Pressuremeters in Geotechnical Design

The pressuremeter is a versatile piece of ground investigation equipment that can be used to test any type of soil or rock in situ. It quantifies in situ stress, stiffness, strength and permeability – the essential properties needed to design geotechnical structures. The results are used in pressuremeter-specific design methods, empirical-design methods and numerical analyses.

This reference book covers the types of pressuremeter and control equipment, methods of installation, test procedures, methods of analysis including direct and indirect methods of interpretation, and application in design. This is supported by an exemplar specification for field operations with the interpretation of the results. Engineers are given enough detail to apply the results confidently.

This comprehensive and thorough discussion of pressuremeter testing in geotechnical design draws on over 40 years' experience in geotechnical engineering. It is essential for professional and academic engineering geologists and geotechnical, civil and structural engineers involved in ground investigation and geotechnical design.

Pressuremeters in Geotechnical Design

Second Edition

B.G. Clarke

CRC Press is an imprint of the
Taylor & Francis Group, an **informa** business

Second edition published 2023
by CRC Press
4 Park Square, Milton Park, Abingdon, Oxon, OX14 4RN

and by CRC Press
6000 Broken Sound Parkway NW, Suite 300, Boca Raton, FL 33487-2742

First edition published by CRC Press 1994

CRC Press is an imprint of Informa UK Limited

British Library Cataloguing-in-Publication Data
A catalogue record for this book is available from the British Library

Library of Congress Cataloging-in-Publication Data
Names: Clarke, B. G. (Barry Goldsmith), 1950– author.
Title: Pressuremeters in geotechnical design / B.G. Clarke.
Other titles: Pressure meters in geogechnical design
Description: Second edition. | Boca Raton : CRC Press, [2023] |
Includes bibliographical references and index.
Identifiers: LCCN 2022021398 | ISBN 9780367464684 (hbk) |
ISBN 9781032321479 (pbk) | ISBN 9781003028925 (ebk)
Subjects: LCSH: Geotechnical engineering–Equipment and supplies. |
Rock mechanics. | Soils–Testing. | Pressure–Measurement–Instruments.
Classification: LCC TA706.5 .C63 2023 | DDC 624.1/5132–dc23/eng/20220906
LC record available at https://lccn.loc.gov/2022021398

ISBN: 978-0-367-46468-4 (hbk)
ISBN: 978-1-032-32147-9 (pbk)
ISBN: 978-1-003-02892-5 (ebk)

DOI: 10.1201/9781003028925

Typeset in Sabon
by Newgen Publishing UK

Contents

Preface to the first edition

In the 1950s, independent developments of pressuremeters took place in France and Japan, which led to the successful use of these probes in ground investigation for foundation design. Design rules were developed from observations of full-scale structures, which are continually being updated and expanded as greater use is made of pressuremeters. In the 1970s, self-boring pressuremeters were developed in France and the UK to determine ground properties directly, especially in situ stress, stiffness and strength. This led to significant developments in the analysis and interpretation of pressuremeter tests. The third major technological advance took place in the 1980s with the development of the cone pressuremeter.

The principal attraction of the pressuremeter test is that it most closely models an ideal condition in which the ground is positively loaded from the in situ stress conditions. By observing the deformation with applied pressure, the in situ stress-strain response of the ground can be obtained. The parameters obtained from a pressuremeter test, which include in situ stress, stiffness and strength, are a function of the probe, the method of installation and the method of testing, as well as the chosen method of analysis and interpretation.

The aim of this book is to explain how pressuremeters can be used in geotechnical design. A form of pressuremeter can be used in all ground conditions, from soft clays to hard rocks. The parameters obtained, either directly, theoretically or empirically, represent realistic estimates of ground properties. Detailed descriptions are given of the major developments in pressuremeter technology, highlighting the most common and significant probes. Site operations, including calibrations, installation, testing and data reduction, are described, highlighting national standards. Attention is drawn to the need for careful installation and correct calibrations if the results are to be of value. A brief summary of the theories of cavity expansion, as they apply to pressuremeter tests, is given in order to describe the development and the limitations of the theoretical or empirical methods of interpretation. Design rules and examples of the use of interpreted values of horizontal stress, stiffness and strength are described to show that foundation behaviour can be predicted.

B.G. Clarke

Preface to the second edition

Pressuremeter testing has continued to develop since the first edition of this book was published. Four significant developments have taken place:

- International standards for all types of pressuremeters, the methods of installation and testing, and methods of interpretation.
- Improvements in instrumentation allowing tests to be automatically controlled and monitored.
- Use of pressuremeter test results with results of other tests to create appropriate ground models and advanced constitutive models.
- Use of pressuremeter test results to validate numerical methods using advance constitutive models.

The use of pressuremeter tests in geotechnical design has continued to expand with improvements in the direct method of design as the database of case studies has expanded, and in the use of numerical methods to enable studies of geotechnical structures to be performed.

The aim of this edition is to reflect on the advances that have taken place to show how the fundamental principles of pressuremeter testing and the use of the results in design have developed to the extent that pressuremeters are used in the design of all types of geotechnical structures in any type of ground condition, and pressuremeter tests are an important tool in the development of constitutive models and numerical methods.

B.G. Clarke

About the author

B.G. Clarke has over 40 years' experience in geotechnical engineering. He worked in ground investigation before joining the University of Cambridge to undertake research into pressuremeters. This led to the formation of a company specialising in pressuremeter testing, and which worked internationally. Dr Clarke is currently a professor in Civil Engineering Geotechnics at the University of Leeds after a time at Newcastle University, where he was Head of Civil Engineering. He is an active member of the geotechnical and civil engineering communities, having been chair of the British Geotechnical Society and president of the Institution of Civil Engineers.

Symbols

A	Pile base area
A	Parameter controlling relative proportions of distortional and volumetric destructuration for a kinematic hardening model
A	Radius of the membrane
a, b	Constants
A, B, C	Constants
A'	Angle for inclined loads and slopes
a_A, a_B	a_A is the radius of the cavity at a pressure of p_A and aB is the radius of the cavity at p_B
a_b, b_b	Model factors to calculate settlement of axially loaded piles
a_e	Cavity radius at end of reloading
a_f	Cavity radius at yield pressure
a_i	Radius of probe during expansion
a_L	Radius of cavity at p_{lm}
a_L	Radius at the modified limit pressure, p_{lm}
a_{max}	Maximum displacement reached at the end of loading or displacement of cavity wall at start of consolidation
a_o	Radius of the probe or cavity
a_p	Initial radius of pocket for a ground anchor
a_r	Radius at the start of reloading
a_s, b_s	Model factors to calculate settlement of axially loaded piles
a_u	Cavity radius at the start of unloading
b	Radius to solid material in which an expanding cavity is creating fractures
b	Membrane stiffness coefficient
B	Width of foundation
B	Radius to solid rock
B'	Corrected width to allow for load eccentricity
B_e	Equivalent dimension of foundation
B_o	Reference width or diameter
c	Constant
c	ratio of major/minor semi-axes of bounding surface ellipse for the bounding surface model (MIT-E_3)
C	Depth to centre of test section
c'	Cohesion

c_h	Coefficient of consolidation
CRR	Cyclic resistance ratio
c_u	Undrained shear strength
CU	Coefficient of uniformity
c_u^6	Undrained shear strength based on a linear elastic perfectly plastic model
c_{uc}	Corrected undrained shear strength
c_{um}	Measured undrained shear strength
D	Depth of foundation
D	Diameter of pressuremeter membrane
d	Drainage path
d_{12}	Internal diameter of calibration cylinder
d_{22}	Probe diameter when undertaking a calibrations
D_a	Diameter of ground anchor
d_e	Equivalent embedment depth
D_r	Relative density
E	Modulus of elasticity
e	Void ratio
e	Eccentricity of load
E_d	Deviatoric stiffness
EI	Pile stiffness
E_m	Ménard modulus
E_m-	Modulus of elasticity from an unloading curve
E_m^+	Modulus of elasticity from a reloading curve
E_{ms}, E_{ma}	Stiffness of soft soil layer and stiffness of layer above
e_o	Initial void ratio
E_o	Slope of the initial portion of prebored pressuremeter curve
E_p	Pile modulus
E_{ro}	Maximum modulus of elasticity from the reloading curve
E_{uo}	Maximum modulus of elasticity from the unloading curve
E_v	Isotropic stiffness
F	Factor of safety
F	Shear load
F	Degree of fixity
F'_c, F'_q	Cavity expansion factors
F_b	Base shear load
FC	Fines content
F_c	Correction factor for length of pressuremeter
g	Intercept of the Horslev surface in q' – p' space
G	Shear modulus
G_i	Initial shear modulus
G_{max}	Maximum shear modulus at small strains
G_o	Small-strain shear modulus obtained from a seismic test or maximum modulus
G_r	Secant-shear modulus from a reloading curve
G_s	Incremental secant modulus
G_u	Secant-shear modulus from an unloading curve

G_{ur}	Shear modulus from an unload/reload cycle
G_{uro}	Modulus at the in situ stress
H	Horizontal load
H	Thickness of compacted fill
h	Slope of the Horslev surface in q' – p' space
h	Material constant for bounding surface plasticity for the bounding surface model (MIT-E_3)
h	Length of element of pile
H_e	Relative depth of foundation
h_k	Spacing between pressuremeter tests
I'_{rr}	Rigidity factor
i_i	Internal radius of membrane during expansion
i_o	Internal radius of membrane
i_δ	Reduction factor for inclined load or adjacent excavations
K	Constant
k	Bearing capacity factor
k	Coefficient of permeability
K	Bulk modulus
K	Ratio of net Ménard limit pressure to true net limit pressure
k	Parameter controlling rate of loss of structure with damage strain for a kinematic hardening bubble model
k	Coefficient of permeability
K_{acv}	Constant volume principal stress ratio in the active state
k_d	Design value of modulus of subgrade reaction
k_e	Equivalent modulus of subgrade reaction
K_g	Modulus number
K_o	Coefficient of earth pressure at rest
K_{ONC}	K_o value for virgin normally consolidated clay for the bounding surface model (MIT-E_3)
k_p	Pile end bearing factor
K_{pcv}	Constant volume principal stress ratio in the passive state
K_r	Modulus coefficient
k_s	Modulus of subgrade reaction
k_s, k_b	Initial stiffness of soil around the shaft of a pile and the base of a pile
L	Length of ground anchor
L	Length of pressuremeter membrane
l_c	Length of measuring cell
l_i	Length of membrane during expansion
l_o	Initial length of membrane
M	Critical stress ratio for Cam Clay
m	material parameter linking M_θ and M for a Kinematic hardening bubble model
m	Number of pressuremeter tests
$m_h{}^e$	Coefficient of volume compressibility
M_o	Moment
M_s	System stiffness

m_v	Coefficient of volume compressibility
M_θ	Dimensionless scaling function for deviatoric variation of M for a kinematic hardening bubble model
N	Constant related to angle of friction
N	Specific volume of normal compression line at unit pressure for Cam Clay
n	Constant related angle of dilation
N_{60}	Standard penetration blow count
OCR	Overconsolidation ratio
p^*	Net cavity pressure
p^*_l	Net limit pressure
p^*_{lm}	Net modified (Ménard) limit pressure
p, p'	Total and effective applied pressure
p'_u	Applied effective cavity pressure at the start of unloading
p_1	Reference mean stress
p_{20}, p_5	Applied pressure at 20% and 5% cavity strain
p_A, p_B	Applied pressure at points A and B
p_c	Pressure in the measuring cell
p_c	Stress variable controlling size of the surfaces for a kinematic hardening bubble model
p_{co}	Initial value of p_c for a kinematic hardening bubble model
p_{cv}, p'_{cv}	Total and effective applied pressure at onset of constant volume conditions
p_d	Pressure on a prebored pressuremeter curve where there is no point of inflexion
p_d	Resistance to horizontal deflection of a pile
p_f	Applied pressure at end of constant creep
p_i	Cavity pressure when radius expanded to radius of a_i
p_{ims}	Pressure at 10% cavity strain from PIP test
p_k	Pressure in the guard cells
p_l	Limit pressure
p_l, p'_l	Total and effective limit pressure
p_{lm}	Modified (Ménard) limit pressure
p_{lme}	Equivalent limit pressure
p_m	Pressure loss
p_{max}	Maximum pressure reached at the end of loading
p_o	Internal pressure
p_o	Initial cavity pressure or pressure at reference volume
p_r	Applied pressure at the start of reloading
p_{ref}	Pressure required to inflate the membrane to reference volume
p_u	Applied pressure at the start of unloading
p_y, p'_y	Total and effective applied pressure at yield
Q	Intersection of hyperbolic model with vertical pressure axis
Q	Normal load
Q	Constant
q	Total bearing pressure
q_c	Cone resistance

Q_f	Ultimate shaft capacity of a pile
Q_m	Compression of pile due to ultimate load
q_n	Net bearing pressure
Q_p	End bearing capacity of axially loaded pile
q_s	Unit skin friction
q_u	Ultimate bearing capacity
R	Radius of an element
R	A factor defining the radius of hyperbola that fits to the pressuremeter curve
R	Parameter describing ratio of sizes of structure and reference surfaces for a kinematic hardening bubble model
$R_{b,k}$	Characteristic base capacity
$R_{c,d}$	Design resistance
$R_{cal:sp}$	Calculated ultimate bearing capacity of piles
R_f	Ratio of the actual strength to the asymptotic value given by the Duncan and Chang model
R_{fs}, R_{bs}	Reduction factors for soil-pile stiffness
R_k	Correction factor for test depth
RMR	Rock mass rating
R_o	Isotropic overconsolidation ratio
r_o	Initial value of r for a kinematic hardening bubble model
r_p	Radius of the plastic zone at the start of unloading
RR	Relative rigidity
$R_{s,k}$	Characteristic shaft capacity
S	Settlement
S	Slope of ln (p') versus ln (ε_{curr})
S, R	Coefficients from reaction curves to calculate pile settlement
s^6	Slope of ln (p') versus ln (ε_{curr}) for length to diameter ratio of six
s_b	Base settlement of axially loaded pile
s_c	Creep settlement of soils subject to dynamic compaction
s_i	Settlement of pile element
S_t	Material constant, related to strain softening of NC clay for a bounding surface model (MIT-E_3)
T	Tension load on ground anchor
T	Time factor
T	Time
T_{50}	Time factor to reach 50% dissipation
t_{50}	Actual time to reach 50% dissipation
t_i	Reduced thickness of membrane on expansion
t_o	Initial membrane thickness
u_o	Ambient pore pressure
V	Current volume
v	Poisson's ratio
V_0^r	Cone module volume correct for system volume increase
V_A, V_B	Volume at points A and B
V_c	Volume of measuring cell

$V_{cav}{}^{u}$	Volume required to inflate FDP membrane to the cone diameter
V_{corr}	Increase in system volume
V_l	Volume of cavity at p_{lm}
V_{mo}	Volume of membrane
V_o	Initial volume of the cavity equivalent to radius a_Q
V_{offset}	Sum of initial cavity volume and maximum cone module volume corrected for the system volume increase
V_p	Initial volume of pocket and when prebored pressuremeter membrane is in contact with pocket wall
V_r	Volume correction
V_{ref}	Reference volume corrected for the volume change required to inflate the membrane to the cylinder wall
$V_{ref}{}^{u}$	Reference volume of calibration cylinder
V_s	Shear wave velocity
V_T	Volume of slotted tube
y	Horizontal displacement
y	Displacement
z_i	Thickness of soil layer
α	Measure of the size of the yield surface in Cam Clay
α	Settlement factor
α	Constant for a power law stiffness profile
α_P, α_D	Constants dependent on stiffness, in situ stress and disturbance
α_s, α_b	Pile factors that depend on soil type
α_s, α_n	Shape factors for horizontally loaded pile
α_v, β_v, γ_v	Volume change parameters dependent on the mean normal stress and relative density
β	Ratio of undrained strength on unloading to that on loading
β	Correction factor for settlement of shallow foundations
β	Soil identification coefficient
β	Constant for a power law stiffness profile
β^*	Pressuremeter constant
β'	Angle for slopes
Δa	Change in radius of probe
Δu	Excess pore pressure
ΔV	Change in volume of the test section
$\Delta V/V_{cv}$	Volumetric strain at onset of constant volume conditions
ΔV_p	Volumetric strain within the sand once yield occurs
$\Delta \varepsilon_c$	Difference between the maximum and minimum cavity strain during a cycle
γ	Shear strain
γ	material constant for bounding surface mapping of flow direction for the bounding surface model (MIT-E_3)
Γ	Specific volume of the critical state line at unit pressure for Cam Clay
Γ	Void ratio at the reference mean stress
γ_{av}	Average shear strain
γ_c	Shear strain at cavity wall

γ_d	Dry density
γ_{peak}	Shear strain at peak shear stress
$\gamma_{R,d}$	Model factor for piles
γ_w	Unit weight of water
δ	Inclination of load
δr	Thickness of soil element
δy	Change in thickness of element
ε	Measured cavity strain
ε^*_c	Cavity strain during unloading
ε_c	Cavity strain
ε_{corr}	Cavity strain corrected for reference datum
ε_{cu}, ε_{cr}	Current cavity strains on unloading and reloading
ε_{curr}	Current cavity strain
ε_{cv}	Cavity strain at onset of constant volume conditions
ε_{max}	Maximum cavity strain
ε_o	Cavity strain at reference datum
ε_{o1}	Offset to allow for disturbance
ε_r	Radial strain
ε_R	Strain at the onset of yield
ε_s	Triaxial shear strain
ε_{um}	Maximum cavity strain during unloading
ε_v	Volumetric strain
ε_{vcv}	Volumetric strain at onset of constant volume conditions
ε_z	Axial strain
ε_θ	Circumferential strain
η_o	Scalar anisotropic structure in a kinematic hardening bubble model
κ	Slope of the swelling line in ln(v) – ln(p) compression plane
κ_o	Compressibility parameter at load reversal for the bounding surface model (MIT-E_3)
λ	Slope of the normal compression line in ln(v) – ln(p) compression plane
Λ	Plastic volumetric strain ratio
λ, λ_D, λ_M	Correction factors for ultimate bearing capacity for shallow foundations
λ_d, λ_v	Shape coefficients
λ_s	Resistance mobilisation factor
ξ	State parameter
ξ	Displacement of membrane
σ'_{av}	Average mean effective stress
σ'_p	Preconsolidation pressure
σ'_R	Onset of yield defined as [σ'_h (1 + sinϕ')] for a yield criterion conforming to the Mohr-Coulomb criterion
σ_h, σ'_h	Total and effective in situ horizontal stress
σ_{hb}	Total horizontal stress at base of pile
σ_m	Minimum value of circumferential stress
σr, σ'r	Radial stress and effective radial stress
σt	Tensile strength
σ_v, σ'_v	Total vertical stress and effective vertical stress

σ_z	Total vertical stress
σ_θ, σ'_θ	Circumferential stress and effective circumferential stress
T	Shear stress
τ^*	Shear stress during unloading
τ^*_{ult}	Ultimate undrained shear strength during unloading
τ_{max}	Maximum shear stress
τ_{ult}	Ultimate undrained shear strength during loading
ϕ'	Angle of shearing resistance
Φ_{cv}'	Angle of shearing resistance at constant volume
ϕ_{TC}'	Friction angle at large strain, triaxial compression test for the bounding surface model (MIT-E_3)
ϕ_{TE}'	Friction angle at large strain, triaxial extension test for the bounding surface model (MIT-E_3)
Φ_{tx}'	Angle of shearing resistance from a triaxial test
χ	Adjustment for limit pressure
ψ_o	Material constant controlling rate of rotation of bounding surface for the bounding surface model (MIT-E_3)
ω	Material constant, controlling nonlinear shear behaviour for the bounding surface model (MIT-E_3)
ψ	Angle of dilation
Ψ	Bearing capacity coefficient

Chapter 1

Introduction

1.1 INTRODUCTION

A ground investigation is undertaken to determine vertical and horizontal variations in ground type and ground properties, which include the in situ stress conditions, deformation and strength parameters and parameters defining time-dependent behaviour. A design engineer can then make a prediction of the behaviour of the ground when it is subject to any change such as that caused by unloading or loading.

There are many techniques used to assess ground conditions and changes in ground conditions, ranging from the simple standard penetration test (SPT) to the most sophisticated computer-controlled true triaxial test. Tests can be divided into laboratory tests and *field* tests which can be further subdivided as shown in Figure 1.1.

Laboratory tests include those that test elements of ground, such as triaxial tests, and those that test prototype models, such as centrifuge tests. Field tests include borehole (in situ) tests, full-scale tests such as preloading trials, and non-destructive tests such as surface geophysical testing. Borehole (in situ) tests include those that penetrate the ground (e.g., the SPT), those that statically load the ground (e.g., pressuremeters) and those that dynamically load the ground (e.g., crosshole geophysics). Robertson (1986) proposed a list of applicability and usefulness of in situ tests for soils and rocks shown in Table 1.1.

He stated that most in situ soil tests are used in sands, silts and clays and concluded that the best in situ tests for soils are electric piezocone tests and self-boring pressuremeter (SBP) tests. In situ tests, commonly used in rocks, include SPT tests (in weak rock), dilatometer or pressuremeter tests, plate tests and geophysical tests.

Results obtained from field tests vary in quality, quantity and applicability. Parameters are test-dependent. A parameter obtained from one test may not have the same value as that obtained from another despite having the same name. Atkinson and Sallfors (1991), in their general report to the 10th European Conference of International Society of Soil Mechanics and Geotechnical Engineering, discussed methods used to determine stress-strain time characteristics of soils and concluded that parameters obtained from in situ tests could be classified as follows:

(a) Primary correlation	sound theoretical basis to analyse a test with few assumptions
(b) Secondary correlation	theoretical basis together with major assumptions and approximations
(c) Empirical correlation	no theoretical basis.

DOI: 10.1201/9781003028925-1

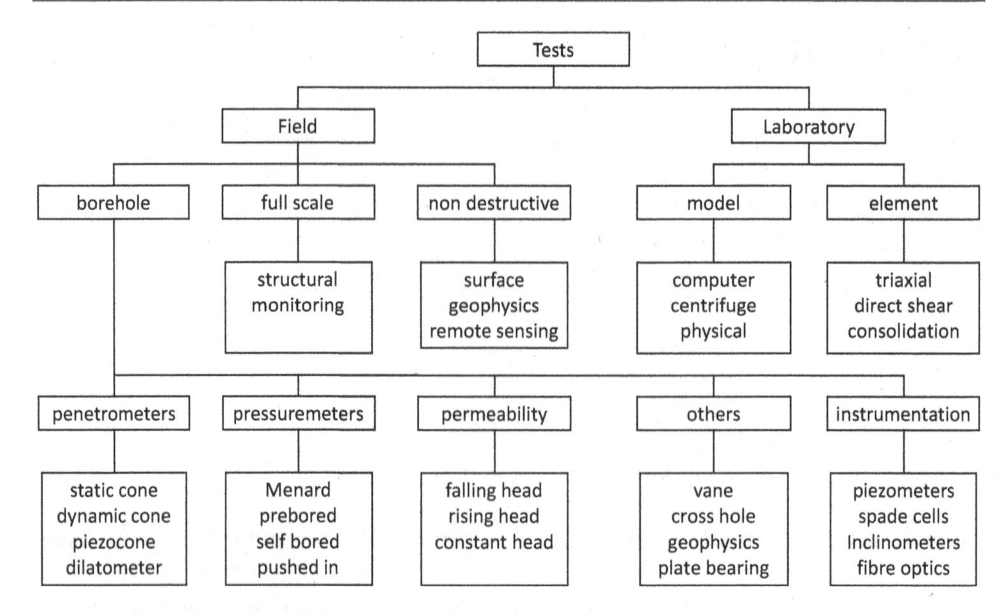

Figure 1.1 Examples of types of in situ and laboratory tests.

Tests are carried out as part of a design exercise and the results can be used either directly or indirectly in design (Figure 1.2).

Data from an in situ test can be converted to ground properties using primary, secondary or empirical correlations. The properties are used in a design method, the choice of which depends on the parameter used, the method by which it was obtained and the ground model chosen. Jamiolkowski et al. (1985), in their report to the 11th International Conference on Soil Mechanics and Foundation Engineering (ISSMFE), outlined advantages and disadvantages of in situ tests and suggested the applicability of in situ tests to soil modelling shown in Table 1.2.

A conclusion that can be drawn from this is that Jamiolkowski et al. suggested that the self-boring pressuremeter has the most potential of all in situ devices since ground properties are derived using primary correlations. Alternatively, in situ test data can be used directly in design methods developed specifically for that test. Examples of this include SPT, cone and pressuremeter design rules. Baguelin (1989), in his report to the 12th ISSMFE, considered that the theoretical interpretation of in situ tests to derive basic soil parameters could only be justified if the application of those parameters in theories of design could be calibrated and validated against an extensive database of the behaviour of full-size structures. He concluded that direct methods of design are more efficient in many cases.

Inspection of Table 1.1 shows that the only type of in situ test that can be used in all ground conditions is the pressuremeter. There are different types of pressuremeter designed for different ground conditions, and it is for that reason that this instrument is so versatile. A further advantage of pressuremeter testing is that the expansion of the pressuremeter can be likened to the expansion of a cylindrical cavity, which can be simply modelled using close-form solutions.

Table 1.1 The applicability and usefulness of in situ tests

Group	Device	Parameters													Ground type						
		Soil	Profile	U	ϕ'	s_u	D_r	m_v	c_v	k	G	σ_h	OCR	σ–ε	Hard rock	Soft rock	Gravel	Sand	Silt	Clay	Peat
Penetrometers	Dynamic	C	B	–	C	C	C	–	–	–	C	–	C		–	C	B	A	B	B	B
	Mechanical	B	A/B	–	C	C	B	C	–	–	C	C	C	–	–	C	C	A	A	A	A
	Static (CPT)	B	A	–	C	B	A/B	C	–	–	B	B/C	B	–	–	C	C	A	A	A	A
	Piezocone (CPTU)	A	A	A	B	B	B	B	A/B	B	B	B/C/C	B	C	–	C	C	A	A	A	A
	Seismic (SCPTU)	A	A	A	B	B	A	B	A	B	A	B	A	B	–	C	C	A	A	A	A
	Flat dilatometer (DMT)	B	A	C	B	B	C	B	–	–	B	B	B	B	–	C	C	A	A	A	A
	SPT	B	B	–	C	C	B	–	–	–	C	–	C	–	–	C	B	A	B	A	B
	Resistivity probe	B	B	–	B	C	A	C	–	–		–	–	–	–	C	–	A	A	A	A
Pressuremeters	PBP	B	B	–	C	B	C	B	C	–	B	C	C	C	A	A	B	B	B	A	B
	SBP	B	B	A*	B	B	B	B	A*	B	A*	A/B	B	A/B*	–	B	–	B	B	A	B
	FDP	B	B	-	C	B	C	C	C	-	A	C	C	C	–	C	–	B	B	A	A
	Cone FDP	B	B	-	C	B	C	C	C	-	A	C	C	C	–	C	–	A	B	A	A
Others	Vane	B	C	–	–	A	–	–	–	–		–	B/C	B	–	–	–	–		A	B
	Screw plate	C	C	–	C	B	B	B	C	C	A	C	B	B		C	–	A	A	A	A
	Plate Load	C	–	–	C	B	B	B	C	C	A	C	B	B	B	A	B	B	A	A	A
	Borehole permeability	C	–	A	–	–	–	–	B	A	–	–	–	–	A	A	A	A	A	A	B
	Hydraulic fracture	–	–	B	–	–	–	–	C	C	–	B	–	–	B	B	-	-	C	A	C
	Crosshole/downhole/ surface seismic																				
	Crosshole/downhole/ surface seismic	C	C	–	–	–	–	–	–	–	A	–	B	-	A	A	A	A	A	A	A

Source: BS 5930:2015+A1:2020, Lunne et al. 1987, Robertson, 1986 and Wroth, 1984.

Note: Applicability: A–high; B–moderate; C–low; – not; * if appropriate sensor used.

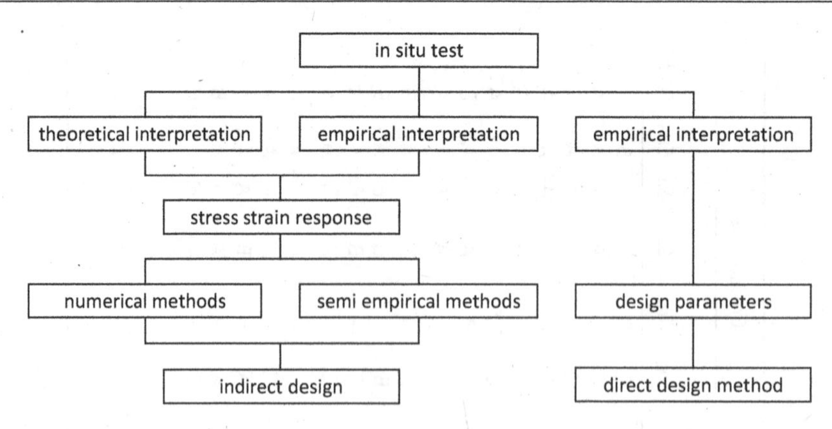

Figure 1.2 The use of in situ tests in design.

Table 1.2 The applicability of in situ tests to soil modelling

Soil behaviour-parameter	*Equipment and/or procedures*	*Comments*
Soil profiling and identification	1.1 Piezocone (CPTU)	1.1a Simultaneous measurement of q_c and u_{max} during penetration has great potential for soil profiling and identification
		1.1b Essentially rigid and extremely well de-aired system with a very quick response for reliable u_{max} measurements
		1.1c Correction of q_c and f_s for unequal end areas effects
	1.2 Static cone (CPT)	1.2a Good for soil profiling but less sensitive to strata changes in comparison to CPTU
		1.2b Friction ratio f_s/qc a poor soil type identifier in especially in sensitive clays
		1.2c Potential may be increased by improving resolution of q_c measurements and more reliable and repeatable f_s measurements
	1.3 Flat dilatometer (DMT)	1.3 I_D a sensitive soil identifier but, since performed discontinuously, generally every 20 cm, less sensitive to strata changes
	1.4 Acoustic cone	1.4 Mainly for soil profiling and identification; needs further field and laboratory validation
	1.5 Electric conductivity	1.5a Measures non-dimensional electrical 'formation' factor which reflects sand structure, hence its anisotropy, particle shape, void ratio and cementation may be relevant for liquefaction studies
		1.5b Needs further validation, especially in the field
In situ σ_h	2.1 SBP	2.1a 'Proven' to be successful in soft clays and stiff clays; less experience in sands
		2.1b Greatest potential among in situ methods
	2.2 Flat dilatometer (DMT)	2.2 Based on empirical correlations; promising, but requires further research to assess reliability

Table 1.2 Cont.

Soil behaviour-parameter	*Equipment and/or procedures*	*Comments*
	2.3 Spade total stress cells	2.3a Limited positive experience only in soft to stiff clays; successfully use in other soils unlikely 2.3b In stiff clay overestimates σ_h; requires correction for bedding error 2.3c Vertical installation essential
	2.4 Hydraulic fracturing	*2.4a Applicable only to cohesive soils having K_o <1* *2.4b Interpretation uncertain*
In situ vertical effective yield stress	3.1 Plate tests	3.1a Limited experience 3.1b Possible applications limited to relatively homogeneous cohesionless deposits at shallow depths in which tests are performed under fully drained conditions *3.1c For screw plate tests the influence of plate shape and disturbance due to its installation on the load/settlement relationship not well understood*
	3.2 Piezocone	3.2a $(u_{max} - u_o)/(q_c - \sigma'_v)$ may be correlated to OCR; in homogeneous deposits, reflects OCR changes 3.2b Possible applications limited to cohesive deposits, further laboratory and field validation needed
Deformability characteristics	*4.1 Plate tests*	4.1a Application limited to shallow depth 4.1b Proven in cohesionless deposits to determine average drained Young stiffness E' within the depth of influence of the plate 4.1c In cohesive soils, despite uncertainty about drainage conditions, it is assumed to yield average undrained Young stiffness E_u 4.1d Since E is obtained from load displacement measurements, an a priori assumption regarding soil constitutive model is necessary 4.1e Very difficult to relate the E obtained from plate tests to the behaviour of a soil macro element, hence to strain or stress levels
	4.2 SBP tests	4.2a Great potential for direct measurement of shear modulus G_h in horizontal direction 4.2b G_h describing the elastic soil behaviour can be assessed from small unloading/ reloading cycles whose role is to minimise the soil disturbance due to probe insertion 4.2c Potential to measure non-linear stiffness profiles in clays, sands and weak rocks
	4.3 Flat dilatometer	4.3a Empirical correlations yield values of tangent constrained modulus in sands and clays 4.3b Presently available correlations have been obtained mainly for predominantly quartz sands and marine and alluvial clays; further laboratory and field validation in a wider range of soils needed

(continued)

Table 1.2 Cont.

Soil behaviour-parameter	*Equipment and/or procedures*	*Comments*
	4.4 Static cone	4.4a Empirical correlations between q_c and E of questionable reliability and not generally valid except for NC sand 4.4b Only applies to predominantly quartz clean uncemented sands in which penetration occurs under fully drained conditions
	4.5 Shear wave velocity	4.5a Proven potential to evaluate small strain ($\gamma < 10^{-3}\%$) G from measurements in horizontally layered soil deposits 4.5b The value of G is calculated after assumptions are made concerning the soil model, the travel path and the soil homogeneity
Flow and consolidation	5.1 Borehole	5.1 Outflow tests at constant head preferred; interpretation above water level extremely complex
	5.2 Large-scale pumping	5.2 Very reliable but also very expensive test; accurate well installation and drawdown measurements with piezometers are required
	5.3 Piezometers	5.3 Constant head tests with Δu small to avoid fracturing are preferred; parameters from outflow tests relevant to OC conditions, inflow tests appropriate for NC conditions
	5.4 Self-boring permeameter	5.4a Careful installation required 5.4b Potential to measure parameters from inflow and outflow tests
	5.5 Holding test	5.5 Careful installation required; difficult interpretation due to non-monotonic changes of effective stress
	5.6 Piezocone	5.6 Very economical and great repeatability, great care required when piezometers probe performing test and interpreting field measurements
	5.7 Back analysis of full-scale structures	5.7 Uncertainties related to initial excess pore pressure or to final consolidation settlement; methods based on consolidation rate need careful analysis of experimental data

Source: Jamiolkowski et al., 1985.

Pressuremeters are now used to

(a) produce design parameters directly
(b) produce in situ values of horizontal stress, shear modulus, shear strength and coefficient of permeability
(c) produce, with results from other tests, parameters for constitutive models
(d) assess improvement in soil properties as part of ground treatment operations
(e) validate numerical methods

There are a number of specifications and guidelines available, including the international standards (BS EN ISO 22476 Geotechnical Investigation and Testing (BS EN ISO 22476 Parts 4, 5, 6 and 8)) and national standards (e.g., ASTM 4719-20:2020).

1.2 DEFINITION OF A PRESSUREMETER

The term pressuremeter was first used by Ménard to describe the testing equipment he developed in 1955. Baguelin et al. (1978) referred to the pressuremeter probe as a device that applies hydraulic pressure through a flexible membrane to the borehole wall. It is further restricted to the definition of a pressuremeter is a cylindrical device, and this definition is recognised internationally by the ISSMFE (Amar et al., 1991). The definition used here, and shown in Figure 1.3, is as follows:

> *A pressuremeter is a cylindrical probe that has an expandable flexible membrane designed to apply a uniform pressure to the walls of a borehole.*

The term pressuremeter can refer to the probe in the test pocket or to the probe, drill rods and testing equipment. The latter definition is used here.

Others have used the term dilatometer to describe pressuremeters. Strictly, the term pressuremeter is more correct since pressuremeter refers to the application of a pressure whereas dilatometer is a term more correctly used to describe expansion due to temperature. There are also several types of instruments, given a name other than pressuremeter, which are actually pressuremeters. Examples of this are the high-pressure dilatometer (Hughes and Ervin, 1980) and the flexible dilatometer (BS EN ISO 22476-5:2012).

A pressuremeter test is defined as the expansion of the membrane after the probe is installed in the test pocket. Pressure and displacement or volume change are monitored during a test, and these data are used to produce a stress–strain curve from which design parameters or ground properties are determined.

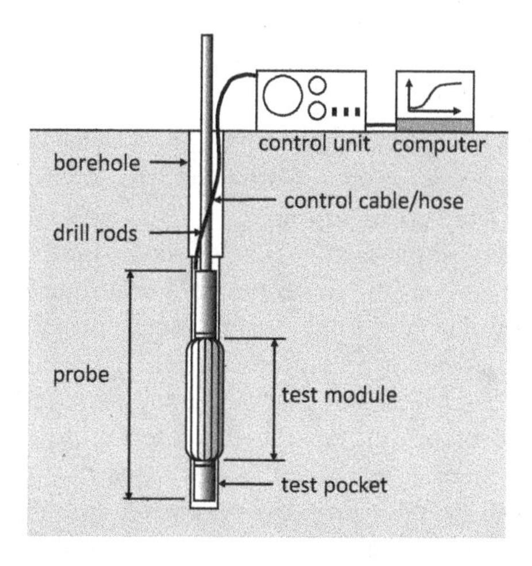

Figure 1.3 The definition of a pressuremeter.

1.3 THE DEVELOPMENT OF THE PRESSUREMETER

The first documented evidence of a pressuremeter is that of Kogler in 1933. In 1954, Fang at Purdue University and Ménard at the University of Illinois, independently began the development of the first of the modern pressuremeters. The instrument developed by Ménard was patented in 1955 and became known as the Ménard pressuremeter. It was first used in Chicago (Ménard, 1957) to obtain ground properties for the design of structures and has since become one of the most widely used types of pressuremeter. In France, Louis Ménard is referred to as the father of pressuremeters since he not only developed the instrument, but also developed a design method based on that instrument which, almost uniquely, became a national standard. This approach is referred to as the Ménard method. Independently, in the 1950s, Fukuoka of Japan developed a lateral load tester to assist in the design of horizontally loaded piles. This form of pressuremeter led to the development of the OYO Corporation's range of pressuremeters, the other most widely used type of pressuremeter.

The Ménard pressuremeter (MPM), lowered into a test pocket of slightly larger diameter than the probe, is a form of prebored pressuremeter (PBP). Ménard recognised in 1955, while developing the instrument, that it would be very difficult, if not impossible, to analyse the tests since the process of preboring changes the properties of the soil adjacent to the pocket. This implies that a PBP test cannot be simply modelled as an expanding cavity in a homogeneous soil. Additional interpretation is required to allow for the effects of installation. To overcome this, Ménard developed a series of design charts based on relationships between foundation performance and pressuremeter test results. This empirical direct design approach is still widely used and is continually being updated as further case studies become available. This approach, similar to that used with other in situ testing equipment, is recognised internationally (e.g., BS EN 1997-1:2004+A1:2013).

In 1968 it was argued (Jézéquel et al., 1968) that it should be possible to install a pressuremeter into the ground without changing the ground properties so that the intrinsic properties, using simple cavity-expansion theories, could be obtained. This led to the development of self-boring pressuremeters (SBP), both in France and in the UK. The SBP, theoretically, causes no disturbance to the surrounding soil since it can be drilled into the ground. The parameters obtained should be properties of the soil rather than of the installation procedure, thus they can be used in any analyses that correctly model soil. These instruments were primarily developed for research but are now used in commercial site investigations.

In practice, there must be some disturbance of the ground during the installation of an SBP, though it should be possible to keep it to a minimum. Self-boring instruments require careful installation to minimise disturbance. Pressuremeters that are pushed into the ground (full displacement pressuremeters (FDP)) were developed in the 1980s to overcome this need for careful installation. They produce repeatable disturbance and an increase in speed of testing though the interpretation is no longer based on simple cavity-expansion theories as the soil is disturbed during installation.

Advances in the analysis and interpretation of pressuremeter tests have taken place in parallel with the development of the equipment. Lamé (1852) was the first to produce an analysis for the expansion of a cavity, but no further substantial developments took place until the 1960s, following the development of the MPM. Since then, there has been an increasing use of pressuremeters, both in design and research.

The theories of cavity expansion do not always apply to PBP and FDP tests because of the disturbance during installation. The results from SBP tests cannot be used in the Ménard design methods, since the installation and testing techniques are different. This is no different from other tests, since many soil parameters are a function of the sampling/installation technique and the testing technique. The interpretation of pressuremeter tests depends on the parameters required, the pressuremeter used, the installation procedure and the test procedure. For these reasons it is necessary, when considering a test, to specify clearly the purpose of the test and the type of pressuremeter to be used.

Design rules, covering bearing capacity and settlement of shallow foundations, pile foundations, caissons and grouted anchors were developed in the 1960s by Ménard Techniques. In 1971 the Laboratoire Central des Ponts et Chaussée (LCPC) produced a standard for the Ménard test and, in 1972, recommendations for design methods. This has since been superseded by LCPC–Setra–1985.

Several countries, including France, Russia and the United States, have developed national specifications, and recommended procedures have been published – for example, by the International Society for Rock Mechanics (ISRM). Institutions and companies have also developed specifications that have since become recognised as good practice. The International Standards Organisation (ISO), an independent, non-governmental international organisation with a membership of 165 national standards bodies, has produced standards for the Ménard pressuremeter test (ISO 22476-4:2012), flexible dilatometer test (ISO 22476-5:2012), self boring pressuremeter test (ISO 22476-6:2018) and full-displacement pressuremeter test (ISO 22476-8:2018).

There have been five international conferences devoted to pressuremeters. The first two – in 1982 in Paris and in 1986 in Texas – were concerned with pressuremeters and their marine applications, the remaining conferences, Oxford (1990), Sherbrooke (1995) and Paris (2005), covered all aspects of pressuremeter testing.

1.4 THE PRESSUREMETER TEST

1.4.1 The probe

The different installation techniques give rise to three groups of pressuremeters: prebored (PBP), pushed-in (FDP) and self-bored (SBP). The Ménard pressuremeter is a form of prebored pressuremeter. Prebored pressuremeters, such as the MPM, high pressure dilatometer and flexible dilatometer are placed in prebored pockets. Pushed-in pressuremeters displace the soil as they are installed. Pushed-in pressuremeters include full-displacement probes, such as the cone pressuremeter, that completely displace the soil during installation, and thick-walled probes that allow some soil to pass up into the probe. Self-bored pressuremeters replace the ground using a drilling system within the probe; that is, the probe is drilled to the test position.

One of the first probes, the Ménard pressuremeter, had three expanding sections, or cells, and is known as a tricell probe (Figure 1.4a).

The central cell, or test section, is connected to a volume-measuring device. The other two cells, known as guard cells, are designed to ensure that the central cell

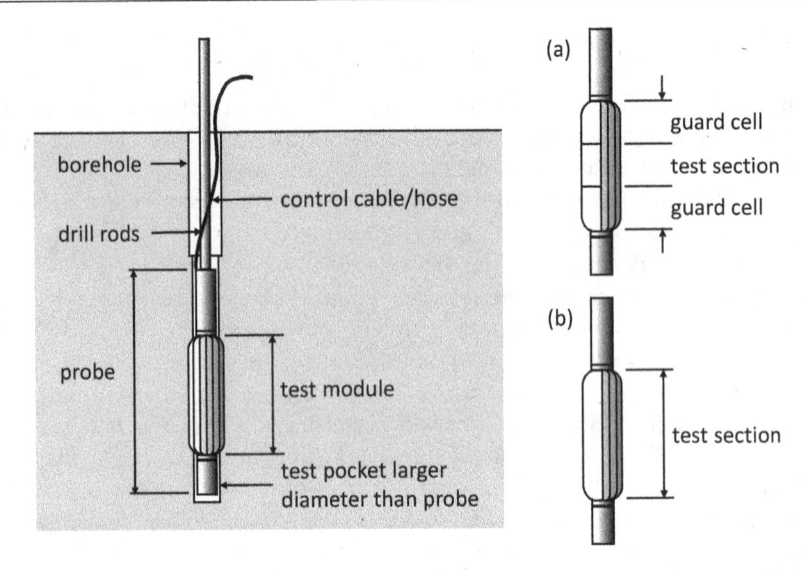

Figure 1.4 Types of prebored pressuremeters: (a) a tricell probe; (b) a monocell probe. The probe is lowered into the predrilled test pocket.

remained a cylinder and, hence, ensured that a true cylindrical cavity expansion is measured. Developments of this probe are still widely used.

In the 1950s, a single-cell, or monocell, probe (shown schematically in Figure 1.4b) was developed by OYO Corporation of Japan. The expansion of the membrane was measured directly using displacement transducers. Many modern pressuremeters, independent of the installation technique, include some form of transducer, both pressure and displacement.

In the late 1960s and early 1970s self-boring probes were developed independently in France (Jézéquel et al., 1968) and the UK (Wroth and Hughes, 1973). The main difference between these single-cell probes is the measuring system. The French probe had a volume-measuring system like the original Ménard pressuremeter, while the UK probe had displacement transducers. Figure 1.5 shows the principle of this instrument.

Pressuremeters specifically developed to be pushed into the ground were first considered in the late 1970s in order to overcome the problems (a) of installing prebored pressuremeters in ground that could collapse and (b) of drilling self-boring pressuremeters in difficult conditions. Ménard pressuremeters are pushed into some soils using slotted casing, but they are not considered as FDPs as the Ménard pressuremeter generally refers to the probe, the test method, the method of analysis and the application of results in design. The first true FDPs were developed from the SBP and static cone for use offshore. These instruments are single-cell probes, as shown in Figure 1.6.

The Ménard pressuremeter is the only tricell probe; all others are monocell probes. The Ménard pressuremeter has a volume displacement measuring system, while the other pressuremeters may have either a volume or a radial displacement measuring system. Thus, within each category there are subdivisions depending on the measuring system, as shown in Figure 1.7.

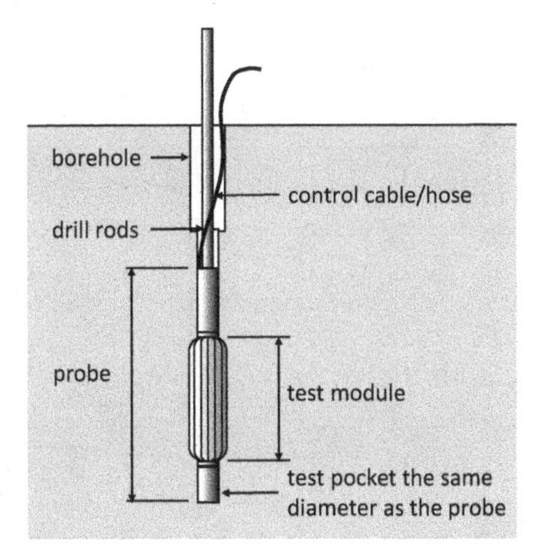

Figure 1.5 The self-boring pressuremeter which creates its own test pocket by drilling.

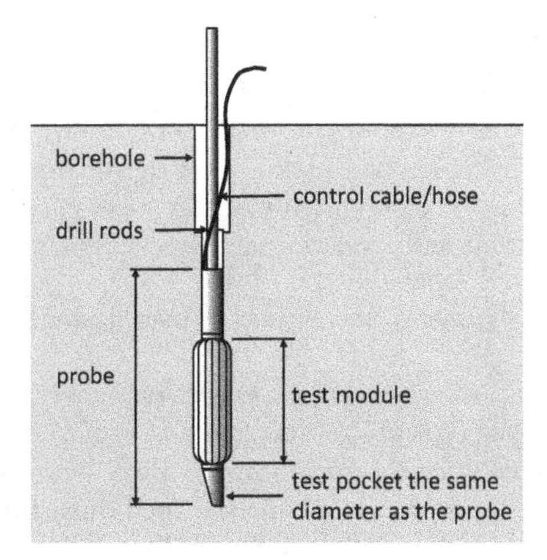

Figure 1.6 The full-displacement pressuremeter showing on the left an external cone for the full-displacement probe, and on the right an internal cone for the push-in, thick-walled probe. The probe is pushed into the ground.

There are numerous pressuremeters, examples of which are given in Table 1.3.

Some pressuremeters have been developed for research purposes only, while others have been developed for commercial use. Only those used extensively in site investigations are described here in detail. The details of the pressuremeters are continually changing as improvements are made, therefore the reader is referred to the manufacturers for exact details of current models.

1.4.2 The expansion curve

The membrane is expanded against the ground, and measurements are taken of the applied pressure and displacement of the membrane. These measurements are shown

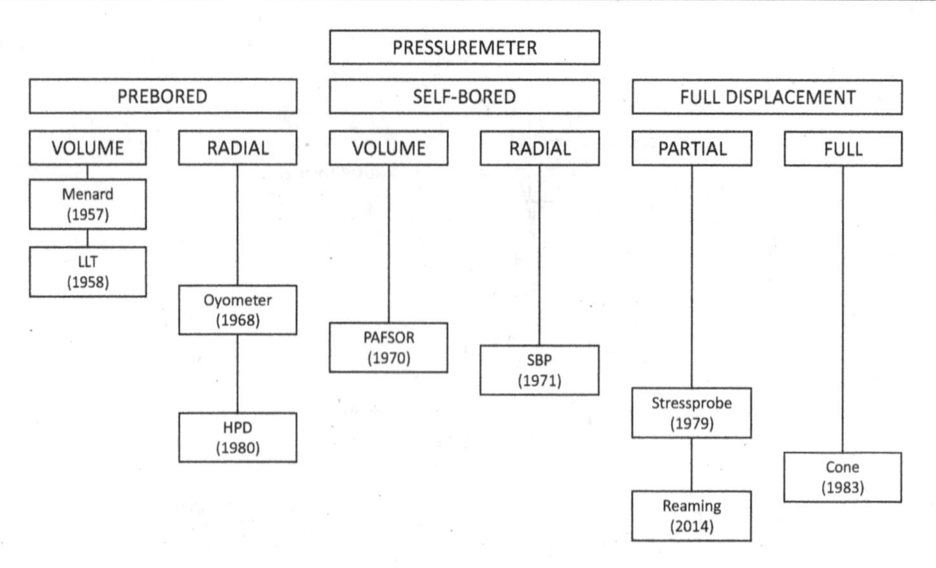

Figure 1.7 Key developments of pressuremeter technology according to the method of installation and the method of measuring deformation.

plotted in Figure 1.8 with the applied pressure on the vertical axis and the displacement on the horizontal axis. The pressuremeter test curve can be used to derive the in situ stress state, deformation and strength parameters. The interpretation of the test and the parameters derived from the test (Table 1.4) are dependent on the ground conditions, the type of instrument, the method of installation, the type of test and the method of interpretation.

The three groups of pressuremeters produce three distinct types of expansion curves, as shown in Figure 1.8. The typical PBP test curve is S-shaped. The first part, OA on Figure 1.8a, is the expansion of the membrane within the mud-filled borehole. The second part, AB, is the deformation of a zone softened during drilling. The third part of the curve, BC, is a measure of elastic behaviour. Point C marks the onset of yielding of the ground adjacent to the membrane.

There are two parts to an SBP expansion curve. Point B (Figure 1.8b), the point at which the membrane begins to move, is equal to the total horizontal stress. Point C is the onset of yielding of the ground. There are also two parts to an FDP expansion curve. Theoretically, during installation an infinite expansion occurs so that there should be very little increase in pressure needed to expand the membrane. This is not the case. As the probe is pushed into the soil the soil is unloaded as it flows past the shoulder of the cone of a full-displacement pressuremeter. Thus point C, Figure 1.8c, represents a yield point. An FDP that partially displaces the soil gives an expansion curve between that of an SBP and a full-displacement pressuremeter.

1.5 SUMMARY

There is a variety of pressuremeters, each of which is designed for a range of ground conditions. A form of pressuremeter can be installed in any ground condition ranging from soft organic clays to hard rocks.

Table 1.3 Examples of specifications of typical pressuremeters

Group	*Name*	*Symbol*	*Manufacturer*	*Pressure capacity (MPa)*	*Strain capacity (%)*	*Diameter (mm)*	*Total length (m)*	*Expanding length (mm)*	*L/D*	*Displacement measurement system*
Prebored	Ménard pressuremeter	GC	Apageo (Fr), Bonne Esperance (Fr), Geomatec (Fr), Roctest (USA)	4	53	74			6.5	Surface volume
	Ménard pressuremeter	GB	Apageo (Fr), Bonne Esperance (Fr), Geomatec (Fr), Roctest (USA)	20	55	74			6.5	Surface volume
	Oyometer	LLT	OYO Corporation	2.5		80				Surface volume
	Oyometer	Elastometer 100	OYO Corporation	10	12	66		520	7.4	Diameter
	Oyometer	Elastometer 200	OYO Corporation	20		66		520	7.4	Diameter
	High-pressure dilatometer	HPD	Cambridge Insitu	20	25	73	1.5	455	6.1	Six displacement transducers
		PA-108	NII Osnovanii	10		108		460		Surface volume
		D-76	NII Osnovanii			76				Surface volume
		CSM	Colorado School of Mines	100		37				
		LNEC		15		74	0.75	540	7.2	Four displacement transducers
	Mazier 60		Mazier	10	55	60			6.6	Three potentiometers
	Mazier 95		Mazier	20	53	95			10.5	Three potentiometers
Self-bored	Cambridge self-boring pressuremeter	CSBP	Cambridge Insitu	4.5	15	84	1	500		Three displacement transducers
	Weak rock self-boring pressuremeter	RSBP	PM Insitu Techniques	20	10		1	400		Three displacement transducers
	Pressiometre autoforeur	PAF-76	Mazier			132	2		2	Surface volume
Full displacement	Stressprobe			3.5		78	>3		4.2	Volume
	Cone pressuremeter	FDCP	Cambridge Insitu				1			Three displacement transducers
	Wireline expansometer	WILE	Syminex	3.0		66	2.75		5	
	LPC-TLM pressio-penetrometer		LCPC	2.5		89			4	Volume

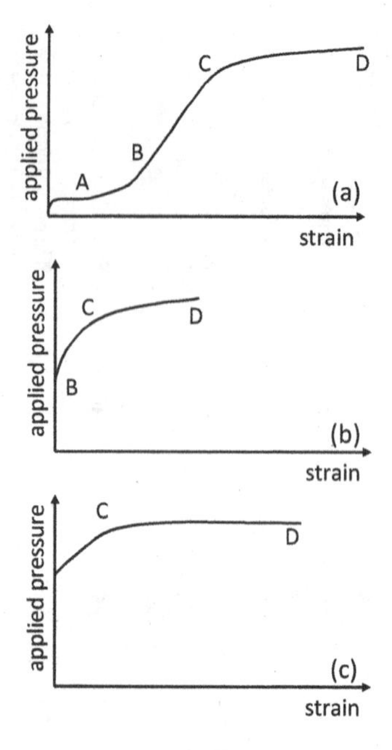

Figure 1.8 Typical expansion curves for the three types of pressuremeters: (a) the prebored pressuremeter; (b) the self-bored pressuremeter; and (c) the full-displacement pressuremeter.

The parameters obtained from a pressuremeter test are a function of the probe, the method of installation, the method of testing as well as the chosen method of analysis and interpretation. Pressuremeter test results have been used directly in design methods based on pressuremeter tests and have been used to produce fundamental properties.

It is important to understand the differences in types of pressuremeters, the site operations and the limitations of the methods of interpretation. This book covers the equipment, the site operations (including test procedures), the theory of cavity expansion, methods of interpreting tests and the application of the results in design.

Each instrument will be supplied with an instruction manual that should describe the probe, control unit and calibrations. However, pressuremeter testing is often specified by engineers who do not necessarily have access to the manuals. Further, the exact type of pressuremeter is often not specified. The purpose of this book is to allow engineers to become familiar with the types of instrument and the method of operation, thus ensuring that they can guarantee quality workmanship and the results they require.

Improvements in equipment, data monitoring, analysis and interpretation have created a robust system that provides a basis for consistency in the operation of pressuremeters to produce either intrinsic ground properties or design parameters. Manufacturers' guidelines and international standards are available.

Table 1.4 Parameters obtained from pressuremeter tests

	Clay						Sand						Rock								
	Soft			*Stiff*			*Loose*			*Dense*			*Gravel*			*Weak*			*Strong*		
Parameter	*PBP*	*SBP*	*PIP*	*PBP*	*SBP*	*PIP*	*PBP*	*SBP*	*PIP*	*PBP*	*SBP*	*PIP*	*PBP*	*SBP*	*PIP*	*PBP*	*SBP*	*PIP*	*PBP*	*SBP*	*PIP*
σ_h		A	CE	C	A	CE		B			C						C	N		N	N
$\mathbf{c_u}$	BE	A	BE	BE	A	BE										CE	B	N	CE	N	N
c'																	B	N		N	N
ϕ'		B			B		CE	A	CE	CE	A	CE	CE	N	N		B	N		N	N
G_i		A			A			A			A			N	N		B	N		N	N
$\mathbf{G_{ur}}$	A	A	A	A	A	A	A	A	A	A	A	A	C	N	N	A	A	N	A	N	N
p_l	BE	A	BE	BE	A	BE	CE	A	CE	CE	A	CE	CE	N	N	CE	B	N	CE	N	N
c_h	B	A	A	B	A	A															

Note: A, excellent; B, good; C, possible; N, not possible; E, empirical.

σ_h, total horizontal stress; c_u, undrained shear strength; c', cohesion; ϕ', angle of friction; G_i, initial shear modulus; G_{ur}, secant shear modulus from an unload/reload cycle; p_l limit pressure; c_h, coefficient of consolidation.

REFERENCES

Amar, S., Clarke, B.G., Gambin, M., and Orr, T.L.L. "The application of pressuremeter test results to foundation design in Europe." *European Regional Technical Committee 4, Pressuremeters* (1991): 1–24.

ASTM Standard D4719–20 (2020) *Standard Test Method for Pressuremeter Testing in Soils.*

Atkinson, J.H., and G. Sallfors, G. "Experimental determination of soil properties. General Report to Session 1." In *Proc. 10th ECSMFE,* Florence, Vol. 3, pp. 915–956. 1991.

Baguelin, F., Jezequel, J.F. and Shields, D.H. (1978) "The Pressuremeter and Foundation Engineering." Trans Tech Publications, W. Germany.

Baguelin, F. (1989) Discussion leader's report: Direct versus indirect use of in situ test results, *Proc. 12th Int. Conf.* SMFE, Rio de Janeiro, Brazil, Vol. 3, pp. 2799–2803.

BS EN 1997-1:2004+A1:2013, *Eurocode 7. Geotechnical design - General rules,* British Standards Institution.

BS 5930:2015+A1:2020, *Code of practice for ground investigations*, British Standards Institution.

BS EN ISO 22476-4:2021, *Geotechnical investigation and testing. Field testing – Prebored pressuremeter test by Ménard procedure*, British Standards Institution.

BS EN ISO 22476-5:2012, *Geotechnical investigation and testing. Field testing – Flexible dilatometer test*, British Standards Institution.

BS EN ISO 22476-6:2018, *Geotechnical investigation and testing. Field testing – Self-boring pressuremeter test*, British Standards Institution.

BS EN ISO 22476-8:2018, *Geotechnical investigation and testing. Field testing – Full displacement pressuremeter test*, British Standards Institution.

Hughes, J.M., and Ervin, M.C. "Development of a high pressure pressuremeter for determining the engineering properties of soft to medium strength rocks." In *Australia-New Zealand Conference on Geomechanics, 3rd, 1980*, Wellington, New Zealand, vol. 6, no. 1 (G), PART1. 1980.

Jamiolkowski, M., Ladd, C.C., Germaine, J.T. and Lancellotta, R. (1985) New developments in field and laboratory testing of soils, *Proc. 11th Int. Conf.* SMFE, San Francisco, Vol. 1, pp. 57–154.

Jézéquel, J.F., Lemasson, H. and Touzé, J. (1968) Le pressiomètre Louis Ménard quelques problèmes de mise en oeuvre et leur influence sur les valeurs pressiométriques, *Bull, de Liaison du LCPC,* No. 32, 97–120.

Lamé, G., (1852) Leçons sur la théorie mathématique de lélasticité des corps solides, *Bachelier,* Paris.

Lunne, T. and Lacasse, S. (1987) Use of In Situ Tests in North Sea Soil Investigations, *Norges Geotekniske Institutt*, No. 169.

Ménard, L. (1957) "An apparatus for measuring the strength of soils in place." Thesis, University of Illinois.

Robertson, P.K. (1986) "In situ testing and its application to foundation engineering." *Canadian Geotechnical Journal* 23, no. 4 (1986): 573–594.

Wroth, C.P. (1984) The interpretation of in situ soil tests. 24th Rankine Lecture, *Geotechnique, 34*(6), 449–489.

Wroth, C.P. and Hughes, J.M.O. (1973) An instrument for the in situ measurement of the properties of soft clays, *Proc. 8th Int. Conf.* SMFE, Moscow, Vol. 1.2, pp. 487–494.

Chapter 2

Pressuremeter probes and testing equipment

2.1 INTRODUCTION

Pressuremeter probes and testing equipment vary between manufacturers and operators, but the principles on which they are based are similar. Specifying a particular manufacturer's probe can often impose an unnecessary restriction on an operator, since other probes may be similar to that specified. The operation of the equipment, including maintenance and its limitations, should be understood. This ensures the correct and full use of the equipment and minimises operational problems due to ground conditions. Installation and test procedures are a function of the pressuremeter specified.

Features common to most pressuremeters are described below in detail. Further descriptions are also given of the major types of probe and their testing equipment, including prebored, self-bored and full-displacement pressuremeters. Specialist probes, such as pavement and offshore pressuremeters, are briefly described.

2.2 KEY FEATURES OF PRESSUREMETERS

2.2.1 The probe

The key features of a typical pressuremeter are shown in Figure 2.1.

A pressuremeter includes the probe (A), the control unit (B) and the connections (C) from the probe to the surface. The probe can be divided into three modules that are connected by the central core tube (1), which is usually made of steel. These are the installation section (D), the test section (E) and a section (F), which can be either a void, or an instrumentation housing or a drilling module, depending on the pressuremeter. Section (F) also helps prevent the membrane expanding into the borehole above the probe. The probe is connected to the surface by drill rods and a control cable (C). The control unit (B) at the surface is often, but not always, an integral part of the pressuremeter equipment.

The type of installation module (D), at the base of the probe, depends on whether the probe is lowered, drilled or pushed to the test location. It can take the form of an internally chamfered shoe, a blunt shoe or a cone, depending on the type of pressuremeter. These shoes are described in more detail under the three groups of pressuremeters.

The probe is shown in greater detail in Figure 2.2.

The test section (E in Figure 2.1), the expanding section, can either be a *monocell* or *tricell* and may or may not include transducers. The probe comprises an expanding

DOI: 10.1201/9781003028925-2

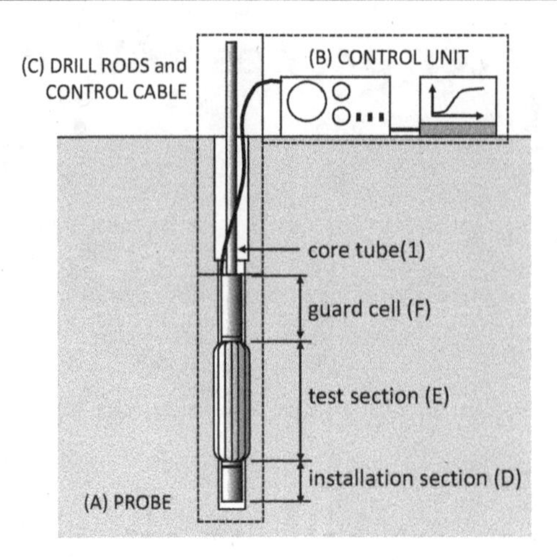

Figure 2.1 Key features of pressuremeters include (a) a probe, (b) a control unit and (c) an installation system.

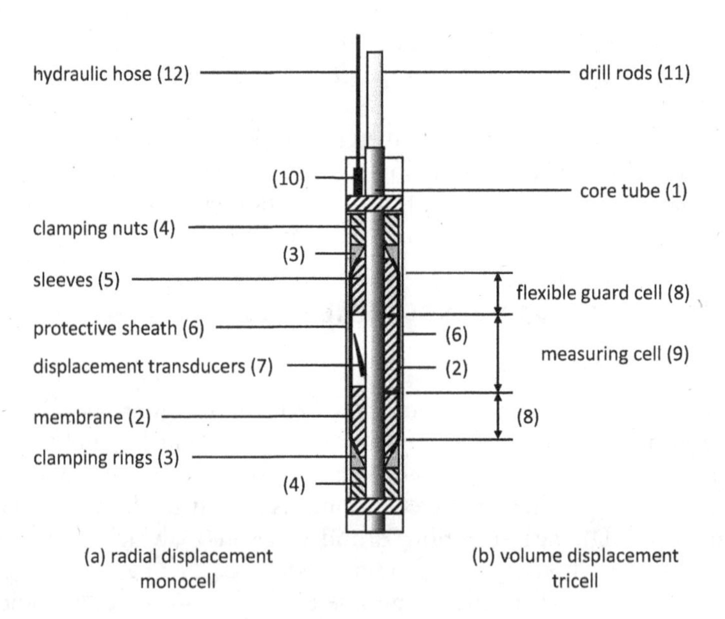

Figure 2.2 Details of the test section of a probe showing the difference between (a) monocell and (b) tricell probes.

membrane (2), which can be made from natural rubber, adiprene, neoprene or metal. Metal membranes are used where small displacements are anticipated and, hence, are not common. However, metal strips, known as a Chinese lantern, may be used to protect the flexible membrane. The membrane is held in place by clamping rings (3), and possibly rubber O rings, which seal the membrane onto the core tube (1) to prevent it from being pulled off and permit the membrane to be pressurised. The clamping rings are held in place by clamping nuts (4) screwed onto the core tube. The membrane is supported on sleeves on the body of the probe (5) during installation, and during a test the membrane is expanded by forcing oil, water or gas into the probe. Measurements are taken of the applied pressure and displacement of the membrane.

Burst membranes can occur during installation if the probe gets caught on the casing shoe when casing is used to support the borehole. However, the most common reason for membrane failure is the presence of existing fissures or voids, ground of varying stiffness and the development of fissures during a test. In these instances, part of the membrane is forced into the fissure or void causing it to be stretched to such an extent that it tears or pulls out of the clamping system. The membrane has to be protected during installation and expansion to reduce potential damage. Various techniques have been used to provide a protective sheath (6). These include thick membranes, reinforced membranes, metal strips covering membranes and metal membranes. As the combined stiffness of the membrane and sheath increases, the pressure sensitivity of the probe reduces, since greater pressure is required to inflate the protective sheath and membrane than that to inflate the membrane alone. Corrections are applied for this effect.

The movement of the membrane can be monitored either by using displacement transducers in the probe or by observing the amount of water or oil forced into the probe. The former are described as radial displacement type probes, the latter as volume-displacement type probes. Volume-displacement systems are almost exclusively used with the Ménard type pressuremeters described below.

It is often assumed that the volume of the membrane remains constant as it expands thus the measured change in volume of fluid injected is equal to the change in volume of the pocket. In that case, the volume-displacement type probes give direct readings of the average displacement of the pocket wall. However, when testing stiffer ground, it is necessary to correct for changes in membrane thickness as it expands.

Radial displacement type probes generally give the displacement of the inner surface of the membrane at a point or points. Displacement transducers (7) include linear variable differential transformers, Hall Effect gauges and strain gauges. The thickness of a membrane reduces as the membrane expands, hence, the measured displacement must be corrected. Ideally a membrane should be as thin as possible to reduce the need for corrections. There are some membranes, designed to operate at very high pressures, which contain buttons that pass through the membrane. In that case, the movement of the ground is monitored directly, and no correction is applied for the change in thickness of the membrane. The buttons, however, are a point of weakness and can lead to membrane rupture. These special membranes tend to be more expensive than rubber membranes and, since membranes are treated as consumables, will inevitably increase the cost of a test.

The test section has a finite length. It is assumed that it expands as a right circular cylinder. In practice it is only cylindrical in homogeneous ground over the middle third since the ends are restrained. Volume-displacement type probes usually contain flexible guard cells (8), which are inflated with the measuring cell (9) to ensure that the measuring cell retains its cylindrical shape. These probes are known as *tricell* probes. Radial displacement type probes usually have only one expanding section, the test section, and are known as *monocell* probes. The test section of a monocell probe is usually longer than the measuring cell of a tricell probe, but the total length of the expanding section is similar.

It is usual with monocell probes to extend the length of the probe so that there is, in effect, a rigid guard cell above the test section. This is to minimise the gap between the probe and the pocket above the test section and, hence, reduce the possibility of the membrane being forced up between the probe and the pocket wall. If this does happen the membrane can burst. Additional measures, examples of which are given below, have to be taken to prevent this happening when using high-pressure probes.

The probe is connected to the surface with rods (11) and a hydraulic hose (12). The rods are used to position the probe in the test pocket. They are either purpose-made drill rods, cone rods or standard drill rods. In all cases they are attached to the core tube. The hydraulic hose is used to transmit the pressurising fluid to the membrane. If the probe contains transducers there is also an electric cable connecting the probe to the surface. This cable may pass through the hose.

If an oil- or water-filled probe is lowered into a dry borehole, the membrane will expand due to the pressure of the weight of fluid in the hydraulic line once the resistance of the membrane is exceeded. These probes have two hydraulic lines. The probe, without fluid, is lowered into the test pocket. The pressurising fluid is pumped down one line, and the air forced out of the other prior to a test. At the end of a test in a dry hole, the membrane will remain expanded due to the weight of the fluid. When that occurs air is forced into the probe down one hydraulic line to push some of the pressurising fluid back to the surface, thus allowing the membrane to contract owing to its elasticity. This expansion can be prevented by filling the borehole with mud.

2.2.2 The control unit

The control unit, Figure 2.3, is used to control and monitor a test. In its simplest form it consists of a pressure supply (13), which can be either a gas supply or a hydraulic pump with oil or water, a displacement and pressure measurement unit (14) and a pressure or displacement control system (15).

Tests can either be stress- or strain-controlled, or a combination of stress- and strain-controlled – the choice depends on the probe design and the test procedure required. In stress-controlled tests the volume or displacement of the membrane is measured and the pressure controlled; in strain-controlled tests the applied pressure is measured and the displacement controlled.

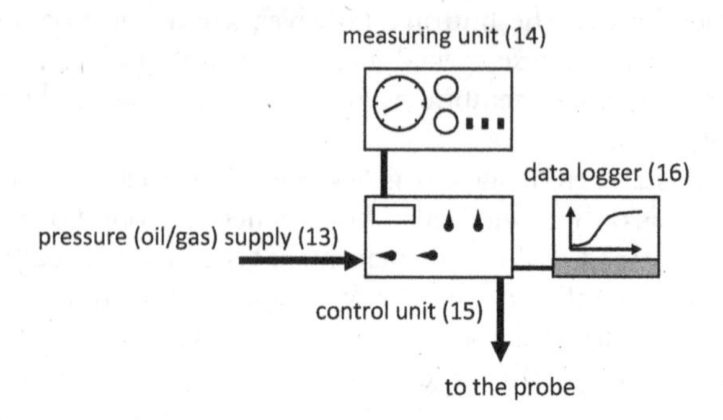

Figure 2.3 The control and monitoring system showing the pressure supply, the measuring unit, the control unit and data logging unit.

In their simplest form, pressuremeter tests are manually operated stress-controlled tests using visual displays to take readings of pressure and displacement. The pressure of the fluid is increased in increments either by adjusting a precision regulator if gas is used as the pressure source or by operating a hand pump if an oil supply is used. The pressure is measured with a Bourdon gauge. The change in volume of the test section is monitored using a volume change unit comprising measuring cylinders.

Some probes are pressurised manually, but the pressure and displacements are recorded automatically using transducers mounted in the test section. The tests are still stress-controlled but it is no longer necessary to monitor the volume change, since the displacement of the membrane is measured directly. Transducers require an electronics unit, either surface mounted or within the probe. The unit contains the source to power the transducers and a module to amplify the signals and convert them into digital form. It is advisable to have a visual display of the pressure and displacement to ensure that a test is proceeding satisfactorily.

The most advanced control units are those that use a feedback system to control either the rate of displacement or the rate of increase of pressure. The feedback system can be either purpose-built hardware or software. The use of computers to monitor and control pressuremeter tests is normal.

2.3 PREBORED PRESSUREMETERS

A prebored pressuremeter (PBP) is designed to be lowered into a prebored or precored pocket. There are many versions of this type of instrument and they are the most common type of pressuremeter used in ground investigations. They were first developed in France and Japan but are now found in a variety of forms throughout the world (for example, Ruppeneit and Bronshtein, 1972; Hughes and Ervin, 1980; Rocha et al., 1966; Briaud and Shields, 1980; Arsonnet et al., 2011; Shaban and Cosentino, 2017; and Aissaoui et al., 2021).

The key features of a prebored radial displacement type probe are shown in Figure 2.4. The probe consists of a test section, described in detail above, and an installation section, which can be either an internally chamfered shoe or a hollow closed-end cylinder. The dimensions of probes vary, depending on the manufacturer. The diameter of a probe is often chosen so that it can be lowered into pockets of standard drilling rod diameters. For example, an NX probe will have a diameter of about 73 mm so that it can be used in a pocket created using an NX core barrel.

There are two groups of PBPs: volume-displacement types and radial-displacement types. The first group is typified by the Ménard pressuremeter; the second group includes the OYO LLT and high-pressure dilatometer. These three probes are described in detail below since they contain features that are common to many of the other PBPs, and are the probes most commonly used in ground investigation.

2.3.1 The Ménard pressuremeter

The Ménard pressuremeter, referred to as the MPM, was developed by the Centre d'Etudes Ménard (Ménard, 1957). The 60 mm diameter G probe, covered by BS EN ISO 22476-4:2021 (Table 2.1), is limited to 5MPa for use in soil.

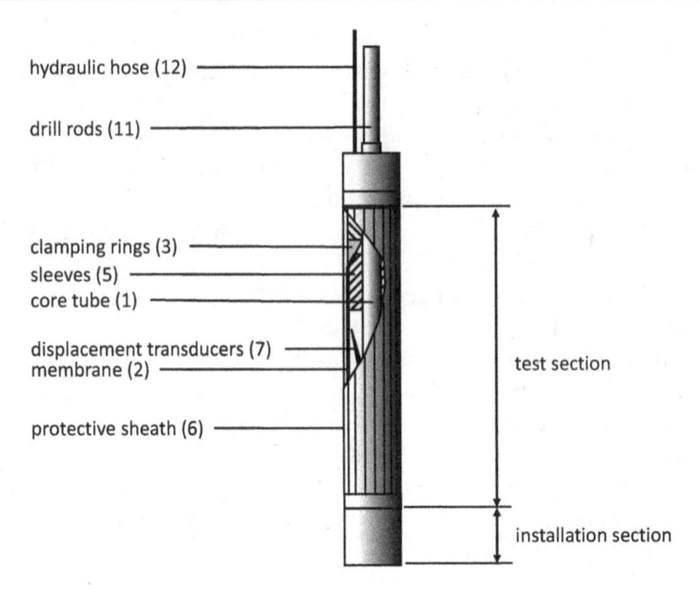

Figure 2.4 Details of a prebored radial displacement pressuremeter.

Table 2.1 Dimensions of Ménard pressuremeters

Parameter			*Value [mm]*	*Tolerance [mm]*
Probe with flexible cover		Central measuring cell length	210	+ 5; 0
		Guard cell length	120	± 15
		Outside diameter	58	± 2
Probe with slotted tube	Inner part: short central measuring cell	Central measuring cell length	210	+ 2; 0
		Guard cell length	200	± 5
		Central measuring cell outside diameter	44	± 2
	Inner part: long central measuring cell	Central measuring cell length	370	± 5
		Guard cell length	110	± 5
		Central measuring cell outside diameter	44	± 2
	Slotted tube	Outside diameter	59	± 5
		Slot length (along tube axis)	≥ 800	-

Source: SO22475-4.

There are a number of MPM devices that are very similar (see Table 1.3). The MPM has evolved as part of a standard procedure to give design parameters directly. This procedure covers the probe, and methods of installation, testing and interpretation. It is referred to as the Ménard method. The MPM can also be used to determine the properties of the ground directly using methods of interpretation based on cavity-expansion theory.

The test section of an MPM consists of three cells: a central measuring cell and two guard cells (Figure 2.5).

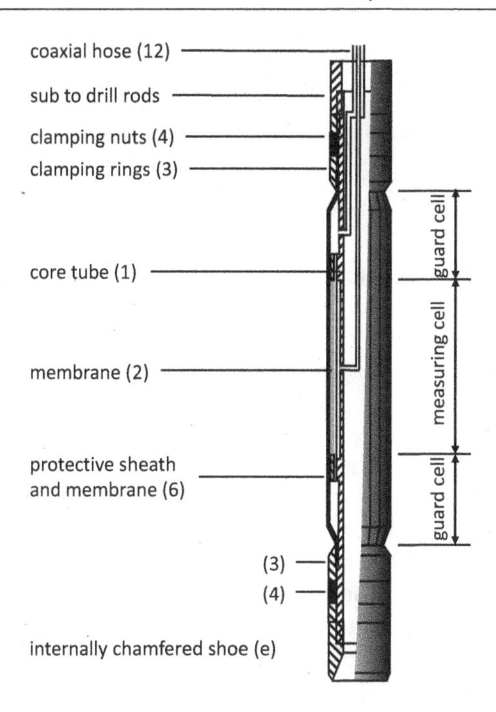

Figure 2.5 The features of a Ménard probe used for testing soil.

The guard cells are inflated to ensure that the central measuring cell expands as a right circular cylinder. The guard cells also prevent the central measuring cell from expanding up into the borehole and down into the pocket. Modern versions of this tricell probe have a sheath, which runs the length of the expanding section and, within that, there is a separate membrane, which forms the central measuring cell. The sheath forms the guard cells. The central cell is inflated with water, and the sheath with gas. Water is used to allow volume changes to be measured. The volumetric capacity if the measuring cell is 800 cm^3 to ensure that the volume of the probe doubles during a test; a requirement of the Ménard test procedure. Versions of this probe include the Apageo standard Ménard pressuremeter with a capacity of 100 bar for testing soil and the Apageo Ménard THP (Very High Pressure) pressuremeter with a capacity of 200 bar for testing rock.

The membranes are made of natural rubber, and in certain conditions are protected on the outside by steel strips, which run the length of the test section. A variant of this is the probe within a slotted tube. The purpose of the tube is to protect the probe when it is pushed into the ground.

The guard cells are inflated with gas. The gas should be at the same pressure as the water so that the central cell expands as a right circular cylinder. However, in order to ensure that the expansion of the central measuring cell is the expansion of the soil – that is, there is no gas between the sheath and the membrane – the gas pressure is actually maintained at a slightly lower pressure than the water pressure.

The key features of the probe are shown in Figure 2.5. The membrane forming the central measuring cell is clamped into place by the wedging action of two metal rings forced against shoulders on the central tube. A rubber sheath (6) covers the central measuring cell and forms the two guard cells. It is held in place by clamping rings (3), which are tightened onto the shoulders with clamping nuts (4). Two truncated membranes attached to steel tubes are held in place by the clamping rings at either end of the sheath. These help prevent the sheath from blowing up the borehole during a test since they provide extra resistance to expansion at the top and bottom of the guard cells.

A shoe (e) with an internal chamfer is screwed onto the bottom of the probe. This ensures that any excess soil is forced into the shoe as the probe is lowered into the pocket. This is necessary should the pocket diameter be less than the probe diameter or if there is a possibility of debris from drilling in the pocket.

A coaxial hose (12) connects the instrument to the surface. Water is passed down the inner tube of the coaxial tubing and gas down the outer tube. Coaxial tubing is used to limit the volume changes of the water line to ensure that the volume changes measured at the surface are equal to the volume changes in the probe.

The control unit at the ground surface is used to pressurise the guard cells and measuring cell and to measure the volume change of the measuring cell. Figure 2.6 is a schematic diagram of the G control unit in which the gas in the guard cells is maintained at a slightly lower pressure than the water in the measuring cell. The compressed air supply (1) is continuously monitored by observing a gauge (2). A volumetric strain gauge, comprising a reservoir (3) and a graduated tube (4), is connected to the measuring cell. The water reservoir has a capacity of 800 cm^3 but as the diameter is too great to give sensitive readings of volume change a small diameter graduated tube, known as a volumeter, is connected in parallel to the reservoir. A change in the water level represents a change in the volume of the coaxial tubing and measuring cell. The volume can be read to the nearest cm^3 – that is, approximately equivalent to an increase in the diameter of the membrane of 0.018 mm (0.02% cavity strain). The pressure in the reservoir is set manually using a regulator (5) and is measured with a Bourdon gauge (6). It is usual to have two gauges, the more sensitive of which is used for lower pressures. The gas pressure in the guard cells is regulated separately (7). It is maintained at a

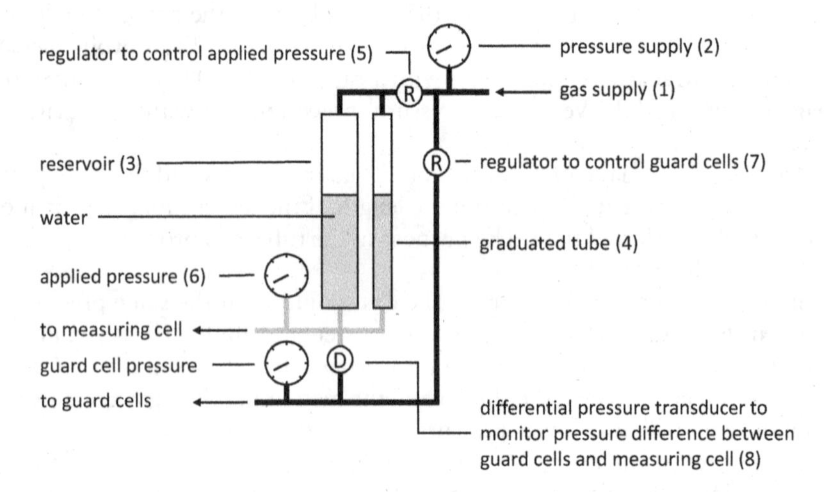

Figure 2.6 The manual control unit for volume-displacement probes for testing soils (e.g., the G probe).

slightly lower pressure than the water in the measuring cell by measuring the difference between the two pressures with a differential pressure transducer (8).

It is possible to monitor the pressure and volume changes automatically using pressure transducers. Figure 2.7 shows a volumetric strain gauge in which the graduated tube is replaced by a reference tube (9). The water level in the reference tube remains constant. The pressure in the reservoir and reference tube are the same but the water level in the reservoir varies during a test. The differential pressure transducer (10) monitors the difference in pressure due to the difference in the height of the columns of water in the reference tube and reservoir.

If the guard cells and measuring cell are maintained at the same pressure in a true tricell probe such as the high pressure volume-displacement probes, and, if water or oil is used in these cells, two reservoirs are required to allow volume changes of the measuring cell to be made. A schematic diagram of this unit is given in Figure 2.8.

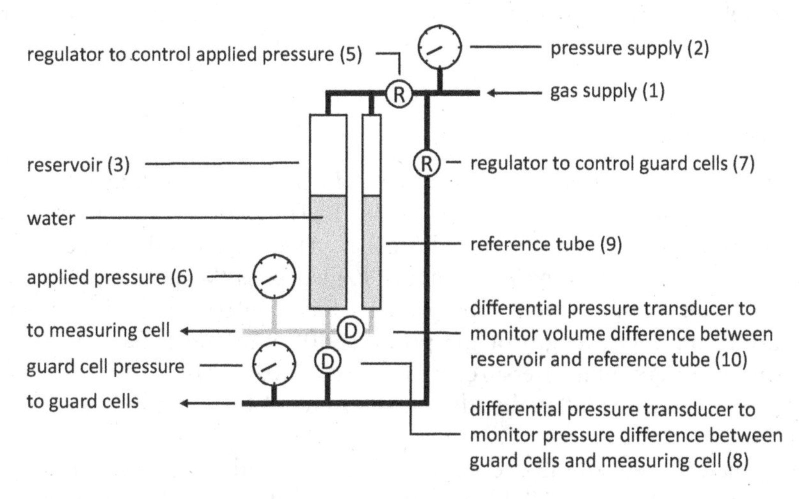

Figure 2.7 A control unit for volume-displacement probes for testing soils that can be used to log the data.

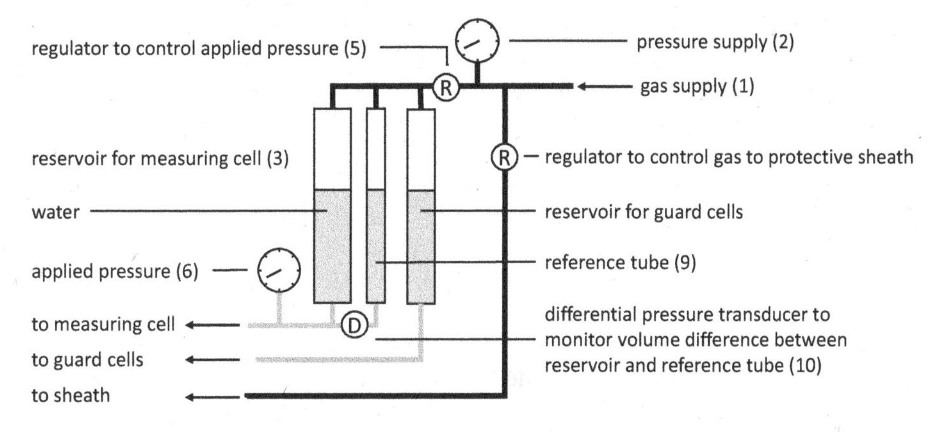

Figure 2.8 The control unit for high-pressure volume-displacement pressuremeters.

This unit is designed to operate up to 20MPa. Two regulators are used to control the applied pressure, one for low pressures the other for high pressures. Similarly, gauges with different ranges are used to monitor the pressures to obtain increased sensitivity. Another line connects the probe to the control unit; this allows the protective sheath to be pressurised, thus forcing the probe to deflate.

It has been possible to control the MPM test automatically since the procedure is fixed. The operator sets the maximum pressure required, and the automatic control unit sets the pressure increments, maintains the pressure within the guard cells and test section and records the volume and pressure. The data are then stored for further analysis. A printout of the data is available at the time of testing in case there is any damage to the storage medium.

Several companies are now manufacturing Ménard type pressuremeters, including, Apageo (Ménard Pressuremeter) in France and Roctest (G-AM II Ménard Pressuremeter) in America. The probes, which vary in detail and size, include monocells, tricells and radial and volume-displacement types. Table 2.1 provides the dimensions of prebored pressuremeters that can be used in the Ménard procedure.

2.3.2 The Oyometer

The Oyometer is the general term for those instruments developed by the OYO Corporation, Japan. The three probes – the LLT M, Elastmeter HQ Sonde and Elastmeter Sonde 4 inch – are monocell probes. The LLT was developed in the late 1950s specifically for designing horizontally loaded piles (Suyama et al., 1966). It is a monocell volume-displacement type probe, which can be supplied in diameters of 60, 70 and 80 mm. The length of the probe is 900 mm and it has a 600 mm long test section. The maximum pressure capacity is 2.5MPa. The LLT-M is a manually operated probe, which is inflated with gas; the Auto LLT2 uses a pump to automatically carry out a test.

The Oyometers were the first commercial probes with displacement transducers. The Elastmeter HQ Sonde, available as a 60 mm or a 70 mm probe, and the Elastmeter 4 inch Sonde are designed for testing rock and can operate up to 20MPa. Displacement transducers are used because of the problems of obtaining accurate information of the cavity expansion with remote volume-measuring systems. This is particularly important for tests in rock where displacements are likely to be small.

The probes have a length/diameter ratio of between 7.2 and 8.4. Membranes used vary in thickness and type, depending on the ground to be tested. The membrane is prevented from expanding into the annulus between the probe and the pocket above and below the test section by its thickness, which can be up to 8 mm, and its stiffness.

The components of the probe are shown in Figure 2.9.

The membrane (2) is in the form of a packer with metal ends bonded onto the rubber. This is slid over the core tube (1) and clamped in place with a set screw (a), which is located on the bottom of the core tube. The packer is prevented from rotating by a plug (b) located on keys on the core tube. The displacement of the inner surface of the membrane is monitored with an LVDT (c), which is mounted coaxially within the core tube. Two spring-loaded arms (7) are connected to the LVDT so that the average movement of the membrane is recorded. A pressure transducer (d) and

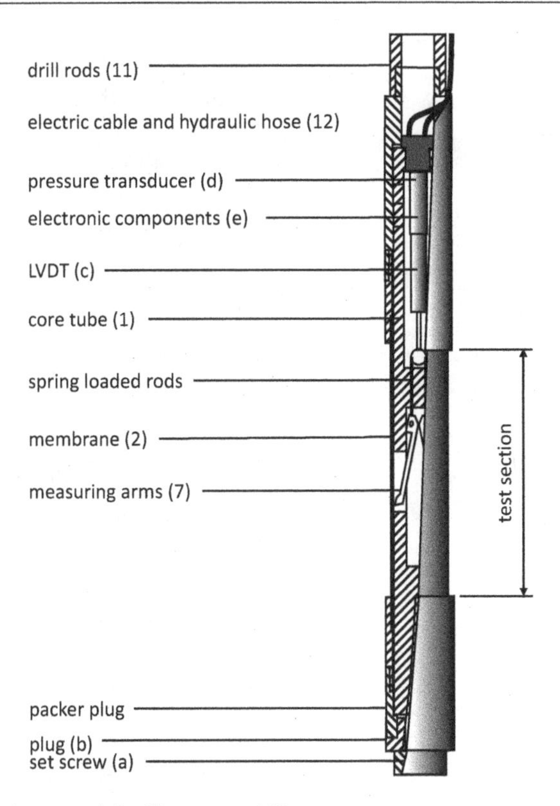

Figure 2.9 Schematic diagram of the Elastmeter 100.

electronic components (e) are mounted above the test section. Two cables connect the probe to the surface, the hydraulic hose and a four-core control cable.

The axis of the probe may move during the expansion of the membrane and it is therefore not possible to measure directly the change in radius of the pocket. A transducer measuring the change in radius will measure it with reference to the axis of the probe. It is for this reason that radial displacement transducers are always mounted in pairs in PBPs, as they detect any change in diameter.

2.3.3 The high-pressure dilatometer

The high-pressure dilatometer, HPD, was developed by Hughes and Ervin (1980). The version described here is manufactured by Cambridge Insitu, UK. It is 73 mm in diameter with a 455 mm long expanding section. It is intended for use in weak rocks, though it can be used in dense sands or stiff clays. The maximum capacity of the instrument is 20MPa.

The components of the probe are shown in Figure 2.10.

It consists of a hollow steel tube (1) on which sits the body of the probe, which contains displacement transducers (7) and a pressure cell (a). A rubber membrane (2) is slid over the core tube (1). Two sets of steel strips (b), known as the Christmas Trees, are placed at the top and bottom of the membrane. They are there to help prevent

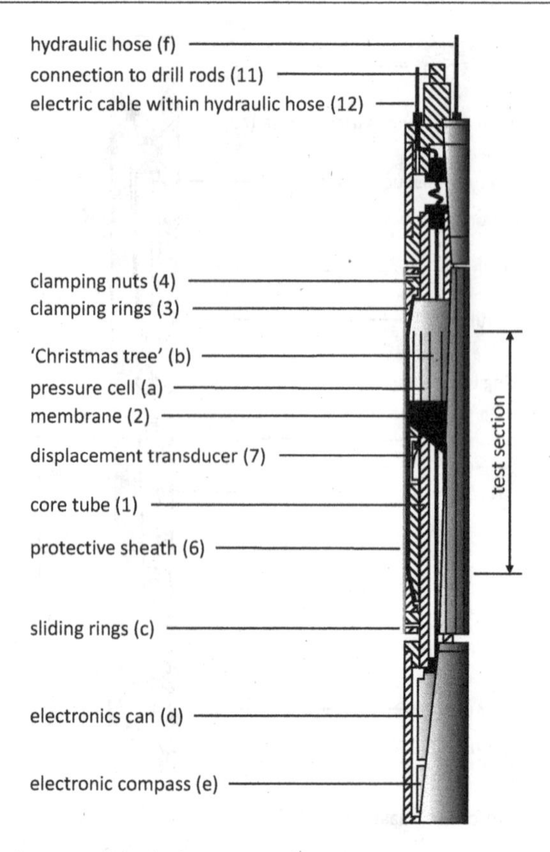

Figure 2.10 Schematic diagram of the high-pressure dilatometer.

the membrane expanding up the borehole. The membrane and Christmas Trees are clamped in place with clamping rings (3) held in place by clamping nuts (4). A Chinese lantern (6), consisting of stainless steel strips attached to sliding rings, is used to protect the membrane when installing the probe in a pocket and to help prevent damage to the membrane as it is expanded in ground containing fissures.

The displacement of the inner surface of the membrane is monitored with six strain arms, each of which consists of a strain follower, which rests on a strain-gauged cantilever spring. The strain followers give an average measurement of displacement rather than a point measurement. Point measurements may be affected by discontinuities within the rock. The six strain arms give independent readings of displacement relative to the central axis of the instrument. The sum of a diametrically opposite pair is used to denote the change in cavity diameter.

The signals from the displacement and pressure transducers are processed in the probe such that digital signals are sent directly to the surface through the control cable (12). The processing unit is mounted below the test section together with an electronic compass (e), which indicates the orientation of the probe. The probe is inflated with oil, which is passed down a second hydraulic hose (f). The first hydraulic hose, which contains the electric cable, is used to pump air into the probe to deflate it when working in a dry borehole.

2.3.4 Other prebored pressuremeters

The probes described above represent key developments in pressuremeter technology. Prebored pressuremeters have been developed throughout the world, including those in France, the UK, United States, Russia, Canada, Portugal, Japan, China and the former Yugoslavia and Czechoslovakia. Many of them are similar to the Ménard pressuremeter and were developed for testing soil for foundation design. These include Apageo probes, the OYO probes (Elastmeter HQ Sonde, Elastmeter 4 inch Sonde) the Roctest probes (DMPe Dilatometer, Trimod, Texam), and the Cambridge Insitu probes (HPD73, HPD95, RPM).

There are further groups of prebored pressuremeters; for example, probes for testing non-homogeneous ground, miniature pressuremeters and model pressuremeters. Some have been developed for specific purposes and may not be widely available.

There are commercially available mini pressuremeters, which are hammered into the ground or placed in shallow hand-augured holes. They can be as small as 22 mm in diameter (e.g. Briaud and Shields, 1980; Shaban and Cosentino, 2017). Briaud and Shields (1980) describe a PBP designed to obtain the stiffness of road pavements. An example is the PENCEL pressuremeter (Figure 2.11).

It is a simple, small, volume-displacement type monocell probe, which is expanded with water at a constant flow rate. The probe is inserted in a hole created by driving E rods with a conical tip through the pavement or by driving the probe into the soil. The 2.5 MPa 32 mm diameter probe is a monocell with an expanding section of 240 mm. A 35 mm diameter shoe is used to limit the friction on the membrane during

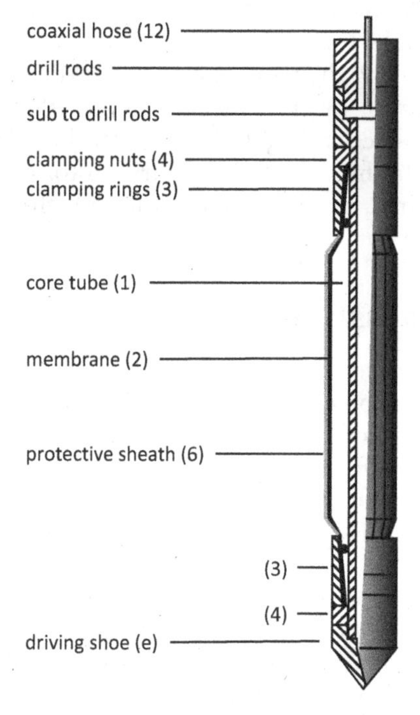

Figure 2.11 Schematic diagram of the pavement pressuremeter, the Pencell.

installation. This means the soil is unloaded prior to a test. Anderson et al. (2005) suggested that a friction reducer be added to the conical tip to reduce damage to the membrane during installation. They concluded that it has an effect, but without further studies the effect could not be quantified. They also investigated stress-controlled and volume-controlled inflation and concluded that stress-controlled tests gave more consistent results. Volume-controlled tests showed a stiffer response.

Cour et al. (2005) undertook a review of the standard MPM, which led to an improved design, yet maintained the principles upon which the original design was based. The features included a downhole volume-control unit pressurised by gas to minimise errors due to connections between the measuring unit and cell, a more robust membrane to minimise bursting, downhole volume sensor and a fully software-controlled test so as to minimise operator influence.

While the principles of a pressuremeter probe have remained the same, the control and monitoring systems have changed such that many pressuremeter tests are controlled and monitored automatically. This includes downhole equipment to reduce corrections for pressure and volume loss. Arsonnet et al. (2011) developed a control system to be used with an MPM to increase its stability up to 25MPa. It consists of motorized piston that controls the volume to a resolution of 1.9×10^{-3} cm^3, The Apageo Gopal is a control unit that automatically controls MPM tests according to BS EN ISO 22476-4:2021 and ASTM D4719-20 standards. The OYO (Elast Logger-3i), Cambridge Insitu (electrical interface unit and strain control unit) and Roctest (D/P box) are examples of units developed to monitor and control pressuremeter tests. Modern pressuremeters are increasingly being used to measure stiffness, which requires accurate measurements of small displacements. For example, Aissaoui et al. (2021) describe the development of a Ménard pressuremeter that incorporates Hall Effect transducers to enable small displacements to be measured up to a volume change of 50%, thus satisfying the requirement of MPM testing yet allowing accurate measurements to be taken to determine small strain stiffness. Frikha and Varaksin (2017) describe an electro-mechanical device to control the expansion of an MPM to an accuracy of 0.01cm3. In this case, a DC electric motor of 24 V and a reducer with a normal torque of 24 N/m drives a ball screw using a high-ratio gearbox with an epicycloid stage gear train.

2.4 SELF-BORING PRESSUREMETERS

In 1957, Ménard used the theory of cavity expansion to determine parameters from PBP tests. He very quickly appreciated that the installation process changes the ground response and therefore it is not possible to obtain a true stress–strain response of the ground using a PBP. The stiffness of the ground can be obtained from an unload–reload cycle from any pressuremeter test provided that the pocket has been expanded to test undisturbed ground. Hence, PBPs are still used extensively to determine the stiffness of rock.

Jézéquel et al. (1968) proposed that a self-boring probe could be used to measure a true ground-reaction curve from the in situ conditions. They developed the *pressiomètre autoforeur* (PAF). Wroth (1973) proposed that there was a need to measure in situ stress because soil behaviour is dependent on that stress. This led to the development of the Camkometer, a self-boring load cell, which later became the Cambridge self-boring pressuremeter (CSBP) (Wroth and Hughes, 1973).

The French self-boring system is used with several different types of instruments, including pressuremeters, load cells and permeameters. The UK system originally developed for the Camkometer, is now used with the self-boring pressuremeter. Other self-boring pressuremeters have developed from these two systems. For example, the Roctest Boremac is a 4MPa 73 mm diameter, 1.8 m long pressuremeter, which incorporates the Texam pressuremeter module and is designed to operate with a drilling rig.

The principle of self-boring, used to install the probe, is to minimise disturbance to the surrounding ground. Consider Figure 2.12a in which a thin-walled tube is inserted into the ground, while soil within the tube is removed as the tube is advanced. The total vertical stress at the base of the tube is zero. There must be a change in total stress within the ground, therefore there will be extension of the ground in front of the tube. This will be accompanied by a reduction in horizontal stress, which may result in a change in the ground properties.

Consider Figure 2.12b in which a solid rod is pushed into the ground. This will cause an increase in the total vertical stress within the ground, leading to compression of the ground in front of the rod. Hence, there will be a change in horizontal stress leading to a change in ground properties. There must be some point between these two extremes at which there is no stress change at the leading face of the probe (Figure 2.12c). This is the principle of self-boring.

The key features of an SBP are shown in Figure 2.13.

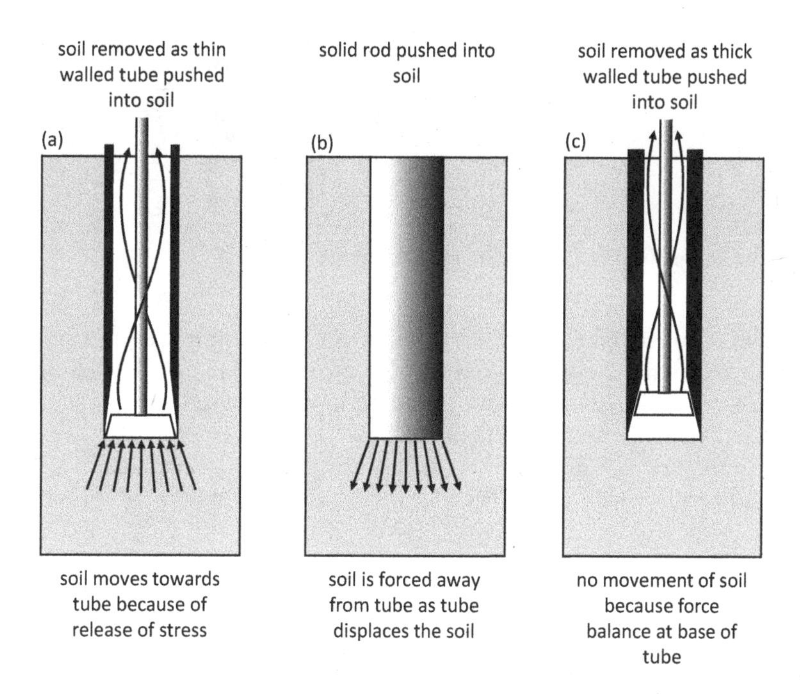

Figure 2.12 The principle of self-boring and the justification for minimum disturbance.

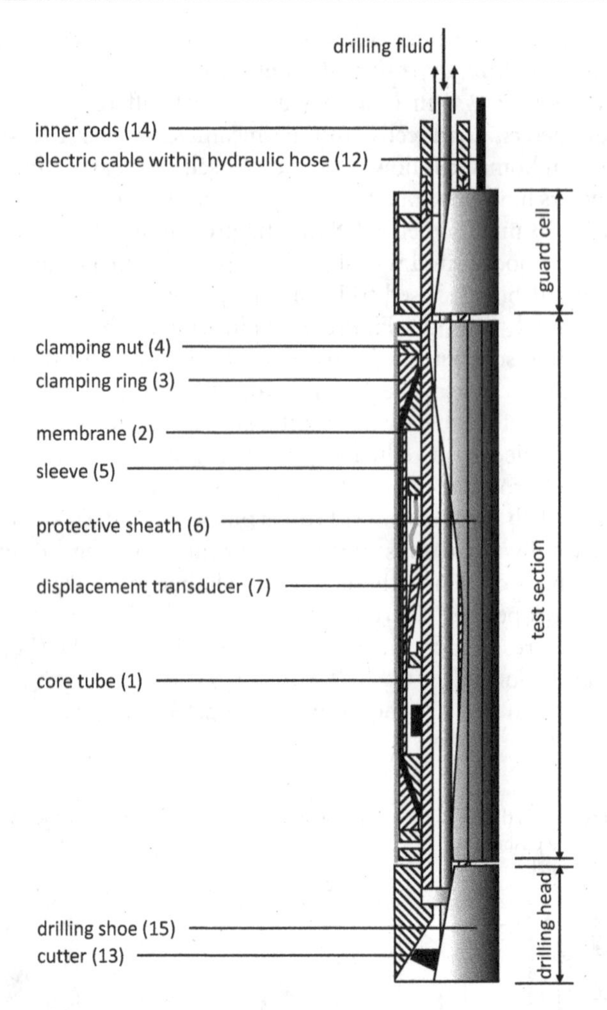

Figure 2.13 The key features of a self-boring pressuremeter.

The key used in the labels conform to Figure 2.2. The test section can either be a volume-displacement type or a radial displacement type. Figure 2.13 shows a radial displacement type. Different self-boring systems have been used, which include drilling, jetting and replacing, as shown in Figure 2.14.

The core tube (1) in Figure 2.13 is hollow to permit the passage of drilling fluid and inner drill rods where necessary. The core tube also transmits the vertical force through the probe to overcome friction between the probe and the ground and provide the thrust to drill the ground. The membrane (2), clamping system, (3) and (4), and protective sheath (6) are similar to those described in section 2.3.

The drilling head (C) is an internally chamfered shoe (15), which contains a rotating cutter. The cutter (13) is driven by rods (14), which can be turned either by a motor in the drilling head or by a motor behind the test section (self-boring module (A)) or by a motor at the surface depending on the type of probe.

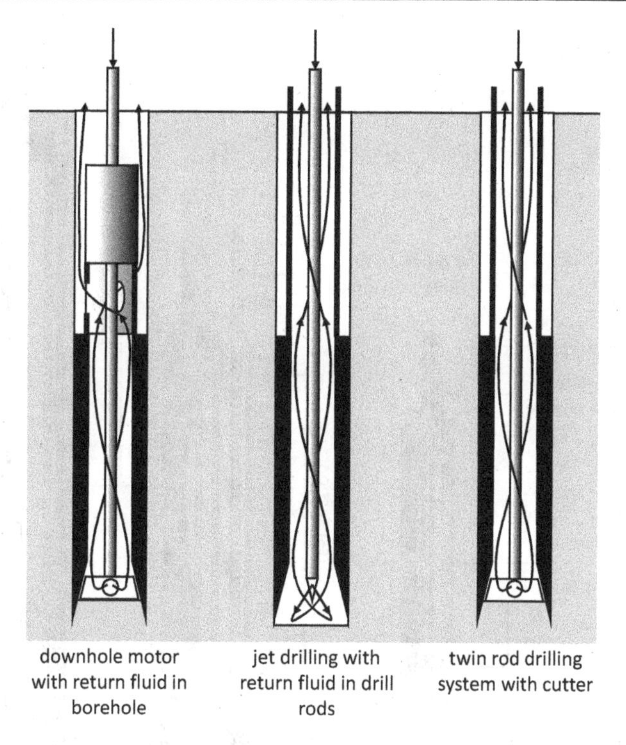

Figure 2.14 Examples of self-boring systems.

The probe is connected to the surface by drill rods, which are used to lower and raise the probe and push the probe into the ground as the soil is removed. They do not rotate. Drilling fluid is pumped down the inner rotating rods or hydraulic hose to the cutter. This flushes away the chippings of the ground as the probe advances.

2.4.1 The pressiomètre autoforeur

In 1968, Jézéquel et al. discussed the effect that the installation of the MPM has upon the test results and concluded there was a potential for the development of a self-boring probe. The first pressiomètre autoforeur (PAF-68) was developed in 1968. The PAF-68 is a surface volume-displacement monocell 89 mm diameter probe with a single rotating rod connecting the probe to the surface. A more robust 132 mm diameter probe (PAFSOR) was developed in 1972. The cutter in this probe is turned by a hydraulic motor mounted behind the test section. A modular probe, PAF-76, was developed in 1976. The test section is mounted above a self-boring unit, which contains a hydraulic motor. Different drilling heads are used for different ground conditions. Outlines of the three probes are shown in Figure 2.15.

The concept of separating the instrument into modules was developed by LCPC so that different measurement modules could be used. LCPC developed a permeameter, friction probe, horizontal penetrometer, geophysical logger and shear meter (Amar et al., 1977).

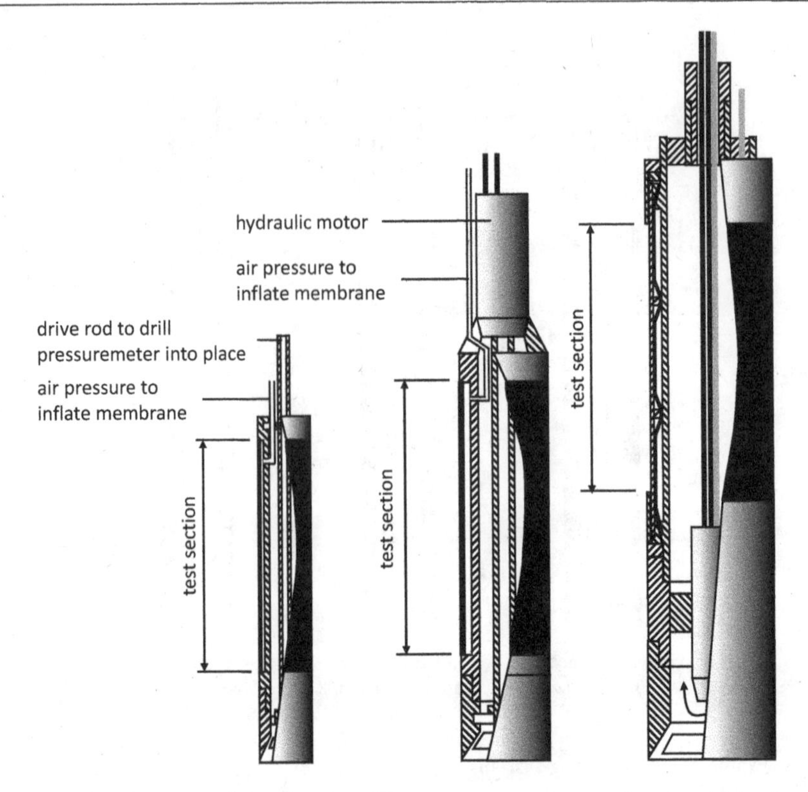

Figure 2.15 **The development of the French self-boring pressuremeters, showing the implementation of downhole hydraulic motors and the development of the test section to incorporate downhole measuring systems.**

The 132 mm diameter PAF-76 (Figure 2.16) has a pressure capacity of 3 MPa.

The two modules, the drilling head containing a hydraulic motor, and the measurement module form a 2m long unit. The drilling head is formed of a thick-walled steel tube (15), which has an internal chamfer to cut into soil. A hydraulic motor (a), mounted within this module, drives the cutter (13). This motor is driven from the surface using hydraulic lines (b) attached to the drill rods (11). Drilling mud is pumped to the cutter through a hydraulic hose (c). The return fluid passes up through the measurement module, through an adapter (d) and through the annulus between the drill rods and the borehole wall.

The measurement module consists of a thin-walled tube (1) into which is milled a recess the length of the expanding section. This recess is covered by a rubber membrane (2), which is clamped top and bottom. The expanding section can also be subdivided to create a tricell by using steel clamps (e) to restrict the membrane at two positions ((a) and (b)). The recess is filled with water and isolated from the surface by a valve (f). The volume of the recess remains constant during drilling, thus a direct reading of the horizontal stress is obtained by measuring, with a transducer, the pressure in the membrane. At the start of a test the valve is opened and water is pumped into the recess forcing the membrane to expand. Strips (6) are glued to the membrane to protect it

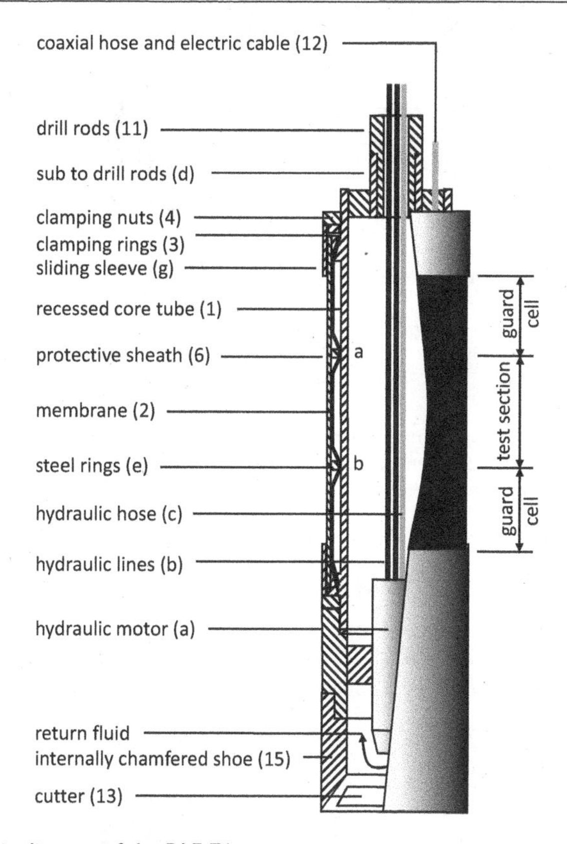

Figure 2.16 Schematic diagram of the PAF-76.

during installation. The ends of the strips are allowed to slide beneath a shield (g), thus permitting expansion of the membrane.

2.4.2 The Cambridge self-boring pressuremeter

The Cambridge self-boring pressuremeter, referred to as the CSBP, was developed in 1971 at Cambridge University, UK. The present instrument, first produced in 1975, is approximately 84 mm in diameter, the exact diameter depending on the membrane being used, which includes adiprene, reinforced rubber and adiprene protected by steel strips. The overall length of the instrument is approximately 1 metre, and the expanding section is 500 mm long. The capacity is 4.5 MPa though, after modification, they have been used to 10 MPa. The key features of the probe are shown in Figure 2.17.

It consists of a hollow steel tube (1) on which sits the body of the probe, which contains displacement transducers (7) and a pressure transducer (a). A membrane (2) is stretched over the body of the probe and clamped in place with clamping rings (3) held in place by clamping nuts (4). The membrane is supported on the brass sleeves (5), which cover the electrical components (b) mounted on the core tube.

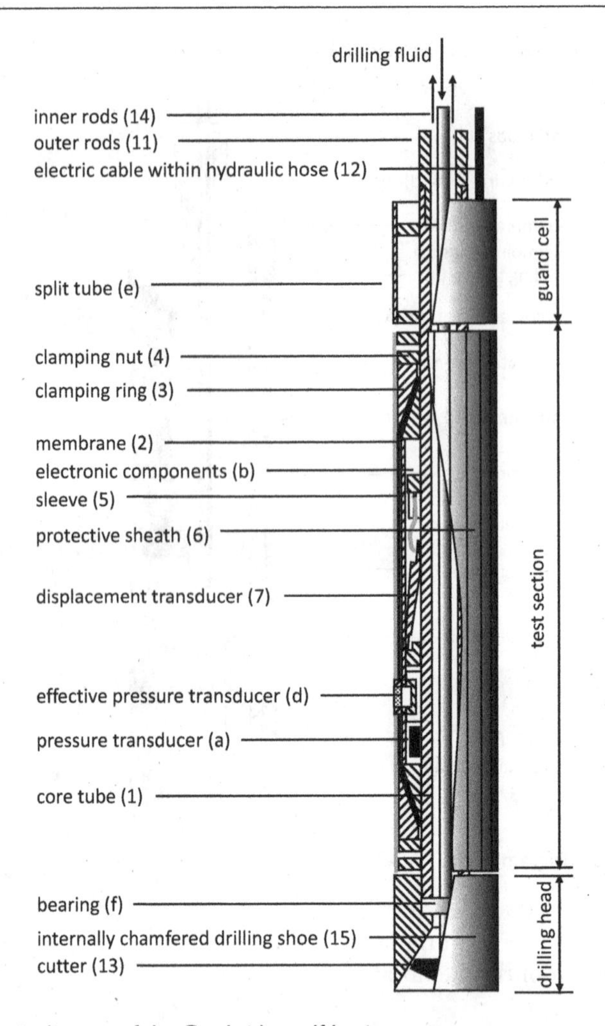

Figure 2.17 Schematic diagram of the Cambridge self-boring pressuremeter.

A Chinese lantern (6), consisting of stainless steel strips bonded onto a rubber sheath, can be used to protect the adiprene membrane when lowering in a borehole and drilling to a test position. It also helps prevent damage to the membrane, as it is expanded in soil containing discontinuities. The strips are connected to sliding rings (c) to allow the lantern to deform during a test. The rubber sheath helps prevent soil particles from becoming trapped between the lantern and the membrane.

Transducers are used to measure displacement and stress. There are three feeler arms (7) mounted at 120° spacing at the centre of the expanding section. An arm consists of a beam (a), which is pivoted about point A (see Figure 2.18).

The arm is loaded by a leaf spring (b), which keeps the end of the arm in contact with the inside of the membrane. The movement of the arm is detected by a change in the resistance of strain gauges (c) mounted on the leaf spring. These gauges are connected in a Wheatstone bridge circuit and are compensated for temperature changes.

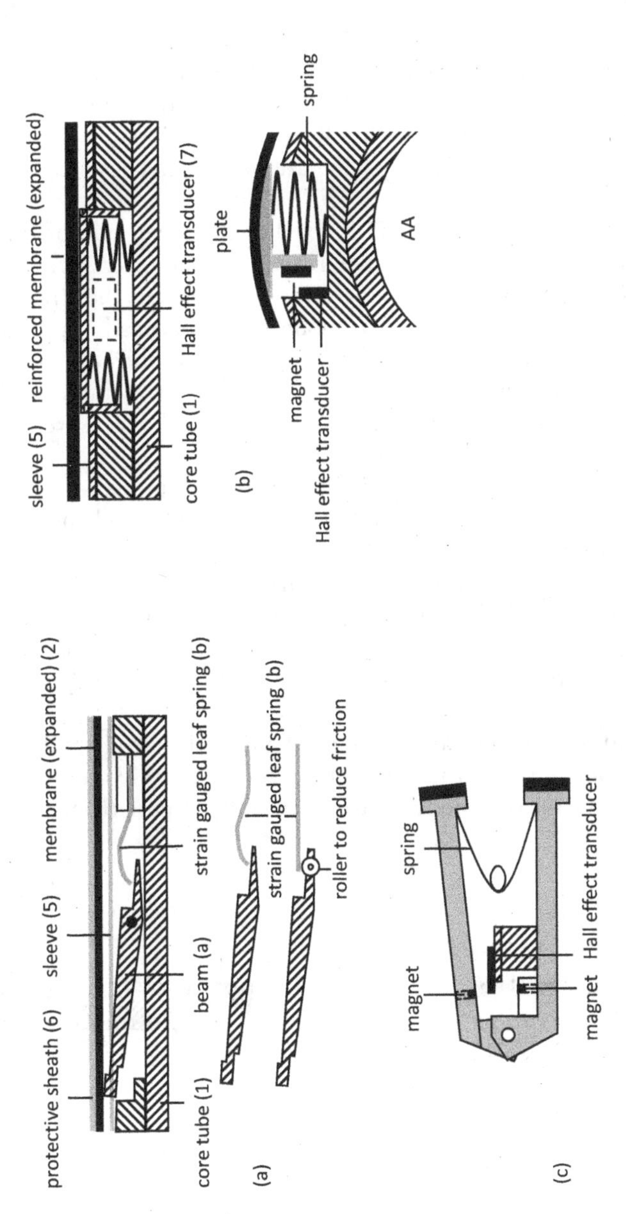

Figure 2.18 Displacement arms used in Cambridge type instruments (a) spring-loaded feeler (and with rollers to reduce friction), (b) Hall Effect transducer, and (c) spring-loaded arms.

Fahey and Jewell (1990) undertook a careful set of calibrations to investigate the effect of system compliance on the interpretation of results. They changed the design of the arms to incorporate rollers on the end to reduce friction between the arms and the membrane, as shown in Figure 2.18a.

Benoit et al. (1990) describe a modified CSBP, which includes nine displacement transducers mounted at three levels within the test section. This has the advantage that a better definition of the true deformed shape of the membrane can be measured. Transducers in PBPs are mounted diametrically opposite so that the change in diameter is measured, thus compensating for possible movement of the probe in the pocket. This has not been considered a problem with SBPs but there is some evidence that it does happen. Probes with six arms at 60° spacing are available. These measure the change in cavity diameter.

A total pressure cell (a) (Figure 2.17) is fixed at the lower end of the test section. It is a thin hollow disc with a strain gauge rosette on one face. A change in pressure deforms the diaphragm and, hence, changes the resistance of the strain gauge. There are two effective pressure transducers (d) mounted diametrically opposite just below the feeler arms. These cells measure the difference in the applied pressure and the pore pressure within the soil – that is, the effective stress at the membrane–soil interface. Figure 2.19 shows a section through a cell to demonstrate the measurement of effective stress.

Fahey et al. (1988) describe a pressure transducer that measures the pore pressure directly (Figure 2.19b). A miniature piezoelectric cell is sealed into the transducer housing so that it is isolated from the applied gas pressure.

A split tube (e) (Figure 2.17) is fixed above the pressuremeter section to protect the connections of the hydraulic hose (12) to the probe and to reduce the possibility of the membrane expanding up the annulus between the pocket and the drill rods (11). This tube can be extended to ensure that the top of the probe remains within a cased borehole.

The drilling head is formed of an internally chamfered shoe (15) into which the soil is forced as the probe is pushed into the ground. A cutter (13), mounted within the shoe, is rotated by means of inner drill rods (14), which pass through the core tube.

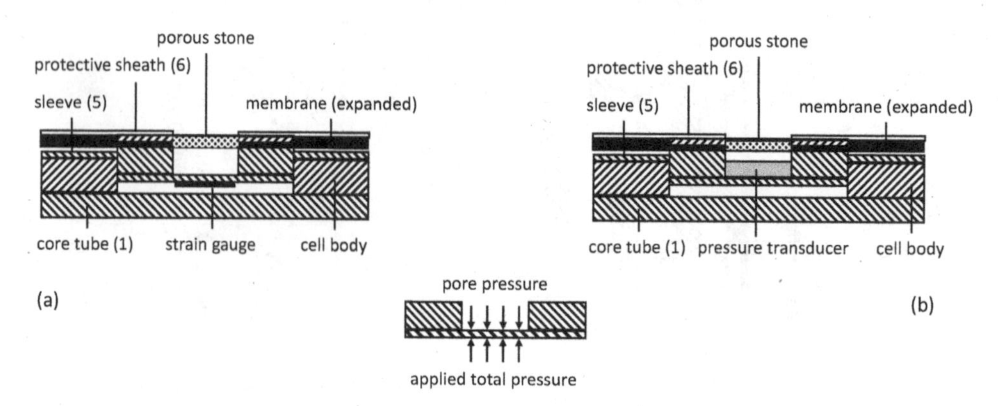

Figure 2.19 Effective stress cells using (a) a purpose-built strain-gauged diaphragm and (b) pressure transducers.

The soil is cut by the rotating cutter and the chippings are flushed to the surface by drilling fluid, which is pumped down through the inner rods and back up the annulus between the inner rods and the core tube. The inner rods are supported laterally and vertically within the drilling head by a bearing (f) mounted behind the cutter.

The CSBP is connected to the control unit (see Figure 2.3) by a hydraulic hose (12) (Figure 2.17), which contains an air passage and electric cable. A gas/electric separator unit is connected to the end of the hose so that the electric cable can be connected to the electronics unit and the hose to the strain control unit. The probe is pressurised with gas, usually nitrogen, though compressed air can be used providing it is dry.

A regulated DC voltage, usually 5 V, powers the transducers through the electric cable. It is normal to monitor the output from the transducers with a computer, which allows data to be reviewed on screen as a test proceeds, thus checking that the system is working. Further, it allows a test to be analysed on site. More recent developments of the probe include the amplifiers within the probe.

A test is controlled by a strain-control unit, which can either be purpose-built hardware or a computer. It is used to monitor the average output from the displacement transducers and to compare that with a preset rate of displacement. If the actual rate during expansion is less than that required, a solenoid valve is opened to allow pressurised gas into the test section. Alternatively, if the rate is greater than that required, another valve is opened and gas is vented to atmosphere. During contraction the roles of the valves are reversed.

Tests are described as strain-controlled but, in practice, they are displacement-controlled with an overriding limit on the rate of change of pressure. Various stress rate limits and strain rates can be set from 2.5 to 40 kPa/min and 0 to 2%/min respectively. They can be changed during a test if required.

2.4.3 The weak rock self-boring pressuremeter

Clarke and Allan (1989) described a development of the Cambridge type SBP for use in weak rocks: the weak rock self-boring pressuremeter (RSBP). It is not possible to drill the CSBP into rock because the shoe cannot penetrate the ground. It is necessary to use a cutter which protrudes from the drilling shoe. Rock roller and drag bits have been used, but they lack the quality of control required to ensure minimum disturbance drilling. Clarke and Allan (1989) proposed a full-face bit ((a) in Figure 2.20), which includes tungsten carbide tips arranged in such a way that one rotation of the bit ensures that the rock is removed over the full face of the shoe. The tips are set at a negative rake to reduce overbreak, thus the rock is ground away rather than chipped away. Tungsten carbide tips are mounted on the edge of the bit to ensure that the wall of the pocket is cut to an exact size determined by the diameter of the bit, which is set to the same diameter as that of the probe. The bit is turned by an inner rod (14) passing through the probe and connected to a rotary rig.

The 73 mm diameter probe can operate from the base of NX boreholes though it is more usual to operate from H or P holes. The membrane (2), with a length to diameter ratio of 5, can be expanded by up to 10% cavity strain at a pressure of 20MPa. It is made of natural rubber reinforced with nylon wire, which prevents axial stretching due to the friction arising from installation but allows radial expansion. The possibility

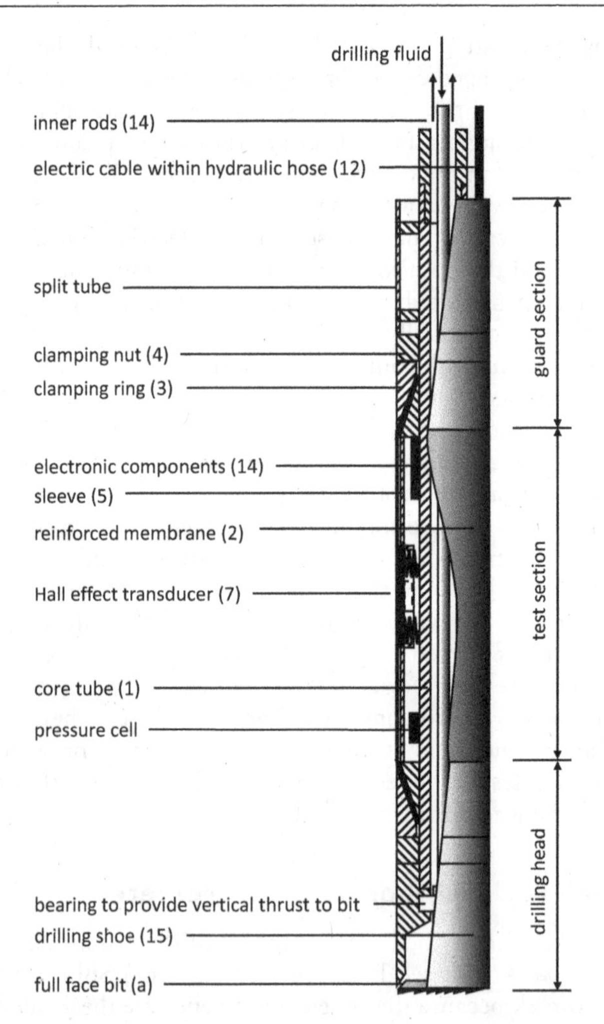

Figure 2.20 Schematic diagram of the weak rock self-boring pressuremeter.

Source: After Clarke and Allan, 1989.

of a burst during expansion is reduced as the nylon reinforces the membrane stretched over any discontinuity. The membrane is sealed onto the core tube with conical wedges (3), which are designed to provide a pressure seal of 20MPa and prevent the membrane being pulled out during installation.

Two hoses (12) are attached to the probe. One is used to pressurise the membrane with oil; the other permits air to be forced into the probe to remove oil when retrieving the probe from a dry borehole. Oil is used for safety reasons. As the expansion of the membrane is measured with transducers (7), it is not necessary to guarantee an air-free system. The second hose also contains an electric cable, which connects the pressure and displacement transducers to a microcomputer at the surface.

The displacement of the membrane is measured using spring-loaded plates (7), shown in Figure 2.18b. Plates are used to allow the transducer to bridge over any discontinuities in the rock occurring naturally or arising during expansion of the membrane (cf. HPD). Two springs are used, thus, the measured movement is the average displacement of the plate. The movement of a plate is monitored with a Hall Effect transducer.

2.4.4 Other self-boring pressuremeters

A number of self-boring probes have been developed, often for research purposes and specific environments such as offshore. Lushnikov and Bystrykh (1980) described a form of self-boring pressuremeter using an inflated PBP with a sacrificial sheath. A thin-walled casing is pushed 300 mm ahead of the borehole and the soil inside the casing is removed. The probe is lowered into the thin-walled casing and inflated, and the casing is withdrawn while the probe is further inflated to ensure that the pocket wall does not contract. The sacrificial sheath is used to overcome the friction between the casing and the PBP membrane as the casing is withdrawn.

Mori (1981) described two volume-displacement type self-boring pressuremeters developed specifically for testing soft clay. They are both 85 mm in diameter and have an expanding length of 440 mm. One is a tricell probe with a 210 mm measuring cell; the other is a monocell probe with a 440 mm long test section. The probes are drilled into the ground using the twin-rod system. Expansion of the membrane is controlled and monitored in a similar manner to that used with the MPM.

Fay and Le Tirant (1982) and Brucy et al. (1982) described a seabed module and probe designed to operate in 700 m of water. The seabed module contains a mud motor and houses a flexible drill stem, which is unreeled as the SBP drills itself down to a maximum depth of 60 m. The probe, based on the PAF-76, has a self-boring module below the test section, which has a length/diameter ratio of 2. In soft soils, the weight of the pressuremeter, drill stem and a drill collar give sufficient reaction for drilling, but in stiffer soils it is necessary to use a packer. This is pressurised against the borehole wall (or casing) and an internal jack uses that for reaction, as shown in Figure 2.21. The probe is 160 mm in diameter, 6m long and weighs 5kN. The seabed unit weighs 160 kN. Brucy et al. (1982) described the operation of this system at several offshore sites in a maximum water depth of 625 m.

Fay and Le Tirant (1990) described a modified version of the PAF-76, which is used with a wireline system. This 75 mm diameter probe, shown in Figure 2.22, has an expanding section 450 mm long. It can operate to a pressure capacity of 2 MPa and up to a cavity strain of 25%. The 7.5 m long system is contained in a housing, which is lowered inside the drill pipe on an umbilical cable. The system is latched onto the base of the drill pipe, which is used to push the probe into the soil as the soil is removed by a rotating cutter driven by an electric motor mounted immediately behind the test section. The membrane is expanded using an electrically driven pump and piston contained at the top of the system.

Campanella et al. (1990) described the University of British Columbia 73 mm diameter probe, which is similar to the CSBP. It is drilled into the ground using a jetting system rather than the twin-rod drilling system. Figure 2.23 shows two forms of jetting systems: one in which the jets are mounted on a central inner rod (a), the other in which the jets are mounted in the wall of the shoe (b). In the former case the position

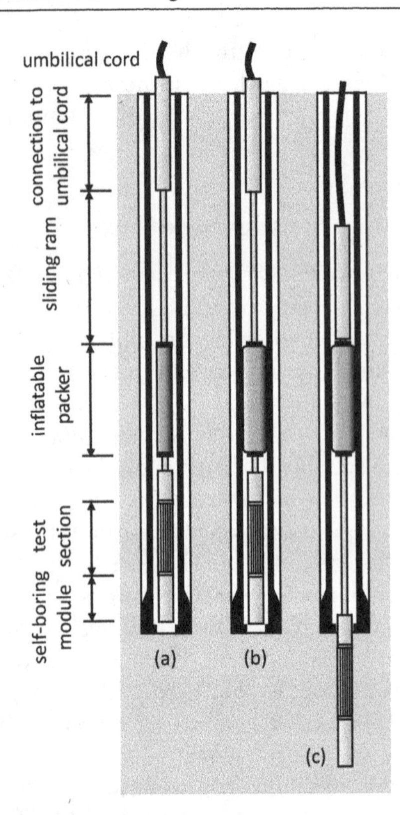

Figure 2.21 A seabed probe shows the use of an inflatable packer to provide reaction for drilling (after Brucy et al., 1982); (a) probe lowered to base of drill string, (b) packer inflated and (c) probe advanced to test location.

of the jet is adjusted by moving the rod. This has the advantage that obstructions due to collection of debris can be removed during installation. In the latter case the shoe is changed to adjust the jet position. Jetting systems have also been used by Benoit et al. (1990) with the CSBP, Jézéquel (1982) and Noret (1976) with the PAF, and Ménard (1974) with the MPM.

2.5 FULL-DISPLACEMENT PRESSUREMETERS

Pressuremeters pushed into soil are known as *pushed-in* pressuremeters and, if the soil is completely displaced, they are known as *full-displacement* pressuremeters (FDP). The jacking forces required are a function of the end resistance and the friction on the probe and, therefore, they can be used in the same soil as static penetrometers, depending on the reaction system. A FDP is pushed in from the ground surface or the base of a borehole in the same way as a penetrometer is pushed into soil. Partial displacement pressuremeters (such as the Stressprobe) are modified versions of PBPs and SBPs and were developed in the late 1970s primarily for use offshore. The FDP (or cone pressuremeter) was originally developed for use offshore, but recent versions are designed for onshore use with cone trucks.

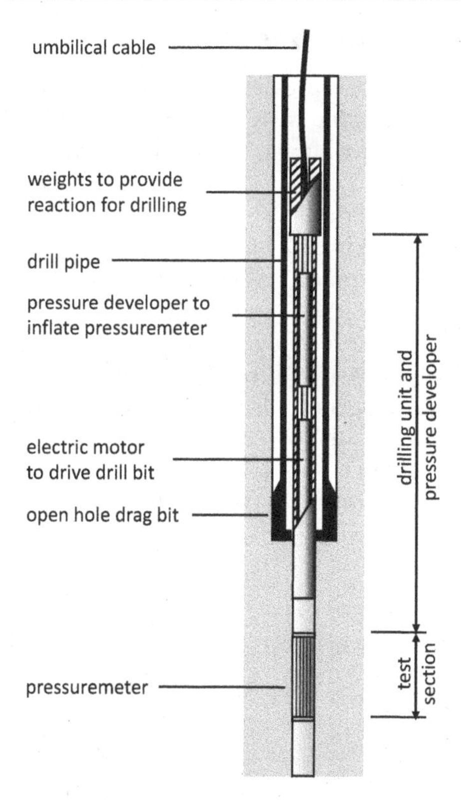

Figure 2.22 Schematic diagram of a wireline self-boring pressuremeter.

Source: After Fay and Le Tirant, 1990.

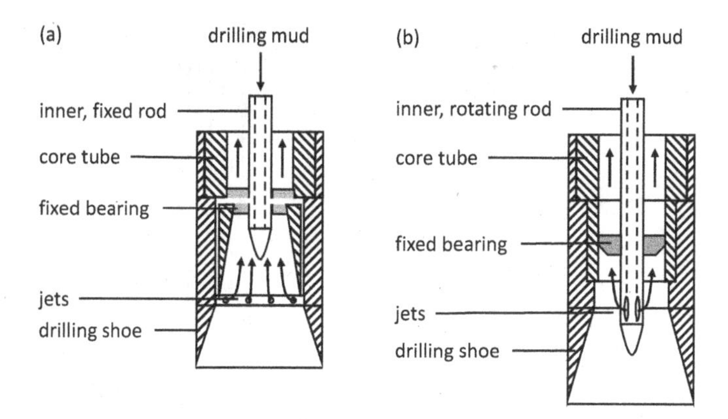

Figure 2.23 Forms of jetting systems: (a) with fixed jets and interchangeable shoe; (b) with inner rotating and moveable jets.

2.5.1 Full-displacement or cone pressuremeter

Jézéquel et al. (1982) described a pressio-penetrometer developed for offshore use, which comprises a 10 cm^2 piezocone with a pressuremeter of the same diameter as the cone mounted above the piezocone. It was a volume-displacement type monocell probe with an operating pressure of 2.5 MPa and a capacity of 100% volumetric strain.

Withers et al. (1986) described the development of a full-displacement pressuremeter, which was 44 mm diameter and over 1 metre long. The expanding section is 450 mm long giving a length/diameter ratio of 10. The general arrangement and form of construction are modelled on the CSBP.

The components of the probe are shown in Figure 2.24.

The probe consists of a hollow steel tube (1) over which is stretched the membrane (2). Three displacement transducers (7) formed of strain-gauged springs are mounted in recesses in the body of the probe at 120° spacing at mid height of the expanding section. The membrane is clamped in place with clamping rings (3) held in place by clamping nuts (4). A Chinese lantern (6), consisting of stainless steel strips backed by

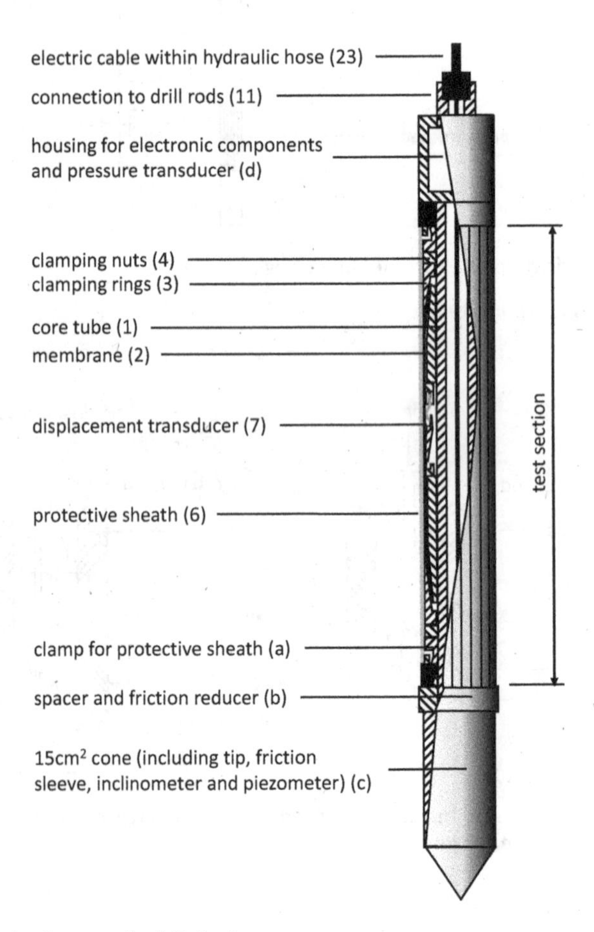

Figure 2.24 Schematic diagram of a full-displacement pressuremeter.

rubber, is used to protect the membrane. The probe is designed to expand by 50%, therefore it is necessary to allow the ends of the Chinese lantern to move. Contraction rings (a) are fitted at the top and bottom. A spacer (b), which can include a friction reducer, is attached to the bottom of the measurement module. The purpose of this is to ensure that uniform conditions exist around the expanding section after the probe is pushed to the test depth. A 15 cm^2 cone (c) is attached to that spacer. An instrument module (d), including a pressure transducer, is attached to the top of the probe, and the probe is connected to cone rods, which are used to push the probe into the ground.

Clarke et al. (2005) and Rehman et al. (2016) describe low-cost pressuremeters that could be driven into place using a dynamic probe rig. The probe, a full-displacement pressuremeter, included Hall Effect transducers (Figure 2.18c) to measure the change in diameter during expansion. The probe can be used in all soils including glacial tills.

2.5.2 The Stressprobe

Henderson et al. (1979) described a partial displacement pushed-in pressuremeter, the Stressprobe, developed from the CSBP for use offshore. Legier (1982) described a similar wireline device. The 78 mm diameter Stressprobe has a total length in excess of 3 m and can operate up to a pressure of 3.5MPa above hydrostatic pressure. There are three sections: a cutting shoe with a stress-relieving step, a pressuremeter module and a pressure developer (Figure 2.25). The probe is forced into the soil by the weight of the drill string. Some soil enters the hollow pressuremeter section; some is displaced. A core of soil can be taken prior to installing the probe to reduce the jacking forces and possibly the amount of disturbance.

The test section is similar to that of the CSBP, but it is a volume-displacement type probe. There is a hollow spacer behind the probe to ensure that the probe can be pushed far enough beyond the base of the borehole to test a zone unaffected by drilling.

The membrane is inflated with oil, which is pumped into the probe using a pressure developer mounted behind the spacer and within the drill pipe. The oil is forced into the probe by a piston (a) driven by an electric motor (b). The position of the piston is monitored by counting the number of turns electronically (c) of the electrically driven screw (d). The pressure in the probe is monitored with a transducer (e) within the pressure developer. A flexible membrane within the pressure developer ensures that the pressure in the outer case is the same as the external hydrostatic pressure. The applied pressure is the sum of the hydrostatic pressure and the pressure developed. Fyffe et al. (1986) described their experience with this probe at 14 sites in water depths up to 220 m. The maximum test depth, in calcareous sands off the coast of Australia, was 127 m below the seabed.

An updated version of this probe is the 47 mm diameter reaming pressuremeter supplied by Cambridge Insitu. This radial displacement monocell can operate up to 12MPa. It is designed to be used in medium to stiff clays, loose to dense sands and weathered or soft rock and can be lowered into a prebored hole slightly larger than the probe (cf. prebored pressuremeter) or pushed into a prebored hole slightly smaller than the probe.

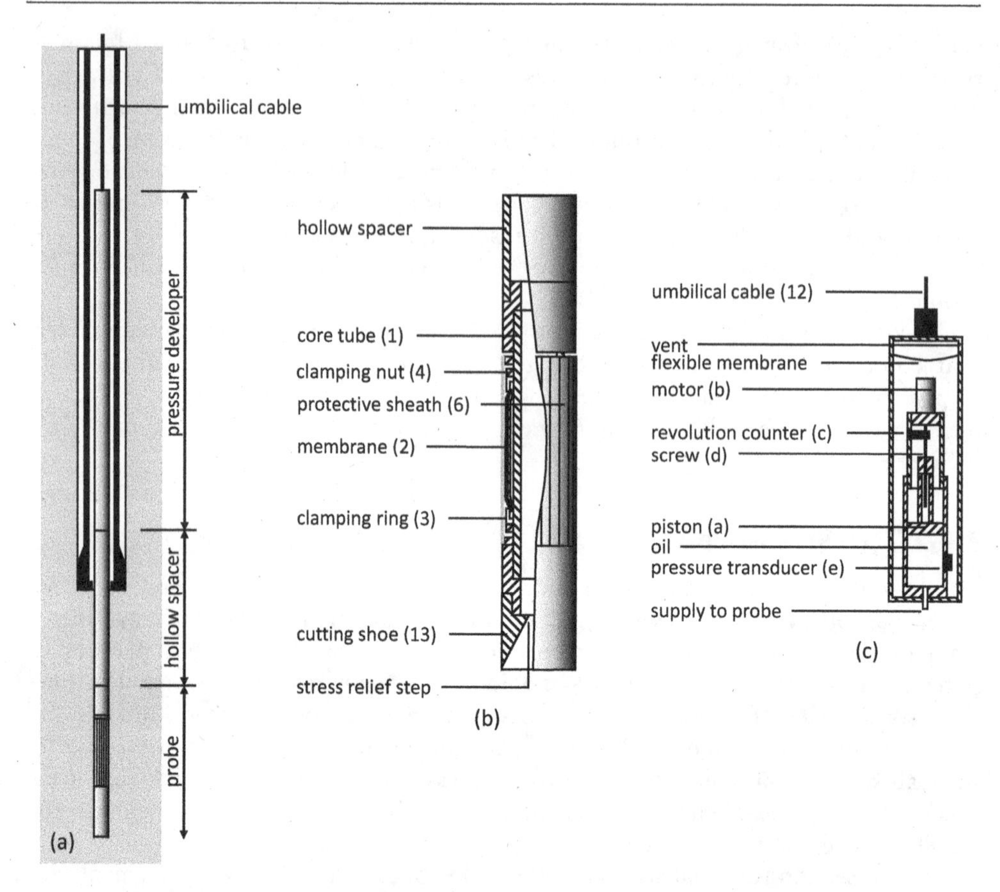

Figure 2.25 Schematic diagram of the Stressprobe, a pushed-in pressuremeter developed for offshore use (after Reid et al., 1982) showing (a) the probe being inserted in a pocket below the drill string, (b) the volume displacement probe, and (c) the means of pressurising and monitoring the expansion of the probe.

2.6 SPECIALIST PROBES

The use of model pressuremeters to support methods of analyses is described by Anderson et al. (1987), Huang et al. (1988), Haberfield and Johnston (1989) and Thorel et al. (2005). Anderson et al. (1987) and Haberfield and Johnston (1989) used a triaxial sample containing an inner membrane to represent the pressuremeter. Huang et al. (1988) described a miniature pressuremeter that was 'cast' in place in a clay slurry, which was allowed to consolidate around the probe. This model probe is 11.1 mm in diameter and 111 mm long. The latex membrane is expanded by pumping water into the probe either at a constant rate of flow or at a constant rate of pressure increase.

Deutsch et al. (1989) described a modified MPM probe used to determine acoustic emissions. They demonstrated that a PBP fitted with transducers in the membrane can measure the mechanical waveforms generated when soil is stressed. In particular, they confirmed the Kaiser effect – that is, acoustic emission increases dramatically when the soil is reloaded to the in situ stress. Other probes have been modified to include devices to measure pore pressure changes and changes in void ratio.

2.7 STANDARDS

National and international standards covering Ménard (BS EN ISO 22476-4:2021), prebored (BS EN ISO 22476-5:2012, ASTM D4719-20, Norme NF P 94-110 2000) self-bored (BS EN ISO 22476-6:2018), and full-displacement pressuremeters (BS EN ISO 22476-8:2018), exist.

Table 2.1 provides the dimension of the 58 mm diameter Ménard pressuremeter (known as the G type probe). This used in a BX pocket. 44 mm and 74 mm diameter probes for AX and NX pockets are available. The probe has to be able to expand by at least 700 cm^3 or 550 cm^3 for a probe with a short measuring cell within a slotted tube.

The dimensions of other prebored pressuremeters are not specified in the standards but, given that the pressuremeters are used in predrilled pockets, they are typically some 3 mm to 6 mm smaller in diameter than standard borehole diameters of 76 mm (NX), 96 mm (HX) and 101 mm. The expansion of the probe is monitored with three or more electrical transducers, either with transducers that measure the diameter of the expanding pocket or the internal diameter of the membrane. The expanding length of the probe must be at least 5.5 times the probe diameter plus the length of the measuring section, which is the length over which the transducer functions. The capacity of prebored pressuremeters for rock testing shall be at least 20MPa. There is no guide given on the range of the displacement measurements but when testing rock they should be at least 10% of the pocket diameter (not the probe diameter); it is unlikely that the probe will expand that much because of the stiffness of the rock. In soils, an expansion of 10% of the pocket diameter allows a full-stress–strain curve to be obtained, from which it is possible to derive the stiffness and strength of the soil.

The dimensions of self-boring probes are not specified in the standard, but the length of the expanding section should be at least four times the probe diameter and the probe should be capable of being expanded by at least 25% of the initial volume of the expanding section. The capacity of the probe should be at least 4MPa.

Full-displacement pressuremeters used with conical tips are also known as *cone pressuremeters*. There are two types, depending on whether the pressuremeter is attached to a cone penetrometer (TC probe) or not (T probe). In the former case, the probe diameter must be between 25 mm and 50 mm. No specification is given for the T probe other than it must have a 60-degree conical tip. The distance between the cone module and the bottom of the expansion section must be less than ten times the maximum cone diameter. The external diameter of the probe must be equal to or less than the cone diameter. It should be possible to increase the radius by 30% or the volume by 75%. Volume and radial displacement probes are possible with the expansion of radial displacement probes monitored at three equidistant points or across a single diameter. No pressure is specified but, given that the probe will be used in soils, it suggests a maximum pressure of 4 MPa.

Apart from the Ménard pressuremeter, there is some flexibility in the size of other pressuremeter probes. However, there are certain principles, which must be satisfied if the results of the tests are to be of value. These are listed in Table 2.2.

Table 2.2 The principles of the pressuremeter probe

Component	*Parameter*	*Comment*
Probe	Length	The expanding section of a probe fits between two sections of a similar diameter. The lower section may be part of the installation system in the case of pushed and self-bored probes. In the case of prebored probes, the lower section may contain part of the monitoring system. In either case, the lower section ensures that the expanding section is beyond the zone of influence of the base of the test pocket. The upper section prevents the membrane expanding into the test pocket or borehole above the probe. It may also contain elements of the control and/or the monitoring system.
Expanding section	Diameter	The probe may be used in a predrilled pocket or drilled from the base of a borehole. In that case the diameter of the probe must be smaller than the pocket or borehole. For prebored pressuremeters, the ratio of the pocket diameter to the probe diameter is 1.1.
	Length to diameter ratio	There is no restriction to the maximum length but the minimum length should be at least five.

2.8 SUMMARY

Pressuremeters in commercial use include monocells and tricells; and radial and volume-displacement types. They cover a range of diameters and lengths, giving length/diameter ratios of between 2 and 10. The pressure capacities vary up to 100 MPa depending on whether they are used to test soil or rock. In addition to the commercially available pressuremeters, a number of pressuremeters have been developed as research tools, such as those used in calibration chambers.

Pressuremeters are designed to simulate the expansion of an infinitely long cylindrical cavity. However, for practical reasons there is a limit to the length of the expanding session. Tricells have inflatable guard cells to ensure that the central section used to measure the volume change expands as a cylinder; monocells rely on the fact that the end restriction has little effect on the assumption of a cylindrical expansion if the length to diameter ratio is at least five.

Thus, pressuremeters are very similar in appearance. However, the membranes, and the systems to hold the membrane in place and protect the membrane during installation vary between manufacturers. Further, the control and monitoring systems vary depending on whether changes in volume or displacement are measured; the test is stress- or strain-controlled; the monitoring and control system are downhole or at the surface; the fluid used to expand the membrane; and the means of recording the data. Examples of commercially available pressuremeters are used to highlight these differences.

Pressuremeters are grouped according to the installation system. Prebored pressuremeters can be used in any ground condition, though the sensitivity of the probe must be changed to suit the strength and stiffness of the ground. Some probes are designed for specific ground conditions and should not be used in other materials without major modification. Self-boring pressuremeters were developed for soils, though they can be used in weak rocks if sufficiently robust. Full-displacement pressuremeters are used in soils.

REFERENCES

Aissaoui, S., Zadjaoui, A. and Reiffsteck, P. (2021) A new protocol for measuring small strains with a pressuremeter probe: Development, design, and initial testing. *Measurement, 169*, p. 108507.

Amar, S., Frank, R., Baguelin, F. and Jézéquel, J.F. (1977) The self-boring placement method and soft clay investigation, Int. Symp. Soft Clay, Bangkok, pp. 337–357.

Anderson, J-B, Townsend, F.C., Horta, E. and Sandoval, J. (2005) Effects of modified probes and methods on PENCEL pressuremeter tests, Symposium International ISP5/PRESSIO 2005, Gambin, Magnan & Mestat, eds., pp. 21–30.

Anderson, W.F., Pyrah, I.C. and Ali, F.H. (1987) Rate effects in pressuremeter tests in clays. *Journal of Geotechnical Engineering, 113*(11), pp. 1344–1358 ASTM Standard D4719–20 (2020) *Standard Test Method for Pressuremeter Testing in Soils.*

Arsonnet, G., Baud, J.P., Gambin, M. and Heintz, R. (2011) HyperPac 25 MPa fills the gap between the Ménard pressuremeter and the flexible dilatometer. *Geotechnics of hard soils and weak rocks, XV ECSMGE,* Athens.

Benoit, J., Oweis, I.S. and Leung, A. (1990) Self-boring pressuremeter testing of the Hackensack Meadows varved clays, *Proc. 3rd Int. Symp. Pressuremeters,* Oxford, pp. 85–94.

Briaud, J.-L. and Shields, D.H. (1980) Special pressuremeter and pressuremeter test for pavement evaluation and design, *Geotech. Test. J., ASTM,* 2(3), 143–151.

Brucy, F., Fay, J.B. and Marignier, J. (1982) Offshore self-boring pressuremeter: the LPC/IFP probe, *Proc. Int. Symp. Pressuremeter and its Marine Appl.,* Paris, pp. 127–142.

BS EN ISO 22476-4:2021, *Geotechnical Investigation and Testing. Field testing – Prebored pressuremeter test by Ménard procedure*, British Standards Institution.

BS EN ISO 22476-5:2012, *Geotechnical Investigation and Testing. Field testing - Flexible dilatometer test*, British Standards Institution.

BS EN ISO 22476-6:2018, *Geotechnical Investigation and Testing. Field testing - Self-boring pressuremeter test*, British Standards Institution

BS EN ISO 22476-8:2018, *Geotechnical Investigation and Testing. Field testing - Full displacement pressuremeter test*, British Standards Institution

Campanella, R.G., Stewart, W.P. and Jackson, R.S. (1990) Development of the UBC self-boring pressuremeter, *Proc. 3rd Int. Symp. Pressuremeters,* Oxford, pp. 65–72.

Clarke, B.G. and Allan, P.G. (1989) Self-boring pressuremeter for testing weak rock, *Proc. 12th Int. Conf. SMFE, Rio de Janeiro, Brazil,* Vol. 1, pp. 211–213.

Clarke, B.G., Allan, P.G., Akbar, A. and Irvine, J. (2005) A simple, robust pressuremeter to tests glacial till. In *ISP5 – Pressio 2005 – Volume 1: Symposium International – 50 ans de pressiomètres, Marne La Vallee, 22-24 AOUT 2005:* (Vol. 1).

Cour, F., Puech, A. and Durand, F. (2005) Un pressiomètre de nouvelle génération. 2005. Proc. ISP5-PRESSIO (1), 63–73.

Deutsch, W.L., Koerner, R.M. and Lord, A.E. (1989) Determination of prestress of in situ soils using acoustic emissions, *J. Geotech. Engng Div., ASCE,* 115(2), 228–245.

Fahey, M. and Jewell, R. (1990) Effect of pressuremeter compliance on measurement of shear modulus, *Proc. 3rd Int. Symp. Pressuremeters,* Oxford, pp. 115–124.

Fahey, M., Jewell, R.J. and Brown, T.A. (1988) A self-boring pressuremeter system, *Geotech. Test. J., ASTM,* 11 (3), 187–194.

Fay, J.B. and Le Tirant, P. (1982) Offshore self boring pressuremeter for deep water, *Proc. Int. Symp. Pressuremeter and its Marine Appl.,* Paris, pp. 305–323.

Fay, J.B. and Le Tirant, P. (1990) Offshore wireline self-boring pressuremeter, *Proc. 3rd Int. Symp. Pressuremeters,* Oxford, pp. 55–64.

Frikha, W. and Varaksin, S. (2017) Auto-Controlled Ménard Pressuremeter: A Novel Tool for Optimal Use of the Pressuremeter. In *International Congress and Exhibition: "Sustainable Civil Infrastructures: Innovative Infrastructure Geotechnology"* (pp. 252–268). Springer, Cham.

Fyffe, S., Reid, W.M.. and Summers, J.B. (1986) The push-in pressuremeter; 5 years of offshore experience, *Proc. 2nd Int. Symp. Pressuremeter Marine Appl., Texam,* United States, ASTM STP 950, pp. 22–37.

Haberfield, C.M. and Johnston, I.W. (1989) Model studies of pressuremeter testing in soft rock, *Geotech. Test. J., ASTM,* 12(2), 150–156.

Henderson, G., Smith, P.D.K. and St John, H.D. (1979) The development of the push-in pressuremeter for offshore site investigation, *Proc. Conf Offshore Site Invest.,* Society for Underwater Technology, London, pp. 159–167.

Huang, A.B., Holtz, R.D. and Chameau, J.L. (1988) A calibration chamber for cohesive soils, *Geotech. Test. J., ASTM,* 30–35.

Hughes, J.M.O. and Ervin, M.C. (1980) Development of a high pressure pressuremeter for determining the engineering properties of soft to medium strength rocks, *Proc. 3rd Aus.-NZ Conf. Geomechanics,* Wellington, Vol. 1, pp. 243–247.

Jézéquel, J.F. (1982) The self-boring pressuremeter, *Proc. Int. Symp. Pressuremeter Marine Appl.,* Paris, pp. 111–126.

Jézéquel, J.F., Lemasson, H. and Touzé, J. (1968) Le pressiomètre Louis Ménard quelques problèmes de mise en oeuvre et leur influence sur les valeurs pressiométriques, *Bull, de Liaison du LCPC,* No. 32, 97–120.

Jézéquel, J.F., Lamy, J.L. and Perrier, M. (1982) The LPC-TLM pressio-penetrometer, *Proc. Int. Symp. Pressuremeter and its Marine Appl.,* Paris, pp. 275–287.

Legier, A. (1982) The wireline Expansometer. In *Symp. on the Pressuremeter and Its Marine Applications,* Paris, pp. 263–274.

Lushnikov, V.V. and Bystrykh, V.F. (1980) Pressuremeter for testing soft soils, *Soil Mech. Found. Engng,* 19, 262–265.

Ménard, L. (1974) Intérêt technique et économique du vibro-marteau hydraulique annulaire pour le prélèvement d'échantillon en mer et la réalisation d'essais géotechniques in-situ, *2e Colloque Int. l'Exploitation des Oceans,* Bordeaux, pp. 1–16.

Ménard, L. (1957) Mesures in situ des propriétés physiques des sols, *Annales des Ponts et Chaussées,* Paris, No. 14, 357–377.

Mori, H. (1981) Study on the properties of soils in the northern coast of Tokyo Bay using a self-boring pressuremeter, *Soils and Founds,* 21 (3), 83–98.

Noret, H. (1976) *Procédé de penetration et d'instrumentation simultane des sols dit pénétro-statique,* Centre d'Etudes et de Construction de Prototypes d'Angers, No. AER 05–22.

Norme NFP 94-110-1 2000, Essai pressiométrique Ménard, partie 1: essai sans cycle, AFNOR, Paris

Rehman, Z., Akbar, A., Khan, A.H. and Clarke, B.G. (2016) Correlations of Pressuremeter Data with SPT, CPT and Laboratory Tests Data. *Pakistan Journal of Engineering and Applied Sciences.*

Reid, W.M., St. John, H.D., Fyffe, S. and Rigden, W.J. (1982) The push-in pressuremeter, Proc. Int. Symp. Pressuremeter and its Marine Appl., Paris, pp. 247–261.

Rocha, M., Silveira, A.Da., Grossman, N. and Oliveira, E.De. (1966) Determination of the deformability of rock masses along boreholes, *Proc. 1st Congr. ISRM, Lisbon,* Vol. 1, pp. 697–704.

Ruppeneit, K.V. and Bronshtein, M.I. (1972) Dilatometer determination of soil deformation and strength properties, *Soil Mech. Found. Engng,* 15, 318–322.

Shaban, A. and Cosentino, P. (2017) Characterizing structural performance of unbound pavement materials using miniaturized pressuremeter and California bearing ratio tests. *Journal of Testing and Evaluation*, *45*(13), pp. 1–18.

Suyama, K., Ohya, S. and Imai, T. (1966) Studies of transverse K-value of ground, *Soils and Founds,* 14(10).

Thorel, L., Rault, G., Gaudin, C., Garnier, J. and Favraud, C. (2005) Un pressiométre miniature pour la caractérisation des massifs de sol en centrifugeuse. In *ISP5 – Pressio 2005 - Volume 1: Symposium international – 50 ans de pressiomètres, Marne La Vallee, 22–24 AOUT 2005:* (Vol. 1).

Withers, N.J., Schaap, L.H.J. and Dalton, J.C.P. (1986) The development of a full displacement pressuremeter, *Proc. 2nd Int. Stmp. Pressuremeter Marine Appl., Texam, USA,* ASTM STP 950, pp. 38–56.

Wroth, C.P. (1973) A brief review of the application of plasticity to soil mechanics. In *The Proceedings of the Symposium on the Role of Plasticity in Soil Mechanics,* September, 13–15, 1973, Cambridge, UK.

Wroth, C.P. and Hughes, J.M.O. (1973) An instrument for the in-situ measurement of the properties of soft clays, *Proc. 8th Int. Conf. SMFE,* Moscow, Vol. 1.2, pp. 487–494.

Chapter 3

Site operations

3.1 INTRODUCTION

The majority of published work on pressuremeters has concentrated on the analyses of tests, comparison of the results with those from other tests and use of the results in design. However, the results depend on the probe, its installation and test procedure. It is necessary to follow correct installation and testing procedures, ensure that the probe and testing equipment are functioning correctly and calibrations are regularly checked if there is to be confidence in the results.

A correct installation procedure ensures either minimum or repeatable disturbance. There are a variety of test procedures including stress- and strain-controlled tests. The data, including the applied pressure and displacement of the membrane, can be recorded either manually, taking readings from gauges at the surface, or automatically with a data logger. Calibrations are used to convert output from transducers to stress and strain. Corrections have to be applied for the stiffness and compression of the membrane and system compliance.

There are national and international standards, recommended procedures and manufacturers' guidelines. Standards and recommendations for site operations include BS 5930:2015+A1:2020, the French (NF P 94-110 2000), the Russian (GOST 20276-12:2012), International Society for Rock Mechanics (ISRM (1987)), the American (ASTM D4719–20:2020) and International Standards (BS EN ISO 22476-4:2021, BS EN ISO 22476-5, BS EN ISO 22476-6 and BS EN ISO 22476-8).

3.2 INSTALLATION TECHNIQUES

3.2.1 Introduction

Table 3.1 is a summary of the applicability of pressuremeters in a variety of ground conditions, showing that a pressuremeter can be used in all types of ground.

A pressuremeter probe is installed in a test pocket, either by lowering, drilling or pushing. A pocket designed for a pressuremeter test is created by a drilling rig or a probe, depending on the type of probe. It can be created from the base of a borehole and, in that case, the borehole diameter will be greater than the pocket diameter. Drilling techniques used to take samples are different from those used to create a test pocket. When creating a test pocket for any in situ test, it is important to minimise disturbance to the surrounding soil; when taking a sample, it is important to minimise disturbance to the sample. Boreholes have to be stable.

DOI: 10.1201/9781003028925-3

Table 3.1 The applicability of pressuremeters to ground conditions

Ground type	*PBP*	*SBP*	*FDP*
Soft clays	A	A	A
Stiff clays	A	A	A
Loose sands	B with support	A	A
Dense sands	B with support	B	C
Gravels	C by driving	N	N
Weak rock	A	B	N
Strong rock	A	N	N

Note: A, very good; B, good; C, moderate; N, not possible.

Table 3.2 Categories of weak rocks for testing for mass strength and static modulus

			In situ *tests*				
Category	*Borehole stability*	*Sample quality*	*SPT*	*Stress cells*	*Plate*	*PBP*	*SBP*
A	Good	a	a	A	a	a	d
B	Fair	b/c	a	b/c	b	b	c
C	Poor	d	b	D	c/d	c/d	a

Note: a, very good; b, good; c, moderate; d, poor.

Prebored pressuremeters (PBPs) can be used in any soil or rock in which it is possible to create a test pocket that remains stable, either with or without water or mud. self-boring pressuremeters (SBPs) can be used in soils containing little to no gravel and in some weak rocks. Full displacement pressuremeters (FDPs) can be used in soils in which it is possible to push a cone. Pocket stability is not critical for SBPs and FDPs, since the probe supports the pocket wall during installation and testing, though care has to be taken to prevent a borehole collapse above the probe.

Clarke and Smith (1992b) proposed that weak rocks could be grouped according to their suitability for in situ tests and sampling, as shown in Table 3.2.

Category A is competent rock. Weathered rock usually falls within categories B and C. Category B is defined as a rock from which it is possible to obtain core for representative laboratory testing and stable pockets can be created for in situ tests. It is neither possible to obtain representative core sufficient for laboratory testing nor to create a stable pocket in category C type rocks. PBPs can be used in category A and B rocks and, possibly in category C rocks if using slotted casing. The rock SBP may be used in category B and C rocks depending on the strength of the rock and the design of the drilling system.

3.2.2 Ground conditions

A pressuremeter can be used to give an assessment of the in situ stress, stiffness and strength of the ground. The quality of the information depends, in part, on the installation technique which, in turn, is affected by the ground conditions. For example:

- The strength of unweathered rock means that a test pocket has to be formed to carry out a pressuremeter test. Thus, it is only possible to carry out PBP tests in unweathered rock. Further, the PBP can only be used to determine the stiffness of the rock because of the strength of the rock;
- PBP and possibly SBP tests can be carried out in weak and weathered rock;
- Any boulders and cobbles (very coarse-grained soils) will obstruct the drilling process to the extent that it is unlikely that pressuremeter tests will be carried out if very coarse-grained soils form the primary fraction. Occasional, very coarse grained particles would prevent a test pocket being formed but, depending on their frequency, it may be possible to create test pockets between the very coarse particles. It is possible to carry out a PBP test in slotted casing driven into place in very coarse-grained soils;
- Gravels and sands (coarse-grained soils) are generally not self-supporting. Therefore, test pockets and boreholes have to be supported. A borehole may be cased or drilled using mud. PBP tests can be carried out in slotted casing driven into place or in mud-filled test pockets. A test pocket is supported by the probe when using SBP and FDP probes in sands;
- Silts and clays (fine-grained soils) can be self-supporting though it is prudent in deeper boreholes to support the boreholes above the test pocket to prevent collapse.

3.2.3 Prebored pressuremeters

Prebored pressuremeters, including the Menard pressuremeter, are usually lowered into a predrilled test pocket. There will inevitably be some disturbance to the surrounding ground as the pocket is formed. This disturbance is due to:

1. *Change of total in situ stress.* Forming a pocket reduces the total stress as the ground is removed. The change will depend on the support to the test pocket used during drilling. The stress on a pocket wall can vary from zero in a dry pocket to the pressure of the flushing medium such as water or drilling mud.
2. *Collapse of the pocket.* The pocket may collapse if the total stress is reduced sufficiently to cause the ground to fail in extension. Pieces of the pocket wall may fall into the pocket, especially if the ground contains discontinuities.
3. *Erosion of the pocket walls.* The sides of the pocket may be eroded during drilling by the passage of the drilling fluid and ground particles.
4. *Softening of the pocket walls.* There is likely to be a reduction in pore pressure in the ground adjacent to the pocket. This may give rise to suction pressures in fine grained soils and, in the presence of drilling fluid, reduce the in situ effective stress causing swelling and, therefore, softening.
5. *Mechanical disturbance.* The drill bits may disturb the pocket walls due to either vibration or eccentric loading.

The borehole should be designed for pressuremeter tests only. The design criteria are hole size, minimum disturbance during drilling and minimum disturbance during removal of drill rods and drill bit. The pocket and borehole diameters are governed by the choice of probe, while the type of ground influences the drilling method chosen. Holes can be formed by either rotary or percussive drilling methods: rotary drilling includes hand auguring, continuous flight augers, open-hole techniques and core drilling; percussive drilling includes driven samples and downhole hammers.

Recommended methods for forming test pockets are discussed by Baguelin et al. (1978), Briaud and Gambin (1984), Finn et al. (1984), Mair and Wood (2013) and BS EN ISO 22476-4:2021, -5, -6 and -8. They are summarised in Tables 3.3 and 3.4.

BS EN ISO 22476-4:2021 describes in some detail the placing of the Ménard probe in the ground. The first test in the ground must be at least 0.75 m below ground level; the first test in a test pocket below a borehole must be at least 0.75 m below the base of the borehole. Subsequent tests have to be at least 0.75 m apart. Test pockets have to be such that the top of the expanding section is at least 0.5 m from the top of the pocket. There must be at least 0.3 m below the expanding length of the probe to the bottom of the test pocket.

Test pockets can be created by open-hole drilling, continuous flight auguring, rotary core drilling; rotary percussive drilling; by pushing, driving or vibrodriven tubes; or by pushing a pressuremeter in with a slotted tube. The probe shall be inserted into the test pocket as soon as possible after drilling, using one of the techniques shown in Table 3.4. Normally there should be one test per pocket, but it is feasible to create a pocket long enough to carry out several tests (Table 3.5).

The techniques given in Tables 3.4 and 3.5 can be used as a guide for installing prebored pressuremeters. Recommended diameters for test pockets for the Menard pressuremeter (MPM) are given in Table 3.6.

The installation procedure for the flexible dilatometer is described in BS EN ISO 22476-5. In this case the pocket shall be drilled in accordance with BS EN ISO 22475-1:2021. The borehole diameter should be either 76 mm, 96 mm, or 101 mm, depending which probe is used, and the probe diameter when deflated should be 3 mm to 6 mm smaller than the borehole diameter. The base of the borehole shall be at least 1 m above the test depth, and the test pocket should be about 3 m. The probe shall be installed within two hours of coring. The top of the expanding section must be at least 0.5 m below the base of the borehole and the bottom of the expanding section 0.5 m above the base of the pocket.

The quality of the test pocket has a significant effect on the quality of a test. The quality of a test pocket depends on the type of ground, the method of creating the pocket and the operator's experience.

The following factors have to be considered when selecting the most suitable method for given ground conditions:

(a) diameter of the pocket and borehole
(b) inclination of the pocket
(c) possible pocket and borehole collapse
(d) erosion of the pocket wall due to uphole velocity of flushing medium and size of cuttings
(e) softening of wall due to soil absorbing water from the flushing medium
(f) presence of gravel leading to irregular holes
(g) depth and spacing of tests.

Table 3.3 Recommended methods for creating test pockets

Soil type		*Hand auger*	*Hand auger with mud*	*Flight auger*	*Driven sampler*	*Driven slotted tube*	*Pushed sampler*	*Pilot hole and pushed sampler*	*Pilot hole and shaving*	*Core barrel*	*Rotary percussion*	*Open hole drag bit*
Clays:	Soft	2B	1	NR	NR	NR	2B	2	2	NR	NR	2B
	Firm to stiff	1B	1	1B	NR	NR	1	2	2	NR	NR	1B
	Stiff to hard	NA	NA	1B	2	NR	2	1	1	1B	NR	1
Silts:	Above GWL	1	2	NR	2	NR	2B	2	2B	NR	NR	1B
	Under GWL	NR	1	NR	NR	NR	NR	NR	2B	NR	NR	1B
Sands:	Loose and above GWL	2	1	2	2	NR	NR	NR	2	NA	NR	1B
	Loose and below GWL	NR	1	NR	NR	NR	NR	NR	2	NA	NR	1B
	Medium to dense	1	1	2	2	NR	NR	NR	2	NR	2B	1B
Sand and gravel:	Loose	NA	NA	NA	NR	2	NA	NA	NA	NA	2	2
	Dense	NA	NA	NR	NR	1D	NA	NA	NA	NA	2	NR
Rock:	Weathered	NA	NA	NA	1	NR	NA	2B	NA	1	2	1
	Strong Rock	NA	NA	NA	NA	NA	NA	NA	NA	1	2B	2B

Source: After Finn et al., 1984, ASTM D4719–20:2020, Amar et al., 1991.

Note: 1, recommended; 2, acceptable; NR, not recommended; NA, not applicable; B, conditional; D, pilot hole drilled first.

Table 3.4 Guidelines for installing the MPM, which can also be applied to prebored pressuremeters

	Rotary Drilling				*Rotary Percussion*			*Sampling Tube*			*Slotted Tube*
Boring Technique	*OHD*	*HA/HAM*	*CFA*	*CD*	*RP*	*RPM*	*STDTM*	*PT*	*DT*	*VDT*	*DST*
Sludge and soft clay	B	A	D	D	NA	D	D	A	NA	NA	C
Soft to firm clayey soils	A	A	B	B	D	C	C	C	C	D	NA
Stiff clayey soils	A	B	A	A	C	B	B	D	C	D	NA
Silty soils above gwl	B	A	B	B	D	C	B	C	C	C	D
Silty soils below gwl	C	B	D	C	D	C	B	D	D	D	C
Medium dense/ dense sandy soils	A	A	A	C	C	B	B	D	C	C	B
Gravel and cobbles	B	D	D	D	C	B	B	NA	C	C	A
Cohesive non-homogeneous soils	B	C	C	B	C	A	C	NA	C	C	NA
Loose non-homogeneous soils	B	C	C	C	C	B	B	D	C	C	B
Weathered rock	A	B	B	B	C	B	B	NA	C	C	NA

Source: After BS EN ISO 22476-4:2021.

Note: OHD – open hole drilling; HA – open hole drilling with a hand augur; HAM – open hole drilling with a hand augur and mud; CFA – continuous flight augur; CD – core drilling; RP – rotary percussion; RPM – rotary percussion with mud; STDTM – slotted tube with integrated drill bit; PT – pushed tube; TWT – pushed thin wall tube; DT – driven tube; VDT – vibro driven tube; DST – driven slotted tube

A – recommended; B – possible; C – acceptable; D – not acceptable; NA – no details

Table 3.5 Maximum length of pocket for multiple MPM tests which could be applied to PBP tests

	Maximum length of pocket for multiple testing [m]		
Soil Type	*Adapted rotary drilling*	*Rotary percussive drilling*	*Smooth tube pushing, driving and vibrodriving*
Sludge and soft clay	1	-	1
Firm clayey soils	2	2	3
Stiff clayey soils	5	4	4
Silty above gwl	4	3	3
Silty soils below gwl	2	1	-
Loose sandy soils above gwl	3	2	-
Loose sandy soils below gwl	1	1	-
Medium dense and dense sandy soils	5	5	4
Coarse soils; gravels, cobbles	3	5	3
Coarse soils with cohesion	4	5	3
Loose non-homogeneous soils (e.g. till)	2	3	2
Weathered rock, weak rock	4	5	3

Source: After BS EN ISO 22476-4:2021.

Table 3.6 Recommended diameters of test pockets for the Menard pressuremeter

Probe		*Diameter of the test pocket [mm]*	
	Diameter [mm]	*Min*	*Max*
AX	44	46	52
BX	58	60	66
NX	70/74	74	80

Source: After BS EN ISO 22476-4:2021.

Table 3.7 Recommendations for drilling test pockets for a prebored pressuremeter

Drilling a test pocket	*Drilling between test locations*
The pocket diameter is critical.	The borehole diameter is not critical.
Disturbance of the soil surrounding the test pocket should be kept to a minimum.	Disturbance of the soil surrounding the test pocket is not critical.
Slow rotation to control borehole diameter (typically 60 rpm).	Fast rotation to advance to the next test location.
Ensure that the disturbance of the pocket walls is kept to a minimum.	The condition of the wall of the borehole is not critical but it is prudent to ensure that the borehole is stable.
The pressuremeter is suspended in the test pocket to allow any cuttings created when forming the test pocket to settle in the base of the pocket.	The base of the borehole should be clean to prevent cuttings collecting in the test pocket.
The pocket should not be cleaned after drilling to prevent damage to the test pocket walls.	All cuttings should be removed once the borehole has reach the depth to the top of the test pocket to remove any cuttings suspended in the drilling mud.

Source: After Briaud, 2013.

The soil profile must be ascertained before carrying out tests so that the correct drilling technique can be chosen. Briaud (2013) recommendations for creating a borehole and test pocket are given in Table 3.7.

The initial tests should be analysed on site to assess the result of the drilling methods, which is to ensure minimum disturbance. Once the correct drilling technique is established for a typical soil on a particular site, it should be used so that the amount of disturbance is a common factor. It is important that the soil is uniform along the length of the membrane to prevent irregular expansion of the membrane. Tests should not be carried out at the interface of two stratum.

Hand auguring gives the most control over the dimensions of a pocket. It is limited to shallow depths and to soils which will remain standing unsupported. It may be used in stiff clays and, possibly, in sands above the water table. It is not practical to use it in weak rocks. Practically it is limited to about 5 m because of the need to remove the auger at frequent intervals. It is time-consuming, and the presence of gravel can lead to refusal, borehole collapse or non-circular holes. Hand auguring can be used to form pockets from the base of boreholes drilled using another rig, but this becomes expensive because of the cost of standing time for the rig while hand auguring.

Continuous flight augers are only used in stiff clays. The rate of advance must be such that the soil is cut and not pushed to the side. The auger must be removed slowly to minimise suction forces being set up below the auger within the test pocket, which can lead to pocket collapse. The augers are rotated as they are withdrawn, the direction of rotation being the same as that used during drilling. Gravel particles within the clay tend to be plucked out from the walls of a pocket, leaving voids.

Open-hole techniques, using drag bits or tricone bits, are not suitable as they can lead to irregular pockets of varying diameter. This is due to rod shake and the erosion of the pocket walls by the passage of large cuttings. However, if tests are required at depth or the spacing between tests is large, these techniques can be used to advance the borehole between test positions. A pocket for the probe is then formed from the base of the borehole using a preferred method.

Percussive rotary bits are not suitable as they lead to irregular pockets but, like open hole techniques, they may be suitable for rapidly advancing the borehole between test positions.

Core drilling is the preferred rotary method as it can lead to a regular pocket of uniform diameter, which may be supported by the flushing medium. Core drilling has the further advantage that the cuttings are small and therefore erosion of the pocket wall is reduced.

The importance of using the correct flushing medium should not be underestimated. The uphole velocity of the medium should be low enough to prevent erosion of the wall but sufficient to remove the cuttings. The medium should not be absorbed into the ground as a result of the suction pressures set up due to the release of total stress. Water flush is not recommended for clays, since it causes erosion and softening. Water tends to erode granular soils because of its low viscosity and high uphole velocity. It is only suitable for competent rocks. Air flush is generally unsuitable for all materials other than very competent rock. Mud flush is the preferred medium for all soils and rocks. A mud flush forms a lining on the pocket wall. This reduces the absorption of water by the surrounding ground, thus reducing the possibility of softening and helps stabilise the wall, preventing collapse. However, if the thickness of the mud lining is too great, it can affect the test (similar to the effect of increasing the disturbed zone). Polymer muds are preferred to bentonite muds, as the mud lining formed is thinner.

Percussive, or pushed, sample tubes can be used to provide a pocket for a test. They can only be used in soils that are self-supporting – that is, clays or possibly sands above the water table. These can be driven from the base of a larger diameter borehole drilled by other techniques. The chamfer on the leading edge of a sampling tube is on the outside so that the soil displaced by the tube wall is pushed outwards to minimise disturbance to the sample (Figure 3.1).

Ideally, the chamfer should be on the inside of a tube used to create pockets for pressuremeter tests, thus compressing the sample and minimising disturbance to the surrounding ground. This means a disturbed sample is obtained. Pushed tubes are preferred to percussive tubes, as they further minimise the disturbance. The tube must be withdrawn slowly to reduce the suction pressures in the cavity formed below the tube.

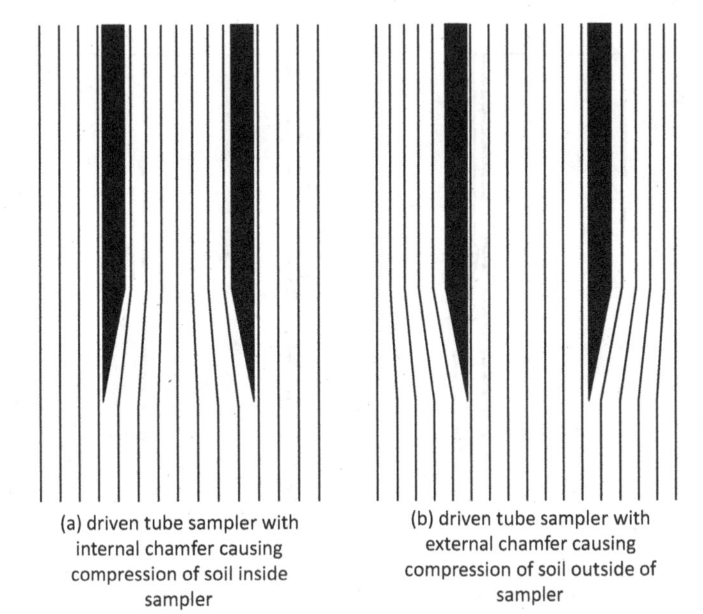

Figure 3.1 The effect of the chamfer (a) to minimise disturbance to the surrounding soil during the installation of the probe, and (b) to minimise disturbance to the soil during sampling.

Pressuremeters with a hollow core and internally chamfered leading edge can be pushed into the ground to minimise the disturbance to the surrounding soil. In this case the diameter of the predrilled test pocket should be a few millimetres smaller than the diameter of the probe.

Tests can be carried out in clays, silts, sands and rocks. It is difficult to carry out tests in gravels because of the size of particles in relation to the probe, the difficulty in supporting the pocket and the difficulty in forming a regular pocket. Sometimes the probe is inserted in slotted casing and the casing driven into the gravel. This inevitably causes a great deal of disturbance, but design rules have been developed for this method of installation. Gravel in other soils can cause irregular holes to be formed as the gravels are plucked out of the side of the borehole. The Russian standard specifically excludes pressuremeter testing in soils containing more than 25% gravel.

Figure 3.2 shows the method developed by Arsonnet et al. (2005) to minimise disturbance.

The STAF system is a slotted tube technique which conforms to BS EN ISO 22476-4:2021. The casing is installed by rotary drilling with mud flush. The bottom of the casing is slotted to allow a pressuremeter test to be carried out. The casing is installed to the maximum depth of the borehole. The drilling bit is removed and replaced by the prebored pressuremeter. A test is carried out. The casing and probe are raised to the next test position. This is repeated until the required number of tests are completed.

Masoud and Khan (2019) describe a sampling technique to minimise the disturbance of soils. A thin wall casing is pushed into the soil. The soil is removed from the casing with a thin-walled sampling tube. The PBP lowered into the cased hole and the casing

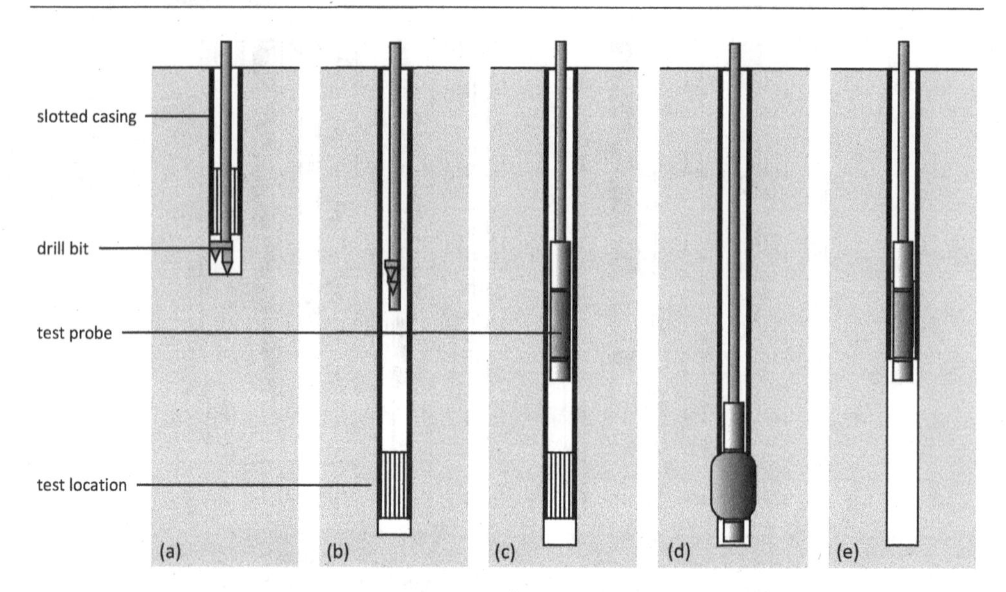

Figure 3.2 The STAF procedure showing (a) a cased hole being created; (b) the drill bit being removed leaving the casing place; (c) lowering the test probe into the slotted base of the casing; (d) the test being carried out; and (e) the casing and test probe being raised to the next test position.

Source: After Arsonnet et al., 2005.

raised to allow a PBP test to be carried out in the pocket created at the base of the borehole. They found that this reduced the disturbance of the ground compared to augured holes. Further, the unload/reload cycles were similar for both types of installation.

The techniques described above are for forming pockets for the probe. It is expensive to prepare the pockets unsuccessfully. If the test positions are at a depth or at sufficient spacing it is often economical to use open-hole techniques between test positions as the quality of hole between test positions is unimportant providing the borehole does not collapse.

During preboring, a disturbed annulus of soil is formed around the pocket because of the reasons given above. The expansion of the probe is limited (see Table 1.3 for examples) and therefore if the disturbed annulus is too large (B in Figure 3.3), the only material tested is the disturbed material.

The pressuremeter curve represents an average stiffness response; therefore, the initial slope will be reduced as the thickness of the disturbed annulus increases.

The probe diameter has to be smaller than the pocket diameter to enable it to be lowered into place. The maximum expansion of the probe is a critical factor when deciding the permitted size of the pocket. As a guide, the ratio of the diameter of the pocket to that of the probe must not be greater than 1.10 (Mair and Wood, 2013). ASTM D4719–20:2020 recommends the ratio should be between 1.03 and 1.20 (Briaud and Gambin, 1984). GOST, referring to 76–127 mm diameter probes, recommends that the pocket diameter should be no more than 2 mm greater than the probe diameter, thus the ratio is between 1.03 and 1.02 depending on the probe. The ISRM recommended procedure suggests that the hole diameter should be 0.5–3 mm

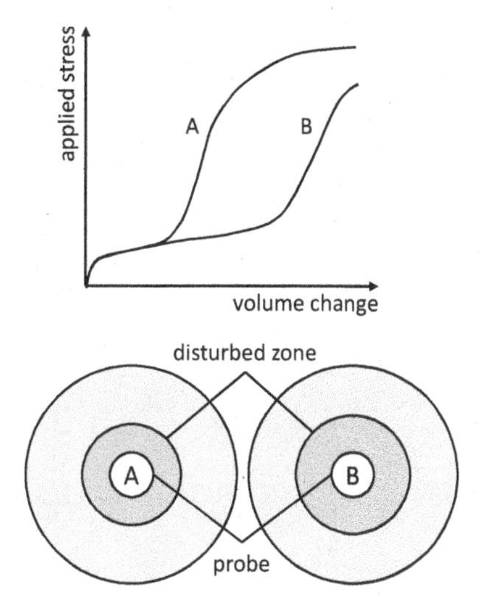

Figure 3.3 The effect of disturbance created during preboring on a pressuremeter test.

larger than the probe diameter. BS EN ISO 22476-4:2021 states that the tool diameter must be not more than 1.08 times the probe diameter.

Once the borehole is prepared, the probe is lowered into the pocket. Two techniques are used. The first is to drill a borehole the same diameter as the required pocket to its maximum depth. The probe is lowered down to the base of the borehole, and tests are carried out as it is withdrawn. There are a number of disadvantages to this method: the borehole has to be completed in one shift; the hole may soften, cave in or spall, thus making it difficult to lower and withdraw the probe; a delay between drilling and testing affects the results obtained. This method, therefore, is not recommended.

The second method is to drill to the test position then carry out the test before advancing the hole to the next test position. This is preferred, though it leads to more time over the borehole as the probe and drill rods are lowered and raised for every test. It is recommended that the probe should be lowered down the hole immediately after the pocket is drilled to reduce the effects of softening and collapsing. Figure 3.4 shows the sequence of operations.

Care has to be taken to prevent the probe acting as a piston or scraping the sides when lowering the probe into a pocket as this could cause further disturbance. If the borehole diameter is similar to that of the probe, then the probe will act as a piston, pressurising the drilling fluid within the pocket causing, in effect, a cavity expansion. The sides of the pocket may be scraped off by the probe, leading to debris collecting at the base if the pocket is smaller than the probe, inclined, or irregularly shaped. There may be debris from drilling in the base of the pocket. For these reasons, the probe may not reach the test position. One way of overcoming this is to use an internally chamfered shoe attached to the base of the probe (Figure 3.5) to allow debris to enter the probe.

Alternatively, the pocket should be longer than the minimum required to allow debris to collect at the base of the pocket and beyond the test location. In pockets full

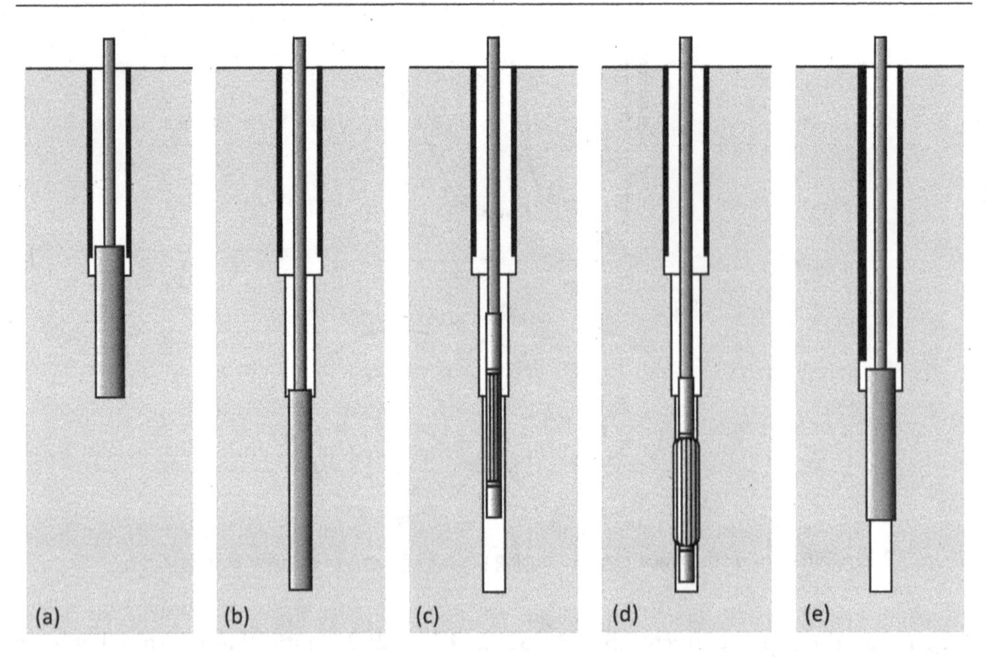

Figure 3.4 The sequence of operations for a PBP test showing (a) a cased hole being created; (b) a core barrel being used to create the test pocket; (c) the test probe being lowered into the test pocket; (d) the test being carried out; and (e) the borehole being advanced to the next test pocket.

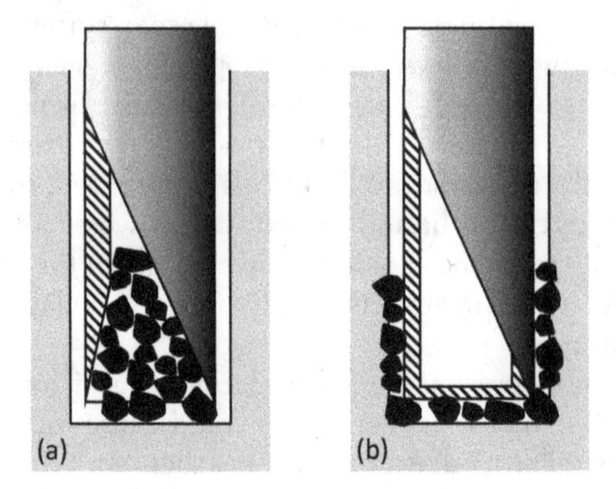

Figure 3.5 The types of shoes for PBPs and the effect they have on positioning the probe; (a) an internally chamfered shoe collects debris within the probe; (b) a solid shoe displaces the debris.

of drilling mud (or water) and of larger diameter than the probe, care has to be taken when lowering the probe to prevent the sides of the pocket being washed away as the mud moves past the probe. Ideally the probe should be lowered slowly into the pocket.

Fluid-filled systems, such as the MPM, can lead to problems in dry holes. The hydrostatic head of the measuring fluid within the control unit and tubing causes the membrane to expand as the probe is lowered. This can be limited by closing off the control unit but, below 10 m, the air pressure in the borehole can no longer support the head in the tubing. This will lead to expansion of the membrane, making it difficult to move the probe in the hole. Further, with water-filled systems, air will come out of solution unless de-aired water is used. Any air in the system will affect readings of volume change. A stiffer membrane may help overcome this problem but this reduces the sensitivity. Some systems have a remote-controlled valve at the probe which locks off the measuring cell thus eliminating this problem. It is preferable, with surface volume-measuring systems, to install a probe in a mud- or water-filled borehole.

Air in fluid-filled radial displacement type probes is not critical. Generally, these systems have twin hoses. The fluid is drained from the system before the probe is lowered into or removed from dry boreholes. Fluid is pumped down one hose, expelling the air through the other hose prior to a test. After the test is complete the fluid is forced out of the probe by pumping air down to the probe.

3.2.4 The self-boring pressuremeter

There are a number of self-boring drilling systems, but all the systems use similar techniques to remove the ground as the probe is advanced. The problems of self-boring are common to all types of SBP. The French system has a paddle mounted within a thin-walled shoe which is designed to break up the soil as it enters the shoe. The Cambridge self-boring pressuremeter (CSBP) has a cutter within an internally tapered shoe, which shaves the soil as it is forced into the shoe. The Rock self-boring pressuremeter (RSBP) has a full face bit mounted just outside the shoe and flush with the probe, which grinds the rock. There are CSBP-type probes which contain rotating or fixed jets (see Figure 2.23). A jetting system is based on the principle that water under pressure will break up the soil as it enters the shoe. This can only be used in soils and is possibly further restricted to soft to firm clays and sands.

Table 3.8 are recommended positions of the drill bit for different ground conditions to minimise disturbance to the surrounding ground when creating a test pocket by self-boring.

The CSBP is drilled into the ground using rotary drilling techniques – that is, the soil is cut by a rotating cutter, and the cuttings are flushed to the surface with a drilling fluid (Figure 3.6).

The probe (1) is connected to the surface by 50 mm outer drill rods (2) which transmit the thrust to the probe and carry the drilling fluid to the surface. The outer rods are connected to a pair of pull-down rams (3). Reaction is required for drilling to overcome the friction between the probe and soil and to push the cutting edge of the shoe into the soil. Typically, it ranges between the self-weight of the probe in very soft clays to several tonnes in hard clays. This reaction is provided by ground anchors or kentledge (4). A hydraulic motor (5), mounted on top of the outer rods, is used to rotate the inner rods (6). A hydraulic power pack (7) provides the power for the rams and motor. The control panel (8) allows the rate and pressure of flow of hydraulic oil to be regulated so that the speed of, and pressure within, the rams and the speed of the motor are controlled. This is referred to as the twin-rod system.

The pressiométrique autoforeuse (PAF-68) is operated in a similar way to the CSBP. Later versions of the PAF, including offshore devices, have a hydraulic motor mounted

Table 3.8 Distance between the cutting edge of the probe and the internal cutting tool for self-boring pressuremeters

Soil type	*Distance between leading edge of the probe and the internal cutting tool [mm]*	*Cutter type*				
		Jet	*Disc*	*FB*	*FFC*	*RRB*
Low and extremely low strength clay	5 to 25	X	X	X		
Medium strength clay	3 to 5			X		
High and very high-strength clay and marl	0 to 3				X	X
Silt above gwl	25 to 50	X		X		
Silt below gwl	25 to 100	X				
Loose sand above gwl	25 to 50	X		X		
Loose sand below gwl	25 to 100	X		X		
Medium dense and dense sand	25 to 50			X		
Highly weathered rock; extremely weal rock	-5 to 3				X	X

Source: After BS EN ISO 22476-6-2012.

Note: Disc – stirring paddle; FFC – full face cutter; FB – flat blade; Jet – high pressure water jet; RRB – rock roller bit

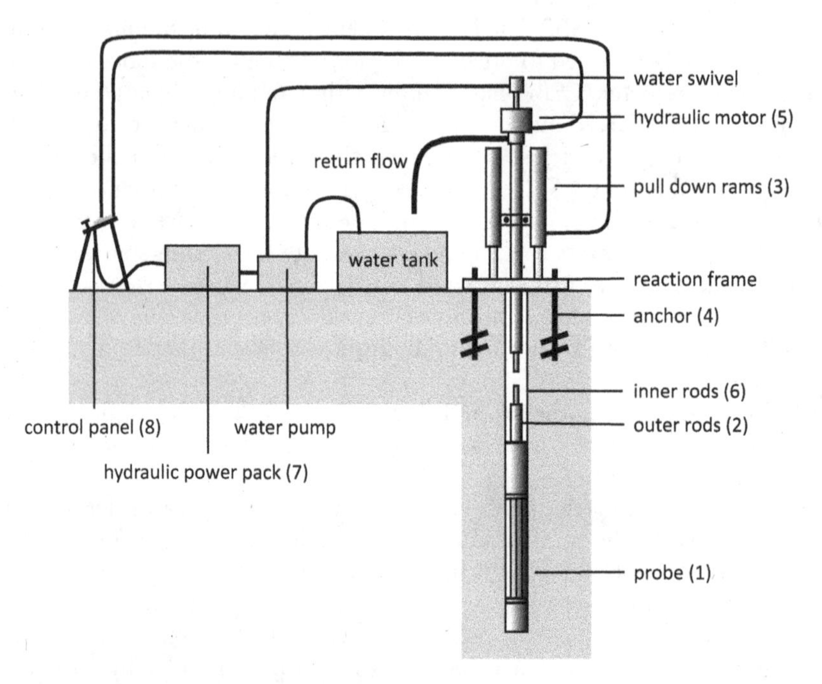

Figure 3.6 The principles of a twin rod self-boring system.

in the probe, which is connected to a hydraulic power pack at the surface. The speed of the motor and the torque can be controlled by altering the pressure and flow rate. This is a downhole system. The probe is pushed into the ground by a rotary rig.

It is possible to drill a SBP continuously in suitable soils, but not in rocks. Clarke and Smith (1992b) describe two drilling systems for the SBP, shown in Figure 3.7.

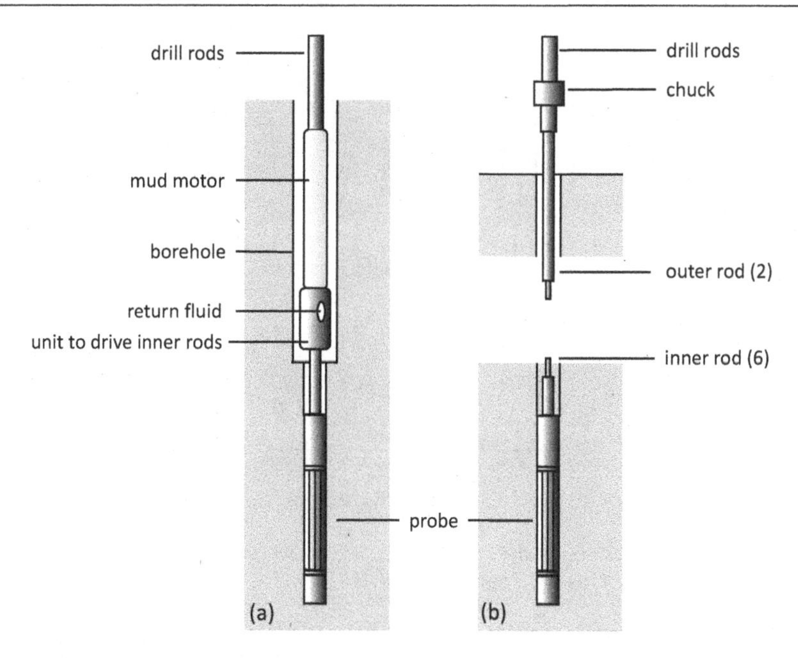

Figure 3.7 The (a) downhole mud motor and (b) twin rod systems used to install a SBP with rotary rigs. The downhole mud motor drives the inner rods and the drilling fluid is recirculated through the boreholes. The twin rod system uses the inner rods driven directly by the rotary rig and the drilling fluid is recirculated through the outer rods.

They both operate from a rotary rig and use the thrust of the rig to push the probe into the ground. The first system, based on the twin-rod system developed for soils, uses the rotary rig chuck to turn the cutter. It is necessary to separate the rotation of the inner rod and the thrust on the outer rod using a special chuck, which is attached to the rotating kelly rod of the rotary rig. The other system uses a downhole mud motor driven by the flushing fluid. This is also used in the PAF-76.

The SBP is pushed into the soil, forcing the soil into the shoe. The soil in the shoe is removed by the rotating cutter or water jets and the particles are flushed to the surface as the probe is advanced. The return fluid is collected in a tank where the chippings settle out. This allows an assessment of the ground type. The drilling fluid is pumped down to the cutter and returned to the surface in several ways, depending on the probe design as shown in Table 3.9 and Figure 2.14.

The cutter or jet position is critical, as this dictates the amount of pressure generated across the face of the probe. Ideally the pressure at the face of the probe should be equal to the total overburden pressure. The balance is maintained by the extrusion force required to push the soil into the shoe, the friction between the internal surface of the shoe and the soil and the weight of water and inner drive rods. Figure 3.8 shows a theoretical relationship between the cutter position and the overburden pressure, for the CSBP.

This is based on extensive laboratory tests and supported by field observations. It is not possible to allow for the local variation in strength, but it is usual to assume that it is constant for many soil deposits over a limited depth. Therefore, the cutter can be

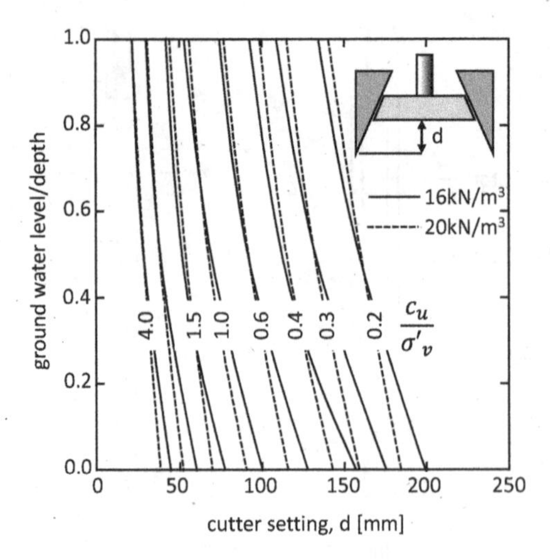

Figure 3.8 The theoretical position of the cutter in the SBP to achieve minimum disturbance drilling in clay based on the unit weight and ground water level.

fixed if the shear strength is known, as suggested in Table 3.8. Without this information the rule for any drilling system with an internally tapered shoe is as follows:

(a) stiff clay and very dense sands: set the cutter near the edge of the shoe;
(b) soft clay and medium dense sands: set the cutter a distance equal to the radius of the SBP back from the leading edge.

The cutter is located within the shoe, the actual position depending on the strength of the soil. The shoe has, therefore, to penetrate the soil and, as the strength of the soil increases, the penetration force increases until it is no longer possible to push the shoe into the soil. It is for this reason that many SBPs are restricted to soils. It is necessary for the cutter to protrude from the probe when self-boring in rock.

There are five parameters that can be varied during drilling: the rate at which the probe penetrates the soil; the thrust acting on the probe; the speed of the cutter; and the pressure and flow rate of the drilling fluid. These have to be balanced to prevent the soil being partially displaced, wash-boring and blockages occurring within the probe.

Ideally, the pressure of the fluid at the drill head should be just slightly greater than that required to lift the column of returning fluid. If it is less, there will be no flow; if it is much greater, hydraulic fracture may occur. The actual mud pressure required depends on the viscosity of the fluid as well as the depth of the probe. Typically, for a CSBP, the flow rate is 12 gal/min, though it may be varied depending on the rate of penetration and type of self-boring.

The speed of the cutter is fixed, usually at about 50 rev/min. This means that the speed of advance is the controlling factor governing the size of chippings. Typically, in soft clays and sands the CSBP can be drilled 1 metre in about 3 min. In stiff clays, the time may be 1 hour, though it is often possible to drill it in considerably less time. As stiff clay chippings swell and tend to block the bearing in the shoe, the speed of advance is reduced to ensure adequate flow and to prevent blockages.

Table 3.9 Methods of delivery and return of drilling fluid in self-boring probes

System		*Down*	*Up*
Twin rod		Inner rod	Outer rod or borehole
Downhole	(a)	Drill rod	Borehole
	(b)	Hydraulic hose	Borehole
Jet	(a)	Hydraulic hose	Drill rod or borehole
	(b)	Inner rod	Drill rod or borehole
	(c)	Drill rod	Borehole

The exact drilling parameters cannot be specified since soil is so variable. As a guide, during drilling, there should be a good return flow and at least a bucketful of soil retrieved for every metre drilled for a nominal 80 mm diameter probe. The first test should be analysed on site. This may help confirm that the drilling parameters are correct. If the interpretation of the test shows that there has been some disturbance during drilling, the probe may have to be removed to alter the drilling system.

The SBP has been used in a variety of soils and weak rocks (Table 3.10).

There are restrictions to the type of material in which it can be used and these are governed by the following:

1. *Strength of material.* The penetration force and probe pressure capacity are limited. The penetration force and pressure capacity required are a function of the strength of the ground.
2. *The horizontal stress.* The friction on the probe is a function of the horizontal stress and strength of the ground. The total penetration force required has to overcome this friction and push the shoe into the ground.
3. *The size of the soil particles.* There is a limit to the size of particles that can pass through the drilling system. The installation disturbance is a function, in part, of the particle size.

The CSBP is designed to work at a maximum pressure of 10MPa. Therefore, a complete test in clay can only be carried out if this exceeds the horizontal stress plus approximately six times the shear strength. For example, for a normally consolidated clay it is theoretically possible to carry out tests at 90 m whereas for a heavily overconsolidated clay the depth limit is 50 m. It is possible to increase the maximum working pressure but this is not recommended for safety reasons as the system is pressurised by gas.

The frictional force on the probe is a function of the effective horizontal stress and the interface friction between the probe and the soil. Typically, in clays, the unit friction is 0.4 c_u (cf. driven piles) which implies, for an 84 mm probe 1 m long, the force increases by approximately 1 tonne for every 100kPa increase in shear strength. Generally, in clays, the speed of the probe remains constant and the reaction required increases with depth. This is not the case for sands. The force on the probe is increased until the sand shears. The probe then moves, and the force is reduced. Thus, the frictional force fluctuates as the sand dilates.

All the soil particles must pass through the probe. This presents no problem when drilling in clays and sands. Tests cannot be carried out in gravels because the gravels

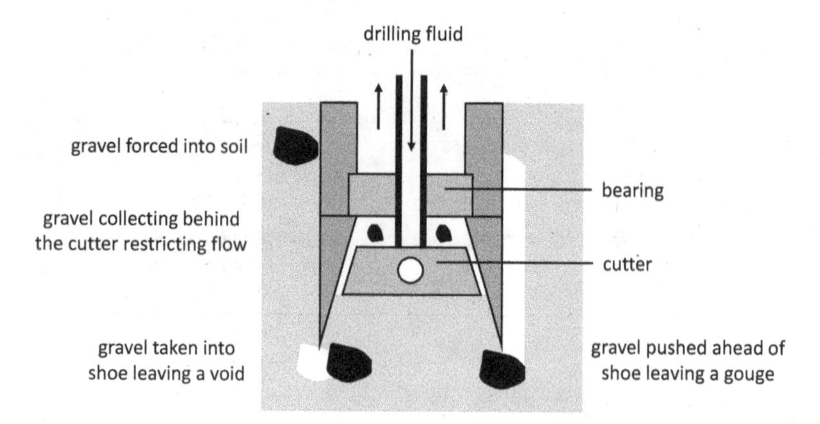

Figure 3.9 The effect of gravels on the soil surrounding a self-boring probe.

cannot be removed. Tests can be carried out in glacial and residual soils which contain occasional particles of gravel. The occasional small piece of gravel in a matrix of sand or clay may not present a problem, though it can lead to some disturbance, as shown in Figure 3.9. It is unlikely that SBP tests will ever be carried out in gravels.

Even if it is theoretically possible to install an SBP within a soil, problems can occur during self-boring. These are summarised in Table 3.11.

The SBP tends to be drilled in by wash-boring in some soils – that is, the drilling fluid is forced out of the bottom of the shoe. This is a common problem in sands and occurs when the pressure of the drilling fluid at the jets in the cutter is too great or when the cutter is too near the leading edge. Reducing the drilling fluid pressure may help reduce the problem though there is a limit since the sand has to be lifted to the surface. This problem is usually solved by either increasing the viscosity of the drilling fluid with mud or setting the cutter further back. Polymer muds have been found to work extremely well, and this is often the best solution. There is a limit to the position of the cutter, since it may lead to overstressing the sand in front of the shoe.

Wash-boring can also occur in clays for two reasons. The clay plug in the bottom of the drilling shoe may be blown out because the drilling fluid pressure is too great. This can happen if the probe becomes blocked. The speed of advance may be too slow, leading to the clay being washed away rather than cut – which is especially the case with soft clays. In soils containing gravels, the hole created by a gravel particle may act as a water channel. Wash-boring may occur when drilling from the base of a predrilled hole because there is a softened layer below the base of the hole due to the action of the other rig. This, however, ceases when undisturbed material is penetrated.

If there is no return of flushing fluid the probe is blocked, usually because the passageways through the bearing are blocked. Should this happen, the ram pressure will increase significantly, and the SBP will be pushed into the ground rather than drilled in. Clays tend to be extruded past the cutter, causing blockages within the probe. This happens when the rate of advance is too great or the probe is pushed in, and in stiff clays due to the swelling of the clay chippings. Clay in suspension tends to settle within the probe either while testing or if the probe is left overnight in the hole. To help prevent this happening, it is necessary to flush the system thoroughly when drilling is complete, ideally with clean water or mud. If this does not cure the

Table 3.10 The range of ground conditions in which the SBP has been used

Ground type	*Examples*	*Geological description*	*Probe*	*Shear strength [kPa]*	*Angle of friction*
Fill	Fly ash		CSBP/RSBP		24–61
Clays	Soft clay		CSBP	5–25	
	Firm to stiff clay		CSBP	25–75	
	Stiff to hard clay			75–150	
		London Clay	CSBP/RSBP	110–3200	
		Boom Clay		75–2750	
		Weald Clay		40–380	
		Woolwich and Reading Beds		80–860	
Sands	Loose		CSBP		<35
	Dense		CSBP/RSBP		>35
		Norwich crag	RSBP		25–64
		Thanet sands			31–61
Rock	Weathered	Chalk	RSBP	340–2400	
		Lower Lias mudstones		90 > 7000	
		Jurassic mudstones		60 > 4800	
		Keuper Marl		130 > 7000	
		Ertrurian Marl		20 > 7000	
		Coal measures		1000 > 2100	
		Carboniferous millstone			35–77
		grit mudstones siltstones			50–60
		sandstones			46–62
		Carboniferous mudstones		1150 > 7000	

Source: After Clarke and Smith, 1992b.

Note: Tests in rock may be drained and rocks exhibit cohesion and friction.

problem, it may be possible to clear the blockage either by the use of air pressure or by increasing the water pressure or by reverse flow. These may lead to spectacular results but often the SBP has to be removed and the blockage cleared manually.

In sands the SBP often becomes blocked while testing, due to either sand settling out of suspension or sand "blowing" up the hole, which can result in the inner rod becoming blocked. As it is very difficult to cure this problem in situ, it is usually necessary to remove the SBP. Increasing the viscosity of the drilling fluid and flushing with clean drilling fluid prior to testing helps to reduce the problem.

On occasions the SBP is advanced with everything appearing to be operating normally, yet the probe is actually being pushed into the soil. This may occur if the inner rods break or if a plug of clay develops in the shoe. In both cases the drilling fluid returns to the surface. The return flow should be checked frequently: fast drilling in clay should bring large pieces of clay to the surface; slow drilling tends to break up the clay, causing the viscosity of the drilling fluid to increase. The forces required to push a SBP into sands are too great once the base of the probe becomes blocked.

The maximum distance a SBP can be drilled depends on the soil type. In soft clays, there is theoretically no limit, but in stiff clays, 30 m is probably the maximum because of wear on the cutter and the bearing. Obstructions, such as gravel, may be found in clay; thus, it is usual in clays to operate the SBP with a drilling rig.

The distance drilled continuously is limited in sands because sand tends to collapse behind the SBP, leading to a ground anchor. It is possible to support the borehole walls with a heavy drilling mud, but success is limited due to the possible collapse of the borehole leading to the loss of the probe. Alternatively, the probe can be lengthened by attaching a cylinder of the same diameter as the probe around the drill rods just above the probe and ensuring that the top of the cylinder always remains above the level of the sand within the casing used to support the borehole.

The SBP drilling system is a rotary drilling system. The twin-rod system was developed for soils where, in some instances, it is possible to use a trailer-mounted rig on its own. In many cases this is not possible because of obstructions, unstable boreholes and distances between test locations. This latter point refers to the economics of drilling a probe continuously without testing. It is often more economic to remove the probe and advance the hole with more conventional drilling techniques, thus ensuring a stable borehole free of obstructions.

Tests can be carried out at similar intervals to those for PBP tests – that is, 1 to 2 metre intervals depending on the probe. BS EN ISO 22476-6 suggests that the minimum spacing between tests is 0.75 m and the minimum depth for a test below the base of a borehole is 0.75 m. There is a delay between drilling to the test depth and the start of a test to ensure that a state of equilibrium is reached. The pressure in the system can be monitored. Alternatively, the delay could be set at 30 mins assuming there is minimum disturbance to the surrounding soil.

The most common technique for drilling a SBP is to use another drilling rig to advance the borehole between test positions and use the same rig to drill the SBP directly. The method of operation is as follows (see Figure 3.10):

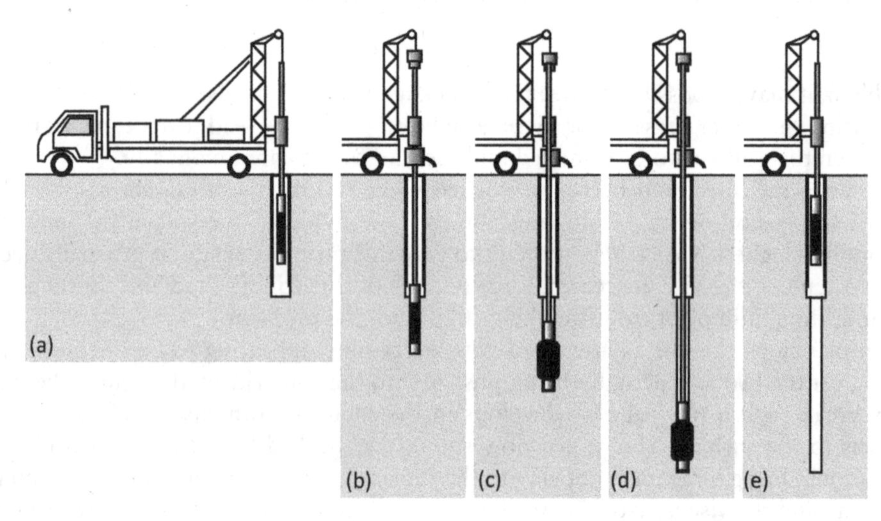

Figure 3.10 Sequence of drilling and testing: (a) lowering to base of hole; (b) drilling to first test position; (c) first test; (d) drilling and testing at adjacent position; (e) removing from hole.

Source: After Clarke and Smith, 1992b.

1. The hole is advanced to within 1 or 2 metres of the test elevation. The borehole should be cased in sands and other unstable ground to prevent collapse. It is important to advance the borehole carefully in sands to prevent sand 'blowing' up into the borehole. Some sand may settle out of the drilling fluid. In either case the probe may become jammed in the sand in the bottom of the casing as the probe is withdrawn. The minimum borehole diameter should be 100 mm.
2. The SBP is suspended in the borehole above the level of the soil. It is prudent to dip the hole prior to lowering the SBP to ascertain this level because if the SBP is inadvertently lowered into the soil it may become blocked. This is common in sands.
3. The SBP is drilled until return flow and reaction are obtained. If there is no return flow it means that water is being lost to the surrounding ground. Lack of reaction confirms this.
4. The SBP will probably be drilling in undisturbed soil provided it is beyond the base of the casing and there is reaction and return flow. The SBP should be drilled a further metre before carrying out the test.
5. The SBP should be withdrawn after the test and the hole advanced with the rig. It is possible to further advance the probe in clays providing the drilling system is still operational.

The CSBP equipment, including 30 m drill rods, weighs about 1 tonne and can be operated from a purpose-built rig or with a rotary rig. The choice depends on the site, though it is more common to operate a SBP directly from a rotary rig in unstable ground and deep boreholes.

3.2.5 The full-displacement pressuremeter

A full-displacement pressuremeter (FDP) should be pushed from the ground surface or the base of a borehole at a rate between 1 cm/s and 5 cm/s. The distance between the top of the expanding section and the base of the borehole (or the ground surface for the first test) shall be greater than ten times the maximum diameter of the cone module. The distance between tests should be at least ten times the maximum diameter of the cone module, and the test should start within two minutes of the end of pushing.

The FDP can be pushed into the ground using a cone truck. Ideally, if the conical tip is instrumented, and it is pushed in at the standard rate, 2 cm/sec, the cone can be used to provide additional data, which are required for empirical correlations. Generally, the probe can be used in soils into which it is possible to push a cone.

The cone pressuremeter is larger than a conventional cone. The force required for installation is greater, therefore the reaction system is often a limiting factor. Increasing the reaction may not be prudent since the increased force on the probe could cause damage.

There are instances where prebored pressuremeters are pushed into the soil using slotted tubes; that is, in effect, they are full-displacement pressuremeters

3.3 CALIBRATIONS

3.3.1 Introduction

A pressuremeter must be calibrated before and after site operations, whenever there is any change to the probe on site and at least weekly when the probe is in daily use. It is imperative that the calibrations are carried out correctly so that the true response of the soil or rock can be determined from a test.

There are four groups of calibrations:

(a) pressure- and displacement-measuring systems (gauges and transducers)
(b) compliance of the probe (membrane stiffness or resistance, membrane compression due to the increase in pressure and membrane thinning due to the expansion of the membrane)
(c) compliance of the system (volume changes)
(d) initial readings.

3.3.2 Pressure gauges

The gauges, used for manually recording pressure, are usually calibrated against standard gauges in the office or laboratory prior to the site visit. These are used with all pressuremeters for either observing the applied pressure or recording the applied pressure or calibrating transducers.

3.3.3 Displacement transducers

The displacement transducers can be LVDTs or purpose-built sensors, which may require a DC voltage power supply. The output from the transducers is a function of the power supply and the displacement but the power supply is regulated, therefore the change in output is a function of the change in displacement.

The sensitivity of the transducers – that is, the change in output with displacement – will change if the cable length changes. However, if the signal processing unit is within the probe, the cable length is not critical. Irrespective of the type of signal and the type of transducer, the technique used to calibrate a displacement transducer is the same.

The membrane is removed from the probe and the probe is placed horizontally. A special jig containing a micrometer is attached to the probe such that the micrometer rests on the end of the displacement arm (the part of the transducer, which is in contact with the membrane). Note that the jig for the High pressure dilatometer (HPD) has two micrometers since the displacement arm is a plate which can yaw. The probe should be connected to the control unit to be used on the project. The arms are calibrated one at a time, keeping the others clamped in place.

The micrometer is adjusted over the full working range of the displacement transducer (for example, 13 mm with the HPD) and readings are taken of the output from the transducer. Typically, readings are taken about every 3 to 5% of the full working range, though it is important to obtain accurate information at small displacements since the interpretation of modulus and horizontal stress are particularly sensitive to this calibration. The displacement increment may vary. A recommendation for

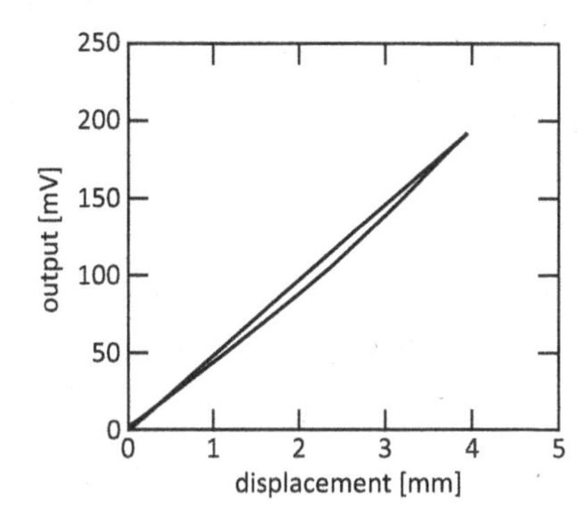

Figure 3.11 A typical calibration curve for a spring type displacement transducer showing the near linear relationship between and voltage output and displacement and the small hysteresis.

self-boring probes, in which the maximum displacement is of the order of 15% of the probe diameter, is to carry out the calibration in steps of 0.1 mm between 0 and 1 mm, and thereafter in 0.5 mm steps up to the maximum displacement. A typical calibration is shown in Figure 3.11.

Readings are taken up to the maximum displacement and then at the same decrements down to zero. Normally the change in output with displacement is linear and shows little hysteresis. It is usual to fit a straight line to the data if this is the case. If either the hysteresis or the coefficient of determination is outside acceptable criteria, and this is confirmed a second time, then the transducer should be checked, cleaned and replaced if necessary. Hysteresis and non-linearity can be introduced if there is dirt in the moving parts of the transducer.

The sensitivity is quoted as mV/mm. It can be quoted in terms of cavity strain defined as the ratio of the change in displacement with the radius of the assembled probe – that is, with a membrane in place. In this case the sensitivity will be quoted as mV/%.

It is important that the total voltage output from the transducers at zero displacement and the average sensitivity of the transducers corresponds to that required by any automatic control system.

3.3.4 Total pressure transducers

Total pressure transducers can either be purpose-built sensors or manufactured sensors. They appear to operate in a similar manner to the displacement transducers; that is, they use the same power supply, and the output range in mV is similar.

The total pressure calibration is carried out with an assembled probe placed upright in a close-fitting thick-walled tube. The pressure within the probe is increased in increments up to the maximum capacity, and readings taken of the output of the transducer. Typically, increments of pressure are about 5% of the full working range.

Normally the change in output with pressure is linear and shows little hysteresis (see Figure 3.12).

It is usual to fit a straight line to the data if that is the case. The sensitivity is quoted as mV/kPa or mV/MPa. If either the hysteresis, or the coefficient of determination is

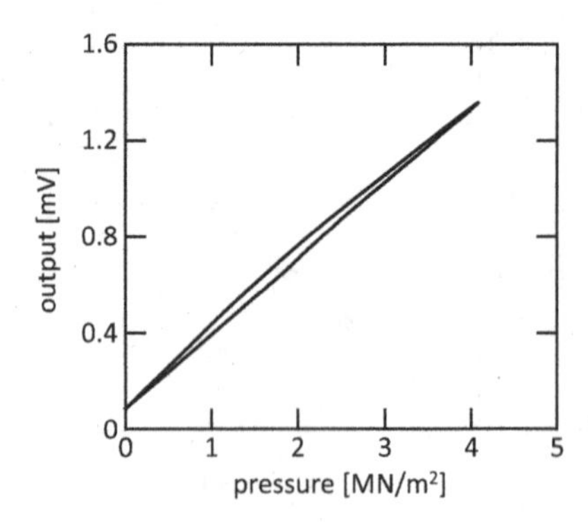

Figure 3.12 A typical calibration curve for a diaphragm type pressure transducer showing the near linear relationship between and voltage output and applied pressure and the small hysteresis.

outside acceptable criteria, and this is confirmed a second time, then the transducer should be checked and replaced if necessary. A new transducer should be pressurised or exercised several times before calibrating. It is recommended that the transducer should be chosen so that the maximum pressure likely to be reached in the test is within the working range of the transducer yet the working range is not too large so that it is possible to achieve maximum sensitivity. For example, in stiff clays a 4 MPa total pressure sensor could be used whereas in soft clays it would be better to use a 1 MPa sensor.

3.3.5 Effective pressure and pore pressure transducers

Some probes are fitted with an effective pressure transducer (Figure 2.19). The simplest calibration for this sensor is similar to that for the total pressure sensor. The thick tube, used in the total pressure calibration, has holes which are aligned with the porous element in the effective stress sensor (see Figure 3.13).

The variation in total pressure in the probe is measured with reference to atmospheric pressure. The calibration procedure is the same as that for the total pressure transducers and the results quoted in the same way. This calibration only applies when the applied pressure is greater than the pore pressure since the exterior pressure during calibrations is atmospheric. This is not the case at the start of a test when the applied pressure is zero. There could be a positive pore pressure. Ideally, the effective pressure sensors must be calibrated by pressurising them externally as well as internally. This is done by placing the probe in a pressure vessel and applying an external pressure to the probe (Figure 3.13). The calibration procedure is the same as before.

True pore pressure transducers are calibrated by applying an external pressure to the probe in a similar way to that shown in Figure 3.13. The calibration procedure is similar to that for a total pressure sensor.

3.3.6 Membrane stiffness

Membrane stiffness or resistance is the pressure required to inflate the membrane, and protective sheath if fitted, in air. The membrane can either be stretched over the body

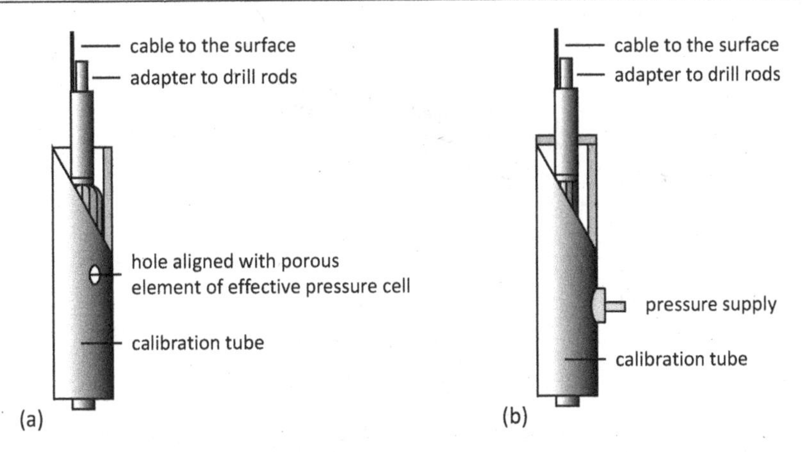

Figure 3.13 Calibrating the total pressure cell and the effective pressure cell using (a) an internal pressure; and calibrating the effective stress cell or pore pressure cells using (b) an external pressure.

of the probe (for example, CSBP) or be slid over the body of the probe (for example, MPM). Pressure is needed to inflate the membrane in both cases. This calibration affects the interpretation of tests in soft ground. It is not critical for tests in stiff clay, dense sand and rock though it is often applied because it is so simple to do so.

The calibration should be carried out at the same temperature as that in the ground since the properties of the membrane are temperature dependent. This is often not possible and, to minimise the problem, the probe and control unit should be kept out of the sun. The ambient temperature must be recorded.

The probe should be completely assembled for its intended use, and tested; for example, if a Chinese lantern or slotted tube is to be used, then they should be there for the calibration. The probe should be vertical. The control unit and probe must be at the same level if a volume-measuring system is used since the weight of fluid within the hydraulic hose will apply pressure to the membrane. New membranes should be exercised at least three times before calibrating.

The membrane is inflated and deflated following a similar procedure to that used in a test, either a stress-controlled or strain-controlled test procedure (see section 3.5). Readings are taken of pressure and displacement up to the maximum displacement at suitable intervals. Typically for strain- controlled tests, in which the membrane is expanded at 1%/min, readings are taken every 10 s. Readings are taken at the end of each increment during stress-controlled tests. Ideally, the rate of inflation should be the same as that for a typical test. This is simple for strain-controlled tests as the strain rate can be set at the required rate, but it may be necessary to carry out several calibrations at different stress rates for stress-controlled tests to ensure that all the strain rates likely to occur during a test are covered. It is not too critical for the majority of tests so it is usual to expand the membrane in increments equivalent to 10% of the maximum expansion and hold each increment for 1 min.

The shape of the calibration curve is dependent on the type of membrane, age of membrane, number of tests carried out with the membrane and ambient temperature. It is essentially composed of two components required to lift the membrane off the

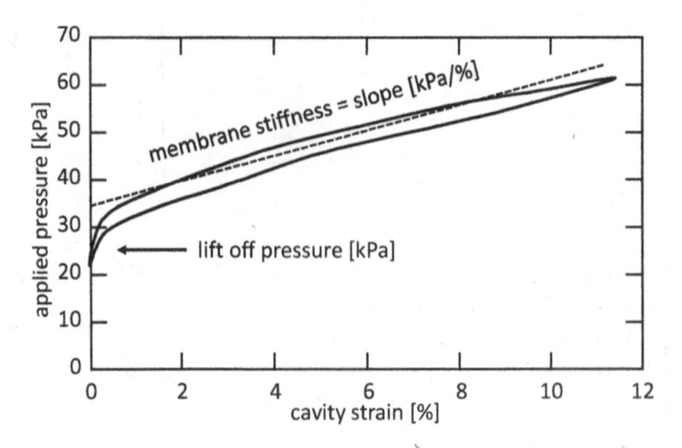

Figure 3.14 The definition of membrane stiffness and lift off pressure when inflating a membrane in air.

body of the probe, known as lift-off pressure, and that are required to cause further expansion. Figure 3.14 shows a calibration curve for a SBP. The lift-off pressure is zero for those probes, such as the MPM, in which the membrane is slid over the body of the probe. The lift-off pressure is typically below 100kPa for those probes, such as the SBP, in which the membrane is stretched over the body. This depends on the membrane and protective sheath used.

The expansion curve is often non-linear and, for those probes that have independent displacement transducers, the response varies around the circumference of the membrane. For tests in soft clays the non-linear curve and the circumferential variation should be taken into account when reducing test data. Otherwise it is normal to take an average best fit line. Typically the pressure required to inflate membranes is between 50 and 150kPa. The membrane stiffness is given as kPa/% cavity strain.

3.3.7 Membrane thinning

Membrane thinning is due to the change in thickness of the membrane as it expands. It is a function of the change in internal diameter of the membrane and not the change in pressure. It is assumed that the volume of the membrane remains unchanged, therefore it is unnecessary to apply this correction to probes with volume-displacement systems.

Consider the membrane and protective sheath shown in Figure 3.15.

The internal diameter, equal to the diameter of the body of the probe, is i_o. The thickness of the membrane is t_o and the external diameter, $(i_o + t_o)$, is a_o. As the pressure is increased the membrane expands and the internal diameter increases to i_i and the thickness reduces to t_i giving a new external diameter of a_i. The length of the expanding section of the membrane increases from l_o to l_i because of expansion at the ends of the membrane. The initial volume of the membrane is V_{mo} and, if the membrane is incompressible, this will remain constant.

$$V_{mo} = \left(\pi a_o^2 - \pi i_o^2\right) / l_o \tag{3.1}$$

$$= \left(\pi a_1^2 - \pi i_i^2\right) / l_i \tag{3.2}$$

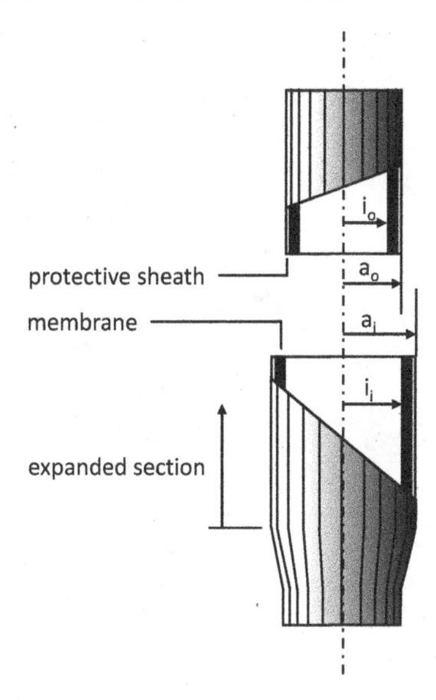

Figure 3.15 The effect of expansion on the thickness of a membrane and the interpreted radius of the pocket wall.

The change in length of the membrane is equal to the change in diameter, which is approximately equal to twice the measured displacement (2ξ).

$$l_i = l_o + 2\xi \tag{3.3}$$

$$i_i = i_o + \xi \tag{3.4}$$

The measured cavity strain, ε, is given by

$$\varepsilon = \frac{\xi}{a_o} \tag{3.5}$$

The actual cavity strain at the pocket wall, ε_c, is

$$\varepsilon_c = \frac{a_i - a_o}{a_o} \tag{3.6}$$

From equations (3.1)–(3.4)

$$a_i = \left[\frac{(a_o^2 - i_o^2) l_o}{l_o + 2\xi} + (i_o + \xi)^2 \right]^{0.5} \tag{3.7}$$

Figure 3.16 shows the effect of probe diameter and membrane thickness on the ratio of the movement of the outside of the membrane to the measured displacement. Equation (3.7) can be approximated to a straight line for a given length, diameter and thickness of membrane. Consider a 500 mm long membrane which is 8 mm thick

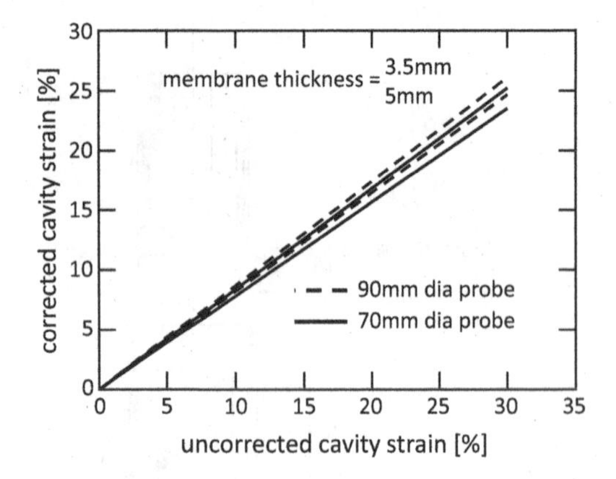

Figure 3.16 The effect of probe diameter and membrane thickness on the difference between the measured movement of the inner surface of the membrane and the actual movement of the ground for a 500 mm long membrane.

when stretched over an NX probe whose external diameter is 73 mm. If the internal diameter of that membrane is expanded by 18.25 mm, equivalent to 25% cavity strain, the corrected cavity strain is 19.96%. This is a significant correction, which can affect derived parameters from a test. Ignoring the change in length produces a corrected strain of 20.53%.

3.3.8 Membrane compression

Membrane compression is the change in thickness of the membrane due to an increase in pressure. It applies to all pressuremeter tests but is usually only determined for radial displacement type probes and, more specifically, for high-pressure pressuremeters – that is, probes that have a pressure capacity of 10–20 MPa. For tests in soils in which the maximum pressure is less than 5 MPa when the membrane is fully expanded, the membrane compression can be ignored. The membrane compression calibration is part of the system compression calibration for those probes with volume-displacement measuring systems. Figure 3.17 shows the apparent change in thickness of the membranes with pressure for the RSBP and HPD. The calibration affects the interpretation of tests in hard ground. It is not critical for tests in soils since thinner membranes are used and the correction is less than 0.1% cavity strain.

Strictly, membrane compression due to pressure includes the change in thickness of the membrane, a change in output of the displacement transducers, the stretching of the membrane in excess of that required to expand the pocket (that is, expansion of the membrane up and down into the pocket) and the lengthening of the probe. It does not include the change in thickness of the membrane simply due to expansion of the membrane, that is, membrane thinning. It is necessary to carry out this calibration in tubes of various diameters for exact calibrations, though this is only important for research purposes. The tube diameters range from one that just slides over the deflated probe to one in which the probe can be expanded up to 90% of its maximum capacity before the membrane reaches the tube. If three tubes are used there will be

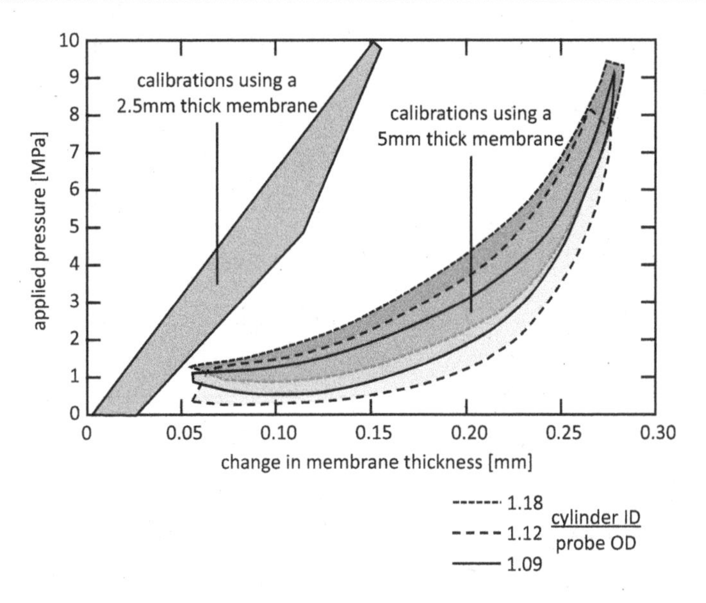

Figure 3.17 The compression of membranes due to applied pressure showing the effect of pressure, initial membrane thickness and ratio of probe diameter to test cell diameter.

Source: After Clarke et al., 1990.

three calibration curves. It is difficult to apply this multiple type of calibration and the increased accuracy of the results may not be important.

The normal membrane compression calibration is carried out by inserting the assembled probe in a steel tube usually as part of a total pressure transducer calibration. The membrane is expanded in air until it comes in contact with the steel tube, as shown in Figure 3.18. The pressure is then increased to the maximum and then reduced to about 20% of that maximum. The calibration is then carried out by increasing the applied pressure to the maximum following a typical test procedure, taking readings of pressure and displacement at suitable intervals. It is prudent to carry out unload/reload cycles and a final unloading since the slope of the curve can vary with the direction of loading.

The change in output of the transducers with pressure is linear and, assuming the membrane and protective sheath are elastic, so is the change in thickness of the membrane. The length of the membrane increases by an amount equal to the difference in diameter between the tube and the probe during initial loading. Further changes in length occur as the ends of the membrane are forced into the annulus between the tube and the probe. Thus, the change in length of the membrane is non-linear. It is usual to assume that the membrane compression curve is linear at pressures greater than the membrane stiffness. That is the reason for the cycle prior to calibrating the membrane.

BS EN ISO 22476-5:2012 suggests that, for high-pressure probes, the probe shall be pressurised in increments up to the maximum capacity of the probe. As the pressure is increased, the cylinder diameter will increase. The increase in cylinder diameter can be measured or calculated, assuming an elastic expansion of a thick-walled cylinder. The

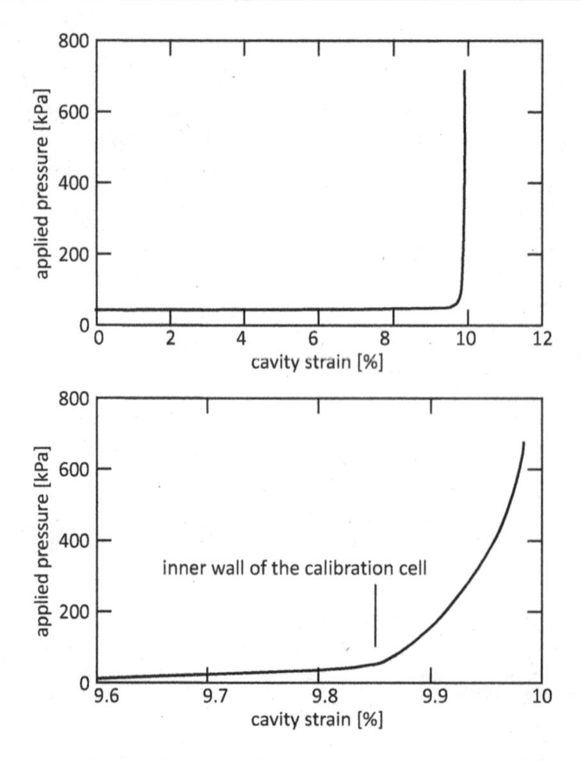

Figure 3.18 The calibration of a membrane to determine its compression characteristics showing the compression of the membrane after contact is made with the calibration cell.

membrane correction is the difference between the two curves shown in Figure 3.19. It is assumed that the cylinder and membrane are linear elastic in their response. Therefore, the membrane compression correction, a, is the slope of the line representing the difference in deformation between the cylinder and the membrane.

$$a = \frac{(d_{22} - d_{21}) - (d_{12} - d_{11})}{(p_2 - p_1)} \tag{3.8}$$

where d_{22} and d_{12} are the probe diameter and internal diameter of the cylinder at an applied pressure p_2; and d_{21} and d_{11} are the values at an applied pressure of p_1. For instruments with multiple displacement transducers, the mean value of a shall be used.

3.3.9 System compression

Membrane compression and membrane thinning referred to above apply to those probes that have displacement transducers. It is unnecessary to carry out such calibrations for probes with volume-displacement systems. A system compression has to be determined instead. This is the change in volume of the probe, the lines connecting the probe to the surface and the control unit. The calibration procedure is very similar to that for the membrane compression – that is, the probe is inflated to its maximum pressure when

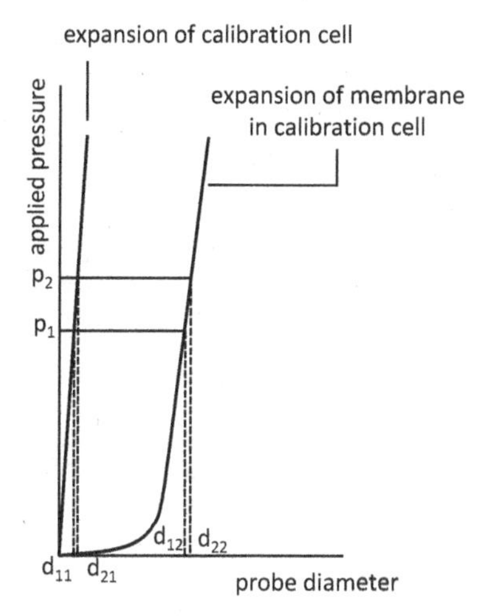

Figure 3.19 Determining the membrane coefficient.

Source: After BS EN ISO 22476-5:2012.

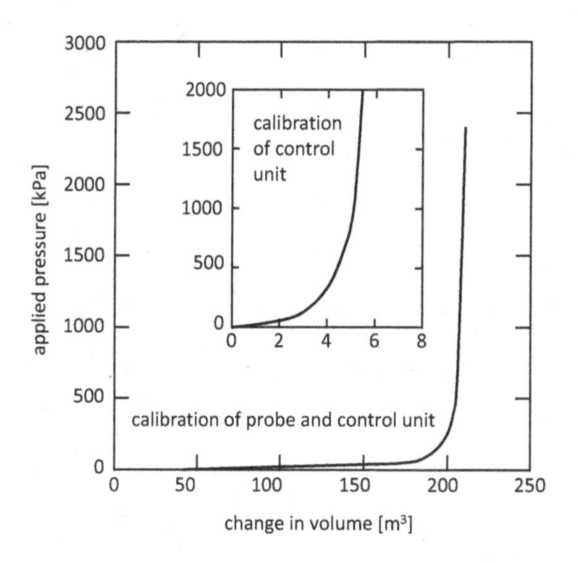

Figure 3.20 The system compliance of a GC probe and its control unit.

Source: After Baguelin et al., 1978.

inside a metal tube. Figure 3.20 shows the result of a system compression test on an MPM. Note the control unit measures the *change in volume*, not the actual volume.

3.3.10 Pressure loss

With volume-displacement systems, it is necessary to determine the pressure loss. This is the difference between the pressure in the probe and the pressure applied to the wall of the pocket. In the case of the MPM, the probe is inflated in increments of 10 kPa

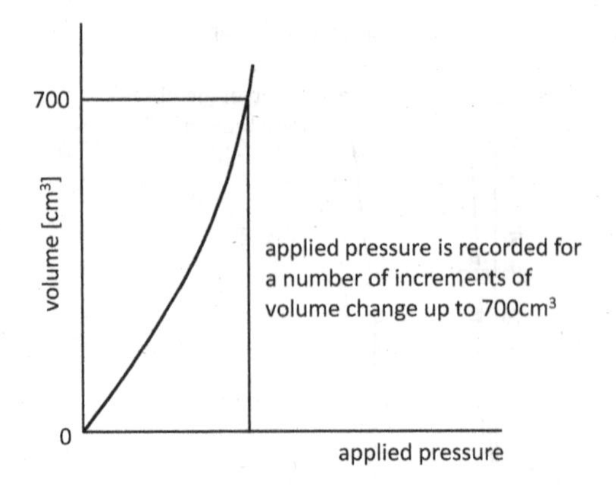

Figure 3.21 Pressure-loss correction for the MPM.

Source: After BS EN ISO 22476-4:2021.

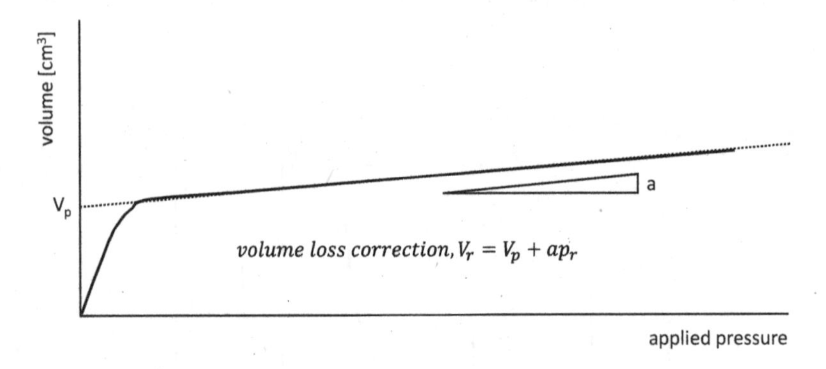

Figure 3.22 Volume-loss correction for the MPM.

Source: After BS EN ISO 22476-4:2021.

(Figure 3.21) until the increase in volume is 700 cm^3, which is the volume used to determine the limit pressure. The probe is placed next to, and ideally level with, the control unit. If a shorter cell is used, as could be the case with a slotted tube, the volume is 550 cm^3. The ultimate pressure loss, p_{el}, is the pressure required to inflate the membrane by 700 (or 550) cm^3.

3.3.11 Volume loss

Volume loss refers to the volume used to pressurise the measuring cell of a volume-displacement type probe. The probe, and slotted tube if necessary, is inserted in a steel cylinder with a wall thickness of at least 8 mm and an internal diameter of 66 mm (for an AX probe). It is pressurised in increments up to the maximum pressure capacity of the probe. Each increment shall be applied within 20 s and held for 60 s.

Initially the probe will expand until the membrane is in contact with the inner wall of the cylinder (Figure 3.22).

The volume correction, V_r, is given by

$$V_r = V_p + ap_r \tag{3.9}$$

where V_p is the intercept on the plot of volume change against applied pressure shown in Figure 3.22, and α is the slope of that line. BS EN ISO 22476-4:2021 suggests that the value of α shall be less than 6 cm³/MPa for a probe fitted with hydraulic hose less than 50 m. Higher α values suggest there is air in the system or there is a leak in the system.

3.3.12 The initial dimension of the probe and readings of the transducers

It is necessary to determine the initial volume or diameter of the probe since tests are analysed with respect to the initial condition. The external diameter of all probes can be measured with calipers. The initial volume can be found from the measured diameter with probes that have displacement transducers even if the membrane expands during installation because it is a fluid-filled probe in a dry borehole. The volume of a volume-displacement type probe at the start of a test is unknown unless the membrane is stretched over the body of the probe and does not expand during installation.

The initial volume of a volume-displacement type probe is defined by testing it in a thick-walled cylinder. The ratio of the internal diameter of the cylinder to the probe diameter should be about 1.005. The probe is placed in the tube, which is longer than the expanding section of the probe. Pressure is applied by pumping fluid into the probe. The point at which the probe can no longer be moved within the tube defines the initial volume. This is best determined by applying the maximum pressure and then allowing the pressure to reduce until the probe can be just slid out of the tube.

With volume-displacement systems where the control system is separate from the probe by the connection lines, it is necessary to determine the initial volume of the probe, V_c. In the case of the MPM,

$$V_c = 0.25\pi l_c d_i^2 - V_p \tag{3.10}$$

where l_c is the length of the measuring cell, and d_i the inside diameter of the calibration cylinder.

For single-cell volume-displacement probes, such as the FDP, it is also necessary to correct for the system volume. If the cone for a FDP exceeds the diameter of the measuring cell, a correction has to be applied to correct for the difference in diameter. Expanding an FDP in a thick-walled cylinder produces a curve shown in Figure 3.23.

Note it is custom to plot volume against pressure for the MPM and pressure against volume for other volume-displacement systems.

For a test carried out in a thick-walled cylinder, the reference volume, V_{ref}^u, is

$$V_{ref}^u = 0.25\pi l_c d_i^2 \tag{3.11}$$

where l_c is the length of the membrane and d_i the internal diameter of the calibration cylinder. V_{ref}^u is not corrected for the system volume increase. The corrected reference

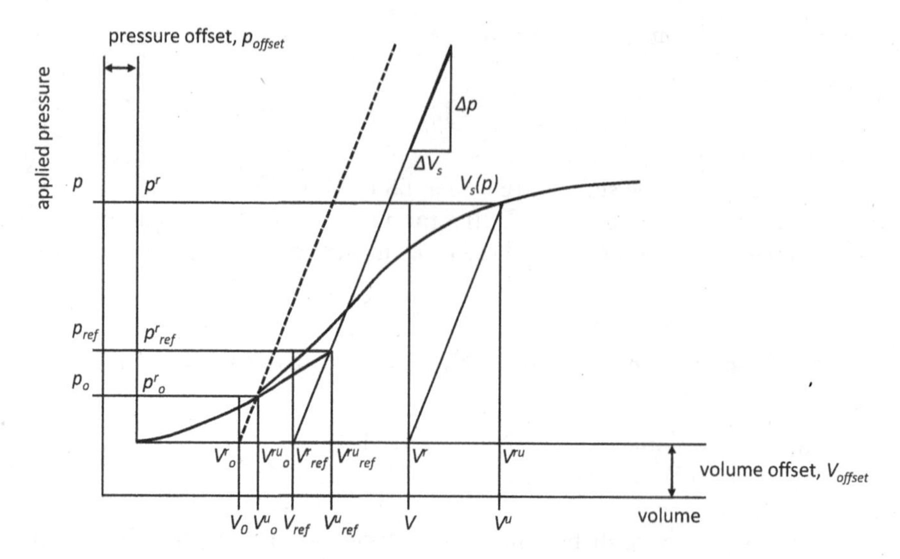

Figure 3.23 Correction for volume-displacement FDP.

Source: After BS EN ISO 22476-8:2021.

volume, V_{ref}, is corrected for the volume change in the system due to the pressure, p_{ref}, needed to inflate the membrane to meet the cylinder wall.

$$V_{ref} = V^u_{ref} - \frac{p_{ref}}{a} \tag{3.12}$$

where a is the slope of the pressure/volume curve after the membrane is contact with the cylinder wall.

If the FDP is attached to a cone with a larger diameter than the membrane diameter, it is necessary to determine the corrected volume allowing for the pressure to initially inflate the membrane to meet the theoretical diameter of the pocket; that is the diameter of the cone. The measured volume, $V_{cav}{}^u$, is the sum of the actual volume increase and the increase in the system volume, V_{corr}.

$$V^u_{cav} = 0.25\pi l_c d_i^2 + V_{corr} \tag{3.13}$$

V_{corr} depends on the pressure, p_o, at which the membrane is expanded to the cone diameter. Note that, the pocket will most likely contract as the soil is unloaded as it passes the cone. Hence, it is an iterative procedure to determine V_{corr} using the pressure, p_o.

$$V^u_{cav} = 0.25\pi l_c d_i^2 + \frac{p_o}{a} \tag{3.14}$$

It is necessary with all probes containing transducers to obtain readings at the surface just before the probe is lowered down the borehole, either to check that the probe is still working or to obtain the zero readings for the pressure sensor in fluid-filled probes (e.g. PAF) and the effective pressure sensors (e.g. CSBP).

3.3.13 Frequency and relevance of calibrations

The number and types of calibrations and the accuracy to which they are carried out must be within reasonable limits. It is important to ensure that calibrations are carried out correctly and frequently or to demonstrate that the calibrations remain constant over a period of time by keeping records of a set of calibrations. Table 3.12 gives a summary of the relevant calibrations for each type of probe.

The ISRM (1987) recommends that calibrations be carried out at least weekly and after major repairs, which includes membrane replacement. Mair and Wood (2013) suggest that it is only necessary to recalibrate transducers after replacing a membrane if the electronics are affected by water. They also suggest that a probe should be calibrated daily or after ten tests.

Table 3.11 Problems of self-boring, their causes and remedies

Problem	*Cause*	*Remedy*
Loss of drilling fluid	Hydraulic fracture of soil in front of probe	Adjust cutter Change drilling fluid Change fluid pressure Change speed of drilling
	Very permeable soils	Adjust cutter Change speed of drilling
	No seal around probe	Adjust cutter Change drilling fluid Change fluid pressure Change speed of drilling
No soil in returned drilling fluid	Probe blocked during drilling	Remove probe and clear blockage Consider different cutter position
	Probe blocked after testing	Flush probe before testing Remove probe after testing and clean
	Inner rods sheared	Remove probe and repair rods
Limit of reaction	Friction on probe too great	Move cutter Cycle drilling force
	Probe blocked	Clean probe
	Obstruction	Remove obstruction with drilling rig
	Hard layer	Advance hole with drilling rig

Table 3.12 Type of pressuremeter and relevant calibrations

Group	*Measuring system*	*Transducers*	*System compression*	*Membrane stiffness*	*Membrane compression*	*Membrane thinning*	*Zero*
Prebored	Volume	If fitted	Yes	Yes	No	No	Yes
	Displacement	Yes	No	Yes	Yes	Yes	Yes
Self-bored	Volume	If fitted	Yes	Yes	No	No	Yes
	Displacement	Yes	No	Yes	Yes	Yes	Yes
Full displacement	Volume	If fitted	Yes	Yes	No	No	Yes
	Displacement	Yes	No	Yes	Yes	Yes	Yes

It is useful to establish a register of the calibrations of the transducers for one probe. Experience has shown that they do not change significantly with time. Therefore, provided their consistency can be demonstrated with historical records, it is only necessary to carry out the calibrations at the start and end of a project and when there is a change to the transducers or cable or control unit.

It is necessary to determine the properties of a new membrane since the properties vary between membranes. This would include an assessment of the membrane's stiffness and compression. Membranes are replaced mainly because either they burst during installation or testing or because the probe is being repaired or cleaned. Should a membrane burst while the probe is in the ground it is prudent to make a simple check on any transducers since they may be affected by water that enters the probe. This can be done simply by inflating the probe in a cylinder of known internal diameter. The output of the transducers is checked against the output expected.

It will be necessary to carry out a full set of calibrations if a probe becomes damaged in any way – for example, if the cable breaks. This should be carried out in warm, dry conditions. It will involve dismantling, drying and cleaning the probe.

Table 3.13 Frequency of calibrations

Calibration	*Frequency*	*Additional calibrations*	*Comments*
Transducers	1. Start and finish of project 2. Regular intervals during project	1. Any change in the transducers 2. Change to the lead connecting the probe to the surface 3. Following damage to probe which involves thorough cleaning and drying	Instrument register
System compression	1. Start and finish of project 2. Regular intervals during project	1. Change to the lead connecting the probe to the surface 2. Change to the control unit	
Membrane stiffness	1. Start and finish of project 2. Regular intervals during project 3. Every time a membrane is replaced		Exercise new membrane before calibrating
Membrane compression	1. Start and finish of project 2. Regular intervals during a project 3. Every time a membrane is replaced		Exercise new membrane before calibrating
Membrane thinning			By calculation
Zero	1. Prior to lowering into borehole		

In summary, calibrations of the transducers should be carried out at frequent intervals and certainly at the beginning and end of an investigation and at intervals during the investigation, depending on the size of the project, as shown in Table 3.13.

Membrane calibrations should be carried out at the same time as the transducer calibrations and whenever a membrane is changed. A new membrane should be taken to the maximum displacement several times before carrying out the calibration. It is important at all times to ensure that calibration tests are carried out safely.

Generally, calibrations of transducers and membranes are, for practical purposes, linear. Typical units are shown in Table 3.14. Typical times taken to carry out those calibrations are also given. A full set of calibrations could take a day, depending on the probe and specification. Specifying a full set every six to ten tests may have a significant effect on the cost of a contract, hence, the importance of a register.

The calibrations required depend on the ground conditions and the probe. Table 3.15 summarises the relevant calibrations according to ground type.

Calibrations for all transducers are critical since their output is used to determine response of the ground to loading and unloading. Membrane stiffness becomes more critical the softer the ground since it affects estimates of strength and lateral stress. Membrane stiffness has little effect on deep tests or tests in hard ground. As the importance of membrane stiffness reduces the importance of membrane compression and thinning or system compression increases. In very stiff ground, membrane compression can become a significant part of the measured ground response and it is, therefore, usual to limit the quoted stiffness of the ground. For example, it is common not

Table 3.14 Summary of form of calibrations showing units and typical times to carry out a complete calibration

Calibration	*Units*		*Time to complete (hours)*
	Slope	*Intercept*	
Pressure transducers	mV/kPa	kPa (only important for probes filled with oil)	2.5–2.75
Displacement transducers	mV/mm	Not important	2.5–2.75
System compression	mm^3/kPa	mm^3	1.5–1.75
Membrane stiffness	mm/kPa	kPa	1.5–1.75
Membrane compression	mm/MPa	Not important	1.5–1.75

Table 3.15 Relevance of calibrations in different ground conditions

Calibration	*Soft ground Max. pressure 1 MPa*	*Most ground Max. pressure 4 MPa*	*Strong ground Max. pressure 20 MPa*
Transducers	Critical	Critical	Critical
System compression	Important	Important	Critical
Membrane stiffness	Critical	Important	Not important
Membrane compression	Unnecessary	Not important	Critical
Membrane thinning	Not important	Not important	Critical
Zero	Critical	Critical	Critical

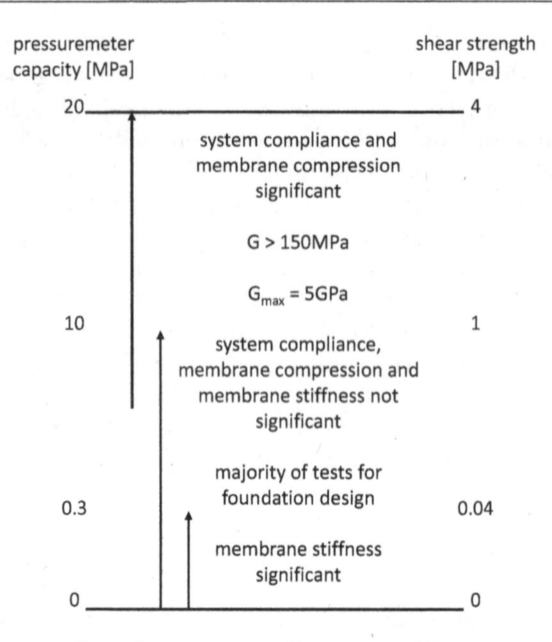

Figure 3.24 The importance of membrane system calibrations in different ground conditions.

to quote moduli from RSBP tests greater than 3 GPa. Figure 3.24 shows the effect of calibrations in different ground conditions.

3.4 ON-SITE SYSTEM CHECKS

The probe and control unit should be checked for leaks regularly. This is not a calibration, but should be carried out at least as frequently as the field calibrations and probably daily, or after five tests, and possibly every time the probe is at the surface. Possible leaks are found by pressurising the probe in a metal tube (e.g., a calibration tube) which is capable of withstanding pressures up the maximum capacity of the probe. For rigorous testing, two different internal diameters of tube should be used, one slightly larger than the probe and the other equivalent to the maximum displacement of the membrane. This is to test the membrane and hose connections at small strains, and the clamping arrangement at larger strains. Leaks in gas systems can be seen by the presence of bubbles when the probe and calibration cylinder are placed under water. Leaks in fluid-filled systems can be detected by a loss of pressure with time.

All volume-displacement type probes must be thoroughly checked for leaks, since any loss of fluid will affect the results. It is important to purge the system of air to ensure that any measured changes in volume are truly a change in the dimension of the test section. Saturation of fluid-filled systems can be determined from the slope of the membrane compression line. The slope for a system containing air will be less than that for a fully saturated system.

Gas leaks are not so critical but if they are appreciable they will lead to loss of supply and may cause damage to pocket walls. If the gas can leak out, water may leak in. This will lead, if electrical transducers are used, to shorting of the circuits and therefore loss of the measuring system. Should any water enter such a probe it is necessary to strip it down, dry it and clean it thoroughly before replacing the membrane. This should be done in a warm dry environment. It is possible to coat the electrical connections with waterproofing compound. This can lead to problems should any rewiring be necessary since the compound will have to be removed.

It is prudent to check the output of any transducer before the probe is lowered into a borehole. If it is not working the probe should be repaired.

3.5 THE TEST

3.5.1 Introduction

In order to obtain repeatable results it is necessary to create test pockets using recommended installation procedures, standardise the time between installation and testing, and follow a recognised test procedure that provides the results required.

Ideally, a test is started once the pocket is created, but this is not always possible. A PBP has to be lowered into the pocket and, in that case, a test should be started immediately the probe is in place. An FDP test can be started once the probe is pushed to a test location, which is common practice.

It is assumed that an SBP has little effect on the pocket wall during installation. There are inevitably some changes due, in part, to the interface shear. Any excess pore pressures will dissipate with time. Windle (1975) conducted tests with a self-boring load cell, the Camkometer, and found that equilibrium conditions were achieved in half an hour. Jamiolkowski et al. (1979) demonstrated that, in soft clays, the time could be much greater. It would be uneconomic to wait too long. Therefore, a period of half an hour is recommended as a compromise between equilibrium and practicality. Using this as a standard ensures that the effects of installation on tests in a similar material will be the same. If effective pressure transducers are fitted, it is possible to monitor excess pore pressures and start the test once those pressures have dissipated.

A test is started when either the pressure in the membrane changes or the membrane moves. Ideally, the membrane is expanded sufficiently to test ground that has been unaffected by installation – that is, undisturbed ground is tested. The movement of the membrane and the pressure required to cause that movement are recorded at intervals and these data are reduced to produce a stress-strain curve. Figure 3.25 shows some typical test curves. Note that the pressure is plotted as a function of the membrane displacement. Sometimes these are reversed, as explained below.

There are procedures, suggested methods and recommendations for tests for different probes, for example (Table 3.16) the Ménard method, and methods described in ASTM D4719-20:2020, GOST 20276-12:2012, and BS EN ISO 22476-4:2021, -5:2012, -6:2018, and -8:2018. The Ménard method is unique in that the probe, control unit, installation, test procedure and interpretation must all follow a specification since the results are used directly in design methods. Other test procedures were developed to determine ground properties, which are subsequently used in design methods not exclusive to the pressuremeter. There are other recommendations for tests for different ground conditions and to obtain different ground parameters (for example, unload/reload cycles for stiffness, holding tests for permeability).

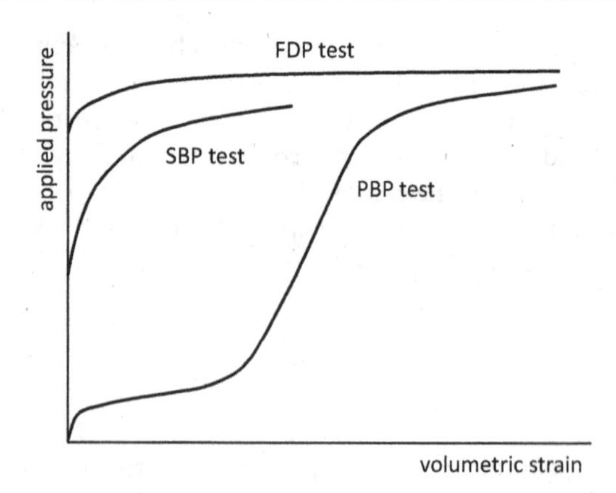

Figure 3.25 Typical stress-strain curves for the PBP, SBP and FDP showing how the installation methods affect the test curve.

Table 3.16 Internationally recognised standards, suggested methods and recommended procedures for pressuremeter tests

Designation	*Country*	*Type of pressuremeter*	*Ground conditions*	*Type of test*
Ménard method (LCPC)	France	Ménard (PBP)	All	Stress
ISRM	International	PBP	Rock	Stress/strain
GOST 20276-12:2012	Russia	PBP	Soils	Stress
ASTM D4719-20:2020	USA	PBP	All	Stress/strain
BS EN ISO 22476-4:2021	International	Ménard (PBP)	All	Stress
BS EN ISO 22476-5:2012	International	PBP	Rock	Stress/strain
BS EN ISO 22476-6:2018	International	SBP	All	Stress/strain
BS EN ISO 22476-8:2018	International	FDP	Soil	Stress/strain

Tests are generally subdivided into stress- or strain-controlled tests, depending on the type of probe and control system. Ideally, the membrane is loaded until it reaches a maximum displacement or until the pressure capacity of the probe is reached. Generally, in soils (provided the correct probe is used), the limit of a test will be the limit of expansion of the membrane. In rocks it is generally the pressure capacity of the probe. Capacities of various probes are given in Table 1.3. Note that these do vary and reference should be made to the manufacturers and operators.

3.5.2 The Ménard method

Ménard realised that the effects of installation produced a disturbed zone around the pocket and the interpretation of the test was complicated by that zone. The standard Ménard procedure covers the method of expansion, which is stress controlled, and the amount of expansion, which, ideally, should double the size of the pocket. The pressure required to achieve this is known as the modified or Ménard limit pressure, p_{lm}. If the initial volume and radius of the measuring cell are V_o and a_o respectively, the initial volume and radius of the pocket are V_p and a_p and the volume and radius of the pocket at the modified limit pressure are V_L and a_L, then the following apply:

$$\text{Volumetric strain} = \frac{\Delta V_p}{V_L} = \frac{V_L - V_p}{V_L} = 0.5 \tag{3.15}$$

$$\text{Cavity strain} = \frac{\Delta a_p}{a_L} = \frac{a_L - a_p}{a_L} = 0.41 \tag{3.16}$$

It is the change in volume, ΔV, of the test section that is measured. Thus

$$\frac{\Delta V}{V_o} = \frac{2V_p}{V_o} - 1 \tag{3.17}$$

V_p is found from the test curve. This can be expressed in terms of the radius of the probe:

$$\frac{\Delta a}{a_o} = \frac{1.41a_p}{a_o} - 1 \tag{3.18}$$

If the radii of the probe and pocket are the same then the membrane has to be expanded by 41% to reach the Ménard limit pressure. Thus, if a pressuremeter is to be used to determine the Ménard parameters, it must be capable of being expanded by at least this amount. If the ratio of the pocket diameter to the probe diameter is 1.1, a typically specified value, then the membrane must be able to expand by 55% to measure p_{lm}.

The Ménard test is a stress-controlled test in which equal increments of pressure are applied at equal intervals of time. It is recommended that there should be ten equal increments during loading though the minimum number of increments can be eight to increase the speed of loading and, to reduce the speed of loading, the maximum number can be 14. An estimate of the ground properties is required before a test is conducted since the Ménard limit pressure must be assumed at the start of a test. Table 3.17 gives guidelines for estimating the Ménard limit pressure for soils. It is unlikely that the modified limit pressure will be reached during tests in rock, therefore it is normal to increase the pressure in increments of between 0.5 and 1 MPa up to the maximum capacity for those tests.

The initial pressure of the oil or water in the measuring cell is equal to the pressure due to the weight of fluid within the hydraulic hoses. The membrane is completely deflated if the pocket is full of mud (because of the weight of the mud) or the test is shallow (because of the membrane stiffness). The membrane is expanded against the pocket wall if the weight of fluid in the hoses exceeds the stiffness of the membrane and the test pocket is dry. In general, it is most likely that tests will be carried out in mud-filled test pockets, therefore the change in volume of the membrane measured at the surface is the change in volume of the test section corrected for system losses.

Table 3.17 Guidelines for estimating the modified limit pressure

Soil	*Description*	*SPT N_{60}*	*Undrained shear strength (kPa)*	*Modified limit pressure (kPa)*
Sand	Loose	0–10		0–500
	Medium	10–30		500–1500
	Dense	30–50		1500–2500
	Very dense	> 50		> 2500
Clay	Soft		0–25	0–200
	Firm		25–50	200–400
	Stiff		50–100	400–800
	Very stiff		100–200	800–1600
	Hard		>200	> 1600

Source: After Briaud, 1992.

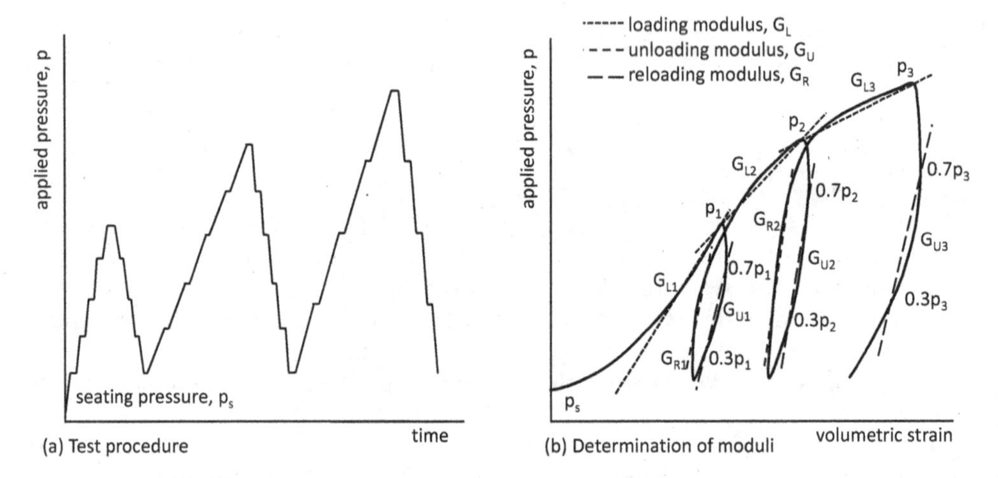

Figure 3.26 Type A test procedure showing the portion of the tests used to calculate the loading modulus (G_L), the unloading modulus (G_U) and the reloading modulus (G_R).

Source: After ISO 22476-5:2012.

Readings of pressure and volume can be recorded at the surface with Ménard probes. Some probes have transducers within the probe. The data recorded are relative to the initial conditions; that is, the values that are measured are the change in pressure and the change in volume.

The pressure in the probe is increased by one increment, and this pressure is maintained for 60 s. Readings of the change in volume are taken at 15, 30 and 60 s after applying the increment of pressure. This is repeated for each increment until the pressure capacity or the volume capacity is reached. The membrane is unloaded in approximately ten decrements, each decrement being held for 1 min. The loading sequence is similar to that shown in Figure 3.26, which also shows two unload – reload cycles not normally required in an MPM test. Note that the stress-strain curve is produced from the data at the end of each 60 s period. This procedure can be used with any prebored pressuremeter that can be expanded by about 55%.

The data recorded from monocell probes with displacement transducers are applied pressure and displacement. The initial pressure and radius are found by taking measurements at the surface prior to lowering the probe into the pocket. The displacement is measured relative to the probe diameter. Provided the membrane stiffness exceeds the pressure of the fluid in the probe or the borehole is filled with mud or water, the measurements at the surface should be the same as those in the pocket. The pressure is recorded relative to atmospheric pressure in gas-filled probes and relative to the borehole mud pressure for fluid-filled probes.

Account must be taken of the membrane stiffness at the start of a test if a tricell probe is used in which the pressure in the guard cells has to be maintained at a different pressure to that in the measuring cell and a test is in a dry pocket. The pressure in the guard cells, p_k, must be in the range:

$$p_c - 3p_m \leq p_k \leq p_c - 2p_m \quad (3.19)$$

where p_c is the pressure in the central measuring cell, p_m is the pressure loss.

For example, consider the G MPM in which a difference in pressure of 110kPa must be maintained between the guard cell gas pressure and the measuring cell water pressure. At the start of a test, the pressure differential may not be 110kPa, depending on the depth of the probe and the mud level in the borehole. Three cases must be considered:

(a) a dry borehole with the probe below 11m
(b) a dry borehole with the probe above 11m
(c) a mud-filled borehole.

There is a pressure in the membrane at the start of the test due to the hydrostatic head in the tubing connecting the probe to the surface. Below about 11 m in dry boreholes the initial pressure in the test section will be greater than 110kPa. Therefore, the difference between the guard cells and measuring cell pressures can be set to 110kPa. Above 11m, the initial pressure in the test section will be less than 110kPa. In that case the gas pressure is maintained at zero until the applied pressure and hydrostatic pressure are equal to 110kPa. The differential pressure is then set. In boreholes full of water or mud, the initial pressure is zero since the hydrostatic pressure within the membrane is equal to that in the borehole. In this case the gas pressure is set when the applied pressure equals 110kPa.

Once the pressure difference is set it is automatically maintained, though it is recommended that a gauge be installed in the circuit to monitor the difference. The differential pressure should be reduced according to the manufacturer's specifications for tests in very soft soils or pressuremeters with low-stiffness membranes. For example, with a low-stiffness membrane and tests in very soft soil the pressure differential should be reduced to 40kPa.

3.5.3 Stress-controlled tests

The Ménard method is a special version of a stress-controlled test. It has been described separately because it is important to distinguish between that test procedure and all other test procedures since the Ménard test forms part of a standard design procedure whereas other stress-controlled tests do not. The procedures for stress-controlled tests are similar to the Ménard method; the main differences are the number of increments, the time intervals and the inclusion of unload/reload cycles.

Briaud et al. (1986b) suggested that increments should be $p_{lm}/10$ and held for 1 min. This is the method specified by ASTM D4719–20:2020. Mair and Wood (2013) suggested that the number of loading increments should be between 15 and 20. BS EN ISO 22476-5:2012 suggests that the pressure increments should be 2% to 5% of the maximum pressure and each increment held for 1min to 3min. Typically, increments vary between 15 kPa for soft clays, 100 kPa for stiff clays and dense sands and up to 500 kPa for weak rocks. The first test not only shows that the correct drilling technique has been used but also gives a check on the chosen pressure increment.

The first increment is applied. If the pressure exceeds the required increment no adjustments are made, and the actual pressure is recorded. The pressure is maintained during the time interval by adjusting the regulator or hand pump. The change in volume or displacement is noted over a period of time during which the pressure is held constant. The usual British practice for PBP tests is to apply increments every 2 min (Mair and Wood, 2013). Windle (1975) used 60 s for CSBP stress-controlled tests; Corke (1990) used 30 s for HPD tests.

The Russian Standard (GOST 20276-12:2012) specifies slow and fast stress-controlled tests. Each increment of pressure (0.025 MPa) is maintained until the rate of displacement is less than 0.1 mm for the time intervals specified in Table 3.18.

At the end of any increment it is possible to undertake an unload/reload cycle by deflating and reflating the membrane using the same stress control procedure. In some instances the decrements and increments of pressure during a cycle are reduced to obtain additional data points.

The ISRM standard recommends a stress-controlled method in which the pressure is increased in at least five equal increments up to the maximum value, but it is different to the tests described above. The test is designed for rock and, therefore, the maximum capacity of the probe is likely to be the limit of the test. No time interval is given for each increment but it suggests that the pressure or volume should be held constant only if dilation information is required. The maximum pressure is held for 10 min. Three complete cycles of unloading and reloading are required to obtain the stiffness properties of rocks.

Table 3.18 Guidelines for the time for a stress increment of 25 kPa

Type of soil	*Properties*	*Type of test*	*Time for each stress increment of 25 kPa*
Sands	Dr <0.8	Slow	15
	Dr > 0.8		30
Clays	LL <0.25		30
	LL > 0.25		60
Organic soils and very soft clays			90
Sands		Quick	3
Clays			6
Organic soils and very soft clays			10

Source: GOST 20276-12:2012.

The membrane is unloaded when the pressure or oil or displacement capacity is reached. It is usual to unload it in a similar way to the loading path. The stress-strain curve and variation in stress and strain rate are very similar to those shown in Figure 3.26. Note that in stress-controlled tests the strain rate varies during a test.

BS EN ISO 22476-5-2012 sets out four procedures for the HPD:

- Procedure A: The membrane is expanded until it contacts the pocket wall when a significant increase in pressure with strain/volume change will be observed. This is the seating pressure. Thereafter, the increments of pressure shall be applied for 1 min to 3 min. At least two unload/reload cycles should be included in which the pressure is reduced in increments down to the seating pressure plus the first increment. Five increments should be used for the first loading. This is followed by four unloading increments. Figure 3.27 shows a loading sequence with three unload/reload cycles.
- Procedure B: This is similar to Procedure A but may include reload/unloaded cycles on the final unloading of the probe.
- Procedure C: This is similar to the MPM test with no unload/reload cycles.
- Procedure D: This is procedure to be followed when time-dependent behaviour is to be assessed.

BS EN ISO 22476-6-2018 allows for a more flexible approach to testing with a self-boring pressuremeter. Tests can be stress- or strain-controlled or a combination of these. Unload/reload cycles on the loading phase and reload/unload cycles on the final unloading are permitted. It is advised that the pressure is maintained prior to unloading to avoid any time-dependent effects. Data should be recorded at least every 10 s to clearly define the curve. An unload/reload cycle should contain a minimum of 20 data points. The pressure range of an unload/reload cycle should be about one third of the pressure at the start of the cycle or, if testing in clays, less than twice the undrained shear strength. The latter is to ensure the soil does not yield on unloading.

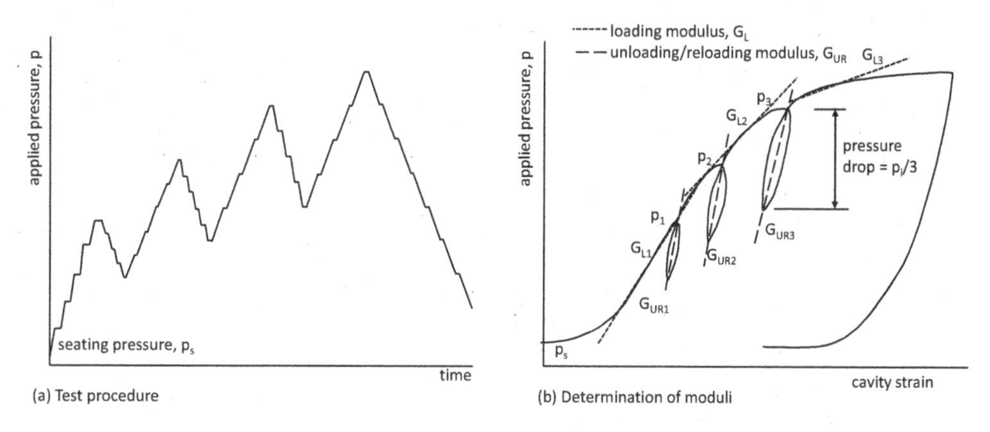

Figure 3.27 Type B test procedure showing the portion of the tests used to calculate the loading modulus (G_L) and the unloading/reloading modulus (G_{UR}).

Source: After ISO 22476-5:2012.

BS EN ISO 22476-8-2018 allows for a similar flexible approach to testing as BS EN ISO 22476-6-2018 for the full-displacement pressuremeter.

The membrane will expand due to the pressure of fluid in the tubing with fluid-filled systems, the amount of expansion depending on the stiffness of the membrane, the level of mud in the borehole and the diameter of the pocket wall. Several possibilities exist for fluid-filled probes.

(a) *a PBP in a dry borehole* – the membrane will expand to the pocket diameter if the pressure of the fluid exceeds the membrane stiffness;
(b) *a PBP in a mud-filled hole* – the membrane will not expand;
(c) *a PBP in a partially mud-filled borehole* – the membrane may expand;
(d) *a SBP in a carefully drilled pocket or a FDP* – the membrane will not expand.

The initial pressure for fluid-filled PBP probes is equal to the hydrostatic pressure, either due to the weight of fluid in the hoses or due to the external pressure of the mud, whichever is greater. With surface-measuring systems this pressure can only be assumed and, in such a case, the pressure recorded is the change in pressure. If transducers are mounted in the probe and readings of the transducers are taken before the probe is lowered into the borehole then the actual pressure is recorded. Gas-filled systems have transducers. The initial readings represent the initial diameter of the probe and absolute pressure.

3.5.4 Strain-controlled tests

The ideal strain-controlled test is one in which the membrane is expanded at a constant rate of displacement. This is relatively simple when using a fluid-filled system. A known quantity of water or oil is pumped into the membrane and the resulting pressure recorded. Briaud et al. (1986b) suggested that each increment of volume, $V_o/40$, should be held for 15 s. The pressure is recorded at the end of each increment. This test procedure can be used with any fluid- filled probe in any ground condition and it gives a complete loading test in 10 min. This method has the disadvantage that the variation in volume during the initial loading of the pocket wall may be too large to obtain any meaningful results. This is especially the case with SBP and FDP tests.

A strain-controlled test is more complicated with gas-filled systems, since they contain displacement transducers. It is necessary to have a feedback control system such as the strain-control unit described in section 2.4.2. The control unit is set to a specified rate, perhaps 1%/min. The outputs from the displacement transducers are monitored and compared to the preset rate. If the rate of expansion is less than that required, gas at pressure is allowed into the probe by manually or automatically adjusting a valve. If either the rate of expansion is too great or the membrane is being unloaded, gas is vented to atmosphere.

A strain-controlled test has the disadvantage that, when loading during an elastic phase (initial loading and reloading), the ground responds very little due to a change in pressure. Thus, too few data, if any, are obtained if the recording intervals are not frequent enough. Further, the test may become unstable because, once the ground yields,

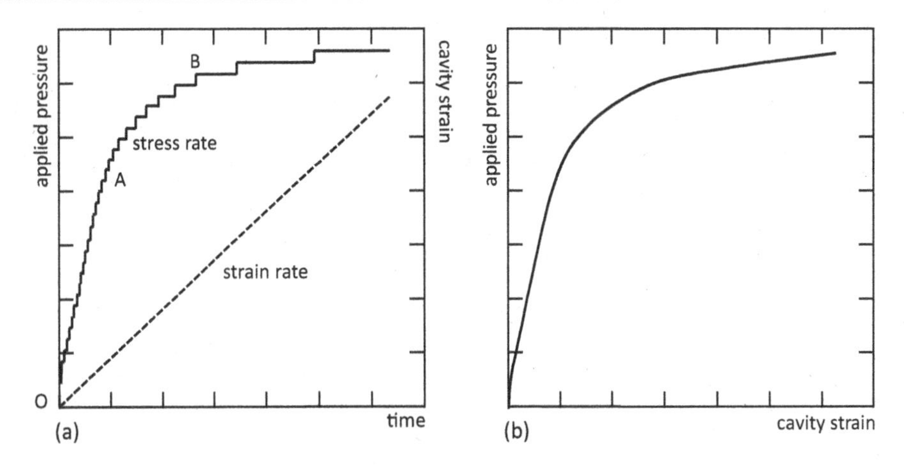

Figure 3.28 A typical CSBP curve showing (a) by varying the time for each stress increment it is possible to obtain a strain-controlled test, which leads to (b) a stress-strain curve.

the membrane will expand in an uncontrolled manner if the feedback system cannot respond quickly enough to close the valves. A compromise is made wherein the elastic portion of a test is stress-controlled and only when the soil begins to yield does the strain control take over. The standard CSBP test is such a hybrid test. The stress rate used is dependent on the soil type and typically ranges from 2 kPa/min for soft soils to 40 kPa/min for stiff soils. The strain rate used depends on the type of test. The usual rate for undrained tests in clay and tests in sands is 1%/min, and the majority of tests reported in the literature have been carried out at this rate.

Figure 3.28 shows test data plotted as a stress-strain curve for a typical CSBP test in clay together with plots of stress and strain rate. From O to A the test is stress-controlled with no strain occurring since the applied pressure is less than the horizontal stress. From A to B the test remains stress-controlled and the strain rate increases. At B the strain rate is 1%/min and thereafter the test is strain-controlled. The transducers are read at 20 s intervals if manual readings are taken or every 10 s if recording automatically. It is important to obtain as many readings as possible between O and B to ensure that there are sufficient data to analyse the test. Once the test becomes strain controlled it is sensible to take readings at 20 s intervals to reduce the amount of data (typically a CSBP test takes 1 hour and with six transducers being monitored, up to 2,160 readings are taken).

3.5.5 Additional test procedures

The procedures described above are those most commonly used. There are variations on these depending on the parameters required. These include unload/reload cycles for determining stiffness and holding tests for determining the coefficient of consolidation.

The pressuremeter is most commonly used to estimate the stiffness of the ground. In the Ménard method a stiffness is obtained from the initial loading of the ground. This is not the stiffness of the ground, but is the stiffness of a zone affected by the

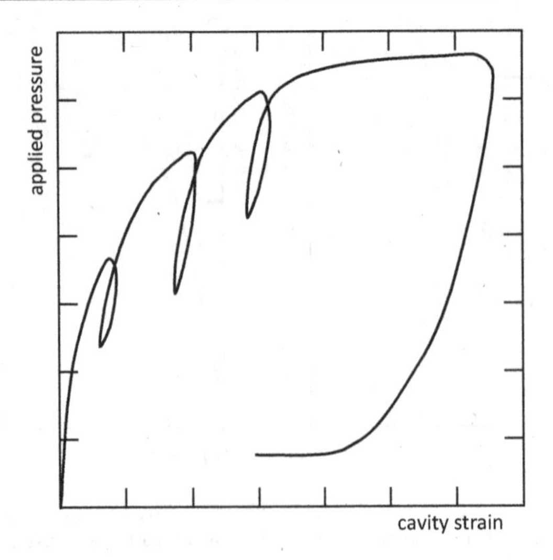

Figure 3.29 A CSBP test showing an unload/reload cycle and final unloading.

installation technique. The stiffness of undisturbed ground is obtained from an unload/reload cycle.

The unloading and reloading of the membrane is either stress- or strain-controlled, depending on the probe and test procedure. In either case there must be sufficient increments to define the curve accurately. Figure 3.29 shows a CSBP test with unload/reload cycles. Note that the membrane is not fully unloaded during a cycle to prevent failure in extension. The stress range is usually defined in terms of the strength of the ground; if it is too large, the ground will fail in extension. The recommended ranges are given in Chapter 6.

The ISRM-suggested method for rock states that the membrane must be fully unloaded prior to reloading. This is acceptable, since it is unlikely that the rock will fail in extension and, therefore, the rock is still responding elastically.

Withers et al. (1989) have shown that, for high-quality, strain-controlled pressuremeter tests in sand, it is necessary to allow the sand to relax before unloading. The displacement of sand at constant pressure is known as *creep*. They recommended that the creep rate should reduce to 0.1%/min before unloading. Figure 3.30 shows the recommended procedure for pressuremeter tests in sand. Note: it is unnecessary to do this with stress-controlled tests as relaxation takes place during loading, since the strain rate varies over a stress increment.

Clarke and Smith (1992a) have shown that, with strain-controlled tests in weak rock, it is necessary to hold the volume of the membrane constant before unloading to allow any effect of creep or consolidation to take place. The membrane will continue to expand even if the pressure is reduced, as shown in Figure 3.31. They showed that, within economic limits, the membrane should be held for 10 min before unloading when testing rock.

It is common practice to carry out more than one unload/reload cycle in a test so that the effects of stress level or cyclic loading can be quantified. It is usual to try to maintain a constant stress or strain range for the cycles, but this is not too critical, since the stiffness can be calculated over a constant stress or a constant strain range. The unload/

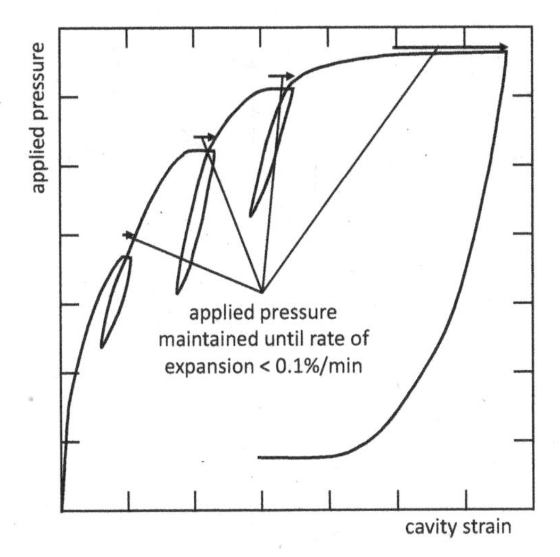

Figure 3.30 A proposed test procedure for tests in sand to obtain the true stiffness of the sand.

Source: After Withers et al., 1989.

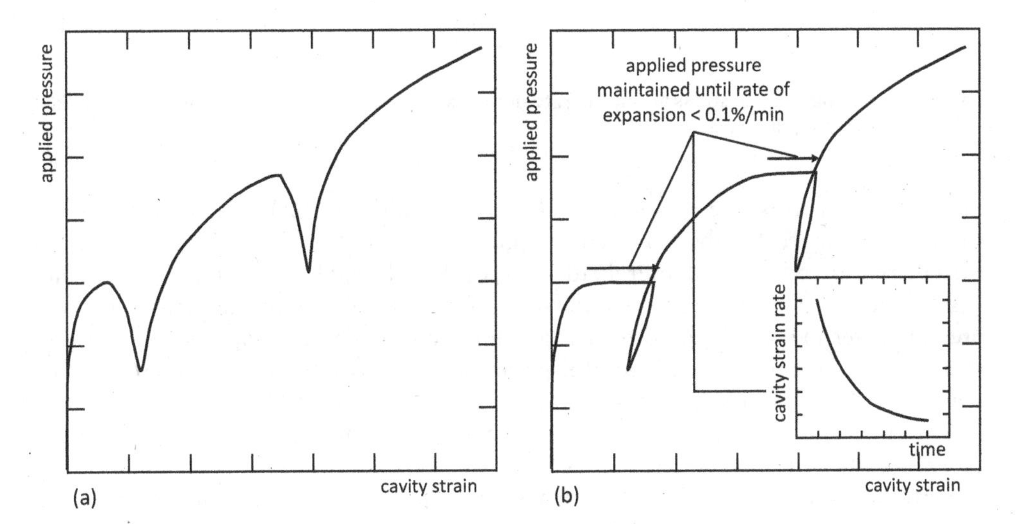

Figure 3.31 The (a) effects of creep on an unloading/reloading cycles in rock (after Clarke and Smith, 1992b) and (b) suggested procedure to eliminate the effects of creep to determine stiffness.

reload cycles indicate hysteretic behaviour – that is, the ground response is non-linear. It is shown in section 4.7.1 that the unload/reload cycle can be used to determine the stiffness over any strain range within the accuracy of the measuring system.

Jézéquel and Le Méhauté (1982) describe several procedures for carrying out cyclic load tests, as these tests have to be automatically controlled to ensure that the amplitude remains constant. They developed a computer-controlled regulator to operate a piston pumping fluid in and out of a PAF, and describe tests that include up to 800 cycles. Note that an unload/reload cycle affects the remainder of the test because of the

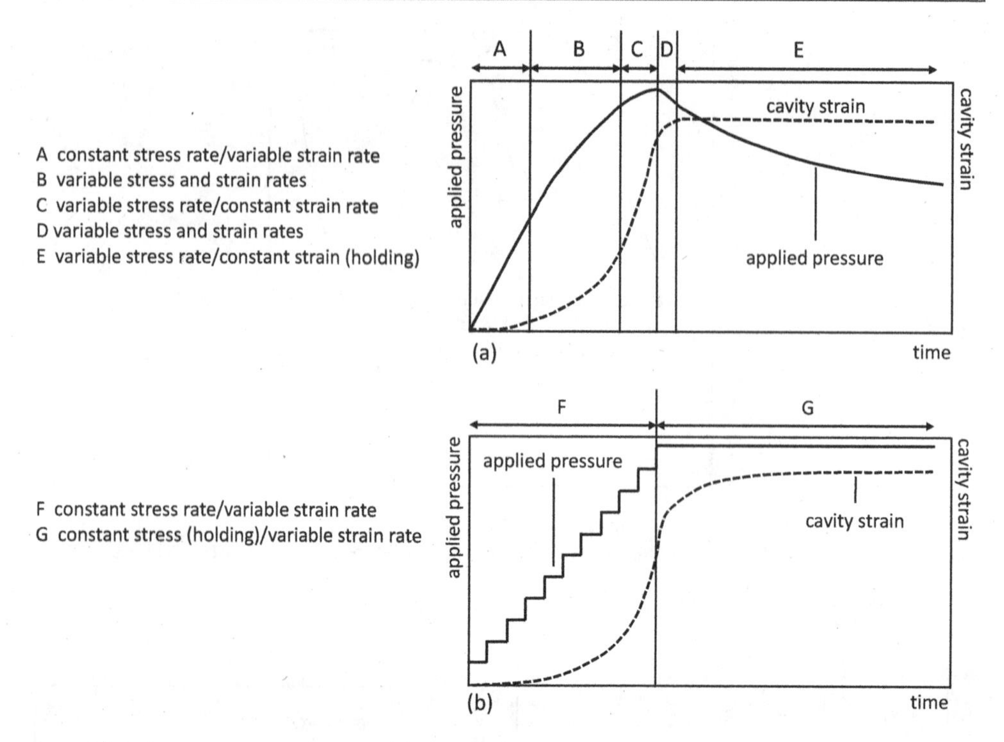

Figure 3.32 The variation in stress and strain rate during (a) a constant strain-holding and (b) a constant stress-holding test.

time it takes. This is one reason that Clarke and Smith (1992a) suggested that a test should be designed for the parameter required. If strength is the required parameter, then there should be no unload/reload cycles or they should be at the maximum strain.

Baguelin et al. (1978) and Clarke et al. (1979) have both suggested that the pressuremeter test can be used to determine the consolidation properties of clays. In this test, the membrane is inflated to, perhaps, 10% cavity strain and thereafter the volume of the membrane is held constant while the change in total and effective stresses (if effective stress transducers are fitted) are observed. Figure 3.32 shows the variation in stress and strain rates for such a test.

Kjartanson et al. (1990) proposed a form of holding test in which the applied pressure is kept constant. This is a form of stress-controlled test in which an increment of stress is maintained until the strain rate becomes constant. The Russian standard describes a similar procedure for soils. Shields et al. (1986) have applied this test in ice to determine the long-term performance of ice structures.

3.5.6 Testing in ice

As a consequence of the development of the Canadian east coast oil and gas reservoirs there is the need to manage icebergs and build on ice islands or ice or ice-rich

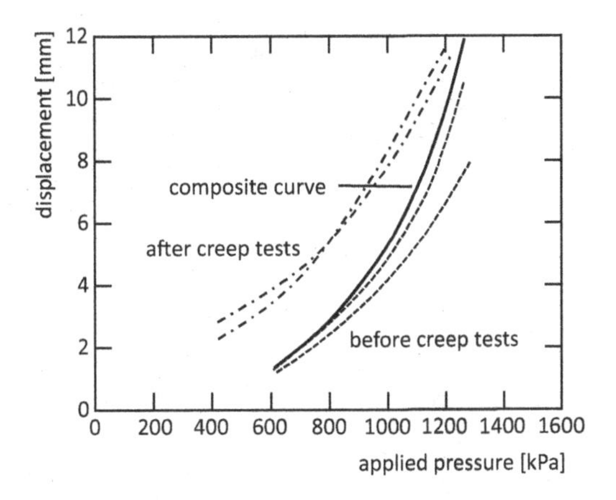

Figure 3.33 Membrane stiffness curves for the Elastometer 100 before and after creep tests.

Source: After Kjartanson et al., 1990.

permafrost. In the former case there is requirement to understand the stress conditions within the iceberg (this is described in detail by Shields et al. (1986)), and in the latter case there is a need to measure the creep of ice. Shields et al. (1986) suggested that the pressuremeter could be used to estimate the stress conditions in icebergs. Ladanyi and Johnston (1973) suggested that the pressuremeter could be used to measure the creep properties of ice.

Shields et al. (1990c) and Kjartanson et al. (1990) recommended that creep tests are best carried out using radial displacement type pressuremeters with gas as the pressurising medium, since this removes the problems of working with water or oil at low temperatures. Small leaks can be tolerated. In addition to the displacement and pressure transducers it is also necessary to measure the temperature. Calibrations have to be carried out on the membrane at the operating temperature because the membrane's compression and stiffness change with time and load. This is particularly important when testing in frozen ground or ice; therefore, any form of creep test must take this into account. Kjartanson et al. suggested that calibrations are carried out before and after a test and a composite curve produced to correct the data as shown in Figure 3.33.

The probe has to be insulated from the surface to prevent heat being conducted to or away from the test section. Stress-controlled tests that last for at least 24 h, and preferably 48 h, are carried out to determine the steady-state creep rate.

3.5.7 Summary of test procedures

Tests can follow a standard procedure, such as the MPM test, or be specified to obtain certain parameters. Table 3.19 summarises possible procedures and gives guidelines for specifications. The appendix describes in detail the specifications for typical test procedures.

Table 3.19 Summary of commonly used test procedures

Name	*Type*	*Probe*	*Ground conditions*	*Recording intervals*	*Stress rate*	*Strain rate*
Ménard	Stress	MPM other PBPs with max strain of 55%	All	15, 30, 60s	$p_{lm}/10$	
ISRM	Stress	PBP	Rock	Not specified	Not specified	
ASTM	Stress	PBP	All	30 s	25–200 kPa	
ASTM	Strain	PBP	All	30 s		0.05–0.1 × V_o
GOST	Stress	PBP	Soils	Varies	25 kPa	
Stress	Stress	All	All	10–30s		Varies
Strain	Strain	All	All	10–30s		1%/min
Holding	Strain	All	All	10–30s		1%/min during loading; 0% during holding

3.6 TERMINATION OF A TEST

3.6.1 Introduction

A test is terminated when the membrane has been deflated after loading to its maximum capacity, either pressure or displacement depending on the capacity of the probe and the ground being tested. There are exceptions:

(a) the maximum-pressure capacity of the probe is reached;
(b) the maximum-volume capacity of the probe is reached;
(c) a membrane bursts;
(d) a displacement transducer reaches its maximum displacement;
(e) the operator is concerned about possible damage to the probe.

Figure 3.34 shows how (a) to (d) can affect a PBP test curve.

3.6.2 Maximum pressure capacity

The maximum-pressure capacity is reached when the pressure in the membrane reaches the pressure capacity of the probe. This is common when testing rock, since the pressure required to reach the maximum cavity strain exceeds 20MPa, the capacity of most probes designed for testing rock. It can occur when testing soil, especially dense sands. As a guide for clays and rocks, if the ratio of the pressure capacity to the undrained shear strength of the ground exceeds 6, then it is possible that the pressure capacity will be reached during a test.

3.6.3 Maximum oil volume capacity

The maximum-volume capacity is reached when the volume of water or oil pumped into a volume-displacement type probe is equal to the theoretical maximum increase in

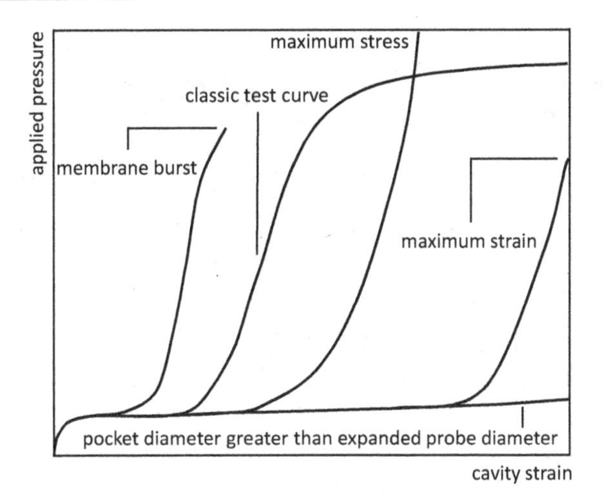

Figure 3.34 Possible shapes of PBP test curves terminated for a reason.

the volume of test section. For example, an MPM has a maximum capacity of 50% of the original capacity. It is possible to pump water or oil into the membrane until, theoretically, the membrane should have reached its maximum displacement yet, in fact, the majority of the membrane does not actually reach the maximum displacement. Thus the maximum capacity does not imply the limit of expansion of the probe. It can be a consequence of either (a) a burst membrane, or (b) a leak or (c) a non-uniform cavity expansion, as described below.

(a) If a membrane bursts there is a sudden drop in pressure and the operator will stop the test and remove the probe.
(b) If there is a leak, the operator will only detect that leak if it is large or, if on deflating the membrane completely, the water or oil in the reservoir does not return to its original level. If the leak is large it is often impossible to develop significant pressures within the membrane. Note that air in the system can give misleading results.
(c) Ground may contain thin layers of varying stiffnesses or discontinuities or cracks may occur during a test. This is especially the case with weak rock. Figure 3.35 shows the possible effects on the expansion of a membrane in layered ground or ground containing fissures.

If displacement transducers are fitted, it is possible, in case (c), to reach the maximum oil capacity of the probe, even though the transducers show that the membrane has not expanded by its full amount. The transducers only measure displacement at a point on the membrane.

The use of oil-filled radial displacement-type probes allows an operator to detect when there may be damage to the membrane. The average expansion, measured by the transducers, should be equivalent to the amount of oil pumped in.

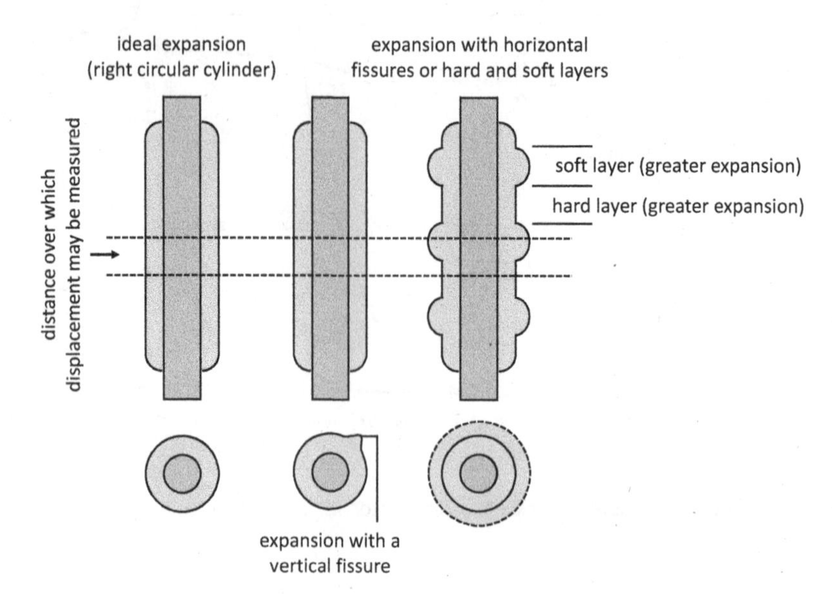

Figure 3.35 The effect of layered ground on the expansion of a pressuremeter.

3.6.4 Burst membranes

Burst membranes can occur for a variety of reasons and are commonly associated with ground conditions similar to those described in section 3.6.3. It is possible for the membrane to be pulled out of the clamping arrangement, giving the effect of a burst membrane. This can occur when the pocket diameter is much greater than the probe diameter or when the top of the membrane is not in the pocket or the top or bottom of the membrane is in softer ground.

A burst membrane is followed by a significant drop in pressure. It is no longer possible to continue a test if a fluid-filled probe is used and, therefore, it must be removed from the ground for repair. It may be possible to continue to expand the membrane if a gas-filled probe is used, provided an adequate gas supply is available. Small gas leaks are often seen during a test, but as they do not affect the ground being tested it is common practice to proceed with the test.

The ISRM recommends that a pocket in rock should be checked for open fissures or voids using a borehole camera. This is not common practice since many probes have protective sheaths to limit damage caused by fissures or voids.

3.6.5 Maximum displacement

The maximum displacement can only be measured when using radial-displacement type probes. The maximum displacement occurs when (a) the ground has yielded sufficiently, (b) there is a void in the ground, or (c) the pocket diameter is too large. If one or more displacement transducers reach their maximum displacement, then the membrane should be unloaded. It is possible to continue to expand the membrane but this is at the risk of bursting the membrane since its position is unknown. However, in some instances, this may be prudent, especially if one transducer shows significantly greater movement than another.

3.7 REDUCTION OF DATA AND INITIAL PLOTS

3.7.1 Introduction

The changes in volume or displacement, and the pressure causing those changes, are monitored at intervals throughout the test, as described in section 3.5. These recorded data may be in units of volts (if transducers are used), volume (if volume-measuring systems are used) or stress (if the pressure is measured directly). The recorded data have to be converted to engineering units and corrections applied for system compliance and membrane properties so that the tests may be interpreted correctly. The calibrations discussed in section 3.4 are used. It is very important that calibrations are applied correctly, since they can radically affect the results (see, for example, Bacciarelli, 1986).

The pressure is measured either directly using Bourdon gauges or indirectly using electrical transducers. The pressure applied to the membrane in a fluid-filled system is equal to the applied pressure plus that due to the hydrostatic head resulting from the difference in level between the control unit at the surface and the level at the centre of the probe. If the pressure is measured relative to the pressure at the start of a test the measured pressure is equal to the change in applied pressure. If it is measured relative to the surface – that is, to the zero conditions taking into account any fluid in the probe – the measured pressure is the actual pressure in the pocket. In gas-filled systems the measured pressure is the actual pressure.

3.7.2 The Ménard pressuremeter test and other volume type pressuremeter tests

Figure 3.36 shows the flow path followed to reduce data from a stress-controlled MPM test.

The pressure in the central measuring cell is equal to the applied pressure plus the hydrostatic head, p_h, between the elevation of the pressure-measuring device and the centre of the probe. The pressure-loss correction, p_e, is a function of the volume of

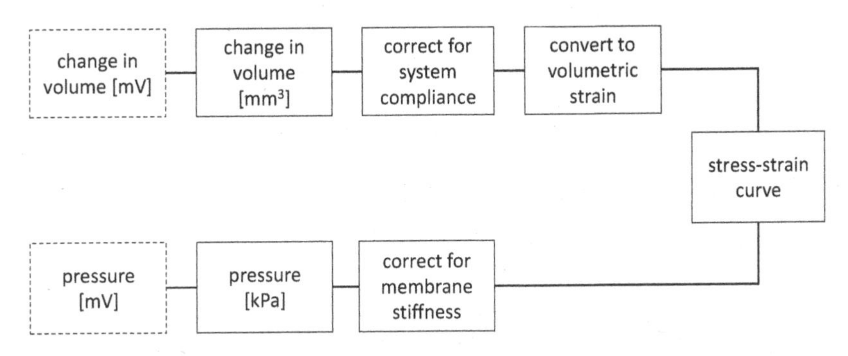

Figure 3.36 The procedure used to reduce the measured data from a volume-displacement type PBP test to a stress-strain curve.

liquid, V_r, at the end of each increment. The pressure-loss correction is the slope of the best fit straight line to the pressure-loss calibration curve. However, the pressure-loss curve is not necessarily linear. Therefore, if a more precise correction is needed, then the best fit line is obtained using a power law.

$$p_e = bV_r^m + c \tag{3.20}$$

An MPM test curve can be approximated by a straight line tangential to two hyperbolic segments. If this curve fitting approach is used, then the corrected pressure, p, is

$$p = p_r - p_e \tag{3.21}$$

The corrected volume allowing for the volume loss correction is

$$V = V_r - \alpha p_r \tag{3.22}$$

where α is the slope of the best fit line to the volume loss calibration curve. This correction is only applied to stiff soils.

Figure 3.37 shows a typical MPM test, which is based on the readings at the end of the increment at 60 secs.

The corrected data are given by

$$p = p_r + p_h - p_e(V_r) \tag{3.23}$$

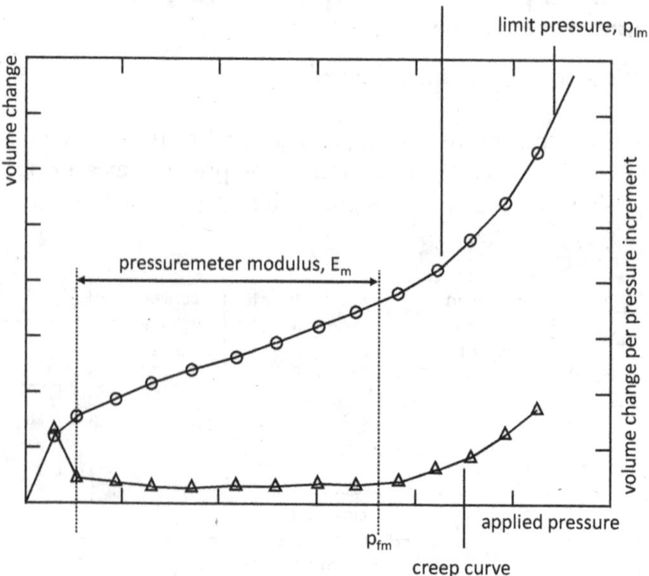

Figure 3.37 Parameters derived from a Menard pressuremeter test include the limit pressure (p_{lm}), the creep pressure (p_{fm}) and the pressuremeter modulus (E_m).

$$V = V_r - V_e(p) \tag{3.24}$$

Note that it is conventional to plot MPM tests with the volume on the vertical axis.

A further plot, the creep plot, is drawn to assist in the interpretation of a test. The volume change between 30 and 60 s for each stress increment is plotted against the corrected applied pressure, as shown in Figure 3.37. The creep curve will be different if the time interval of each increment is changed.

The method to convert data from strain-controlled PBP tests is similar to that described above, but all the data are plotted to produce the pressure-displacement curve. Volumetric strain is calculated by expressing the change in volume with respect to the original volume.

3.7.3 Radial displacement type PBP tests

Converting data from a radial displacement type PBP test is similar to that described for a volume-displacement type probe. It is not necessary to correct for system compliance, since displacement is measured directly, but it is necessary to correct for membrane compression and thinning, as shown in Figure 3.38.

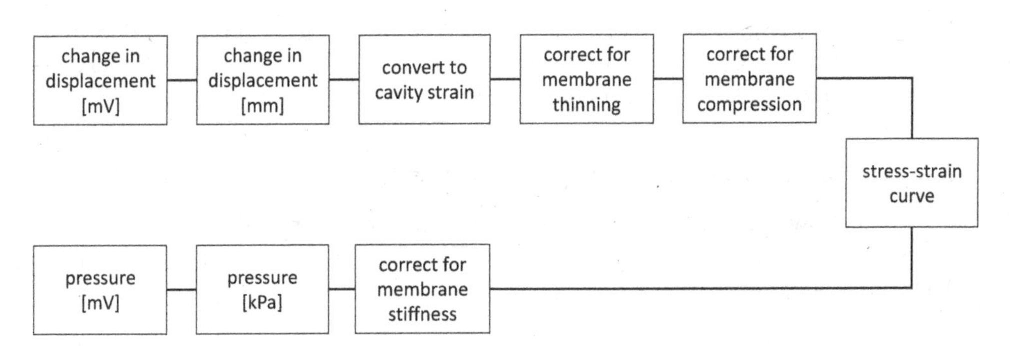

Figure 3.38 The procedure used to reduce the measured data from a radial displacement type PBP test to a stress-strain curve.

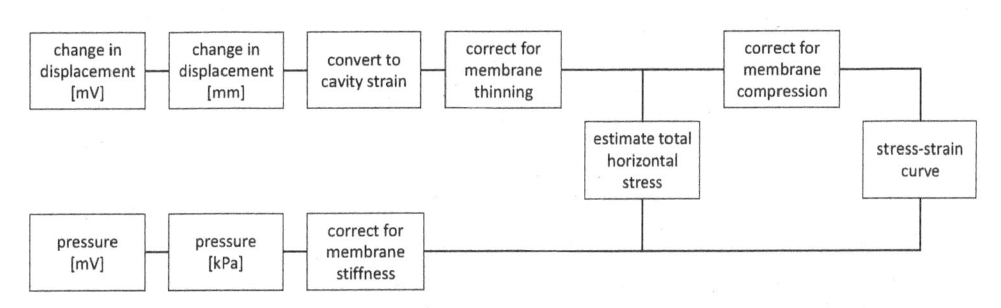

Figure 3.39 The procedure followed to reduce the measured data from a radial displacement type SBP test to a stress-strain curve.

3.7.4 Radial displacement type SBP tests

Figure 3.39 shows the path followed to convert data from a radial displacement type SBP test in which the membrane is under external ground pressure at the start of a test.

The external pressure is less than, equal to or greater than the in situ horizontal stress depending on whether there was any installation disturbance. Therefore, it is necessary to estimate this pressure first and apply the membrane compression and thinning corrections to the displacement once the membrane begins to expand.

The corrected pressure, p, is

$$p = p_r(d_r) - bd_r - p_c \tag{3.25}$$

where d_r the measured displacement, b, the membrane stiffness coefficient, is the best fit line to the membrane stiffness calibration, and p_c the intercept to the membrane stiffness calibration (se).

The corrected displacement d, is

$$d = d_r - ap_r \tag{3.26}$$

where a is the slope of the pressure-displacement curve from the membrane-compression test.

The membrane compression-correction is necessary for all tests in rocks and for tests with thick membranes (such as those used with the OYO pressuremeters). Otherwise, it is generally not used as the errors introduced by ignoring it are small.

3.7.5 Radial displacement type FDP tests

Figure 3.40 shows the path followed to convert data from a radial displacement type FDP tests.

The corrected volume, V, is

$$V = V^r + V_{offset} = V_{r,u} - V_s(p) + V_{offset} = V_{r,u} - \frac{p}{M_s} + V_{offset} \tag{3.27}$$

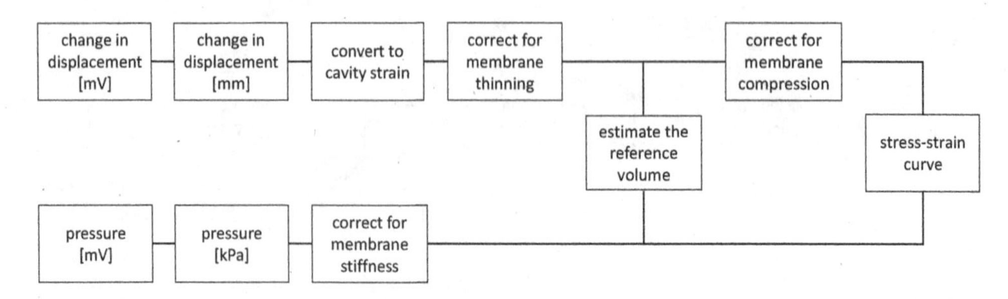

Figure 3.40 The procedure followed to reduce the measured data from a radial displacement type FDP test to a stress-strain curve.

where M_s is the system stiffness, $V_{r,u}$ the readings recorded during a test and V_{offset} is

$$V_{offset} = V_0 - V_0^r = V_{ref} - V_{ref}^r \tag{3.28}$$

where V_0 is the initial cavity volume and V_0^r is the maximum cone module volume corrected for the system volume increase.

3.8 SUMMARY

The results of a pressuremeter test are a function of the instrument used, the installation procedure, the test procedure and the methods of analysis and interpretation. It is important to ensure that the installation procedure produces repeatable minimum disturbance to the ground. The pocket created for a prebored pressuremeter is dependent on the type of drilling used. That created for a self-bored pressuremeter should, theoretically, represent the minimum disturbance to the surrounding ground, but the disturbance is a function of the control used during drilling as well as particle size. The pushed-in pressuremeter creates the same amount of disturbance every time, provided a standard rate of penetration is used.

A pressuremeter has to be calibrated before use so that the test data can be converted to stress and strain to obtain a test curve for analysis. There are calibrations for transducers, the membrane and the system compliance. All transducers have to be calibrated, usually at the beginning and end of a project and when there is a major change to the probe. Membrane calibrations are carried out at the same intervals as the transducer calibrations and whenever the membrane is changed. They include membrane stiffness – which is particularly important when testing soft soils – and membrane compression when testing rocks. The system compliance has to be ascertained for all volume-displacement type probes, and membrane thinning has to be calculated for radial-displacement type probes used in rock.

There are several test procedures some of which are set out in national and international standards. Tests can be stress- or strain-controlled or a combination of these. Tests are designed to measure different parameters, including in situ stress, modulus, strength, permeability and creep, which are derived from the test curve using theories of cavity expansion explained in Chapter 4.

REFERENCES

Amar, S., Clarke, B.G. Gambin, M. and Orr, T.L.L. (1991) *The application of pressuremeter test results to foundation design in Europe*, Report to the ISSMFE.

Arsonnet, G., Baud, J.P. and Gambin, M.P., 2005. Pressuremeter tests inside a self-bored slotted tube (STAF). In *International Symposium 50 Years of Pressuremeters (ISP5-Pressio 2005)* (pp. 31–45).

ASTM D4719–20 (2020) Standard Test Methods for Prebored Pressuremeter Testing in Soils, ASTM International.

Bacciarelli, R.E. (1986) The calibration and use of high capacity pressuremeter to determine rock stiffness, *Proc. 2nd Int. Symp. Pressuremeter Marine Appl., Texam, United States*, ASTM STP 950, pp. 75–96.

Baguelin, F., Jézéquel, J.F. and Shields, D.H. (1978) *The Pressuremeter and Foundation Engineering*, Trans. Tech. Publication.

Briaud, J.-L. (1986b) Pressuremeter and foundation design, *Proc. In situ '86: Use of In Situ Tests in Geot. Engng*, Blacksburg VA, pp. 74–115.

Briaud, J.-L. (1992) *The Pressuremeter*, Balkeema, Rotterdam.

Briaud, J.L., 2013, September. Ménard Lecture: The Pressuremeter Test: Expanding its use. In *Proceedings of the 18th International Conference on Soil Mechanics and Geotechnical Engineering*, Paris (pp. 107–126).

Briaud, J.-L. and Gambin, M. (1984) Suggested practice for the preparation of a pressuremeter test borehole, *Geotech. Test. J., ASTM*, pp. 36–40.

BS 5930:2015+A1:2020, *Code of practice for ground investigations*, British Standards Institution.

BS EN ISO 22475-1:2021 *Geotechnical investigation and testing – Sampling methods and groundwater measurements – Part 1: Technical principles for the sampling of soil, rock and groundwater*, British Standards Institution.

BS EN ISO 22476-4:2021, *Geotechnical investigation and testing. Field testing – Prebored pressuremeter test by Ménard procedure*, British Standards Institution.

BS EN ISO 22476-5:2012, *Geotechnical investigation and testing. Field testing – Flexible dilatometer test*, British Standards Institution.

BS EN ISO 22476-6:2018, *Geotechnical investigation and testing. Field testing – Self-boring pressuremeter test*, British Standards Institution.

BS EN ISO 22476-8:2018, *Geotechnical investigation and testing. Field testing – Full displacement pressuremeter test*, British Standards Institution.

Clarke, B.G., Carter, J.P. and Wroth, C.P. (1979) In-situ determination of the consolidation characteristics of saturated clays, *Proc. 7th Eur. Conf. SMFE*, Brighton, Vol. 2, pp. 207–211.

Clarke, B.G., Newman, R. and Allan, P. (1990) Experience with a new high pressure self-boring pressuremeter in weak rock, *Ground Engng*, 22(5), 36–39; (6), 45–51.

Clarke, B.G. and Smith, A. (1992a) A model specification for radial displacement measuring pressuremeters, *Ground Engng*, *25* (2), 28–38.

Clarke, B.G. and Smith, A. (1992b) Self-boring pressuremeter tests in weak rocks, *Construction and Building Materials*, 6(2), 91–96.

Corke, D.J. (1990) Self-boring pressuremeter in situ lateral assessment in London clay, *Proc. 24th Ann. Conf. of the Engng Group of the Geological Soc.: Field Testing in Engineering Geology*, Sunderland, *pp.* 55–62.

Finn, P.S. (1984) New developments in pressuremeter testing, *Ground Engng*, 17, 17–29.

GOST 20276–12 (2012) updated version of GOST 20276–85 (translated by Foque, J.B. and Sousa Coutinho, G.F.) *Soils Methods for Determining Deformation Characteristics*.

ISRM (1987) Suggested methods for deformability determination using a flexible dilatometer, *J. Int. Soc. Rock Mech.*, pp. 125–134.

Jamiolkowski, M., Lancellotta, R., Marchetti, S., Nova, R. and Pasqualini, E. (1979) Design parameters for soft clays, *Proc. 7th Eur. Conf SMFE*, Brighton, Vol. 5, pp. 27–58.

Jézéquel, J.F. and Le Méhauté, A. (1982) Cyclic tests with the self boring pressuremeter, *Proc. Int. Symp. Pressuremeter and its Marine Appl.*, Paris, pp. 209–222.

Kjartanson, B.H., Shields, D.H. and Domaschuk, L. (1990) Pressuremeter creep testing in ice. Calibration and test procedures, *Geotech. Test. J., ASTM*, No. 13, 3–9.

Ladanyi, B. and Johnston, G.H. (1973) Evaluation of in-situ creep properties of frozen soils with pressuremeter, *Proc. 2nd Int. Conf. Permafrost North American Contributions*, Yakutsk, pp. 310–318.

Mair, R.J. and Wood, D.M. (2013) *Pressuremeter Testing – Methods and Interpretation*, Elsevier.

Masoud, Z. and Khan, A.H., 2019. An improved technique for prebored pressuremeter tests. *KSCE Journal of Civil Engineering*, *23*(7), pp. 2839–2846.

NF P 94-110 2000 (2000) *Sols: reconnaissance et essais – Essai pressiométrique Ménard*, AFNOR.

Shields, D., Ladanyi, B., Murcat, J.R. and Clark, J.I. (1986) Stresses in icebergs – will the pressuremeter work?, *39th Canadian Geotechnical Conf In Situ Testing and Field Behaviour, Ottawa*, pp. 159–166.

Shields, D.H., Domaschuk, L., Kjartanson, B.H. and Azizi, F. (1990) Measuring the creep properties of ice in situ, *Proc. 24th Ann. Conf. Engng Group of the Geological Soc.: Field Testing in Engineering Geology*, Sunderland, pp. 111–116.

Windle, D. (1975) "In situ Testing of Soils with a Self-boring Pressuremeter," PhD Thesis, University of Cambridge, UK.

Withers, N.J., Howie, J., Hughes, J.M.O. and Robertson, P.K. (1989) Performance and analysis of cone pressuremeter tests in sands, *Geotechnique*, 39 (3), 433–454.

Chapter 4

Analysis of expanding cavities

4.1 INTRODUCTION

This chapter deals with the analysis of a pressuremeter test – that is, theories of expanding cavities for which there are numerous references. An ideal pressuremeter test is often modelled as an infinitely long expanding cylindrical cavity in an elastic plastic continuum, the theory of which is well documented. The practical interpretation of a pressuremeter test is more complicated because of the multiphase nature and the stress history of soils and rocks, the impact the type of probe and methods of installing the probe have upon the surrounding ground, and the test procedure. Interpretation, discussed in Chapter 6, is often based on simpler theories, which are described in detail in this chapter. In recent years advances in analysis and numerical methods have enabled parametric studies into the effects that ground properties, the probe geometry and installation, and the test procedure have upon a pressuremeter test.

There are two approaches to the theoretical interpretation of a pressuremeter test. The first is to assume a constitutive model for the ground and derive an equation that will give an approximate fit to different parts of the test curve depending upon the parameters under investigation (e.g., Gibson and Anderson, 1961; Prevost and Hoeg, 1975; Denby and Clough, 1980; Manassero, 1989; Bolton and Whittle, 1999; Cao et al., 2001; Mo and Yu, 2017). The second is to use a constitutive model to generate the pressuremeter curve and undertake parametric studies to determine the most appropriate set of parameters that model the curve using inverse analysis (e.g., Ajalloeian and Yu, 1996; Zentar et al., 2001; Dano and Hicher, 2002; Mecsi, 2011; Abed et al., 2016; Toumi and Abed, 2018; Levasseur et al., 2008; Zheng, 2021). Table 4.1 gives references to some of the methods.

Cavity expansion and contraction solutions have been developed for elastic perfectly plastic material and for elastic strain softening/hardening materials (e.g., Yu, 2000). The perfectly plastic model assumes that the strength of the material remains constant, whereas, in reality, the strength of soil varies with deformation. This produces a further complexity, one which is often analysed using the critical state concept (Schofield and Wroth, 1968). Critical state models are used with numerical methods because strength is a function of effective stress.

It is usual to assume either the soil is fully saturated and the volume remains constant (e.g., undrained expansion in saturated clays) or the volume of the soil varies, and there is no change in pore pressure (e.g., drained expansion in sands). There are some analyses which allow for partial drainage – for example, the analyses of consolidation

DOI: 10.1201/9781003028925-4

Table 4.1 Examples of the theoretical interpretation of a pressuremeter test

Lamé (1852)	Linear elastic material
Bishop et al. (1945)	Cohesive material
Ménard (1957)	Frictional cohesive material
Vesic (1972)	Linear elastic, perfectly plastic material
Gibson and Anderson (1961)	Linear elastic perfectly plastic material with no volume changes
Windle and Wroth (1977)	
Jefferies (1988)	
Houlsby and Withers (1988)	
Denby (1978)	Non-linear elastic perfectly plastic material with no volume changes
Ferreira and Robertson (1992)	Hyperbolic model
Prévost and Hoeg (1975)	Elastic–plastic with strain hardening or softening with no volume changes
Ladanyi (1963)	Linear elastic perfectly plastic material with volume changes
Hughes et al. (1977)	
Robertson and Hughes (1986)	
Houlsby et al. (1986)	
Palmer (1972)	Cohesive material with no volume changes
Manassero (1989)	Cohesionless material with volume changes
Cao et al. (2001)	Modified Cam Clay
Mo and Yu (2017)	Unified state parameter model

tests in clay. It is usual to apply an analysis to a particular ground type. For example, the analysis developed by Windle and Wroth (1977) is widely quoted for the interpretation of undrained SBP tests in clay.

The analyses presented here are based on small strain theory – that is, the strain is defined in terms of the original radius. Pressuremeter tests can cause strains of up to 50% at the pocket wall. Several authors have investigated the expansion of cavities in compressible and dilatant material taking into account the large strains involved; these include Houlsby and Yu (1990); Carter et al. (1986); Sousa Coutinho (1990); Selvadurai (1984); Cao et al. (2001); and Mo and Yu (2017). They found that the solution for incompressible materials (that is, undrained tests in clays) is similar for the large and small strain theories. A comparison for dilatant materials shows that it depends on the angle of dilation and strain level. For practical purposes the small strain theory is often considered adequate because of the spatial variability of the composition and properties of the ground.

Numerical analytical techniques have been used to study installation (e.g., Whittle and Aubeny, 1993; Aubeny et al., 2000), the effects of the dimensions of the test section (for example Laier et al., 1975; Ajalloeian and Yu 1998), the test procedure (e.g., Pyrah et al., 1988), derived parameters (for example, Cambou et al., 1990) and soil characteristics (e.g., Yu and Collins, 1998; Rouainia and Wood, 2000; Russell and Khalili, 2002; Schnaid et al., 2004; Gonzalez et al., 2007; Gonzolaz et al., 2009; Hao et al., 2010; Han et al., 2011; Jiang and Sun 2012; Elwood et al., 2015; Li et al., 2016; Sivasithamparam and Castro, 2018; Qui et al., 2019; Chen et al., 2020; Gong et al., 2021; Rouainia et al., 2020; Yang et al., 2021; Han et al., 2021; Li et al.; 2021). They enable parametric studies to be carried out to investigate the sensitivity of an analysis to the assumptions made (Yu, 2006).

It is important to understand the assumptions made and the methods of analyses to realise the practical limitations of the interpretation of pressuremeter tests. Therefore, the development of the theory is discussed here in outline with references to the relevant publications.

4.2 CONSTITUTIVE MODELS

In order to interpret a pressuremeter test, it is necessary to establish equations of equilibrium and compatibility, and constitutive models linking stress to strain. Soil is a multiphase material with a skeleton of a combination of grains from fine-grained particles (clay and silt), coarse-grained particles (sand and gravel) to very coarse-grained particles (cobbles and boulders). The voids between the particles contain water (fully saturated), air (dry) or water and air (partially saturated). The behaviour of soil depends on the type, shape and properties of the soil particles, the degree of packing, the contact forces between the soil particles, and the properties of the pore fluid. It is theoretically possible to use discrete element methods to gain an insight into the microscopic behaviour of soil (e.g., Geng et al. 2013) but interpretation of pressuremeter tests is normally based on macroscopic behaviour in which soils are assumed to be a continuum. Saturated soils in which there is no drainage, saturated soils in which free drainage is permitted and dry soils can be treated as a single continuum. This is not feasible for partially saturated soil and saturated soils subject to transient flow because there is interaction between the soil particles and pore fluid flow. In that case, the soil is considered as a two-phase continuum to model the transient response.

Ideally, a constitutive model should cover all loading and unloading situations, be dependent on a small number of parameters that can be independently assessed and model the actual behaviour so that the parameters derived from a pressuremeter test can be used in geotechnical design.

The simplest model is the linear elastic model (Figure 4.1), which is often used to determine the shear modulus from pressuremeter tests.

However, soil is not linear elastic and its elastic behaviour is limited. Non-linear models such as the hyperbolic and the power law models (Figure 4.1) are used to derive the modulus degradation curve, but they do not model the failure of soil. Simple elastic plastic models incorporating a yield surface or failure surface (Figure 4.1) such as the Tresca yield criterion for undrained behaviour and Mohr Coulomb yield criterion to limit the elastic behaviour are used for drained behaviour.

Strain hardening/softening models based on the critical state concept (Figure 4.1), which incorporate the influence of stress history to account for the variation in strength with deformation have been used to gain a greater insight into the effects of geometry, installation disturbance, stress history, soil structure and anisotropy on the response of soil around an expanding cavity. The family of critical state models developed from the Cam Clay model to more complex models are often used for research purposes as they require more input parameters from specialist testing. The complexity of these models means that that they are inevitably used with numerical methods.

Other more complex models (Figure 4.2) include extended hardening models with a kinematic hardening surface, double hardening surfaces, multi surface models, and bubble models below the yield surface to more precisely describe the loading and

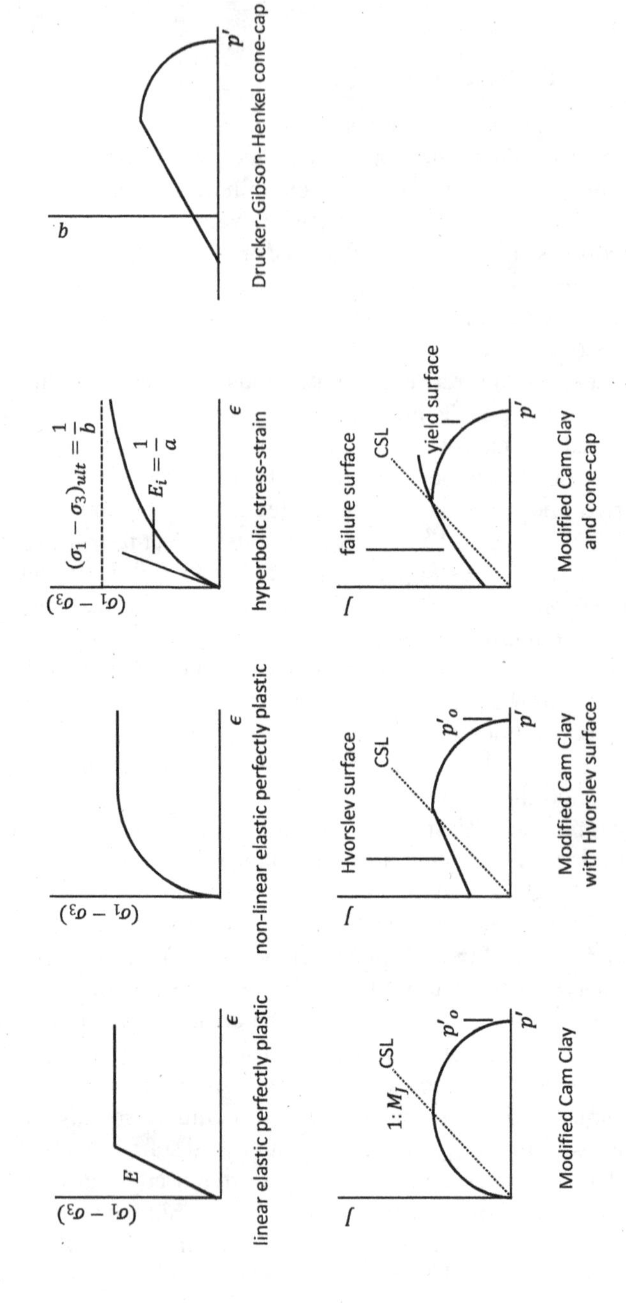

Figure 4.1 **Examples of constitutive models used in the analysis of pressuremeter tests.**

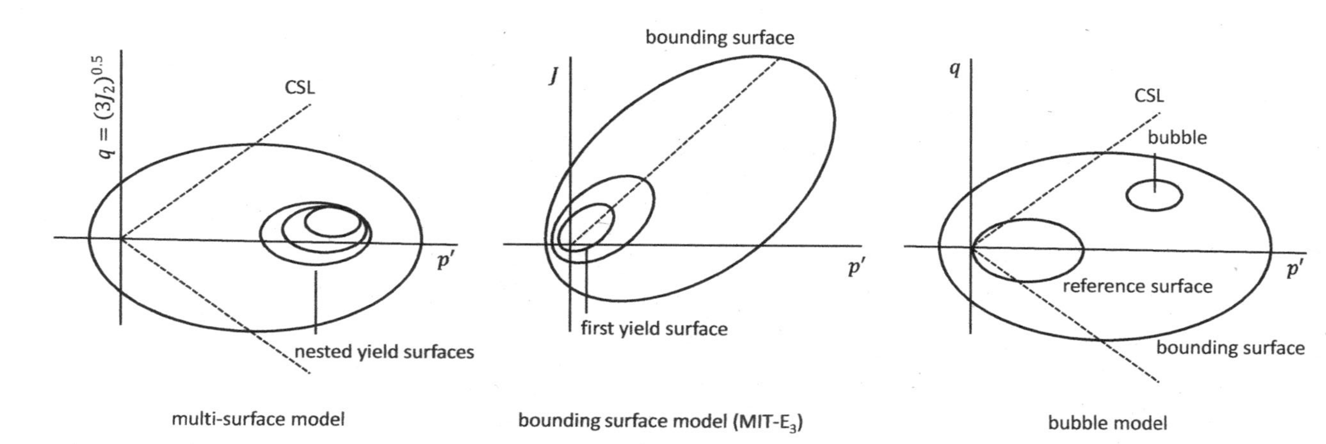

Figure 4.2 Examples of multi- and bounding-surface models used in the analysis of pressuremeter tests.

Table 4.2 A comparison between constitutive models highlighting the number of parameters used to define the models

Model	*Parameters*
Linear elastic	E, υ
Non-linear elastic (e.g. hyperbolic, power law)	a, b
Perfectly plastic	c', ϕ'
Linear elastic perfectly plastic	E, υ, c', ϕ'
Drucker-Gibson-Henkel cone-cap	$\kappa, \lambda, \phi', \theta, b, q_u, q_t$
Cam Clay	$N, \Gamma, \kappa, \lambda, G, \phi'$
Modified Cam Clay	$N, \Gamma, \kappa, \lambda, G, \phi'$
Modified Cam Clay with Hvorslev surface	$N, \Gamma, \kappa, \lambda, G, \phi', g, h$
Modified Cam Clay and cone-cap	$N, \Gamma, \kappa, \lambda, G, \phi', \alpha, M_\theta, \theta$
Kinematic hardening bubble model	$N, \Gamma, \kappa, \lambda, G, K, \phi', M_\theta, \theta, m, \beta, \psi, k, A, p_{co}, r_o, \eta_o$
Bounding surface model (MIT-E_3)	$e_o, \lambda, K_{ONC}, G, K, \phi_{TC}', \phi_{TE}', G_{max}, h, C, n, S_t, c, \omega, \gamma, \kappa_o, \psi_o$

unloading of strain hardening/softening soils and anisotropy, visco-elastic and visco-plastic models to take into account time effects and models of unsaturated soils.

The pressuremeter test is unusual in geotechnical engineering in that the geometry of the test can be simply defined. This has allowed close formed solutions to be developed from which soil properties can be obtained directly from a test. More complex models are used to gain an insight into the pressuremeter test. A comparison of the function and the parameters required of a number of soil models (Table 4.2) shows that the number of parameters increases as the complexity of the model increases. Given the complexity of the models and the uncertainties associated with installation and testing, the simpler models are used in practice to interpret pressuremeter tests to determine soil properties and design parameters.

4.3 DISTRIBUTION OF STRESS AND STRAIN

Consider the ideal situation in which a probe is installed into the ground without disturbing the surrounding material. The ground is assumed to be homogeneous and isotropic. The probe is assumed to be vertical, and the length/diameter ratio of the expanding section is large enough such that the pressuremeter test can be modelled as the expansion of an infinitely long right circular cylinder. At the start of a test the radius of the probe, or cavity, is a_o and the internal pressure, p_o, is equal to the total in situ horizontal stress, σ_h. As p, the applied pressure, is increased to p_i the cavity expands in a radial direction to a_i (see Figure 4.3).

All movements will be in the radial direction as the length of the cavity is considerably greater than its diameter. Axial symmetry applies as the soil is assumed to be homogeneous and isotropic. The expansion is plane strain. If the vertical stress prior to a test is a principal stress, the radial, circumferential and vertical stresses around the cavity are and remain principal stresses.

Consider an element of soil, thickness δr, at radius r, measured from the centre of the cavity, subject to principal stresses σ_r, σ_θ and σ_v. Timoshenko and Goodier (1934) showed that the equation of equilibrium is

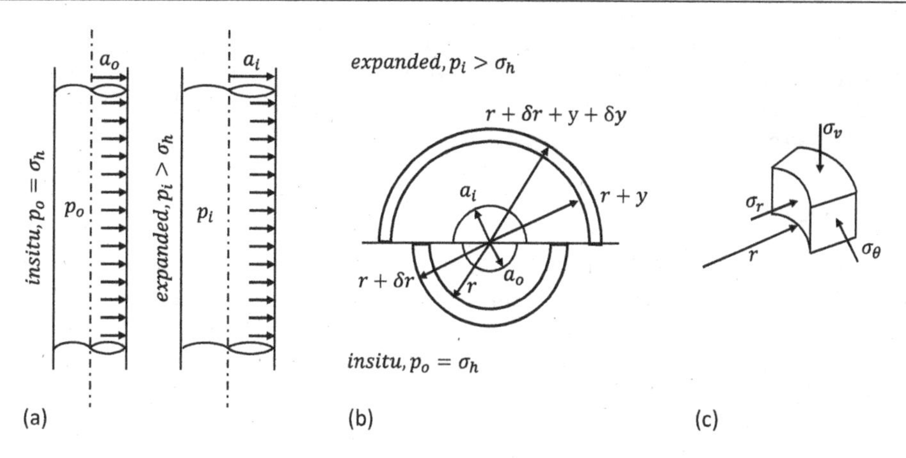

Figure 4.3 The definitions used in the analysis of the expansion of a cylindrical cavity: (a) expansion of a cylindrical cavity; (b) expansion of an element at radius r; (c) stresses on an element at radius r.

$$\frac{d\sigma_r}{dr} = -\frac{\sigma_r - \sigma_\theta}{r} \tag{4.1}$$

The inner radius of the element expands to $r + y$ and the thickness to $\delta_r + \delta_y$ as the pressure in the membrane is increased from p_o to p_i Thus, the tensile circumferential strain, ε_θ, is

$$\varepsilon_\theta = \frac{y}{r} \tag{4.2}$$

since the circumference increases from $2\pi r$ to $2\pi(r + y)$.

The thickness of the element changes by δy, therefore the radial strain, ε_r, is

$$\varepsilon_r = \frac{\delta y}{\delta r} \tag{4.3}$$

The only variables measured in a test are the applied pressure, p, and the radius of the membrane, a. The circumferential strain at the cavity wall is referred to as the cavity strain, ε_c, which is defined as

$$\varepsilon_c = \frac{a - a_o}{a_o} \tag{4.4}$$

There are instances in which the *volume* of the cavity is measured. The change in volume, ΔV, is simply related to the cavity strain by

$$\frac{\Delta V}{V} = 1 - \frac{1}{(1 + e_c)^2} \tag{4.5}$$

where V is the current volume ($= V_o + \Delta V$).

It is assumed that at some distance from the probe the strain and the change in radial stress are zero; that is, $\sigma_r = \sigma_\theta = \sigma_h$ and $\delta y = 0$. This is not enough information to determine the properties of the ground. It is necessary to consider a relationship between stress and strain, both post- and pre-yield.

4.4 ELASTIC GROUND

Consider a cavity expanded in an ideal linear elastic isotropic soil – that is, one which obeys Hooke's Law. The principal radial, circumferential and axial strains, ε_r, ε_θ and ε_z, are related to the principal stress changes by

$$\begin{aligned} E\varepsilon_r &= \Delta\sigma_r - v(\Delta\sigma_\theta + \Delta\sigma_z) \\ E\varepsilon_\theta &= \Delta\sigma_\theta - v(\Delta\sigma_z + \Delta\sigma_r) \\ E\varepsilon_z &= \Delta\sigma_z - v(\Delta\sigma_\theta + \Delta\sigma_r) \end{aligned} \tag{4.6}$$

where E is the modulus of elasticity and v is Poisson's ratio. The vertical strain is zero therefore

$$\Delta\sigma_z = v(\Delta\sigma_\theta + \Delta\sigma_r) \tag{4.7}$$

Combining equations (4.1)–(4.4), (4.6) and (4.7) gives the differential equation

$$r^2\frac{d^2y}{dr^2} + r\frac{dy}{dr} - y = 0 \tag{4.8}$$

The boundary conditions are

$$\begin{aligned} y &= 0 && for\, r = infinity \\ y &= (a - a_0) && for\, r = a \end{aligned} \tag{4.9}$$

The displacements and stresses within the soil are given by

$$y = \varepsilon_c\frac{a_o a}{r} \tag{4.10}$$

$$\Delta\sigma_r = \sigma_r - \sigma_h = 2G\varepsilon_c\frac{a_o a}{r^2} \tag{4.11}$$

and

$$\Delta\sigma_\theta = \sigma_\theta - \sigma_h = -2G\varepsilon_c\frac{a_o a}{r^2} \tag{4.12}$$

where G is the shear modulus and σ_h is the total horizontal stress.

The radial and circumferential strains at r are equal and opposite and, since there is no vertical strain, there must be no volume changes in the ground – that is, deformation takes place at constant volume during the elastic phase.

At the cavity wall, $r = a$, and $\sigma_r == p$. As $(a - a_o)$ is small during the initial expansion, equation (4.11) can be written as

$$p - \sigma_h = 2G\varepsilon_c \tag{4.13}$$

Hence, for a linear elastic isotropic material, the shear modulus at the start of a test can be determined simply by measuring displacement of the cavity wall and the change in pressure causing that displacement. The initial stiffness of the ground is given, in terms of cavity strain, as

$$G = \frac{0.5(p - \sigma_h)}{\varepsilon_c} \tag{4.14}$$

This can be expressed in terms of volumetric strain as

$$G = (p - \sigma_h)\frac{V_o}{\Delta V} \tag{4.15}$$

where V_o is the initial volume of the cavity, equivalent to radius a_o.

The more general forms of equations (4.14) and (4.15), used, for example, to determine a stiffness from an unload/reload cycle, which may be more representative of the in situ conditions than the initial loading, are

$$G = 0.5\frac{a}{a_o}\frac{dp}{d\varepsilon_c} \tag{4.16}$$

$$G = V\frac{dp}{dV} \tag{4.17}$$

Since the change in radial stress is equal and opposite to the change in circumferential stress, and as there is no change in vertical stress, the mean stress remains constant. Therefore, no pore pressures are developed during the cavity expansion in an elastic material. Further, the expansion of the cavity is a shearing process, even though it is the compressive stress that changes. This is significant since the derived modulus will be independent of drainage conditions.

4.5 UNDRAINED EXPANSION OF CYLINDRICAL CAVITY

4.5.1 General analysis

The analysis described in section 4.4 only applies to an elastic material. At the onset of yield at the cavity wall there is a change in behaviour, which results in either the development of excess pore pressures (undrained expansion) or volume changes within the ground (drained expansion). In practice, most tests will be either partially or fully drained but, in order to develop an analysis, it is assumed that a test is either fully undrained or fully drained.

It is assumed that no volume changes take place in the ground during an undrained expansion of a cavity, and the strain rate does not affect the response of the ground. Thus, all elements of soil at all radii are subject to the same mode of deformation, though the magnitude of deformation differs between each element at any moment,

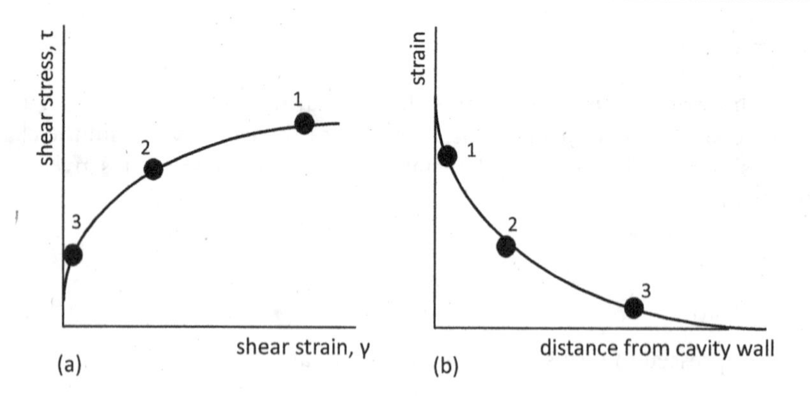

Figure 4.4 The variation in strain magnitude throughout the ground surrounding an expanding cavity: (a) stress–strain curve; (b) variation in strain within a soil mass containing an expanding cavity.

as shown in Figure 4.4. The deformation of each element will vary with the square of the radius.

The stiffness at the membrane/ground interface is the integrated effect of all the elements from the interface to infinity. Palmer (1972), Ladanyi (1972) and Baguelin et al. (1972) independently showed that the shear stress, τ, is given by

$$\tau = 0.5\varepsilon_c (1+\varepsilon_c)(2+\varepsilon_c)\frac{dp}{d\varepsilon_c} \tag{4.18}$$

This equation describes the complete shear stress–strain response at the cavity wall. For small strains equation (4.18) can be approximated to

$$\tau = \varepsilon_c \frac{dp}{d\varepsilon_c} \tag{4.19}$$

that is, at any strain, the shear stress is equal to the slope of the pressure cavity strain curve multiplied by the cavity strain or, more simply, the difference in pressure (ab in Figure 4.5a) between the intercept of the slope of the stress–strain curve and the vertical axis and the pressure at that strain.

This is the basis of the subtangent method, which can be used to determine a shear stress–strain curve (Wroth and Hughes, 1973). This method is very sensitive to installation disturbance as it depends on the datum selected for the initial cavity radius.

Equation (4.19) can be written in terms of volumetric strain

$$\tau = \frac{dp}{d\left(\ln(\Delta V / V)\right)} \tag{4.20}$$

The shear stress is equal to the slope of the pressure volumetric strain curve, as shown in Figure 4.5b.

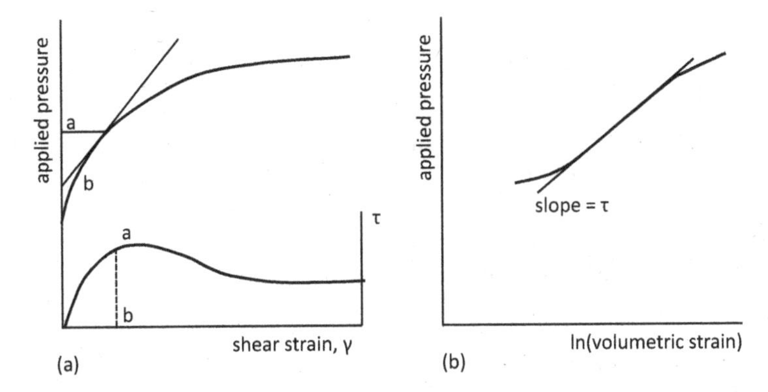

Figure 4.5 The derivation of shear stress from curves of pressure against (a) cavity and (b) volumetric strain.

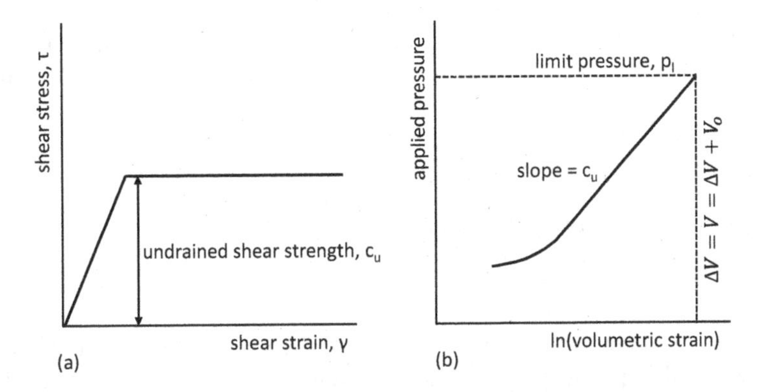

Figure 4.6 The derivation of undrained shear strength from the pressure volumetric strain curve.

4.5.2 Linear elastic perfectly plastic soil

If the ground is linear elastic perfectly plastic, Figure 4.6, then τ will be constant once yield occurs and equal to the undrained shear strength, c_u.

The ground will respond as an elastic material (equation (4.14)) until

$$p - \sigma_h = c_u \tag{4.21}$$

The volumetric strain at the onset of yield is given by

$$\frac{\Delta V}{V} = \frac{c}{G} \tag{4.22}$$

The change in pressure during the expansion of a cavity in an elastic perfectly plastic material can be obtained by integrating equation (4.20) with respect to ln $(\Delta V/V)$ to give

$$p - \sigma_h = c_u \left[1 + \ln\left(\frac{G}{c_u}\right) + \ln\left(\frac{\Delta V}{V}\right) \right] \tag{4.23}$$

Thus, if the results are plotted as cavity pressure against the logarithm of the volumetric strain, the slope of the plastic portion is equal to the undrained shear strength, c_u, of the soil.

There is a limiting condition to equation (4.23), since the maximum value of $\Delta V/V$ is 1; that is, when the change in volume is equal to the current volume, as shown in Figure 4.6. The pressure at that point is known as the limit prèssure, p_1 such that

$$p_1 - \sigma_h = s_u \left[1 + \ln\left(\frac{G}{c_u}\right) \right] \tag{4.24}$$

This equation was proposed by Ménard (1957) who subsequently used it to produce correlations between a modified limit pressure and shear strength.

Thus, equations (4.13) and (4.24) describe the complete loading curve for a test in linear elastic perfectly plastic incompressible material. Equation (4.13) applies between the limits

$$\sigma_h < p < c_u + \sigma_h \tag{4.25}$$

Equation (4.24) applies between the limits

$$c_u + \sigma_h < p < p_1 \tag{4.26}$$

Jefferies (1988) extended this to take into account the unloading portion of the curve. This was developed to produce a method that could be used to select a value of σ_h that would not be subjective. The loading portion is given by equation (4.23), which can be expressed in terms of displacement as

$$p - \sigma_h = c_u \left\{ 1 + \ln\left[\left(\frac{G}{c_u}\right) \left(1 - \frac{a_o}{a}\right)^2 \right] \right\} \tag{4.27}$$

During the unloading phase the clay will unload elastically such that

$$p - p_{max} = 2G\left(\frac{a - a_{max}}{a_{max}}\right) \tag{4.28}$$

where p_{max} is the maximum pressure reached at the end of loading (see Figure 4.7) and a_{max} is the maximum displacement reached at the end of loading.

Failure will occur in extension when

$$p - p_{max} = (1 + \beta) c_u \tag{4.29}$$

where β is the ratio of undrained strength on unloading to that on loading.

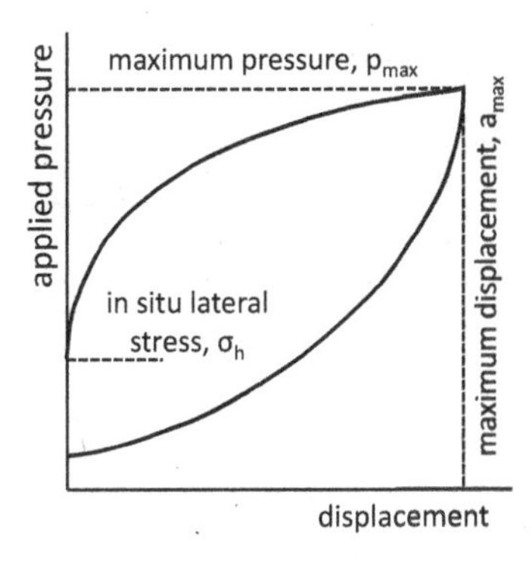

Figure 4.7 Key points on an ideal pressuremeter curve.

The plastic unloading curve is given by

$$p - p_{max} = -c_u\left[(1+\beta) - \ln\left\{\left[1-\left(\frac{a}{a_{max}}\right)^2\right]\left[\frac{G}{(1+\beta)c_u}\right]\right\}\right.$$
$$\left. -\beta ln\left\{\left[\left(\frac{a}{a_{max}}\right)^2 - 1\right]\left[\frac{G}{(1+\beta)c_u}\right]\right\}\right] \tag{4.30}$$

Jefferies (1988) reports that constant mean total stress triaxial tests on specimens of clay showed that the undrained strength on unloading after a peak strength had been obtained during loading gave a value of β of 0.83. For practical purposes, β can be taken as 1, which gives a plastic unloading curve of the form of

$$p - p_{max} = -2c_u\left\{1 + \ln\left[\left(\frac{G}{2c_u}\right)\left(\frac{a_{max}}{a} - \frac{a}{a_{max}}\right)\right]\right\} \tag{4.31}$$

Thus, if the results are plotted as cavity pressure against the logarithm of $\left(\frac{a_{max}}{a} - \frac{a}{a_{max}}\right)$, the slope of the plastic portion is equal to twice the undrained shear strength, c_u, of the soil.

4.5.3 Non-linear material

The general expression (equation (4.18)) for the variation in shear stress with strain has limited practical use because of fluctuations in test data and installation disturbance. This gives rise to unacceptably high values of peak stress and fluctuating stress–strain curves. It is possible to fit a curve to the test data and analyse that curve. This removes some of the fluctuation in shear stress but it does not overcome the effects of installation.

The alternative linear elastic perfectly plastic model allows an average stiffness and strength to be evaluated. It does not take into account non-linear elastic behaviour or

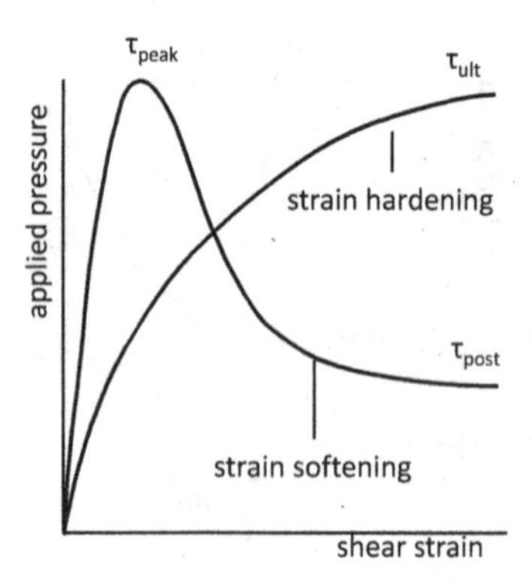

Figure 4.8 Stress-strain curves proposed by Prévost and Hoeg (1975).

strain hardening/strain softening. A number of non-linear models have been proposed, and these include hyperbolic and power law models, which are defined either in terms of constants or engineering parameters.

Prévost and Hoeg (1975) proposed two relationships, one for strain-hardening soils and the other for strain-softening soils, as shown in Figure 4.8.

The shear stress, τ, for strain-hardening soils is a function of the shear strain, γ, such that

$$\tau = \frac{\gamma}{\alpha + \gamma}\tau_{ult} \tag{4.32}$$

where α is a constant and τ_{ult} is the ultimate undrained shear strength during loading. The distributions of stress and strain given in section 4.2 are combined with equation (4.32) to define the cavity expansion curve as

$$p = \sigma_h + \frac{1}{\sqrt{3}}\tau_{ult}\, ln\left[1 + \frac{2\varepsilon_c}{\alpha\sqrt{3}}\right] \tag{4.33}$$

The relationship between τ and γ for strain-softening soils is

$$\tau = A\left[\frac{B\gamma^2 + \gamma}{1 + \gamma^2}\right] \tag{4.34}$$

B is an experimental constant defined by

$$\gamma_{peak} = B + \sqrt{(1 + B^2)} \tag{4.35}$$

where γ_{peak} is the shear strain at the peak shear stress. The peak strength is obtained by substituting equation (4.35) into equation (4.34). The post-rupture strength is equal to AB. The derived cavity expansion curve is given by

$$p = \sigma_h + \frac{A}{\sqrt{3}}\left[\frac{B}{2} ln\left[1 + \frac{4\varepsilon_c^2}{3}\right] + tan^{-1}\left(\frac{2\varepsilon_c}{\sqrt{3}}\right)\right] \tag{4.36}$$

Thus, for strain-hardening soils α and τ_{ult} are adjusted until the best fit to the test curve defined by equation (4.33) is obtained. A and B are adjusted until equation (4.36) fits the test data to obtain engineering parameters for strain-softening soils.

Equations (4.32) and (4.34) include non-engineering constants. The choice of the constants not only has to satisfy the best fit criteria but should give realistic values of stiffness when using equations (4.16) and (4.17) to derive G. The alternative is to define the soil behaviour in terms of engineering parameters. Denby and Clough (1980) and Ferreira and Robertson (1992) proposed hyperbolic models based on stiffness and strength. Denby and Clough assumed a hyperbolic relationship based on that proposed by Duncan and Chang (1970) of the form

$$\tau = \frac{\varepsilon_c}{[1/2G_i]+[R_f/c_u]\varepsilon_c} \tag{4.37}$$

where R_f is the ratio of the actual strength to the asymptotic value given by the Duncan and Chang model and G_i is the initial shear modulus.

The cavity expansion curve is given by

$$\frac{d\varepsilon_c}{dp} = \frac{1}{2G_i} + \frac{R_f}{c_u}\varepsilon_c \tag{4.38}$$

This can be expressed as

$$p = \sigma_h + \frac{c_u}{R_f}\ln\left(1 + 2G_i\frac{R_f}{c_u}\varepsilon_c\right) \tag{4.39}$$

This is the same as the expression developed by Prévost and Hoeg (1975) if $R_f = 1$. A plot of the tangent to the curve against cavity strain will produce a straight line, which has an intercept $[1/2G_i]$ and slope $[R_f/c_u]$, as shown in Figure 4.9.

The ultimate shear strength can never be achieved in a test in a material with hyperbolic rheological behaviour, so Denby and Clough developed a hybrid model in which they assumed a hyperbolic model for the first part of the test and a rigid plastic model for the second part – that is, a non-linear elastic perfectly plastic soil, as shown in Figure 4.9. This is a combination of equations (4.23) and (4.37).

Ferreira and Robertson (1992) extended Denby and Clough's hyperbolic model using an R_f value of 1 to include the unloading curve. The shear stress, τ^*, during unloading is given by

$$\tau^* = \frac{\varepsilon_c^*}{[(1/2G_i)-(\varepsilon_c^*/\tau_{ult}^*)]} \tag{4.40}$$

where τ_{ult}^* is the ultimate undrained shear strength during unloading, ε_c^* is the cavity strain during unloading ($= (\varepsilon_c - \varepsilon_{max})/(1 + \varepsilon_{max})$) in which ε_{max} is the maximum cavity strain during loading. Figure 4.10 shows the definition of the terms used to describe the complete loading and unloading hyperbolic model.

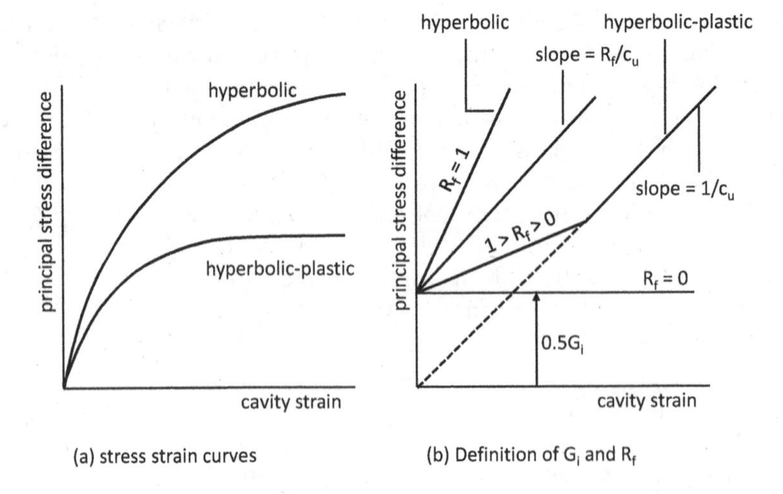

Figure 4.9 The derivation of Gi and Rf.

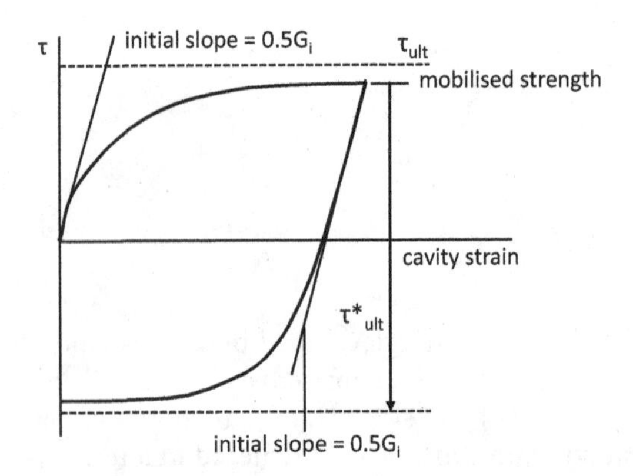

Figure 4.10 A hyperbolic model to describe the loading and unloading of a cavity wall.

Source: After Ferreira and Robertson, 1992.

The distributions of stress and strain given in section 4.2 are combined with equations (4.37) and (4.40) to define the cavity expansion curve. The loading curve is given by

$$p = \sigma_h + \tau_{ult} \ln\left[1 + \left(2G_i \frac{\varepsilon_c}{\tau_{ult}}\right)\right] \tag{4.41}$$

The cavity contraction curve is given by

$$p = p_{max} + \tau^*_{ult} \ln\left[\frac{1}{1 - \left[2G_i\left(\varepsilon_c - \varepsilon_{max}\right)\right] / \left[\left(1 + \varepsilon_{max}\right)\tau^*_{ult}\right]}\right] \tag{4.42}$$

where $p_{\max}$ is the maximum applied pressure during loading (see Figure 4.7).

Equation (4.41) contains three unknown engineering parameters, σ_h, G_i and τ_{ult}; equation (4.45) contains two unknowns, G_i and τ^*_{ult}. Methods used to solve these equations are discussed in Chapter 6.

Briaud (2013) suggested it is possible to determine G_{max} from a prebored pressuremeter test by assuming a hyperbolic model (Baud et al., 2013):

$$\sigma = \frac{\varepsilon}{\dfrac{1}{2G_{max}} + \dfrac{\varepsilon}{p_l}} \tag{4.43}$$

This curve is defined by the pressuremeter test beyond the yield point on the PBP curve such that:

$$\frac{\varepsilon}{\sigma} = \frac{2}{2G_{max}} + \frac{\epsilon}{p_l} \tag{4.44}$$

4.5.4 Critical state models

The self-boring pressuremeter test in clay is usually interpreted using undrained cavity expansion theory based on total stresses. This is reasonably accurate for normally and lightly overconsolidated clays where the shear resistance of the soil does not change significantly during the pressuremeter test. For heavily overconsolidated clay, however, the shear resistance may vary considerably with deformation and this cannot be easily accounted for by the total stress approach with a perfectly plastic soil model.

Collins and Yu (1996) were the first to derive analytical solutions for large strain cavity expansion using the critical state concept. Yu and Collins (1998) using solutions based on large strain cavity theory showed that the direct application of the perfectly plastic model is accurate for soils with low overconsolidation ratio (OCR) values but tends to underestimate the undrained shear strength of heavily overconsolidated soils. As shown in Figure 4.11, the underestimate could be as high as 50% for soils with a very high OCR value.

Cao et al. (2001) developed a close form solution for undrained expansion of a cavity assuming a Modified Cam Clay model which takes into account stress history. The undrained shear strength, c_u, is related to the deviator stress, q_u, at the critical state

$$c_u = \frac{q_u}{\sqrt{3}} = \frac{Mp'_o}{\sqrt{3}}\left(\frac{OCR}{2}\right)^{\Lambda} \tag{4.45}$$

where M is the slope of the critical state line in p'-q plane, Λ the plastic volumetric strain ratio, and OCR the isotropic over-consolidation ratio. The ultimate deviator stress, q_u, is

$$q_u = Mp'_u = Mp'_o\left(\frac{R}{2}\right)^{\Lambda} \tag{4.46}$$

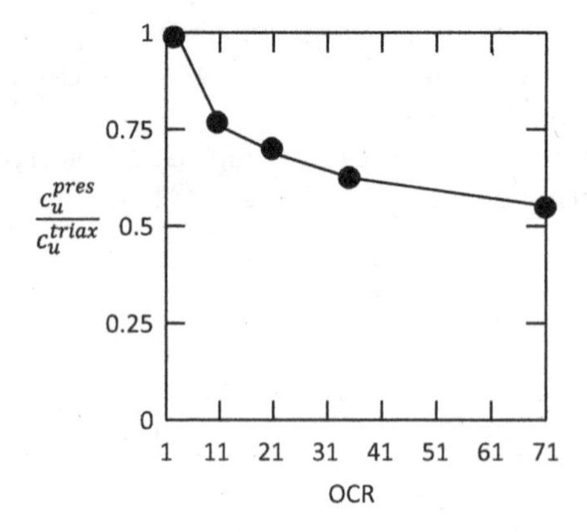

Figure 4.11 The effect of strain softening due to overconsolidation on the undrained shear strength derived from a pressuremeter test.

Source: After Yu and Collins, 1988.

The variation of the effective stress and the radial distance from the centre of the cavity is

$$ln\left(1-\frac{a^2-a_o^2}{r^2}\right)=-\frac{q}{G\sqrt{3}}-2\sqrt{3}\frac{\kappa\Lambda}{\upsilon M}\left[\beta_1-tan^{-1}\left(\frac{\eta}{M}\right)+tan^{-1}\sqrt{(R-1)}\right] \tag{4.47}$$

where

$$\beta_1=0.5ln\left\{\frac{(\eta+M)\left(\sqrt{(r-1)}-1\right)}{(\eta-M)\left(\sqrt{(r-1)}+1\right)}\right\} \tag{4.48}$$

Further

$$\frac{\kappa}{\upsilon}=\frac{3(1-2v')p'_o}{2(1+v')G} \tag{4.49}$$

Solving this equation requires a numerical solution which has been the basis of numerous analyses discussed in Chapter 5 to investigate the factors that control the pressuremeter test.

Cao et al. (2001) proposed an approximate closed form solution assuming the deviator stress, q, in the plastic zone is equal to the ultimate deviator stress, q_u, since, theoretically, the difference is small. The pressure at the cavity wall is

$$p_c=\sigma_h+\frac{1}{\sqrt{3}}q_u+\frac{1}{\sqrt{3}}q_u ln\left\{\left[\frac{G\sqrt{3}}{q_u}\right]\left[\frac{a^2-a_o^2}{a^2}\right]\right\} \tag{4.50}$$

where a is the current radius of the cavity and a_o the original radius.

Mo and Yu (2017) describe an effective stress solution for undrained expansion for cylindrical cavities using a unified state parameter model for clay and sand (CASM), developed by Yu (1998). The state boundary surface of CASM is

$$\left(\frac{\eta}{M}\right)^n = \frac{ln\left(\frac{p'}{p'_y}\right)}{lnr^*} \tag{4.51}$$

Rowe's stress dilatancy relationship is

$$\frac{\dot{\delta}^p}{\dot{\gamma}^p} = 0.5\frac{9(M-\eta)}{9+3M-2M\eta} \tag{4.52}$$

The hardening law is

$$p'_y = \frac{vp'_y}{\lambda-\kappa}\dot{\delta}^p \tag{4.53}$$

Mo and Yu (2017) developed the analytical solution using this model for the elastic, plastic and critical state regions and undertook a parametric study of the a SBP test. It is routine in SBP tests to record the total applied pressure and resulting cavity radius. The critical state models provide a means of determining the variation of radial and tangential stresses within the soil but to obtain the total stresses it is necessary to undertake a numerical integration. Further, a number of parameters are used to define these models; that is, a number of combinations of those parameters will produce a fit to the pressuremeter data. The power of these models lies in their ability to undertake parametric studies to gain an insight into the ground responses to an expanding cavity.

Gaone et al. (2019) describe an optimisation strategy to determine the parameters for the Modified Cam Clay model from undrained pressuremeter tests. The undrained shear strength and lift off pressure were determined from 'standard' pressuremeter tests which had no unload/reload cycles. The MCC parameters, M, Λ and R_o, were adjusted in a systematic way (Figure 4.12) until an acceptable fit to the test data was obtained.

The model was validated against the behaviour of two 1.8 m square foundations with a formation level at 1.5 m in estuarine clay.

4.6 DRAINED EXPANSION OF A CYLINDRICAL CAVITY (TESTS IN SAND)

4.6.1 Volume changes

Tests in sands are drained since no excess pore pressures are developed. Volume changes will occur and if these are not taken into account the interpretation of a test will be in error. The changes in volume within the sand have been represented in different ways by different authors and these include experimental observations, stress dilatancy theory and the state parameter.

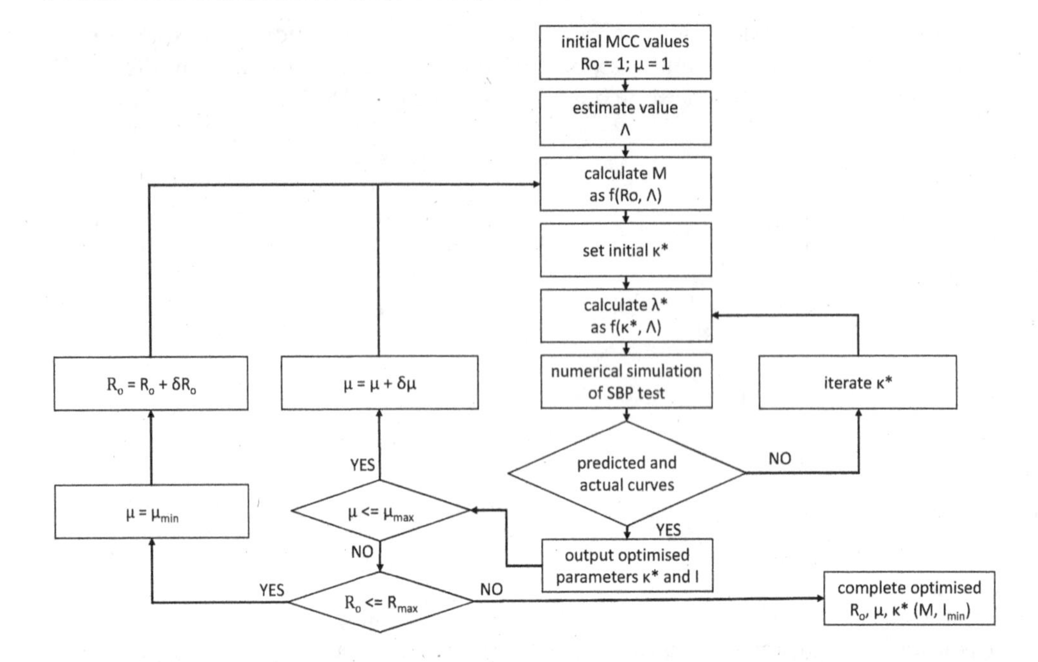

Figure 4.12 **Nested single variable optimisation strategy to derive MCC parameters from SBP test data.**

Source: Gaone et al., 2019.

Ladanyi (1963) proposed that failure occurred at a constant effective stress ratio and at constant volume. The volumetric strain prior to yield is selected by trial and error by altering the initial volume of the cavity until a straight line is obtained for the graph of p' against $\Delta V/V$ to logarithmic scales. Note that this trial and error approach is commonly used with methods of analysis in the interpretation of tests.

Vesic (1972) proposed a general solution for the expansion of a cylindrical cavity in a soil possessing both cohesion and friction. The test curve is given by

$$(p_l' + c'\cot\phi')(I_{rr}'\sec\phi')^{\sin\phi'/(1+\sin\phi')} = (\sigma_h' + c'\cot\phi')(1+\sin\phi') \tag{4.54}$$

where c' is the cohesion, ϕ' is the angle of friction, p_l' is the effective limit pressure and I_{rr}' is a rigidity factor given by

$$I_{rr}' = \frac{\left[G/(c'+\sigma_h'\tan\phi')\right]}{\left\{1+\left[G/(c'+\sigma_h'\tan\phi')\right]\Delta V_p\sec\phi'\right\}} \tag{4.55}$$

The limit pressure is given by

$$p_1 = c'F_c' + \sigma_h F_q' \tag{4.56}$$

where F_c' and F_q' are cavity expansion factors given by

$$F_q' = (1+\sin\phi')(I_{rr}' \sec\phi')^{\sin\phi'/(1+\sin\phi')} \quad (4.57)$$

$$F_c' = (F_q' - 1)\cot\phi' \quad (4.58)$$

A value of volumetric strain at failure has to be assumed or derived in order to apply equation (4.45). Vesic suggested that results of plane strain triaxial compression tests could be used together with an iterative procedure to determine the average volumetric strain in the surrounding ground. Equation (4.54) is the same as equation (4.23) if friction and volume changes are ignored since $F_c' = (1+\ln(G/c_u))$.

Sayed (1989) also proposed a method in which volume changes are measured in the laboratory from tests on specimens of sand prepared to the same density as that in situ. A quadratic equation (equation (4.48)) relates the volumetric strain, ε_v, to the circumferential strain, ε_θ.

$$\varepsilon_v = \alpha_v + \beta_v \varepsilon_\theta + \gamma_v \varepsilon_\theta^2 \quad (4.59)$$

where α_v, β_v and γ_v are volume change parameters dependent on the mean normal stress and relative density. The volume change parameters are found experimentally or are taken from databases of volume measurements from laboratory tests, and these are combined with a non-linear stress-strain function to define the test curve.

4.6.2 General analysis

It is difficult to obtain volume change parameters for sand. They can be determined experimentally from triaxial tests on specimens prepared at different densities but this requires representative samples and time. Hughes et al. (1977) assumed that the rate of volume change is constant during cavity expansion and is related to the peak angle of friction using the concept of stress dilatancy. This is reasonable for very dense sands and is a commonly used to interpret tests in sands. Robertson and Hughes (1986) extended this work empirically to allow for the variation in the rate of volume change which can occur in medium dense and loose sands since the expansion of a typical pressuremeter is insufficient to reach constant volume conditions required for the analysis of Hughes et al. Manassero (1989) proposed a more general solution based on Rowe's stress dilatancy theory which involves a numerical technique to produce a complete stress–strain curve for tests in loose to dense sands. These methods, based on stress dilatancy, are described in detail below.

Elastic strains are neglected since they are negligible in relation to plastic strains. Once yield occurs it is assumed that the behaviour of the sand in plane strain is governed by Rowe's stress dilatancy theory. The stress ratio, $K(=\sigma_1/\sigma_3)$, will vary with shear strain, as shown in Figure 4.13.

The ratio of effective stresses at any radius will change depending on the density of the sand at that radius as the cavity is expanded. This will be a function of the volume changes occurring within the sands, which can be expressed in terms of the angle of dilation, ψ, where ψ is defined in terms of the ratio of the change in volumetric strain, ε_v to shear strain, γ, as

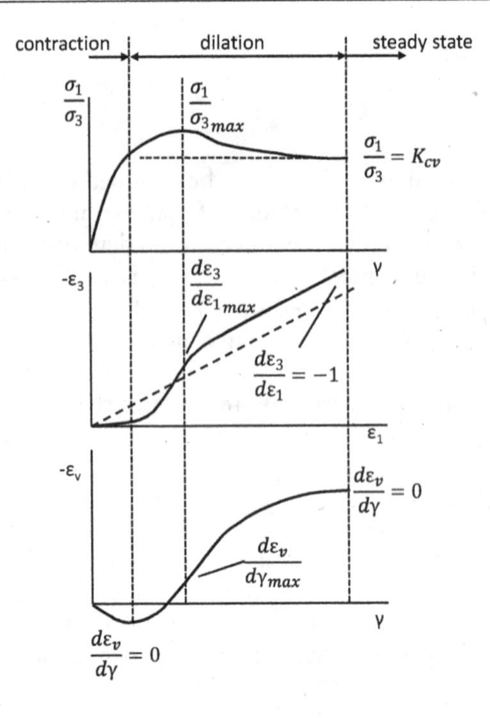

Figure 4.13 Stress-strain relationships developed from dilatancy theory.

Source: After Manassero, 1989.

$$\sin\psi = -\frac{d\varepsilon_v}{dy} \tag{4.60}$$

The change in volume will depend on the density of the sand. If it is loose, the sand will compress and, if it is dense, the sand will dilate. There is a particular density for which no volume changes will occur. The mobilised friction at this density is known as the angle of friction at constant volume, ϕ'_{cv} and the stress ratio for sand at failure at that density is K_{pcv}, where

$$K_{pcv} = \frac{1+\sin\phi'_{cv}}{1-\sin\phi'_{cv}} \tag{4.61}$$

where K_{pcv} is the constant volume principal stress ratio in the passive state.

Provided there is sufficient expansion of the cavity, all tests will reach this constant volume state. The change in the stress ratio with volumetric changes can be expressed in terms of a flow rule such as the stress dilatancy relationship proposed by Rowe, that is

$$\frac{1+\sin\phi'}{1-\sin\phi'} = \left(\frac{1+\sin\psi}{1-\sin\psi}\right)\left(\frac{1+\sin\phi'_{cv}}{1-\sin\phi'_{cv}}\right) \tag{4.62}$$

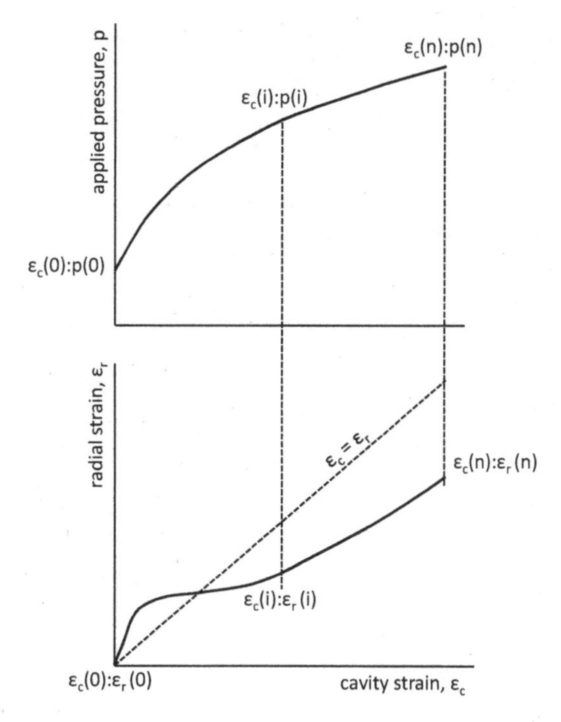

Figure 4.14 The interpretation of a pressuremeter curve in sand.

This can be expressed in terms of the principal stresses, volumetric strains and shear strains by

$$\frac{\sigma_r'}{\sigma_\theta'} = K_{pcv}\left(\frac{1-d\varepsilon_v/d\gamma}{1+dv/d\gamma}\right) \tag{4.63}$$

At large strains $d\varepsilon_v/d\gamma$ is zero, as shown in Figure 4.14.

The governing equations of stress and strain distribution given in section 4.2, combined with equation (4.63), produce a relationship which is valid at any radius

$$\frac{d\sigma_r}{d\varepsilon_\theta} = -\sigma_r'\left(\frac{1+K_{acv}\left(d\varepsilon_r/d\varepsilon_\theta\right)}{\varepsilon_r-\varepsilon_\theta}\right) \tag{4.64}$$

where $K_{acv} = 1/K_{pcv}$. At the cavity wall $\sigma_r = p$ and $\varepsilon_\theta = \varepsilon_c$, thus in order to solve equation (4.64) it is necessary either to make an assumption about the volumetric strains (Hughes et al. (1977) assumed that they are constant) or use a numerical technique since the applied pressure is a function of the cavity strain and this is known. The slope at any point i on the curve shown in Figure 4.14 is approximately given by

$$\frac{dp}{d\varepsilon_c} = \frac{p_{(i)} - p_{(i-1)}}{\varepsilon_{c(i)} - \varepsilon_{c(i-1)}} \tag{4.65}$$

The change of radial strain with respect to cavity strain is approximately given by

$$\frac{d\varepsilon_r}{d\varepsilon_c} = \frac{\varepsilon_{r(i)} - \varepsilon_{r(i-1)}}{\varepsilon_{c(i)} - \varepsilon_{c(i-1)}} \tag{4.66}$$

Therefore, at any point i the radial strain can be deduced by substituting equations (4.65) and (4.66) into equation (4.64), which gives

$$\varepsilon_{r(i)} = \frac{p_{(i)}\left[\varepsilon_{c(i-1)} + K_{acv}\varepsilon_{r(i-1)}\right]}{2\left[p_{(i)}(1+K_{acv}) - p_{(i-1)}\right]} - \frac{p_{(i-1)}\varepsilon_{c(i)}}{2\left[p_{(i)}(1+K_{acv}) - p_{(i-1)}\right]} + \frac{p_{(i)}\left[\varepsilon_{c(i-1)} - \varepsilon_{r(i-1)}\right]}{2K_{acv}p_{(i-1)}} + \frac{p_{(i-1)}\left[\varepsilon_{r(i-1)}(1+K_{acv}) - \varepsilon_{c(i)}\right]}{2K_{acv}p_{(i-1)}} \tag{4.67}$$

At the start of a test (i = 0) the radial strain is zero. The radial strain can be deduced at all other points on the curve using equation (4.67). In order to derive the complete shear strain curve it is necessary to determine the shear and volumetric strains from equation (4.63), again using a numerical technique.

4.6.3 Very dense sands

Hughes et al. (1977) assumed that the sand around the cavity is elastic perfectly plastic with a constant angle of dilation. The relationship between shear and volumetric strains, from equation (4.60), is

$$\varepsilon_v = c - \gamma \sin\psi \tag{4.68}$$

where c is a constant equal to the intercept shown in Figure 4.15.

The relationship between cavity strain and applied effective pressure is given by

$$\varepsilon_c + \frac{c}{2} = \left(\varepsilon_R + \frac{c}{2}\right)\left[\frac{p'}{p'_y}\right]^{[(n+1)/(1-N)]} \tag{4.69}$$

where

$$N = \frac{1-\sin\phi'}{1+\sin\phi'}$$

$$n = -\frac{1-\sin\psi}{1+\sin\psi}$$

p'_y is the onset of yield defined as $\sigma'_h(1+\sin\phi')$ for a yield criterion conforming to the Mohr Coulomb criterion, and ε_R is the strain at the onset of yield. For very dense sands, c is negligible and for practical purposes can be ignored. Jewell et al. (1980), conducting pressuremeter tests using a probe cast into dense sand, confirmed that this

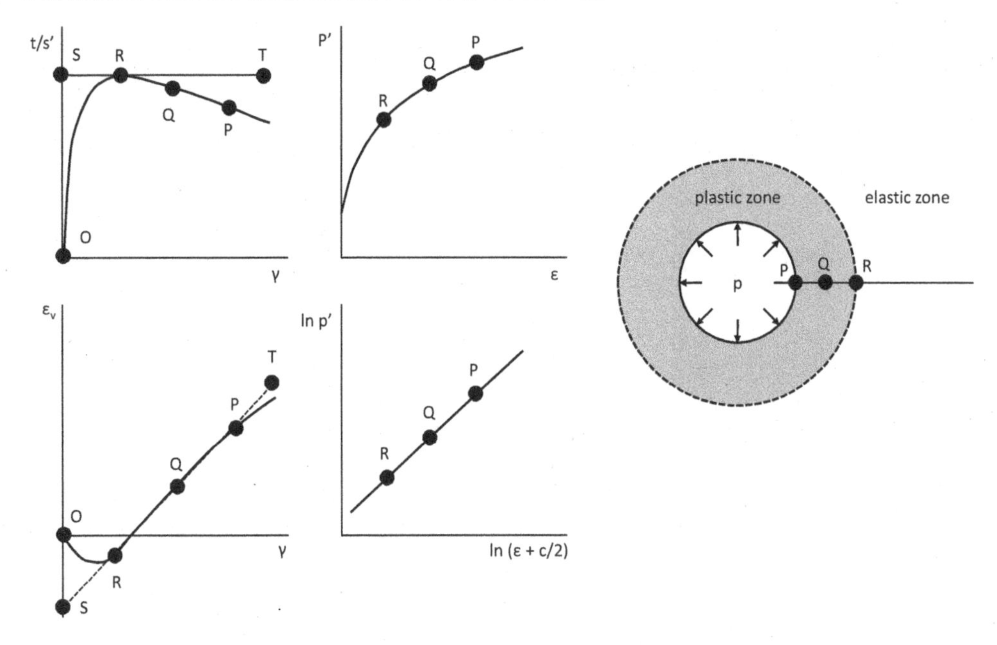

Figure 4.15 Stress–strain behaviour of a sand around an expanding cavity.

After Hughes et al., 1977.

assumption is reasonable. Thus, if equation (4.69) is plotted to a log scale, a straight line is produced, as shown in Figure 4.15.

The gradient of this line, s, is equal to $[(1 - N)/(1 + n)]$. This, together with Rowe's relationship for stress dilatancy (equation (4.62)), gives values of ϕ' and ψ:

$$\sin\phi' = \frac{s}{1+(s-1)\sin\phi'_{cv}} \tag{4.70}$$

$$\sin\psi = s + (s-1)\sin\phi'_{cv} \tag{4.71}$$

Houlsby et al. (1986) extended this analysis to investigate the unloading curve of a cavity expansion test in sand. Unloading is initially elastic until failure in extension occurs at the cavity wall. The stress states during unloading after the onset of failure during loading and unloading are shown in Figure 4.16.

A relationship, similar to that developed by Hughes et al. (1977), is obtained for the effective applied pressure, p', in terms of the cavity strain.

$$p' = np_{max}\left[\frac{\left[\dfrac{2G(1+nN)(\varepsilon_{max}-\varepsilon_c)}{(1-N)p_{max}}\right]-(1-n)}{n(1+N)}\right]^{[(N-1/N)/(N+1/n)]} \tag{4.72}$$

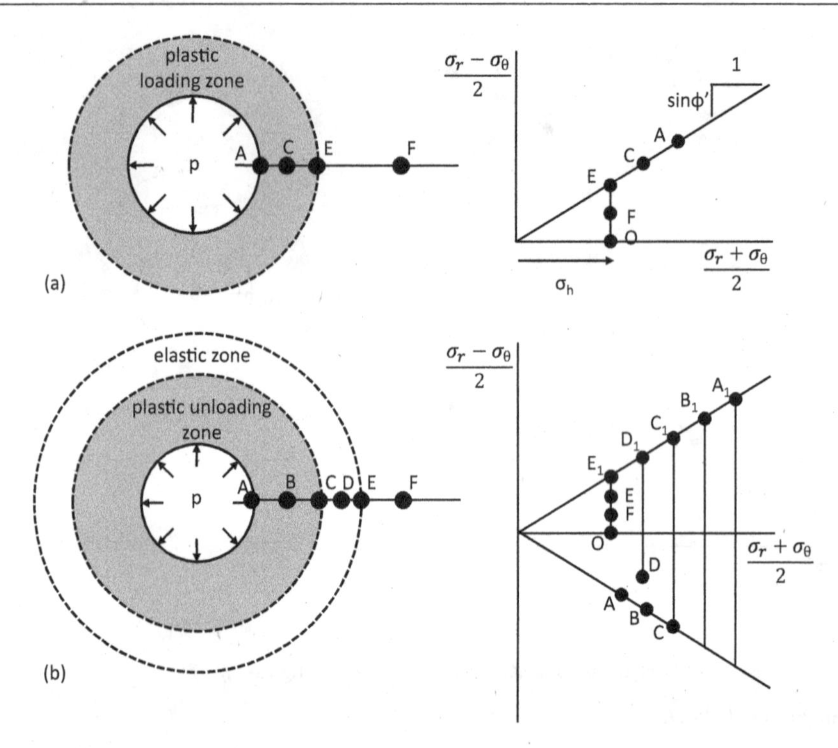

Figure 4.16 The stress states around the wall of a cavity unloading in sand: (a) expansion; (b) contraction.

4.6.4 State parameter

Volume changes can also be defined in terms of the state parameter. The state parameter (Wroth and Bassett, 1965; and Been and Jefferies, 1985) is defined as the difference between the current void ratio and the void ratio at the steady state at the same mean effective stress (Figure 4.17).

It enables the effects of relative density and confining stress to be taken into account and, using Figure 4.17, establish the angle of friction of sand. The state parameter, ξ, is defined as

$$\xi = e + \lambda ln\left(\frac{p}{p_1}\right) - \Gamma \tag{4.73}$$

where e is the void ratio, p the mean effective stress, p_1 the reference mean stress (typically 1kPa) and Γ the void ratio at the reference mean stress.

Yu (1994) showed that there is a unique relationship between the pressuremeter expansion curve and the initial state parameter. The numerical results of self boring pressuremeter tests on sands showed that

$$\xi = 0.59 - 1.85s \tag{4.74}$$

where s is the slope of the pressuremeter curve.

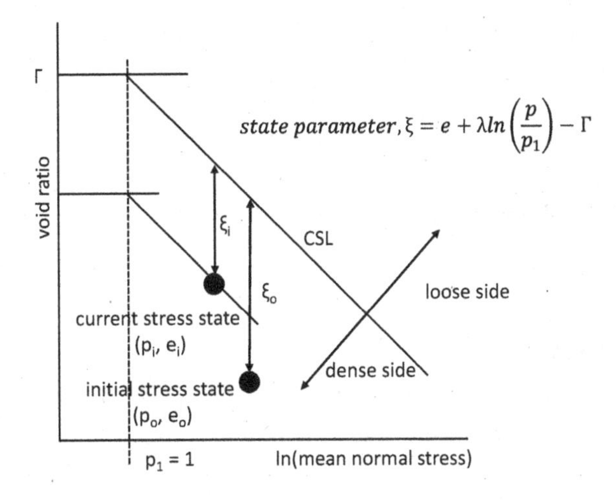

Figure 4.17 The definition of the state parameter.

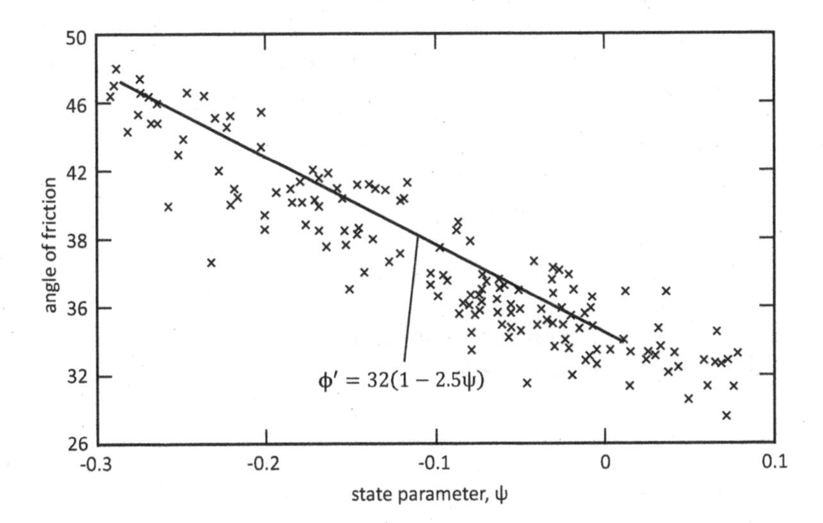

Figure 4.18 Experimental observations of the relationship between the angle of friction and state parameter from triaxial tests in compression and extension.

Source: After Been et al 1987.

The angle of friction, ϕ', based on experimental observations (Figure 4.18) by Been et al. (1987) is

$$\phi' = 0.6 + 107.8s \tag{4.75}$$

Installation effects means that the ground surrounding a probe will be disturbed. For this reason, it is preferable to use the unloading curve to derive soil parameters from a test in sand. Yu (1996) showed that the slope of the applied pressure plotted against the cavity strain relative to the maximum cavity strain plotted to a log scale is a function of the state parameter

$$\xi = 0.53 - 0.33s \quad (4.76)$$

where s is the slope of the unloading pressuremeter curve. The angle of friction, ϕ', based on experimental observations (Figure 4.18) by Been et al. (1987) is

$$\phi' = 6.6 + 18.4s \quad (4.77)$$

4.7 TESTS IN ROCK

The analyses given above are based on the assumption that the soil behaves as a continuum throughout the test. Many soils are naturally fissured and therefore the analyses do not apply. Briaud et al. (1987) demonstrated that cracking could occur in soils if the tests were relatively shallow. Consider Figure 4.19, which shows the stress paths followed at the cavity wall.

The radial and circumferential stresses are equal to the in situ total horizontal stress, σ_h, at the start of the test. The radial stress increases by the same amount as the applied pressure and the circumferential stress reduces by an equal and opposite amount, as described in section 4.3. At yield

$$p_y = \sigma_h + c_u \quad (\text{for } c_u \text{ materials}) \quad (4.78)$$

$$p'_y = \sigma'_h(1 + \sin\phi') \quad (\text{for } \phi' \text{ materials}) \quad (4.79)$$

$$p'_y = \sigma'_h(1 + \sin\phi') + c'\cot\phi' \quad (\text{for } c'\phi' \text{ materials}) \quad (4.80)$$

If the minimum value of circumferential stress, σ_m, is less than the tensile strength, σ_t, cracking will occur – that is, $\sigma_t < (\sigma_h - p_y)$ for cracking not to occur.

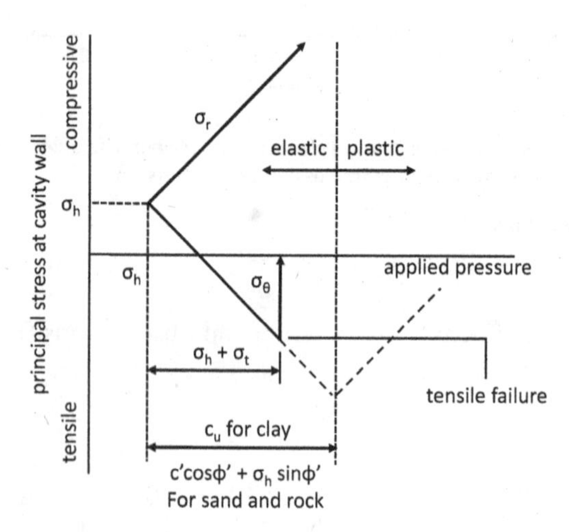

Figure 4.19 Total stress paths for cavity expansion.

In many clay soils $(\sigma_h - p_y)$ will always be positive for tests at depth but near the ground surface, $(\sigma_h - p_y)$ could become negative and less than the tensile strength of the clay causing cracks to form. Haberfield and Johnston (1989) demonstrated, with model tests, that two or three cracks develop in soft rock. The assumption that the ground behaves as a continuum is valid if the tensile strength, σ_t, is less than σ_m. If σ_m is less than σ_t, tensile cracks will propagate from the cavity wall. Figure 4.20 shows the failure processes around an expanding cavity in rock.

Rocha et al. (1966) proposed that a series of wedges would be formed as cracks propagated from the cavity wall and these wedges of rock are contained within an elastic solid as shown in Figure 4.21.

The deformation of a cavity wall within a solid mass is

$$a - a_o = 2(1+v)a\frac{p}{E} \tag{4.81}$$

The deformation of a cavity wall in a material which contains natural discontinuities is

$$a - a_o = 2\left(1 + v + \ln\left[\frac{b}{a}\right]\right)a\frac{p}{E} \tag{4.82}$$

where b is the radius to the solid rock. Equation (4.82) is independent of the number of fissures and the orientation of the displacement transducers relative to the fissures. A numerical analysis, in which the displacements at the centre and edge of a wedge were calculated, was used to show the effect of fissuring on the measured displacement (Table 4.3).

The expansion of the cavity in a solid mass is elastic until tensile failure occurs in the cavity wall. During the elastic phase equation (4.81) is satisfied. Once cracks start to propagate, equation (4.82) becomes valid. The depth of fissure is given by

$$b = a\left[\frac{p}{\sigma_t}\right]^{0.5} \tag{4.83}$$

where σ_t is the tensile strength of the rock. An unload/reload cycle will be linear if tensile cracking occurs because it represents the elastic contraction of the wedges, not the closing of the cracks.

Haberfield and Johnston (1993) showed that the drainage conditions in a pressuremeter test are a function of the coefficient of consolidation, c_h. For a given permeability, the value of c_h will depend on the compressibility of the ground, m_v. The time, t, for dissipation of pore pressure is given by

$$t = \frac{Td^2}{c_h} \tag{4.84}$$

where T is a time factor and d is the drainage path.

The coefficient of consolidation is given by

$$c_h = \frac{k}{m_v \gamma_w} \tag{4.85}$$

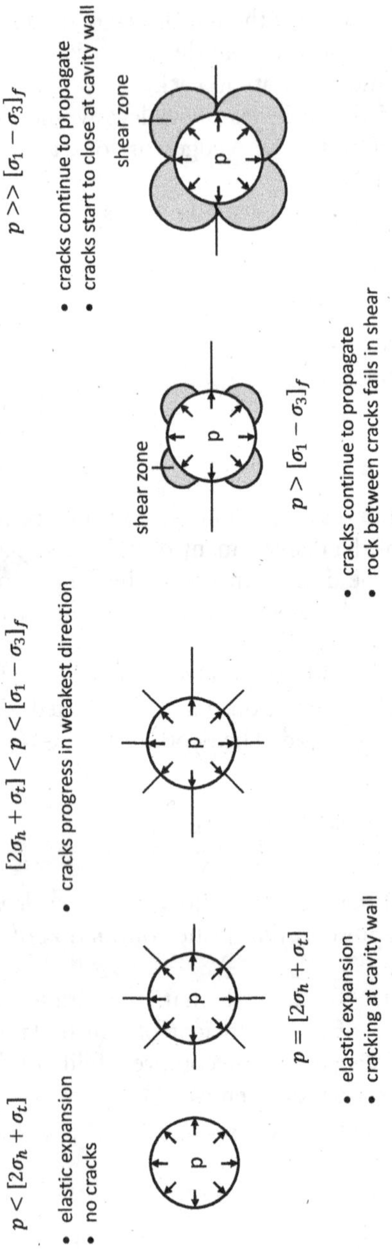

$p < [2\sigma_h + \sigma_t]$

- elastic expansion
- no cracks

$p = [2\sigma_h + \sigma_t]$

- elastic expansion
- cracking at cavity wall

$[2\sigma_h + \sigma_t] < p < [\sigma_1 - \sigma_3]_f$

- cracks progress in weakest direction

shear zone

$p > [\sigma_1 - \sigma_3]_f$

- cracks continue to propagate
- rock between cracks fails in shear

$p >> [\sigma_1 - \sigma_3]_f$

- cracks continue to propagate
- cracks start to close at cavity wall

shear zone

Figure 4.20 Failure processes around an expanding cavity in rock.

Source: After Haberfield and Johnston, 1990.

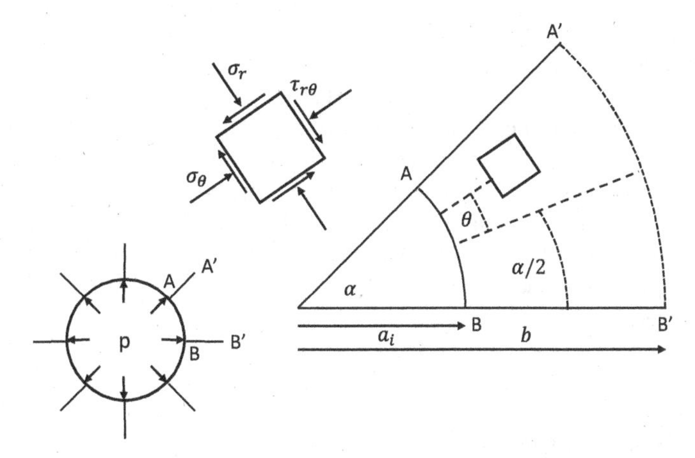

Figure 4.21 The pressure distribution and geometry around an expanding cavity in a fractured rock.

Source: After Rocha et al., 1966.

Table 4.3 The effects of orientation, number and depth of fissures on stiffness (assuming $v = 0.2$ and $E = 100$ MPa)

No. of fissures	*Depth of fissures / Cavity diameter*	*Displacement at point fissures form*	*Displacement midway between fissures*	*Displacement adjacent to fissures*
0	0	90	90	90
2	2	1.58	1.44	0.91
	4	2.16	2.13	0.95
	8	2.73	2.82	1.04
4	2	1.58	1.47	1.12
	4	2.16	2.02	1.49
	8	2.73	2.73	2.22
8	2	1.58	1.58	1.56
	4	2.16	2.16	2.38
	8	2.73	2.62	2.69

where k is the coefficient of permeability. The ratio (k/m_v) is small for clays and, therefore, excess pore pressures remain during loading. The ratio for rocks is larger than that for clays, thus pore pressures will dissipate more quickly in rocks than in clays even if their coefficients of permeability are similar. Dissipation will be quicker if there are discontinuities present, such as those propagated during a test. Thus, it is possible that tests in rock are drained tests.

Haberfield and Johnston (1990) developed a numerical method for fitting a curve to pressuremeter data. This depends on four parameters: the modulus, angle of friction and dilation and in situ stress. The modulus can be obtained from an unload/reload cycle. A wide range of values for the remaining parameters will produce curves that fit the data, therefore it is necessary to undertake other tests on the rock in order to interpret the pressuremeter data. The analysis is further complicated by the presence of discontinuities.

Haberfield and Johnston (1993) suggest that the only parameter that can be obtained from a pressuremeter test in soft rock is the unload/reload cycle, G_{ur}. The initial modulus is affected by any discontinuities and unless the spacing of discontinuities is significantly less than the diameter of the probe, the initial modulus will not represent the mass modulus of a fractured rock. It could represent the modulus of intact rock. The value taken from an unload/reload cycle may be representative since radial cracking will have occurred prior to unloading. The presence of soft inclusions will affect the results.

4.8 SPECIFIC ANALYSES

4.8.1 Non-linear stiffness

Moduli are dependent on strain range and mean effective stress. Reid et al. (1982) suggested that the shear modulus should be chosen for the required stress or strain range. Briaud et al. (1983) showed that, for tests in soils above groundwater level, there was a correlation between the modulus and the total applied pressure, and the unload and reload moduli could be represented by a hyperbolic model. Clarke and Wroth (1985) demonstrated that moduli obtained from tests in sands corrected for stress level reduced the scatter in the data, suggesting that modulus was dependent on the effective stress. O'Brien and Newman (1990) and Newman et al. (1991) suggested that the pressuremeter moduli should be quoted over a constant strain range. Robertson and Hughes (1986) and Newman et al. (1991) showed that moduli should be corrected for stress and strain level. All these authors used a single value of shear modulus from the pressuremeter test.

The variation in modulus with strain can be assessed from the initial loading, the final unloading and both the unloading portion and reloading portion of an unload/reload cycle. The shape of the initial loading curve is sensitive to the installation technique and therefore is unlikely to be of use in the interpretation.

It is possible to show the variation in reload modulus, for example, in terms of cavity strain range thus describing a non-linear stiffness curve. The cavity strain represents the integrated effect of the strains in the ground. In order to apply a constitutive model to predict the performance of a structure it is necessary to determine the incremental modulus. Muir-Wood (1990) developed a relationship between incremental tangent and secant moduli from triaxial tests and the shear modulus from a pressuremeter test using Palmer's analysis.

The incremental secant modulus, G_s, can be simply obtained from a pressuremeter test in clay by using equation (4.19) to obtain the stress–strain curve.

$$G_s = \frac{\Delta \tau}{2\varepsilon_c} \tag{4.86}$$

Equation (4.14), combined with equation (4.89), produces a relationship between the incremental secant modulus and the average unload/reload shear modulus of the form

$$G_s = G_{ur} + \varepsilon_c \frac{dG_{ur}}{d\varepsilon_c} \tag{4.87}$$

where G_{ur} is the secant shear modulus from an unload/reload cycle. In order to determine the non-linear incremental stiffness curve it is necessary to differentiate the G_{ur} versus ε_c curve. The test data will not necessarily lie on a smooth curve because of errors within the measuring systems. Simply applying equation (4.87) to the data will amplify this variation. Alternatively, a polynomial, for example, could be used to produce a smooth curve that fits the data. Muir-Wood suggested a relationship based on the empirical relation developed by Jardine *et al.* (1986) from triaxial tests on clays of the form

$$G_p = A + \left[\frac{B}{\sqrt{(1+\alpha^2)}}\right] \times \left\{\cos\left[\alpha\, ln\left(\frac{\varepsilon_s}{C}\right) - \theta\right] + \alpha\frac{C}{\varepsilon_s}\right\} \tag{4.88}$$

where A, B, C and α are constants and θ is $\cos^{-1}[1/\sqrt{(1 + \alpha^2)}]\}$. Note: the triaxial shear strain, ε_s is $[2\varepsilon_c/\sqrt{3}]$ but for practical purposes Muir-Wood suggests that $\varepsilon_c = \varepsilon_s$.

Jardine (1991) developed this method further by the use of 'equi-stiffness' curves taken from triaxial tests on till, London Clay and chalk, as shown in Figure 4.22.

The curves in Figure 4.22 represent the values of shear and cavity strain at which G_s and G_{ur} are equal which gives a trend defined by equation (4.88). The non-linear secant shear modulus curve from pressuremeter data is obtained by dividing the cavity strain by $[1.2 + 0.8 \log(\varepsilon_c/10^{-5})]$:

$$\frac{\varepsilon_c}{\varepsilon_s} = 1.2 + 0.8 log\left(\frac{\varepsilon_c}{10^{-5}}\right) \tag{4.89}$$

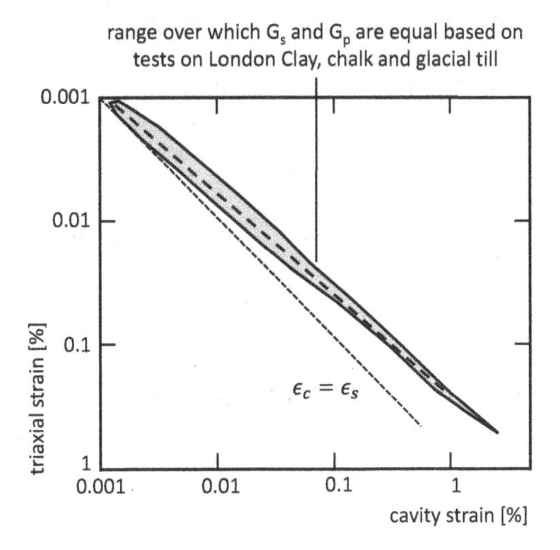

Figure 4.22 Non-linear secant modulus profiles from triaxial tests and the development of the conversion between ε_c and ε_s for $G_{ur} = G_s$.

Source: After Jardine, 1991.

The hyperbolic models described above can be used to define a non-linear profile rather than the experimental observations of Jardine (1991). Ferreira (1992) suggested that the variation in secant shear modulus with shear strain could be obtained from

$$\frac{G_{ur}}{G_o} = \frac{G_i}{G_o}\left[\frac{1}{1+G_i\gamma/\tau_{ult}}\right] \tag{4.90}$$

where G_o is the small strain shear modulus obtained from a seismic test. Note that G_i is not equal to G_o.

A similar non-linear stiffness profile can be developed from tests in sands but, in this case, it is necessary to take into account the variation in stress level as well as strain since the effective stress varies throughout the test unlike that in clays. Janbu (1963) proposed that the modulus at any stress level could be related to the modulus at another stress level which, for pressuremeter tests, gives

$$G_{uro} = G_{ur}\left(\sigma_h' / \sigma_{av}'\right)^n \tag{4.91}$$

where G_{uro} is the modulus at the in situ stress, n is a modulus component and σ_{av}' is the mean effective stress on a horizontal plane, n varies with strain level (Wroth et al., 1979) but, for sands, it tends to be in the range 0.4-0.5.

Bellotti et al. (1989) proposed a framework to enable G_{ur} to be compared to moduli from other tests to take this into account. The method involves

(a) calculating the average mean plane strain effective stress and shear strain amplitude at the start of the unload/reload cycle
(b) determining the in situ conditions
(c) calculating the equivalent modulus at the in situ stress level.

The average mean effective stress, σ_{av}', at the start of unloading is given by

$$\sigma_{av}' = \frac{\left[\dfrac{1}{2\sin\phi'} - \left(\dfrac{1+\sin\phi'}{2\sin\phi'}\right)\dfrac{\sigma_h'}{p_u'}\right]p_u'}{\ln\left[\left(\dfrac{p_u'}{\sigma_h'(1+\sin\phi')}\right)^{(1+\sin\phi')/2\sin\phi'}\right]} \tag{4.92}$$

where p_u' is the applied effective cavity pressure at the start of unloading.

The average shear strain, γ_{av}, during the unload/reload cycle is given by

$$\gamma_{av} = \frac{\Delta\varepsilon_c\left[1-\left(a_u / r_p\right)^2\right]}{\ln\left(r_p / a_u\right)} \tag{4.93}$$

where $\Delta\varepsilon_c$ is the difference between the maximum and minimum cavity strain during a cycle, a_u is the cavity radius at the start of unloading and r_p is the radius of the plastic zone at the start of unloading.

Bellotti *et al.* found, from CSBP tests in calibration chambers and in situ CSBP tests, that equations (4.92) and (4.93) could be approximated to

$$\sigma'_{av} = \sigma'_h + \alpha(p'_u - \sigma'_h) \tag{4.94}$$

$$\gamma_{av} = 2\beta\Delta\varepsilon_c \tag{4.95}$$

where α and β are constants. The range of ϕ' for these tests was 35° to 60°. The average value of α was found to be 0.2 and the average value of β, 0.5, which is similar to that suggested by Robertson and Hughes (1986).

Harden and Drnevich (1972) proposed a hyperbolic relationship expressing the non-linear stiffness profile in terms of the strain.

$$\frac{G}{G_o} = \frac{1}{\left[1 + G_o(\gamma/\tau_{max})\right]} \tag{4.96}$$

where G_o is the maximum modulus and τ_{max} is the maximum shear stress. Bellotti et al. (1989) proposed that a form of this could be used to determine the equivalent maximum modulus for pressuremeter tests. This could then be used to determine a non-linear stiffness profile given by

$$\frac{G_{uro}}{G_o} = \frac{1}{\left[1 + G_o(\gamma/2\sigma'_h \sin\phi')\right]} \tag{4.97}$$

Bolton and Whittle (1999) used the power law function proposed by Gunn (1992) and Bolton et al. (1993)

$$\tau = \alpha\gamma^{\beta} \tag{4.98}$$

The response of the soil within an unload reload cycle is

$$p_c - p_o = \frac{\alpha}{\beta}\left(\frac{\delta A}{A}\right)^{\beta} \tag{4.99}$$

where $\left(\frac{\delta A}{A}\right)$ is the shear strain at the cavity wall, α and β are derived from an unload reload cycle plotted to a log log scale such that the gradient is β and the intercept is $\ln(\alpha/\beta)$.

The secant shear modulus is

$$G = \eta\beta\gamma_c^{\beta-1} \tag{4.100}$$

where $\eta\beta$ is the stiffness constant α.

4.8.2 Undrained analysis assuming entire expansion at the limit pressure

The analyses described above refer to an ideal pressuremeter test in which the probe is installed with no disturbance to the surrounding ground. Gibson and Anderson (1961) and others have proposed analyses which take into account the unloading of the cavity wall prior to testing. In practice they are of little use because the true unloading curve of the ground during installation is unknown since it is not measured.

The installation of a full displacement FDP in an incompressible rigid material is similar to the limiting expansion of a cavity, therefore, theoretically, the pressure at the cavity wall is equal to the limit pressure given by equation (4.101):

$$p_1 - \sigma_h = s_u \left[1 + \ln\left(\frac{G}{c_u} \right) \right] \tag{4.101}$$

Houlsby and Withers (1988) assumed that the clay behaves as a linear elastic perfectly plastic material with no volume changes occurring within the clay during installation and expansion. The pressure at the start of unloading is equal to the limit pressure. The unloading will be elastic until failure occurs in extension at $(p_1 - 2c_u)$, as shown in Figure 4.23.

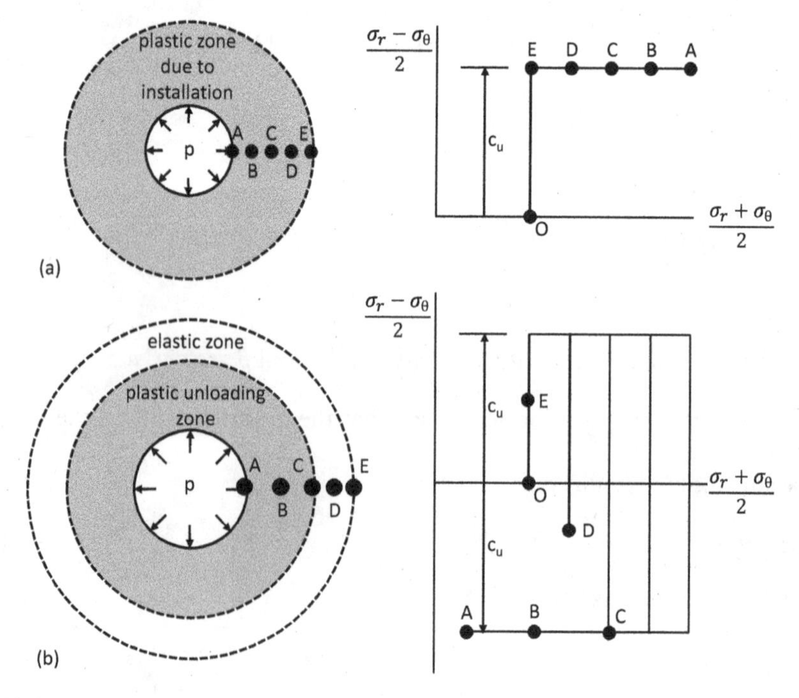

Figure 4.23 Stress paths followed during (a) installation and (b) contraction of an FDP.

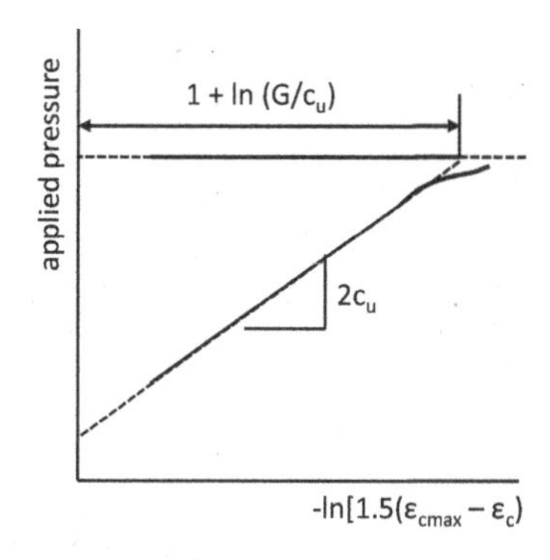

Figure 4.24 The analysis of FDP tests in clay.

The unloading curve is then defined by

$$p = p_1 - 2sc_u\left[1 + \ln\left\{\frac{\sinh(\varepsilon_{max} - \varepsilon)}{\sinh(c_u / G)}\right\}\right] \quad (4.102)$$

where ε_{max} is the maximum cavity strain. Thus, a plot of applied pressure against $\ln(\varepsilon_{max} - \varepsilon)$ will be linear, as shown in Figure 4.24, with a slope equal to the compression strength of the clay.

Further, the intersection between the loading portion (constant stress) and the extrapolation of the linear portion of the unloading curve intersects at $[1 + \ln(G/c_u)]$, which can be used to determine the shear modulus.

The above analysis is based on an incompressible material. It cannot be applied to tests in sands, which will dilate during installation and testing. The analyses described above for an ideal test in sand are based on small strains and cannot be applied to FDP tests because of the large shear strains induced during shearing and testing. Any closed-form solution developed for sands will have to take into account the volume changes occurring during installation and testing.

4.8.3 Coefficient of consolidation

The rapid expansion of a cavity in clay will create excess pore pressures in the clay. Baguelin et al. (1978) and Clarke et al. (1979) proposed that the decay in pore pressure at the cavity wall, while the volume of the cavity wall is held constant, could be used to determine the horizontal coefficient of consolidation, c_h. This is known as a strain-holding test. Fahey and Carter (1986) proposed a pressure holding test in which the pressure is held constant while the probe continues to expand.

Consider an elastic perfectly plastic soil. Up to yield the change in cavity pressure is equal to the change in the effective radial stress at the membrane/ soil interface since the pore pressure remains constant. After yield, the effective stress remains approximately constant and the increase in applied pressure is taken by an increase in pore pressure. The excess pore pressure, Δu, once yield has occurred, is given by

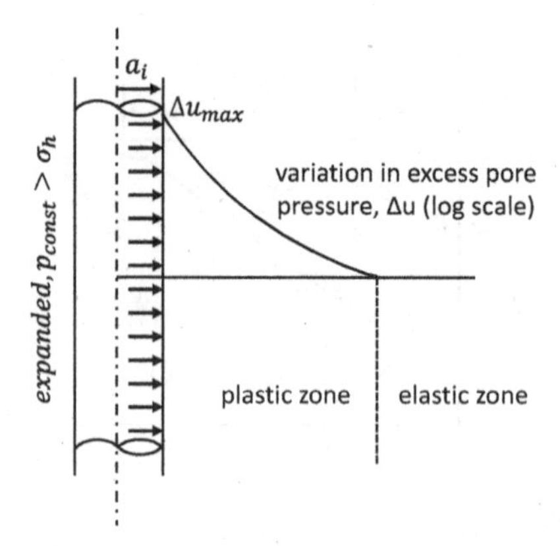

Figure 4.25 The distribution of excess pore water around an expanded cavity.

Source: After Clarke et al., 1979.

$$\Delta u = c_u\left[\ln\left(\frac{G}{c_u}\right)\left(\frac{\Delta V}{V}\right)\right] \tag{4.103}$$

The distribution of pore pressure is logarithmic within the zone of yield, as shown in Figure 4.25. If no further expansion occurs then the excess pore pressures will dissipate.

Numerical solutions (Carter et al., 1979) show that, as consolidation occurs, the total pressure and the pore pressures reduce though the reduction in total pressure is less than the reduction in pore pressure.

A closed form solution was developed by Randolph and Wroth (1979) based on the assumption that the soil behaves elastically during consolidation. This indicates that the total pressure remains constant during consolidation but, within experimental accuracy, the numerical solution indicates that the closed form solution is sufficiently accurate. Clarke et al. (1979) proposed that an analogous method to fitting data to the oedometer test could be used based on 50% dissipation. The curve derived from the work of Randolph and Wroth (1979) is shown in Figure 4.26.

The time factor to reach 50% dissipation, T_{50}, is given by

$$T_{50} = \frac{c_h t_{50}}{a_{max}^2} \tag{4.104}$$

where t_{50} is the actual time to reach 50% dissipation and $a_{\max}$ is the radius of the cavity wall at the start of consolidation (the maximum expansion of the cavity).

4.9 NUMERICAL METHODS

Uncertainties that exist in ground properties and in situ stress field are associated with aleatory uncertainties (e.g., spatial variability) and epistemic uncertainties (e.g., measurement errors, limited information, and model uncertainty) in a pressuremeter test (Nadim 2007). In general, the interpretation of pressuremeter tests uses graphical

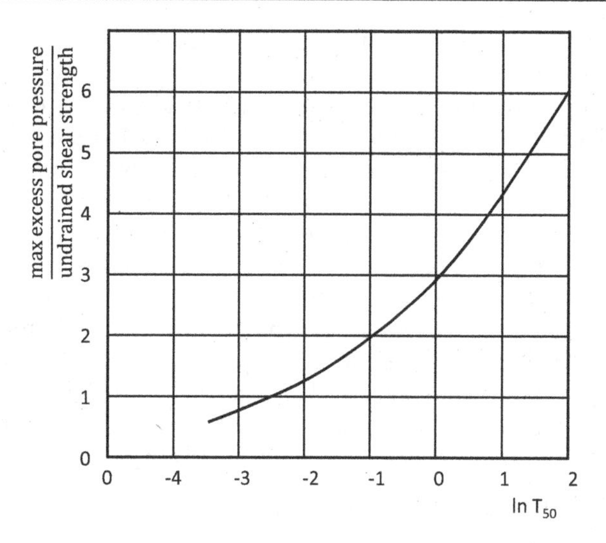

Figure 4.26 Time for 50% pore pressure decay at the cavity wall.

Source: After Randolph and Wroth, 1979.

plotting methods (e.g., Gibson and Anderson 1961; Houlsby and Withers 1988; Marsland and Randolph 1977) or curve fitting methods (e.g., Arnold 1981; Denby and Clough 1980). The graphical plotting methods involve fitting a line to the loading or unloading curve. A computational model, which can be either a closed-form solution or a numerical model, is used to determine the geotechnical properties of the ground. However, since the pressuremeter curve is a function of the in situ stress state and the elastic and plastic properties of the ground, the solution is not unique (Houlsby and Wroth, 1989). A number of techniques have been developed to give more confidence in the results. This includes a comparison of the properties with published results and results from other tests, least square error curve fitting techniques, optimization approaches based on a gradient method and genetic algorithms, neural networks and artificial intelligence.

Inverse analysis is used to determine soil parameters from pressuremeter tests either by optimization iterative algorithms such as gradient method or by techniques from the field of artificial intelligence such as the artificial neural networks (Levasseur et al., 2008). The gradient method is a parameter identification of an a priori known local constitutive law, whereas the artificial neural network is a method which creates, by learning phases, its own structural behaviour law from the geotechnical measurements. Both methods assume there is a unique solution. However, modelling errors and in situ measurement uncertainties are not taken into account by inverse analysis processes to be sure about solution accuracy. There is not one unique exact solution but rather an infinity of approximate solutions around an optimum. Levasseur et al. (2008) compared the gradient method and a genetic algorithm optimisation approach to produce a pressuremeter curve based on a Mohr Coulomb model using a finite element analysis. They concluded that the gradient method was not robust because of the complex nature of soil and cannot produce a unique solution. They observed that the genetic algorithm method was more promising to converge to a unique solution.

Dano and Hicher (2002) developed a semi-analytical solution of the pressuremeter curve using a linear elastic perfectly plastic model with a post-peak strain-softening and a small strain hypothesis and applied the Gauss-Newton algorithm to optimise the chosen soil properties. They found that the simultaneous optimisation of all the parameters by the Gauss-Newton algorithm led to divergent computations or erroneous estimates. In order to reduce the number of parameters that have to be simultaneously optimised, they used a procedure based on a sensitivity study and on laboratory experiments. The horizontal stress was determined from the initial portion of the curve (cavity strain less than 1.5%); the shear modulus from an unload reload cycle; and the angle of friction from the loading curve. The strain softening model takes into account the dilatency characteristics, the reduction in friction angle and the softening rate. Various empirical and theoretical relationships were applied to reduce the number of variables.

Abed et al. (2014, 2016) used a finite element analysis with a non-linear elastic plastic model to find the best fit to experimental data. The test curve was split into three; the initial response dominated by elastic behaviour defined by the stiffness, final response dominated by the yield criteria defined by the angle of friction and an intermediate section defined by the degree of curvature (Olivari and Bahar, 1993).

Dzousa et al. (2021) used a statistical inverse analysis to interpret pressuremeter tests (Figure 4.27).

The framework included (1) a priori knowledge of initial values and uncertainty quantification of input variables; (2) selection of a computational model and coupling with an optimization algorithm; (3) establishing an objective function; (4) identification

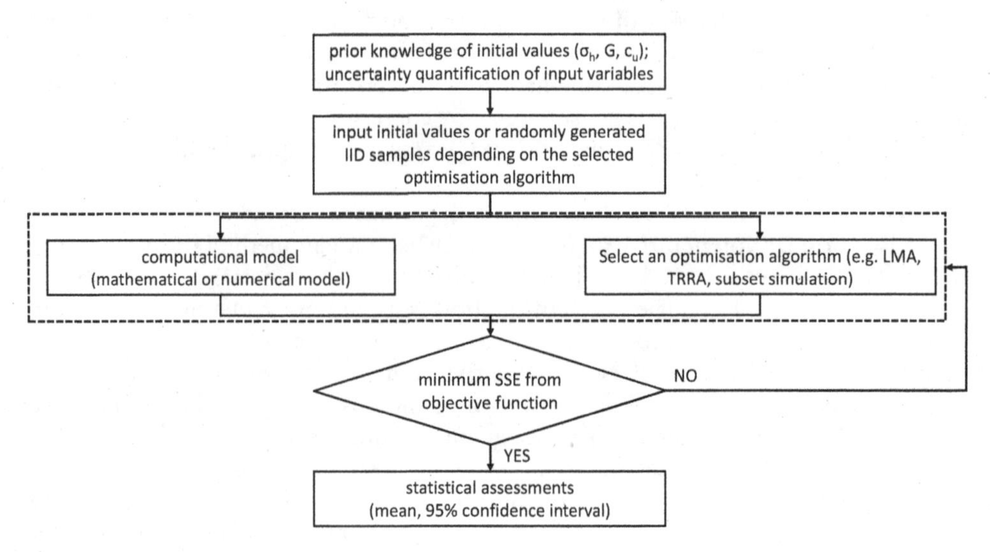

Figure 4.27 Flowchart showing the process to determine soil properties from a pressuremeter test using a statistical inverse analysis.

Source: After Dzousa et al., 2021.

of the optimal dataset based upon the optimization criteria; and (5) execution of a statistical assessment on the identified parameters. The analysis used the closed-form solutions attributed to Gibson and Anderson (1961) and Jefferies (1988) to model the loading and unloading curves for a pressuremeter test in clay and numerical analysis using finite differences with a Mohr Coulomb model. The criteria to select the best fit line was based on the least mean square of the difference between the predicted and actual test curve and the length of the error bar. However, the final selection of soil properties is still subject to an expert opinion.

4.10 SUMMARY

A brief description of the simple theories of cavity expansion, which can be used to interpret pressuremeter tests, has been used to show the assumptions made and the limitations of the methods. Most of them have been developed for ideal pressuremeter tests in which the probe is installed in the ground without causing any change to the ground – that is, the probe is "wished" into place. This is impossible in practice other than for research applications in which probes are cast into soil contained within a calibration chamber.

Analytical methods reviewed include linear and non-linear elastic material, perfectly plastic incompressible material, strain-hardening and strain-softening incompressible material, and dilatant material. Loading and unloading curves can be analysed. Most of the analyses assume the material to be a continuum, but in many practical cases the ground either contains discontinuities or discontinuities are created during a test. This has little effect on stiffness taken from an unload/reload cycle but will affect the derived strength.

More sophisticated methods of analysis use more complex constitutive models that take into account the uncertainties associated with the interpretation of a pressuremeter test, and the limitations imposed by the geometry and spatial variability of the ground properties. These models do not produce a unique solution because of the number of parameters involved. A range of optimisation techniques have developed to lead to an optimum solution but, ultimately, it is expert opinion that decides on the most appropriate ground properties.

The practical interpretation of pressuremeter tests has to take into account installation disturbance, fluctuation in test data, and the difference between actual behaviour and predicted behaviour.

In practice, there are three approaches to the interpretation of pressuremeter tests:

a) Closed-form solutions using linear and non-linear elastic and perfectly plastic models to derive soil properties;
b) Numerical methods with complex constitutive models to gain an insight into installation effects, probe geometry and ground properties on cavity expansion;
c) Empirical rules developed from observations of full and pilot scale structures to produce design parameters.

Therefore, pressuremeter tests undertaken as part of routine investigations are interpreted using empirical rules or linear and non-linear elastic and perfectly plastic models. More complex constitutive models in association with numerical methods are used to study expanding cavities and determine the sensitivity of the expansion to the probe geometry, the installation method, the test procedure, the ground response and the characteristics of the ground.

REFERENCES

Abed, Y., Bahar, R., Dupla, J.C. and Amar Bouzid, D.J. (2014) Identification of granular soils strength and stiffness parameters by matching finite element results to PMT data. *International Journal of Computational Methods*, 11(02), p. 1342001.

Abed, Y., Amar Bouzid, D.J., Bahar, R. and Toumi, I. (2016) Parameters identification of granular soils around PMT tests by inverse analysis. In *Advances in civil, environmental, and materials research World Congress (ACEM 16)*. Jeju Island, Korea.

Ajalloeian, R. and Yu, H.S. (1996) A calibration chamber study of the self-boring pressuremeter test in sand. In *7th Australia New Zealand Conference on Geomechanics: Geomechanics in a Changing World*: Barton, ACT: Institution of Engineers, Australia. pp. 54–59

Ajalloeian, R. and Yu, H.S. (1998) Chamber studies of the effects of pressuremeter geometry on test results in sand. *Geotechnique*, 48(5), pp. 621–636.

Arnold, M. (1981) Empirical evaluation of pressuremeter test data, *Can. Geotech. J.*, 18 (3), 455–459.

Aubeny, C.P., Whittle, A.J. and Ladd, C.C. (2000) Effects of disturbance on undrained strengths interpreted from pressuremeter tests. *Journal of Geotechnical and Geoenvironmental Engineering*, *126*(12), pp. 1133–1144.

Baguelin, F., Jézéquel, J.F., Le Mée, E. and Le Méhauté, A. (1972) Expansion of cylindrical probes in cohesive soils, *J. SMF Div., ASCE*, 98 (SM11), 129–142.

Baguelin, F., Jézéquel, J.F. and Shields, D.H. (1978) *The Pressuremeter and Foundation Engineering*, Trans. Tech. Publication.

Bahar, R. and Olivari, G. (1993) Analyse de la réponse du modèle de Prager généralisé sur chemin pressiométrique. In *Actes du 6ème Colloque Franco-Polonais de Mécanique des Sols Appliquée* (pp. 97–104).

Been, K. and Jefferies, M.G. (1985) A state parameter for sands. *Géotechnique*, *35*(2), pp. 99–112.

Been, K., Crooks, J.H.A., Becker, D.E., and Jefferies, M.G. (1987) The cone penetration test in sand. II: General inference of state. *Geotechnique*, 37(3), 285–299.

Bellotti, R., Ghionna, V., Jamiolkowski, M., Robertson, P.K. and Peterson, R.W. (1989) Interpretation of moduli from self-boring pressuremeter tests in sand, *Geotechnique*, 39 (2), 269–292.

Bishop, R.F., Hill, R. and Mott, N.F. (1945) The theory of indentation and hardness tests. Proceedings of the Physical Society (1926-1948), *57*(3), p.147.

Bolton, M.D., Sun, H.W. and Britto, A.M. (1993) Finite element analyses of bridge abutments on firm clay. *Computers and Geotechnics*, *15*(4), pp. 221–245.

Bolton, M.D. and Whittle, R.W. (1999) A non-linear elastic/perfectly plastic analysis for plane strain undrained expansion tests. *Géotechnique*, *49*(1), pp. 133–141.

Briaud, J.L. (2013) September. Ménard Lecture: The pressuremeter Test: Expanding its use. In *Proceedings of the 18th International Conference on Soil Mechanics and Geotechnical Engineering, Paris, France* (pp. 107–126).

Briaud, J.-L., Lytton, R.L. and Hung, J.T. (1983) Obtaining moduli from cyclic pressuremeter tests, *J. Geotech. Engng Div., ASCE*, 109(NGT5), 657–665.

Briaud, J.L., Cosentino, P.J. and Terry, T.A. (1987) Pressuremeter moduli for airport pavement design and evaluation. Texas Transportation Inst College Station

Cao, L.F., Teh, C.I. and Chang, M.F. (2001) Undrained cavity expansion in modified Cam clay I: Theoretical analysis. *Geotechnique*, *51*(4), pp. 323–334.

Cambou, B., Boubanga, A., Bozetto, P. and Haghgou, M. (1990) Determination of constitutive parameters from pressuremeter tests, Proc. 3rd Int. Symp. Pressuremeters, Oxford, pp. 243–252.

Carter, J.P., Randolph, M.F. and Wroth, C.P. (1979) Stress and pore pressure changes in clay during and after the expansion of a cylindrical cavity, Int. J. Numer. Analyt. Methods in Geo. Mech., *3*(4), 305–322.

Carter, J.P., Booker, J.R. and Yeung, S.K. (1986) Cavity expansion in cohesive frictional soils, *Geotechnique, 36*(3), 349–358.

Chen, H., Li, L. and Li, J. (2020) Elastoplastic solutions for cylindrical cavity expansion in unsaturated soils. *Computers and Geotechnics*, 123, p. 103569.

Clarke, B.G. and Wroth, C.P. (1985) Discussion on 'Effect of disturbance on parameters derived from self-boring pressuremeter tests in sand' by Fahey, M. and Randolph, M.F., *Geotechnique, 35*(2), 219–222.

Clarke, B.G., Carter, J.P. and Wroth, C.P. (1979) In-situ determination of the consolidation characteristics of saturated clays, *Proc. 7th Eur. Conf. SMFE, Brighton,* Vol. 2, pp. 207–211.

Collins, I.F. and Yu, H.S. (1996) Undrained cavity expansions in critical state soils. *International Journal for Numerical and Analytical Methods in Geomechanics*, *20*(7), pp. 489–516.

Dano, C. and Hicher, P.Y. (2002) Evolution of elastic shear modulus in granular materials along isotropic and deviatoric stress paths. In *15th ASCE Engineering Mechanics Conference.*

Denby, G.M. (1978) Self-boring pressuremeter study of San Francisco Bay Mud, PhD Thesis, University of Stanford.

Denby, G.M. and Clough, G.W. (1980) Self-boring pressuremeter tests in clay, *J. Geotech. Engng Div., ASCE,* 106 (GT12), 1369–1387.

Duncan, J.M. and Chang, C.-Y. (1970) Nonlinear analysis of stress and strain in soils, *J. SMFE Div., ASCE, 96*(5), 755–757.

Dsouza, N., Spyropoulos, E., Maakaroun, T. and Kolla, V. (2021) Prebored pressure-meter tests for an offshore project, its interpretation and comparison of deformation modulus. *J Arch Des Cons Tech*, 2(2), pp. 8–17.

Elwood, D.E., Derek Martin, C., Fredlund, D.G. and Ward Wilson, G. (2015) Volumetric changes and point of saturation around a pressuremeter probe used in unsaturated soils. *Journal of Geotechnical and Geoenvironmental Engineering*, *141*(11), p. 04015046.

Fahey, M. and Carter, J.P. (1986) Some effects of rate of loading and drainage on pressuremeter tests in clay. In *Proceedings of the Specialty Geomechanics Symposium: Interpretation of Field Testing For Design Parameters* (pp. 50–55).

Ferreira, R.S. (1992) "Interpretation of Pressuremeter Tests Using a Curve Fitting Technique," PhD Thesis, University of Alberta.

Ferreira, R.S. and Robertson, P.K. (1992) Interpretation of undrained self-boring pressuremeter test results incorporating unloading, *Can. Geotech. J., 29,* 918–928.

Geng, Y., Yu, H.S. and McDowell, G.R. (2013) Discrete element modelling of cavity expansion and pressuremeter test. *Geomechanics and Geoengineering*, *8*(3), pp. 179–190.

Gibson, R.E. and Anderson, W.F. (1961) In situ measurements of soil properties with the pressuremeter, *Civ. Engng Public Wks. Rev.,* 56, 615–618.

Gaone, F.M., Doherty, J.P. and Gourvenec, S., 2019. An optimization strategy for evaluating modified Cam clay parameters using self-boring pressuremeter test data. *Canadian Geotechnical Journal*, *56*(11), pp. 1668–1679.

Gong, W., Yang, C., Li, J. and Xu, L., 2021. Undrained cylindrical cavity expansion in modified cam-clay soil: a semi-analytical solution considering biaxial in situ stresses. *Computers and Geotechnics*, *130*, p. 103888.

González, N., Arroyo, M. and Gens, A. (2007) The effect of structure in pressuremeter tests in clay. *Nu [5] merical Models in Geomechanics,* NUMOG X, Pande, G.N. & Pietrusckzak, S. (Eds), pp. 721–732.

González, N.A., Arroyo, M. and Gens, A. (2009) Identification of bonded clay parameters in SBPM tests: A numerical study. *Soils and Foundations, 49*(3), pp. 329–340.

Gunn, M.J., 1992. The prediction of surface settlement profiles due to tunnelling. In *Predictive soil mechanics: Proceedings of the Wroth Memorial Symposium held at St Catherine's College, Oxford, 27-29 July 1992* (pp. 304–316). Thomas Telford Publishing.

Haberfield, C.M. and Johnston, I.W. (1989) Model studies of pressuremeter testing in soft rock, *Geotech. Test. J., ASTM, 12*(2), 150–156.

Haberfield, C.M. and Johnston, I.W. (1990) A numerical model for pressuremeter testing in soft rock, *Geotechnique, 40*(4), 569–580.

Haberfield, C.M. and Johnston, I.W. (1993) Factors influencing the interpretation of pressuremeters tests in soft rocks, *Proc. Conf. Geotechnical Engng Hard Soils-Soft Rocks,* Athens, Vol. 1, pp. 525–531.

Han, W., Liu, S., Zhang, D. and Zhou, H. (2011) Characteristics analysis of cavity expansion with anisotropic initial stress in a two-dimensional numerical model. In *Geo-Frontiers 2011: Advances in Geotechnical Engineering* (pp. 4186–4194).

Han, Y., Alruwaili, K.M. and Al Tammar, M.J., (2021) Numerical Modeling of Pressuremeter Test (PMT) in Rock Formations. *In 55th US Rock Mechanics/Geomechanics Symposium. OnePetro.*

Hao, D., Luan, M., Li, B. and Chen, R., 2010. Numerical analysis of cylindrical cavity expansion in sand considering particle crushing and intermediate principal stress. *Transactions of Tianjin University, 16*(1), pp. 68–74.

Hardin, B.O. and Drnevich, V.P. (1972) Shear modulus and damping in soils: design equations and curves, *J. SMF Div., ASCE,* No. SM7, 667–692.

Houlsby, G.T. and Wroth, C.P. (1989) The influence of soil stiffness and lateral stress on the results of in situ soil tests, *12th International Conference on Soil Mechanics and Foundation Engineering,* Rio De Janeiro, pp 1–19.

Houlsbv. G. T., Clarke. B. G. and Wroth. C. P. (1986) Analysis of' the unloading of a pressuremeter in sand. *Proc. 2nd Int. Symp. The Pressuremeter and It Marine Applications. Special Technical Publication 950.* New York: ASCE.

Houlsby, G.T. and Withers, N.J. (1988) Analysis of the cone pressuremeter test in clay, *Geotechnique, 38*(4), 575–587.

Houlsby, G.T. and Yu, H.S. (1990) Finite element analysis of the cone pressuremeter test, *Proc. 3rd Int. Symp. Pressuremeters,* Oxford, pp. 221–230.

Hughes, J.M.O., Wroth, C.P. and Windle, D. (1977) Pressuremeter tests in sands, *Geotechnique, 27*(4), 455–477.

Janbu, N. (1963) Soil compressibility as determined by oedometer and triaxial tests, *Eur. Conf. SMFE,* Vol. 1 .

Jardine, R.J. (1991) Discussion on "Strain dependent moduli and pressuremeter tests," *Geotechnique, 41*(4), 621–626.

Jardine, R.J., Potts, D.M., Fourie, A.B. and Burland, J.B. (1986) Studies of the influence of non-linear stress strain characteristics in soil structure interaction, *Geotechnique, 36*(3), 377–396.

Jefferies, M.G. (1988) Determination of horizontal geostatic stress in clay with self-bored pressuremeter, *Can. Geotech. J., 25*(3), 559–573.

Jewell, R.J., Fahey, M. and Wroth, C.P. (1980) Laboratory studies of the pressuremeter test in sand, *Geotechnique, 30*(4), 507–531.

Jiang, M.J. and Sun, Y.G. (2012) Cavity expansion analyses of crushable granular materials with state-dependent dilatancy. *International Journal for Numerical and Analytical Methods in Geomechanics, 36*(6), pp. 723–742.

Ladanyi, B. (1963) Evaluation of pressuremeter tests in granular soils, *Proc. 2nd Pan American Conf. SMFE,* Brazil, Vol. 1, pp. 3–20.

Ladanyi, B. (1972) In situ determination of undrained stress-strain behaviour of sensitive clays with the pressuremeter, *Can. Geotech. J., 9*(3), 313–319.

Laier, J.E., Schmertmann, J.H. and Schaub, J.H. (1975) Effect of finite pressuremeter length in dry sand, *Proc. ASCE Spec. Conf. In Situ Measurement of Soil Properties, Raleigh,* Vol. 1, pp. 241–259.

Lamé, G., 1852. *Leçons sur la théorie mathématique de l'élasticité des corps solides*. Bachelier.

Levasseur, S., Malécot, Y., Boulon, M. and Flavigny, E. (2008) Soil parameter identification using a genetic algorithm. *International Journal for Numerical and Analytical Methods in Geomechanics, 32*(2), pp. 189–213.

Li, L., Li, J. and Sun, D.A. (2016) Anisotropically elasto-plastic solution to undrained cylindrical cavity expansion in K0-consolidated clay. *Computers and Geotechnics, 73*, pp. 83–90.

Li, L., Chen, H. and Li, J. (2021) An elastoplastic solution to undrained expansion of a cylindrical cavity in SANICLAY under plane stress condition. *Computers and Geotechnics, 132*, p. 103990.

Manassero, M. (1989) Stress-strain relationships from drained self-boring pressuremeter tests in sands, *Geotechnique, 39*(2), 293–307.

Marsland, A. and Randolph, M.F. (1977) Comparisons on the results from pressuremeter tests and large in situ plate tests in London clay, *Geotechnique, 27*(2), 217–243.

Mecsi, J. (2011) Determination of the physical parameters of soil from pressuremeter tests. In *Proceedings of the 15th European Conference on Soil Mechanics and Geotechnical Engineering* (pp. 441–446). IOS Press.

Ménard, L. (1957) Mesures in situ des propriétés physiques des sols, *Annales des Ponts et Chaussées,* Paris, No. 14, 357–377.

Mo, P.Q. and Yu, H.S. (2017) Undrained cavity expansion analysis with a unified state parameter model for clay and sand. *Géotechnique, 67*(6), pp. 503–515.

Muir-Wood, D. (1990) Strain dependent moduli and pressuremeter tests, *Geotechnique, 40*(26), 509–512.

Nadim, F. (2007) Tools and strategies for dealing with uncertainty in geotechnics. In *Probabilistic Methods in Geotechnical Engineering* (pp. 71–95). Springer, Vienna.

Newman, R.L. (1991) Interpretation of data from self-boring pressuremeter tests for the assessment of design parameters in sand, *Tech. Sem. Pressuremeters for Design in Geotechics, Soil Mechanics*, No. 3.

O'Brien, A.S. and Newman, R.L. (1990) Self-boring pressuremeter testing in London Clay, *Proc. 24th Ann. Conf. Engng Group of the Geological Soc.: Field Testing in Engineering Geology,* Sunderland, pp. 39–54.

Palmer, A.C. (1972) Undrained plane-strain expansion of a cylindrical cavity in clay: A simple interpretation of the pressuremeter test, *Geotechnique, 22*(3), 451–457.

Prévost, J.-H. and Hoeg, K. (1975) Analysis of pressuremeter in strain softening soil, *J. Geotech. Engng Div., ASCE,* 101 (GT8), 717–732.

Pyrah, I.C., Anderson, W.F. and Pang, L.S. (1988) Effects of test procedure on constant rate of strain pressuremeter tests in clay, *Proc. 6th Conf. Numerical Methods in Geomechanics,* Innsbruck, pp. 647–652.

Qiu, M., Kong, Q., Chen, P., Shi, Z. and Li, C. (2019) Effects of overconsolidation ratio on undrained shear strength of clay considering pressuremeter of limited length: cases study. *In IOP Conference Series: Earth and Environmental Science*, 304, No. 5, pp. IOP Publishing.

Randolph, M.F. and Wroth, C.P. (1979) An analytical solution for the consolidation around a driven pile, *Int. J. Numer. Analyt. Methods in Geomech.,* 3, 217–229.

Reid, W.M., St. John, H.D., Fyffe, S. and Rigden, W.J. (1982) The push-in pressuremeter, *Proc. Int. Symp. Pressuremeter and its Marine Appl.*, Paris, pp. 247–261.

Robertson, P.K., Hughes, J.M.O., Campanella, R.G., Brown, P., and McKeown, S. (1986) Design of laterally loaded piles using the pressuremeter, *Proc. 2nd Int. Symp. Pressuremeter Marine Appl.*, Texam, United States, ASTM STP 950, pp. 443–457.

Rocha, M., Silveira, A.Da, Grossman, N. and Oliveira, E.De. (1966) Determination of the deformability of rock masses along boreholes, *Proc. 1st Congr. ISRM*, Lisbon, Vol. 1, pp. 697–704.

Rouainia, M. and Muir Wood, D. (2000) A kinematic hardening constitutive model for natural clays with loss of structure. *Géotechnique*, *50*(2), pp. 153–164.

Rouainia, M., Panayides, S., Arroyo, M. and Gens, A. (2020) A pressuremeter-based evaluation of structure in London Clay using a kinematic hardening constitutive model. *Acta Geotechnica*, *15*(8), pp. 2089–2101.

Russell, A.R. and Khalili, N. (2002) Drained cavity expansion in sands exhibiting particle crushing. *International Journal for Numerical and Analytical Methods in Geomechanics*, *26*(4), pp. 323–340.

Sayed, S.M. (1989) Alternate analysis of pressuremeter test, *J. Geotech. Engng Div., ASCE*, *115*(12), 1769–1786.

Schofield, A.N. and Wroth, P. (1968) *Critical state soil mechanics* (Vol. 310). London: McGraw-Hill.

Schnaid, F., Kratz de Oliveira, L.A. and Gehling, W.Y.Y. (2004) Unsaturated constitutive surfaces from pressuremeter tests. *Journal of Geotechnical and Geoenvironmental Engineering*, *130*(2), pp. 174–185.

Selvadurai, A.P.S. (1984) Large strain and dilatancy effects in pressuremeter, *J. Geotech. Engng Div., ASCE*, 110 (NGT3), 421–436.

Sivasithamparam, N. and Castro, J. (2018) Undrained expansion of a cylindrical cavity in clays with fabric anisotropy: Theoretical solution. *Acta Geotechnica*, *13*(3), pp. 729–746.

Sousa Coutinho, A.G.F. (1990) Radial expansion of cylindrical cavities in sandy soils. Application to pressuremeter tests, *Can. Geotech. J.*, *27*(6), 737–748.

Timoshenko, S.P. and Goodier, J.N. (1934) *Theory of Elasticity*, McGraw-Hill, New York.

Toumi, I., Abed, Y. and Boufia, A. (2018) Soil parameters identification around Ménard's pressuremeter test by inverse analysis in SNTF El-Harrach site, *14th ASEC Conference in Jordan*, Jordan University of Science & Technology.

Vesic, A.S. (1972) Expansion of cavities in infinite soil mass, *J. SMF Div., ASCE*, 98 (SM3), 265–290.

Whittle, A.J. and Aubeny, C.P. (1993) The effects of installation disturbance on interpretation of in situ tests in clays, *Predictive Soil Mechanics, Proc. Wroth Memorial Symp.*, Oxford, pp. 585–605.

Windle, D. and Wroth, C.P. (1977) Use of self-boring pressuremeter to determine the undrained properties of clays, *Ground Engng*, 10 (6), 37–46.

Wroth, C.P. and Hughes, J.M.O. (1973) An instrument for the in situ measurement of the properties of soft clays, *Proc. 8th Int. Conf. SMFE, Moscow*, Vol. 1.2, pp. 487–494.

Wroth, C.P. et al. (1979) A review of the engineering properties of soil with particular reference to the shear modulus, University of Cambridge.

Wroth, C.P. and Bassett, R.H. (1965) A stress–strain relationship for the shearing behaviour of a sand. *Géotechnique*, *15*(1), pp. 32–56.

Yang, C., Li, J. and Li, L., 2021. Expansion responses of a cylindrical cavity in overconsolidated unsaturated soils: A semi-analytical elastoplastic solution. *Computers and Geotechnics*, *130*, p. 103922.

Yu, H.S. (1994) State parameter from self-boring pressuremeter tests in sand. *Journal of Geotechnical Engineering*, *120*(12), pp. 2118–2135.

Yu, H.S. (1996) Interpretation of pressuremeter unloading tests in sands. *Geotechnique*, *46*(1), pp. 17–31.

Yu, H.S. (1998) CASM: A unified state parameter model for clay and sand. *Int. J. Numer. Anal. Methods Geomech*, 22, 621–653.

Yu, H.S. (2000) *Cavity expansion methods in geomechanics*. Springer Science & Business Media.

Yu, H.S. (2006) The First James K. Mitchell Lecture In Situ Soil Testing: From mechanics to interpretation. *Geomechanics and Geoengineering: An International Journal*, *1*(3), pp. 165–195.

Yu, H.S. and Collins, I.F. (1998) Analysis of self-boring pressuremeter tests in overconsolidated clays. *Geotechnique*, *48*(5), pp. 689–693.

Zentar, R., Hicher, P.Y. and Moulin, G. (2001) Identification of soil parameters by inverse analysis. *Computers and Geotechnics*, *28*(2), pp. 129–144.

Zheng, D., Zhang, B. and Chalaturnyk, R. (2021) Uncertainty quantification of in situ horizontal stress with pressuremeter using a statistical inverse analysis method. *Canadian Geotechnical Journal*, (ja).

Chapter 5

Factors affecting the interpretation of pressuremeter tests

5.1 INTRODUCTION

The analyses described in Chapter 4 are limited to ideal pressuremeter tests in ground that conforms to the constitutive model chosen. The ideal pressuremeter test, in which the probe is installed without disturbing the surrounding ground, is very nearly satisfied by the SBP test and, therefore, these tests are often interpreted directly using theories given in Chapter 4. Preboring for a PBP test inevitably causes disturbance. Semi-empirical methods are often used to interpret these tests, though it is possible in some cases to apply theories of cavity expansion (e.g., determining the shear modulus from an unload/reload cycle). The installation of an FDP is similar to an expanding cavity, and the interpretation of an FDP test can be based either on theory or on correlations.

The assumption that a pressuremeter test can be modelled as the expansion of an infinitely long cylindrical cavity in a continuum affects the results of the analyses; that is, the results of the analyses are a function of the assumptions and the methods of analysis. This is no different from the interpretation of any other test on soil or rock. The development of constitutive models and numerical methods has enabled researchers to investigate the effects of installation, probe geometry and soil properties upon the results of a pressuremeter test. This is the subject of this chapter.

5.2 FACTORS AFFECTING PARAMETERS DERIVED FROM PRESSUREMETER TESTS

5.2.1 Introduction

There are several reasons for discrepancies between theories of cavity expansion and the practical interpretation of tests. These include:

1. Installation affects the initial size of cavity and the properties of the surrounding ground.
2. The probe may not be vertical.
3. The vertical stress may not be the intermediate stress once yield has occurred.
4. The horizontal stress may not be the same in all directions.
5. The ground may not behave as a continuum, especially if it contains discontinuities or discontinuities are formed during a test as the cavity expands.
6. The ground may not be homogeneous both vertically and radially.

DOI: 10.1201/9781003028925-5

7. The ground properties may be anisotropic.
8. The ground response may not conform to the model chosen; for example, a perfectly plastic model does not model strain softening/hardening.
9. Drainage can occur during a test, which means tests in fine-grained soils are likely to be partially drained.
10. Ground properties are test rate-dependent.
11. The cavity may not expand as a cylinder.
12. The probe dimensions do not conform to those of a theoretical pocket.

It is for these reasons that simple models are used for the practical interpretation of pressuremeter tests. Wroth (1984) suggested that a simple model is adequate for the interpretation of most in situ tests, because the increase in accuracy of the model cannot be justified in any saving in the foundation design. Further, the uncertainty associated with the spatial variations of the ground composition and ground properties means that it is necessary to undertake several tests to estimate the characteristic properties.

All types of pressuremeters disturb the ground during the installation process. Aubeny et al. (2000) undertook a review of the effects of installation disturbance on SBP and FDP tests and concluded the following:

1. Probe installation affects both the stress state and properties of soil adjacent to the pressuremeter membrane.
2. Peak shear strengths estimated from SBP tests in soft clays are typically 30–50% higher than those measured in high quality, K_0-consolidated undrained triaxial compression shear tests.
3. In general, the lift-off pressure is not equal to the true in situ lateral stress. Data from Ghionna et al. (1982) and Benoit and Clough (1986) indicate that low SBP contact pressures ($p_0 < \sigma_{h0}$) are associated with higher estimates of shear strengths and vice versa.
4. Full displacement pressuremeter strength estimates are significantly lower than those obtained from SBP tests performed in the same soil (e.g., Lacasse et al. 1990).
5. The length-to-diameter ratio (L/D) of the pressuremeter membrane can have a significant effect on the interpreted shear strength. Ghionna et al. (1982) showed that undrained shear strength from a PAFSOR-type device with L/D = 2 were 100– 250% higher than those obtained from a similar device with L/D = 4. Numerical analyses to investigate the effects of the membrane length on derived undrained shear strength for L/D = 6 showed that the finite membrane length overestimates the theoretical (i.e., infinite cavity) undrained strength by 25–40% (Yeung and Carter 1990; Houlsby and Carter 1993) to 5–20% (Shuttle and Jefferies 1995).
6. The assumption of a unique soil stress-strain curve may be violated due to partial drainage during membrane expansion and contraction, and the fact that soil response is strain-rate dependent. Prévost and Hoeg (1975) showed that, due to rate-dependent behaviour of clays, the derived stress-strain curve from a constant rate of strain pressuremeter test will exhibit strain softening even in materials that are actually strain hardening. Benoit and Clough (1986) suggested

that there is a 30% increase in estimated strength for a 20-fold increase in expansion rate.

5.2.2 Effects of installation

The basis of the theories proposed in Chapter 4 is that the cavity at the start of expansion is unaffected by installation – that is, the ground has not been displaced and the in situ stresses and pore pressure have not changed. However, the installation of a probe *does* have a significant effect on the ground surrounding the cavity, therefore the installation technique will affect the interpreted results of a test. This applies to all pressuremeters though the amount of disturbance varies between pressuremeters, the ability of the operators and the properties of the ground.

Compared to the installation of an ideal pressuremeter, installation can cause

(a) a reduction in the ground pressure acting on the pressuremeter (e.g., the installation of a PBP);
(b) a reduction in cavity diameter as the ground relaxes due to the reduction in cavity pressure (e.g., installation of a PBP);
(c) a change in properties of the ground due to induced shear and a change in cavity pressure (e.g. installation of an SBP);
(d) and an increase in cavity pressure (e.g., installation of an FDP).

The cavity diameter at the in situ total horizontal pressure is referred to as the reference datum, and the interpreted results are dependent on that chosen datum. The installation can change the horizontal stress on the pocket wall, and the pocket diameter is then no longer the reference datum. The pocket diameter can be significantly greater than the reference datum, as in FDP tests, or it can be less, as in PBP tests.

The installation process disturbs the soil, making it difficult to determine the initial reference volume. Using strain-path analysis, Aubeny et al. (2000) showed that it is better to use the unloading curve to determine undrained shear strength of clays. The properties of the ground around the pocket change during installation. It is important to expand the membrane sufficiently to ensure that undisturbed ground is stressed so that in situ parameters are obtained. This may not be possible because of the limitation of the expansion of the test section and, hence, the amount of disturbance must be considered at the time of interpretation.

Theoretically, the SBP produces the least disturbance. If an SBP is installed correctly, it is assumed that the initial pocket diameter, equal to the probe diameter, is the reference datum. A test can be interpreted using the closed formed solutions given in Chapter 4. Many of the numerical studies of cavity expansion are based on the principle that the probe diameter is the reference diameter. Any disturbance during installation, however small, will change the conditions around the probe and therefore a subjective decision regarding the reference datum will have to be taken before a test can be analysed. The selection of the reference datum is subjective and affects the selection of horizontal stress, initial loading stiffness and strength. The unload/reload modulus, provided that the cycle of unloading and reloading is carried out at sufficient strain so that undisturbed ground is being tested, is less affected by the selection of the

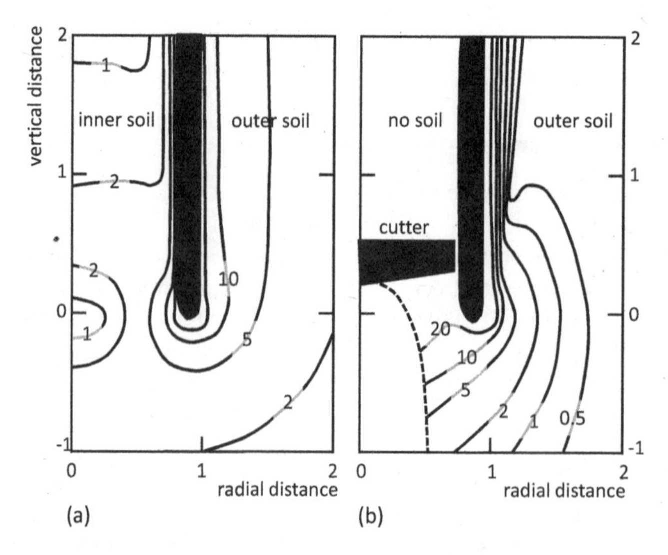

Figure 5.1 The shear strains induced in soil adjacent to a probe during installation: (a) open-ended pile; (b) self-boring pressuremeter.

Source: After Whittle and Aubeny, 1993.

reference datum. This means it is possible to carry out an unload/reload cycle in any pressuremeter test to determine the in situ shear modulus.

Shear strains are induced in the ground during installation of an SBP or FDP, and these can affect the properties of the ground adjacent to the probe. Figure 5.1 shows the strains induced due to installation of (a) an open-ended pile (cf., a partial displacement PBP) and (b) an SBP.

Using strain path analysis, Whittle and Aubeny (1993) showed that the interpretation of an FDP loading test will underestimate the strength, while that for an SBP loading test will overestimate the strength. The unloading curve is unaffected by the installation technique provided there is sufficient expansion to test undisturbed ground.

Zhou et al. (2014) showed that the shear stress at the cavity wall boundary has a significant influence on the pressure expansion curve and should not be neglected when deriving the cavity wall pressure or excess pore pressure. An elastic perfectly plastic soil with large strains was assumed in the plastic zone and small strains in the elastic zone around the cavity in their analysis. They showed that the initial shear stress has little effect on the plastic zone, suggesting that it is feasible to derive a strength from an SBP test, even allowing for shear during installation at the cavity wall.

One consequence of the induced shear strains is to change the stress conditions around an SBP. The implication of this is that the total horizontal stress is no longer a principal stress, but the pressure acting on the probe is equal to the total horizontal stress (Figure 5.2a). The exception to this is when the ground is at a state of failure, such as that encountered in front of a retaining wall. In that case, the measured horizontal stress will be less than the true horizontal stress (Figure 5.2b). Conversely, if the ground is in an active state, the horizontal stress would increase during installation.

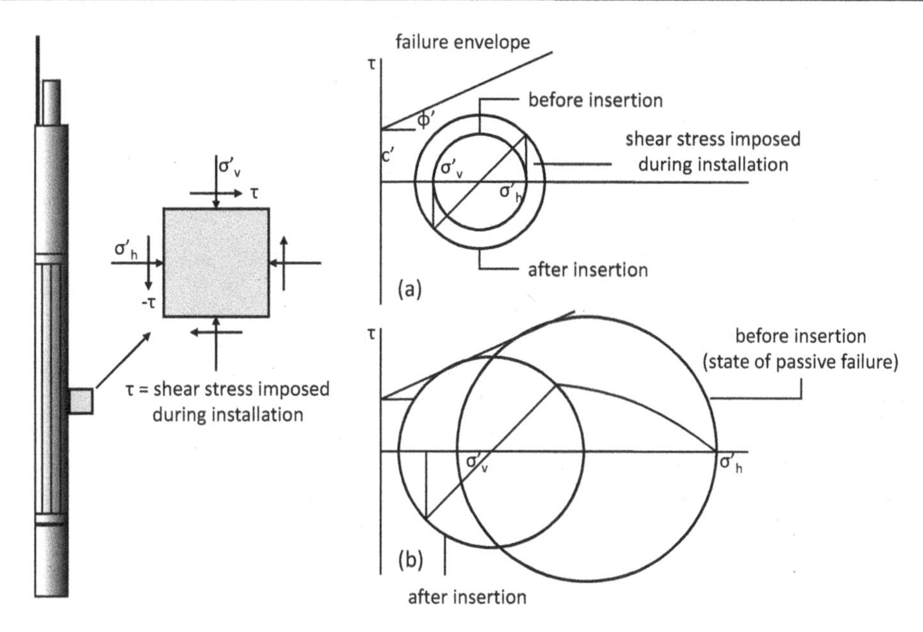

Figure 5.2 Stress conditions in a soil element in contact with a SBP: (a) $1 < K_o < K_p$; (b) $K_o = K_p$.

Source: After Clarke and Wroth, 1984.

It is assumed that the probe is vertical. All expansion should be in the horizontal direction and, provided the horizontal stress and ground properties are the same in all directions, the expansion should be uniform. FDPs and SBPs tend to deviate from a vertical line because the jacking force at the surface, remote from the probe, causes the rods to buckle in the borehole. Stabilisers on the rods can help reduce this effect, but problems with the control cable can arise if that cable becomes caught in the borehole. In the extreme case, the deviation will give rise to a reduction in stress on one side of the probe and an increase on the other, as shown in Figure 5.3.

The increase in stress occurs because the probe is forced into the soil. The reduction in stress arises from the possibility of the rods bending. This will only be apparent when using radial-displacement type probes, since the movement of the transducers will vary depending on their position relative to the axis of the probe. This means the lift off pressure will vary between the transducers, suggesting incorrectly a variation in horizontal stress. This will not be detected with volume-displacement type probes. An inclinometer and compass are installed in some probes (for example, the FDP and HPD) so that the orientation and alignment of the probe can be measured.

Aubeny et al. (2000) undertook strain-path analyses in conjunction with the MIT-E3 effective stress soil model, to simulate the effects of installation disturbance on subsequent pressuremeter expansion and contraction tests in normally and moderately overconsolidated Boston Blue Clay.

1. The installation of a FDP generates large excess pore pressures in the surrounding soil, such that the contact pressures greatly exceed the in situ total horizontal

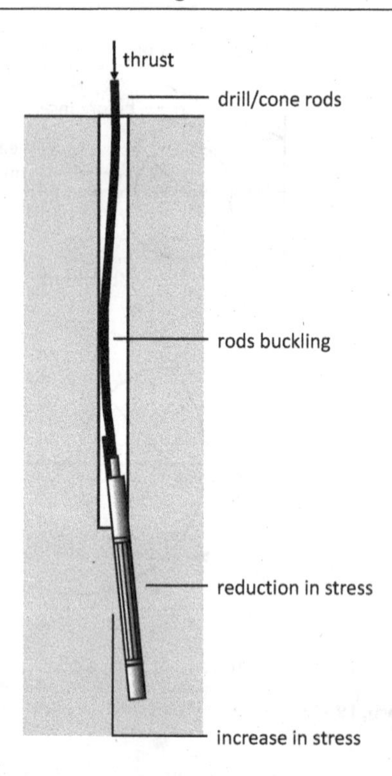

Figure 5.3 The effect of installation on the orientation of pushed-in and self-boring probes and the implications to the in situ stress.

stress. Undrained shear strengths derived from the expansion curves underestimate the strength of the intact clay (by 50–90%).

2. Disturbance induced during ideal self-boring penetration reduces the pressure acting on the probe, and leads to an increase in the derived peak shear strength with post peak strain softening that are inconsistent with the behavior of the intact clay. This is amplified when the finite membrane length is included in the analyses.
3. The analyses of both FDPs and SBPs suggest that undrained shear strength can be estimated reliably from the unloading portion of a test in fine grained soils (assuming prior membrane expansion to large strains).

5.2.3 Effects of the in situ stress

It is assumed that the total vertical stress is a principal stress and does not affect the behaviour of the ground since it remains an intermediate stress once yield is initiated – that is, failure takes place on vertical planes. During an expansion test the radial stress increases, and the circumferential stress decreases until equation (5.1) is satisfied, that is

$$\frac{\sigma_r'}{\sigma_\theta'} = \frac{1+\sin\phi'}{1-\sin\phi'} \tag{5.1}$$

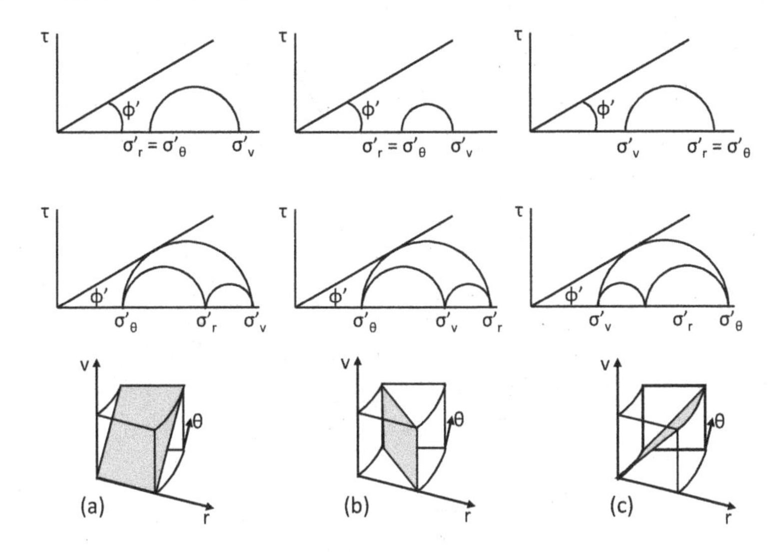

Figure 5.4 Orientation of failure planes assuming Mohr-Coulomb failure: (a) normally consolidated; (b) lightly overconsolidated; (c) heavily consolidated.

Source: After Wood and Wroth, 1977.

This applies for purely frictional materials. Wood and Wroth (1977) showed that the orientation of the failure planes depends on the initial stress conditions. For normally consolidated (a), lightly overconsolidated (b) and heavily overconsolidated (c) soils failure occurs on inclined planes, as shown in Figure 5.4 and, therefore, the vertical stress is no longer the intermediate stress.

In practice, soils with low values of K_0 do not behave elastically prior to yielding, therefore the vertical stress in those soils becomes and remains the intermediate stress. This may not be the case for heavily overconsolidated soils.

Figure 5.5 shows the theoretical limits to the range of K_0 for conventional theories of cavity expansion to apply. It is assumed that the horizontal stress and ground properties are the same in all directions, though in practice the expansion of the membrane is non-uniform. This can be seen when inspecting tests carried out using radial-displacement type probes, an example of which was given by Dalton and Hawkins (1982). The transducers only give measurements at a few points (normally three or six) and, hence, the true shape of the expanding cavity is unknown. It is not possible to determine the true shape of the expanded membrane when using volume-displacement type probes.

It is assumed that the membrane expands as a right circular cylinder. A test can be interpreted using the average expansion, either by taking the actual volume change measured or by taking the average of the displacement transducers. Alternatively, with displacement systems, a test can be analysed as if it were a number of independent tests, each test corresponding to a transducer. This allows the average and range of parameters to be quoted. Further, it permits an assessment of the quality of the data as discussed in Chapter 6. In either case, the assumptions made in the analysis are violated.

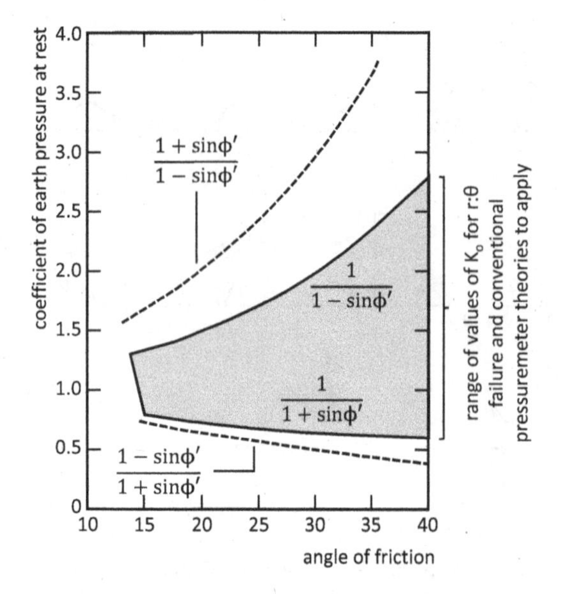

Figure 5.5 Limits to the coefficient for earth pressure at rest for conventional pressuremeter theories to apply.

Source: After Mair and Wood, 2013.

If the minimum value of circumferential stress is less than the tensile strength, cracks will propagate. The minimum value of circumferential stress is a function of the in situ horizontal stress and strength. At shallow depths, where soils are likely to be overconsolidated due to desiccation, for example, the soil may be stiff but the in situ stresses are low and radial cracking does occur. Ménard and others have shown that there is a critical test depth above which the vertical stress will influence the interpreted values of limit pressure and strength. Briaud and Shields (1981) established from laboratory tests that there was no critical depth for stiffness, which confirms the assumption made by Rocha et al. (1966) for tests in fractured rock – that is, unload/reload stiffness of ground containing discontinuities is a function of the intact ground.

5.2.4 Effects of stress history

The effect of stress history (OCR) on soil behaviour is not considered when evaluating undrained shear strength with an elastic perfectly plastic model, but, in reality, the stress history does influence the derived undrained shear strength of soil (Gutierrez et al., 2008; Ching et al. 2010). Yu and Collins (1998) showed that the direct application of the perfectly plastic model is accurate for soils with low overconsolidation ratio (OCR) values but tends to underestimate the undrained shear strength of heavily overconsolidated soils. Min Qui et al. (2019) undertook an Finite element (FE) analysis to investigate the effect of OCR on the undrained shear strength when using a pressuremeter of finite length. They compared the results from the FE analysis using the modified Cambridge model (Cao et al., 2001) to those derived using the analytical methods attributed to Gibson and Anderson (1961) and Cao et al. (2001). They found that the undrained shear strength obtained from the Gibson and Anderson method is significantly smaller than that obtained from FE analysis, and that difference increases as the OCR increases for an infinitely long pressuremeter. The analytical methods over-predict the shear strength when length is taken into account in the pressuremeter analysis. As the OCR increases, the possibility of vertical deformation reduces, which means the expansion is closer to plane strain. Thus, the undrained shear strength increases with an increase in OCR and reduces with a reduction in the length of the pressuremeter test section. There is a point when these two effects cancel each other out. Qui et al. (2019) found that the critical value of OCR depends on the soil type.

Naggar and Naggar (2012) undertook a numerical study of the effect of depth of embedment on a linear elastic perfectly plastic cohesionless soil. They found that the cavity pressure predicted by the close-formed solution exceeded that produced by the numerical analysis if K_o <1. They proposed that the cavity pressure should be multiplied by the correction factor:

$$R_k = (K_0 - 1)\left(0.5 - \frac{1.08}{\frac{C}{D}} + \frac{0.57}{\frac{a}{a_0}}\right) - 0.24\left(K_0^2 - 1\right) \tag{5.2}$$

where C/D is the ratio of the depth to the centre of the test section to the probe diameter.

5.2.5 Effects of discontinuities and bands of hard and soft layers

Discontinuities and layered materials can result in non-uniform expansion and loss of continuum. Figure 3.35 shows the effect on the pocket shape of vertical and horizontal fissures and hard and soft layers which can only be detected, if at all, with radial-displacement type probes. A vertical or horizontal fissure will not be noticed unless a displacement transducer is directly in line with that fissure. In that case, the average of the measured strains will appear to be greater than the true average strain. Volume-displacement type probes will always give an average strain.

The horizontal stress, stiffness and strength of hard layers will be underestimated, and the properties of soft layers will be overestimated if the ground is layered and volume-displacement type probes are used. Radial-displacement type probes will give the properties of the layer in which the transducers sit. If the transducer is aligned with a soft layer, the interpretation will give the parameters for that layer, though the analysis will not be valid since there will be an element of horizontal shear between the soft layer and the hard layers above and below. If the transducer is aligned with the hard layer, the interpretation will give the parameters for that layer.

Thus, the interpretation of tests in such ground – for example, weathered rocks, stiff fissured clays and laminated clays – will produce significant scatter in results, which may be due to natural variations in the ground as well as variations due to the effects of the discontinuities upon the test.

5.2.6 Effects of particle type

Sand particles can crush at high stresses (McDowell and Bolton 1998). Russell and Khalili (2002) proposed a new critical state line for coarse-grained soils to cater for particle rearrangement at low stresses, particle crushing and, at high stresses, further particle alignment of the altered particles. They found with a Mohr Coulomb failure criterion and state parameter model that ignoring particle crushing may lead to an apparent stiffer response especially for dense sands at depth.

5.2.7 Effects of test procedure

Interpreted and mobilised ground properties are dependent on the type of test and the speed of test. The rate of strain in a stress-controlled test varies as shown in Figure 5.6, hence the results obtained will be different from those in a strain-controlled test.

Anderson et al. (1984) developed an apparatus for testing thick hollow cylinders of soil, which they proved, using a numerical analysis, simulated to a high degree of accuracy, a pressuremeter test. This apparatus was used to investigate the effect of stress-controlled tests on the shape of the pressuremeter curve and therefore on the results. A 150 mm diameter cylinder with a 25 mm diameter cylindrical cavity in the centre was formed with normally consolidated kaolin. Figure 5.7a shows the effect of varying the pressure increment on the shape of the pressuremeter curve and the derived shear stress-strain curves. A similar set of experiments was conducted using a constant stress increment of 10kPa but with different 'holding' times. The results of these tests are shown in Figure 5.7b.

Creep occurs in addition to elastic plastic deformation and consolidation. Withers et al. (1989) showed, experimentally, that creep occurred in sand. Clarke and Smith (1992) showed that creep occurred in weak rocks which can, in the extreme case, lead to expansion of the test section even if the applied pressure is reduced.

A numerical study on the effects of creep, consolidation, probe diameter, and test technique on values of undrained shear strength by Anderson and Pyrah (1986) showed

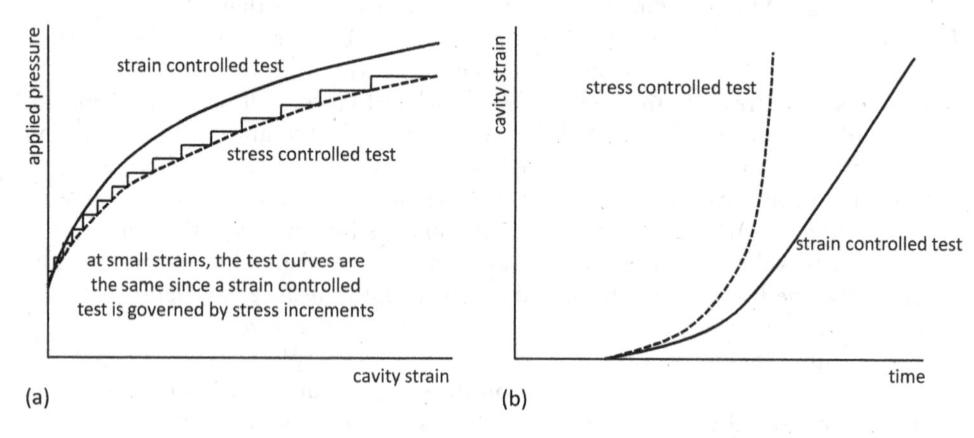

Figure 5.6 A comparison between stress- and strain-controlled tests showing the effects on (a) the test curve and (b) the cavity strain rate.

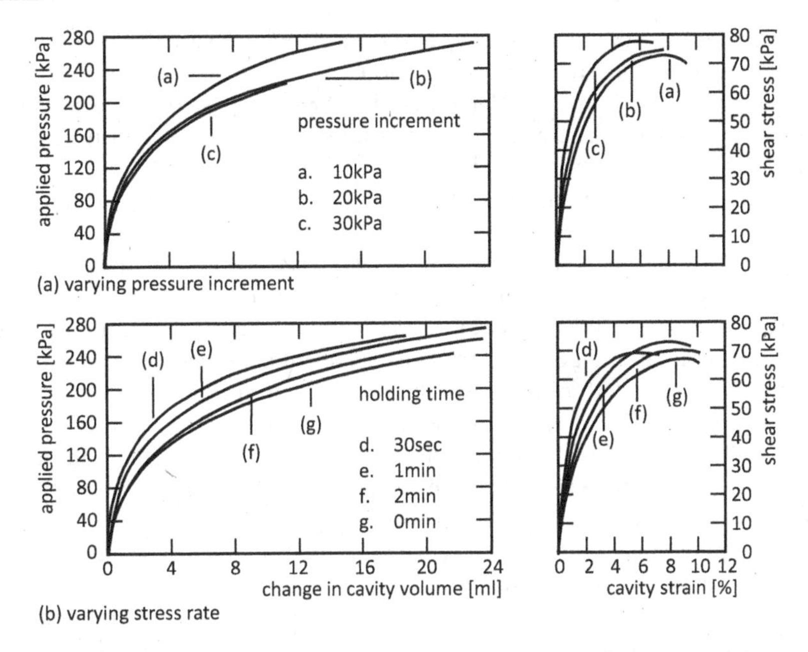

Figure 5.7 An experimental study of the effects of (a) the magnitude of stress increment and (b) the time of maintaining a stress increment on derived shear stress.

Source: After Anderson et al, 1984.

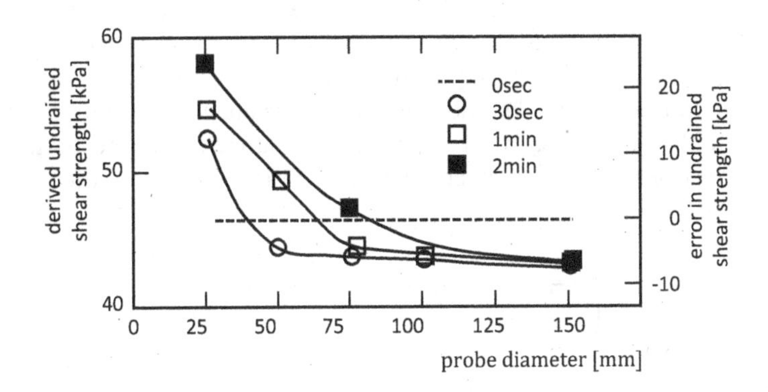

Figure 5.8 The effects of 'holding' time and probe diameter on the derived strength.

Source: After Anderson and Pyrah, 1986.

that, for one normally consolidated clay, the error in derived strength could be up to 30% (Figure 5.8).

Consolidation occurs around the probe as the applied pressure is increased. This results in additional strain to that occurring due to shear. The strength of the clay around the probe increases. The two effects cancel each other out since the cavity expands more because of consolidation, and the applied pressure has to increase as the soil strength increases because of consolidation.

Fukagawa et al. (1990) and Anderson et al. (1987) suggested that drainage would occur in quick tests in clay, affecting the results obtained. Anderson et al. (1990) reported the results of numerical studies (Pyrah et al., 1985, 1988), which showed that the strength obtained, assuming that the soil is perfectly plastic, varies according to the rate of testing and type of test control (stress or strain). They suggested that consolidation is the dominant effect in strain-controlled tests, while creep affects the results of stress-controlled tests. Tests in fine-grained soils are assumed to be undrained. Fioravante et al. (1994) and Jang et al. (2003) showed that this assumption is acceptable if the strain rate is 1% per min, and the coefficient of permeability is less than 10^{-9} m/sec. Thus, it is only valid for clays which are not fissured. The presence of fabric and coarser particles means that tests will be partially drained. Jang et al. (2003) studied the effect of strain rate showing that a strain rate of 1%/min results will produce an overestimate of undrained shear strength compared that from laboratory tests. Note, they did not take into account anisotropy or the difference between the stress paths in the two test types, which will also affect the mobilised strength. The effects were most significant for strain-softening behaviour.

Marcil et al. (2015) undertook a series of stress- and strain-controlled tests to show that strain controlled tests generally yield slightly higher values of modulus and lower values of limit pressure than the stress-controlled method. These differences are, in part, due to the time a stress increment is applied (Briaud, 1992). However, the differences may be small compared those introduced by other factors.

The limit pressure is a function of the shear strength, therefore it will also vary with test procedure. Powell et al. (1983) undertook stress-controlled PBP tests in a glacial till varying the time of holding each increment from 0.5 to 30 min to show the effect that this has upon limit pressure (see Figure 5.9).

The shear modulus should be independent of drainage, but Windle and Wroth (1977) showed that changing the rate of strain changed the interpreted values of moduli. This may be due to the effects of consolidation increasing the effective stress around the probe.

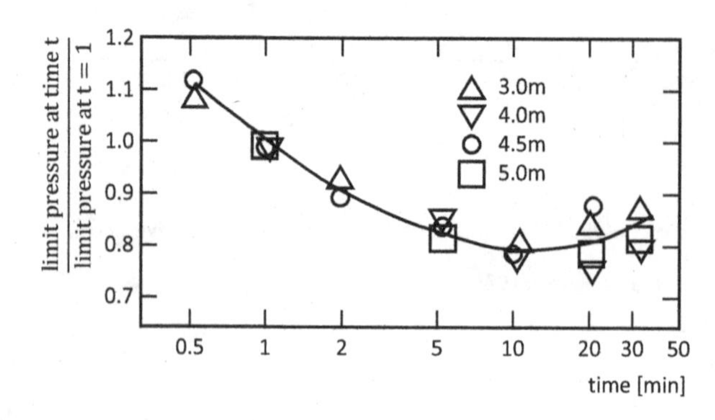

Figure 5.9 The effect of the time of maintaining the pressure increment during a stress-controlled PBP test in till at various depths.

Source: After Powell et al., 1983.

5.2.8 Effects of test cavity shape

A test pocket is unlikely to be perfectly cylindrical unless it is created by an FDP, or an SBP is drilled in correctly. Even if the test pocket created is cylindrical, the expansion of the cavity may not be truly cylindrical due to the limited length of the test section and the anisotropic properties of the ground. In both cases, the assumption that a pressuremeter test can be likened to a cylindrical cavity expansion is not valid.

Suyama et al. (1983) carried out tests using one-third scale models of the LLT and MPM in sand held in a chamber 40 cm wide, 16 cm thick and 40 cm deep. Lead shot of 2 mm diameter was placed at 15 mm intervals on a vertical plane through the axis of the probe that was cast in place. X-rays were used to monitor the position of the shot as the test section was inflated. The results, shown in Figure 5.10, indicate that the cavity does not expand as a cylinder but tends to deform as a flattened sphere. The deformation remote from the probe varied inversely with the radius, but deformation also occurred above and below the test section.

Fragaszy and Cheney (1977) showed that the average displacement of an elliptical cavity in an elastic medium is greater than the displacement of a circular cavity of the same area, as shown on Figure 5.11.

The maximum difference was 8.5% for an ellipse which had a ratio of major to minor axis of 1.3. The effect that this has upon the interpreted value of modulus is small.

Yu (2000), Yu et al. (2005), Silvestri et al. (2008), Mair and Wood (2013) and Isik et al. (2015) suggested that assuming the pressuremeter is of unlimited length is the main reason why undrained shear strengths determined from the conventional interpretation methods of pressuremeter tests are generally higher than those determined from other field or laboratory tests.

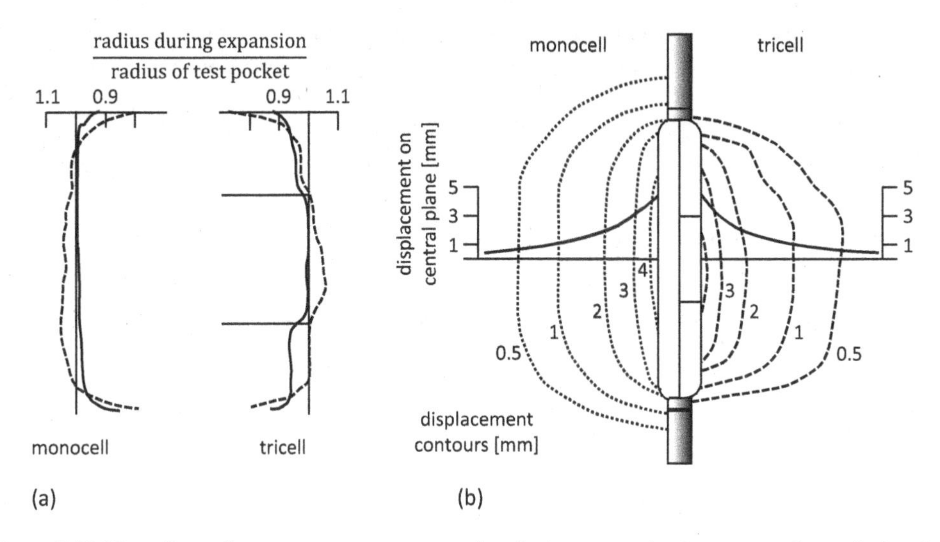

Figure 5.10 The effect of pressuremeter tests on the displacements in the surrounding soil showing (a) the shape of expanded tricell and monocell probes and (b) the deformations within the ground.

Source: After Suyama et al, 1983.

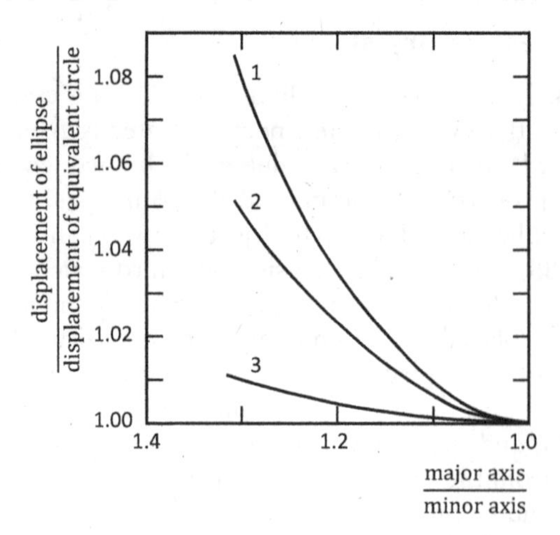

Figure 5.11 The effect of non-cylindrical expansion on a test curve: (1) average displacement at 0° and 90°; (2) average displacement at 11 equal spaced points; (3) displacement at 45°.

Source: After Fragaszy and Cheney, 1977.

Houlsby and Carter (1993) undertook a numerical study to investigate the effect of length/diameter ratio on shear strength and stiffness. They concluded that the interpretation of pressuremeter tests in clays overestimates the shear strength by up to 43% and the stiffness by less than 2% for a typical L/D of 6. This discrepancy arises because of non-cylindrical expansion caused by end effects.

Yu (1990, 1993) and Yeung and Carter (1990) undertook numerical studies to gain an insight into the effect that pressuremeter length assumption has upon derived test parameters. They found that the undrained shear strength of clays would be overestimated. Yu (1990, 1993) suggested a correction factor, F_c, should be applied to self boring pressuremeter test results in which the length to diameter ratio is 6.

$$F_c = \frac{c_u}{c_u^6} = 1 - 0.002 ln\left(\frac{G}{c_u^6}\right) \tag{5.3}$$

where c_u^6 is the undrained shear strength derived using a linear elastic perfectly plastic model. Yu et al. (2005) showed that an effective stress analysis gave a similar result but the effect decreased as the OCR value of the clay increased. Houlsby and Carter (1993) showed that there was little effect on stiffness.

Yu (1990) showed numerically and Ajalloeian and Yu (1998) showed experimentally that the response of sand is 10% to 20% stiffer than that predicted by cavity-expansion theory (for length to diameter ratio of 6). This is dependent on a factor, F_c:

$$F_c = \frac{s}{s^6} = 1.19 - 0.058 ln\left(\frac{G}{p'_o}\right) \leq 1 \tag{5.4}$$

where s^6 is the slope of ln (p′) versus ln (ε_{curr}) for a membrane length to diameter ratio of six.

5.2.9 Effects of probe type

Faugeras et al. (1983) and others suggested that the choice of probe will influence the results obtained. An average expansion is obtained with volume-displacement type probes. It is common to analyse a test with a radial-displacement type probe as if it were three tests represented by each arm and then quote the average of the results.

Fragaszy and Cheney (1977) showed that the stiffness of the ground in the middle to three-quarters of the test zone dominated the behaviour of the test, therefore results from volume-displacement type probes will differ from results from radial-displacement type probes unless a tricell probe is used. Examples of the effects of probe type and the length of test section on strength and limit pressure are shown in Figures 5.12 and 5.13.

Corrections for membrane compression and stiffness are applied to measured data to produce the corrected test curve. Monnet et al. (2016) suggest that further corrections are necessary for the non-linear expansion of the membrane and the change in the membrane thickness as the membrane expands. They suggested, based on numerical studies, that the pressure applied to the soil is 74% of the pressure within the membrane for a standard MPM test, and 66% if a slotted tube is used. They recommend that a correction coefficient of 0.8 should be applied to the pressure when testing with the G probe and, if a slotted tube is used, the correction factor is 0.6. However, it should be noted that this may invalidate the use of design curves from MPM tests, which were developed for the uncorrected pressuremeter test. The correction factor is significant if the intrinsic properties of a soil are derived from the MPM test.

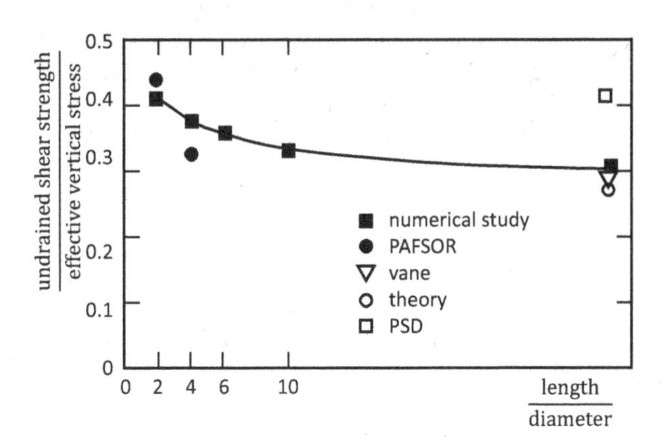

Figure 5.12 The influence of the length/diameter ratio on the derived average undrained shear strength.

Source: After Borsetto et al, 1983.

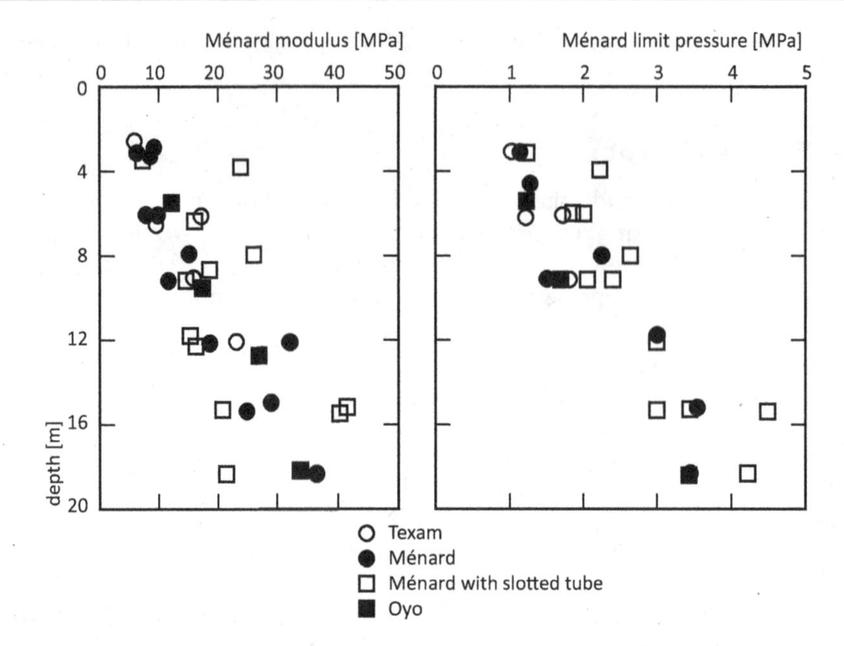

Figure 5.13 The effect of probe type on limit pressure and Ménard modulus from tests in sand and gravel.

Source: After Briaud, 1986.

5.2.10 Effects of depth of embedment

Naggar and Naggar (2012) undertook a numerical study of the effect of depth of embedment and K_o on a linear elastic perfectly plastic cohesionless soil to show that the depth of embedment and anisotropy affect the cavity pressure, which means the closed-form solution of Yu and Houlsby (1995) based on isotopic stress conditions may not be appropriate in many cases. If the overburden depth to probe diameter exceeds 20, then ignoring the depth has little effect. Given that, for most pressuremeter tests, the ratio will exceed 20, the assumption that the expansion is radial is acceptable. They propose that the cavity pressure should be multiplied by the correction factor:

$$R_d = \frac{-2.14}{\frac{C}{D}} - \frac{0.51}{\frac{a}{a_o}} + \frac{0.45}{\frac{C}{D}\frac{a}{a_o}} + \frac{2.32}{\left(\frac{C}{D}\right)^2} + \frac{0.51}{\left(\frac{a}{a_o}\right)^2} + 1.14 \qquad (5.5)$$

where C/D is the ratio of the depth to the centre of the test section to the probe diameter and a/a_o is the ratio of the current cavity diameter to the original cavity diameter.

5.3 SUMMARY

The main assumptions that apply to theories of cavity expansion are that the probe is 'wished' into place, causing no change to the properties of the ground, and the initial pressure acting on the probe is equal to the in situ stress. This is not the case.

Even an SBP, often considered to be the probe representing the nearest to an ideal pressuremeter, does cause some disturbance during installation, though the disturbance is much smaller than that created by preboring or pushing a probe into the ground.

It is also assumed that the ground contains no discontinuities and that the material is uniform over the length of the expanding section. This is unlikely to be correct in many cases. The ground will be variable and will contain discontinuities, and in some situations cracks will propagate due to the expansion of the membrane. Further, the expansion of the probe may not be cylindrical.

Different test procedures will give different results. During a test, drainage and creep can occur, which could affect results.

A review of the effects that installation, probe geometry and ground properties have upon the derived parameters suggests that their influence can be significant. Ideally these facts should be taken into account when interpreting a test but, given the level of uncertainty of the spatial variation of the composition and properties of the ground after a probe is installed, it is better to use the simple models to determine the ground properties or determine design parameters directly. This is no different from the interpretation of other in situ tests and laboratory tests.

REFERENCES

Ajalloeian, R. and Yu, H.S. (1998) Chamber studies of the effects of pressuremeter geometry on test results in sand. *Geotechnique*, *48*(5), pp. 621–636.

Anderson, W.F., Pyrah, I.C. and Haji Ali, F. (1984) Pressuremeter testing of normally consolidated clays – the effects of varying test technique, *Proc. 20th Regional Meeting of the Geol. Soc.*, Guildford, Vol. 1, pp. 21–32.

Anderson, W.F., Pyrah, I.C. and Haji Ali, F. (1987) Rate effects in pressuremeter tests in clays, *J. Geotech. Engng Div., ASCE, 113*(11), 1344–1358.

Anderson, W.F., Pyrah, I.C. and Pang, L.S. (1990) Strength and deformation parameters from pressuremeter tests in clay, *Proc. 24th Ann. Conf. Engng Group of the Geological Soc.: Field Testing in Engineering Geology*, Sunderland, pp. 23–32.

Anderson, W.F. and Pyrah, I.C. (1986) Consolidation and creep effects in the PMT in clay, *Proc. 12th Int. Conf SMFE*, Rio de Janeiro, Vol. 1, pp. 153–156.

Aubeny, C.P., Whittle, A.J. and Ladd, C.C. (2000) Effects of disturbance on undrained strengths interpreted from pressuremeter tests. *Journal of Geotechnical and Geoenvironmental Engineering*, *126*(12), pp. 1133–1144.

Benoit, J. and Clough, G.W. (1986) Self-boring pressuremeter tests in soft clay, *J. Geotech. Engng Div., ASCE, 112*(1), 60–78.

Borsetto, M., Imperato, L., Nova, R. and Peano, A. (1983) Effects of pressuremeters of finite length in soft clay, *Proc. Int. Symp. Soil and Rock Investigations by In-situ Testing*, Paris, Vol. 2, pp. 211–215.

Briaud, J.-L. (1986) Pressuremeter and deep foundation design, *Proc. 2nd Int Symp. Pressuremeter Marine Appl., Texam*, United States, ASTM STP 950, pp. 376–405.

Briaud, J.-L. (1992) *The Pressuremeter*, Balkeema, Rotterdam.

Briaud, J.-L. and Shields, D.H. (1981) Pressuremeter tests at very shallow depth, *J. Geotech. Engng Div., ASCE*, 107 (NGT8), 1023–1040.

Cao, L.F., Teh, C.I. and Chang, M.F. (2001) Undrained cavity expansion in modified Cam clay I: Theoretical analysis. *Geotechnique*, *51*(4), pp. 323–334.

Ching J., Phoon K. K, and Chen, Y.C. (2010) Reducing shear strength uncertainties in clays by multivariate correlations. *Canadian Geotechnical Journal*, *47*(1): 16–33.

Clarke, B.G. and Smith, A. (1992) Self-boring pressuremeter tests in weak rocks, *Construction and Building Materials, 6*(2), 91–96.

Clarke, B.G. and Wroth, C.P. (1984) Analysis of Dunton Green retaining wall based on results of pressuremeter tests, *Geotechnique, 34*(4), 549–561.

Dalton, J.C.P. and Hawkins, P.G. (1982) Fields of stress – some measurements of the in-situ stress in a meadow in the Cambridgeshire countryside, *Ground Engng, 15*(4), 15–22.

Faugeras, J.C., Gourves, R., Meunier, P., Nagura, M., Matsubara, L. and Sugawara, N. (1983) On the various factors affecting pressuremeter test results, *Proc. Int. Symp. Soil and Rock Investigations by In-situ Testing,* Paris, Vol. 2, pp. 275–281.

Fioravante, V., Jamiolkowski, M. and Lancellotta, R. (1994) An analysis of pressuremeter holding tests. *Géotechnique*, *44*(2), pp. 227–238.

Fragaszy, F.U. and Cheney, J.A. (1977) Influence of borehole imperfections on a pressuremeter, *J. Geotech. Engng Div., ASCE,* 103 (GT9), 1009–1013.

Fukagawa, R., Fahey, M. and Ohta, H. (1990) Effect of partial drainage on pressuremeter test in clay, *Soils and Founds, 30*(4), 134–146.

Gibson, R.E. and Anderson, W.F. (1961) In situ measurements of soil properties with the pressuremeter, *Civ. Engng Public Wks. Rev.,* 56, 615–618.

Ghionna, V.N., Jamiolkowski, M. and Lancellotta, R. (1982) Characteristics of saturated clays as obtained from SBP tests, *Proc. Int. Symp. Pressuremeter and its Marine Appl.,* Paris, pp. 165–186.

Gutierrez, M., Nygård, R., Høeg, K. and Berre, T. (2008) Normalized undrained shear strength of clay shales. *Engineering Geology*, *99*(1-2), pp. 31–39.

Houlsby, G.T. and Carter, J.P. (1993) The effects of pressuremeter geometry on the results of tests in clay. *Geotechnique*, *43*(4), pp. 567–576.

Isik, N.S., Ulusay, R. and Doyuran, V. (2015) Comparison of undrained shear strength by pressuremeter and other tests, and numerical assessment of the effect of finite probe length in pressuremeter tests. *Bulletin of Engineering Geology and the Environment*, *74*(3), pp. 685–695.

Jang, I.S., Chung, C.K., Kim, M.M. and Cho, S.M. (2003) Numerical assessment on the consolidation characteristics of clays from strain holding, self-boring pressuremeter test. *Computers and Geotechnics*, *30*(2), pp. 121–140.

Lacasse, S., D'Orazio, T.B. and Bandis, C. (1990) Interpretation of self-boring and push-in pressuremeter tests, *Proc. 3rd Int. Symp. Pressuremeters,* Oxford, pp. 273–286.

Mair, R.J. and Wood, D.M. (2013) *Pressuremeter testing: Methods and interpretation.* Elsevier.

Marcil, L., Sedran, G. and Roger, A.F. (2015) Values of pressuremeter modulus and limit pressure from stress or strain controlled PMT testing. In *Proceedings of the 7th International Symposium on Pressuremeters,* Hamammet, Tunsia.

McDowell, G.R. and Bolton, M.D. (1998). On the micromechanics of crushable aggregates. *Géotechnique*, *48*(5), pp. 667–679.

Monnet, J., Mahmutovic, D. and Boutonnier, L. (2016) Membrane correction for pressuremeter test. In 5th International Conference on Geotechnical and Geophysical Site Characterisation; Gold Coast, Queensland, Australia.

El Naggar, H. and El Naggar, M.H. (2012) Expansion of cavities embedded in cohesionless elastoplastic half-space and subjected to anisotropic stress field. *Geotechnical and Geological Engineering*, *30*(5), pp. 1183–1195.

Powell, J.J.M., Marsland, A. and Al-Khafagi, A.N. (1983) Pressuremeter testing of glacial clay tills, *Proc. Int. Symp. Soil and Rock Investigations by In-situ Testing,* Paris, Vol. 2, pp. 373–378.

Prévost, J.-H. and Hoeg, K. (1975) Analysis of pressuremeter in strain softening soil, *J. Geotech. Engng Div., ASCE,* 101 (GT8), 717–732.

Pyrah, I.C., Anderson, W.F. and Haji Ali, F. (1985) The interpretation of pressuremeter tests: time effects for fine grained soils, *Proc. 5th Int. Conf Numerical Methods in Geomechanics,* Nagoya, pp. 1629–1636.

Pyrah, I.C., Anderson, W.F. and Pang, L.S. (1988) Effects of test procedure on constant rate of strain pressuremeter tests in clay, *Proc. 6th Conf. Numerical Methods in Geomechanics,* Innsbruck, pp. 647–652.

Qiu, M., Kong, Q., Chen, P., Shi, Z. and Li, C. (2019) Effects of overconsolidation ratio on undrained shear strength of clay considering pressuremeter of limited length: cases study. *In IOP Conference Series: Earth and Environmental Science*, 304, No. 5, pp. IOP Publishing.

Rocha, M., Silveira, A.Da, Grossman, N. and Oliveira, E.De. (1966) Determination of the deformability of rock masses along boreholes, *Proc. 1st Congr. ISRM,* Lisbon, Vol. 1, pp. 697–704.

Russell, A.R. and Khalili, N. (2002) Drained cavity expansion in sands exhibiting particle crushing. *International Journal for Numerical and Analytical Methods in Geomechanics,* 26(4), pp. 323–340.

Shuttle, D.A. and Jefferies, M.G., 1995. A practical geometry correction for interpreting pressuremeter tests in clay. *Géotechnique*, *45*(3), pp. 549–553.

Silvestri, V. and Abou-Samra, G. (2008) Analysis of instrumented sharp cone and pressuremeter tests in stiff sensitive clay. *Canadian Geotechnical Journal*, *45*(7), pp. 957–972.

Suyama, K., Ohya, S., Imai, T., Matsubara, M. and Nakayama, E. (1983) Ground behaviour during pressuremeter testing, *Proc. Int. Symp. Soil and Rock Investigations by In-situ Testing,* Paris, Vol. 2, pp. 397–402.

Whittle, A.J. and Aubeny, C.P. (1993) The effects of installation disturbance on interpretation of in situ tests in clays, *Predictive Soil Mechanics, Proc. Wroth Memorial Symp.,* Oxford, pp. 585–605.

Windle, D. and Wroth, C.P. (1977) In-situ measurement of the properties of stiff clays, *Proc. 9th Int. Conf. SMFE,* Tokyo, Vol. 1, pp. 347–352.

Withers, N.J., Howie, J., Hughes, J.M.O. and Robertson, P.K. (1989) Performance and analysis of cone pressuremeter tests in sands, *Geotechnique, 39*(3), 433–454.

Wood, D.M. and Wroth, C.P. (1977) Some laboratory experiments related to the results of pressuremeter tests, *Geotechnique, 27*(2), 181–201.

Wroth, C.P. (1984) The interpretation of in situ soil tests. 24th Rankine Lecture, *Geotechnique, 34*(6), 449–489.

Yeung, S.K. and Carter, J.P. (1990) Interpretation of the pressuremeter test in clay allowing for membrane end effects and material non-homogeneity, *Proc. 3rd Int. Symp. Pressuremeters,* Oxford, pp. 199–208.

Yu, H. S. (1990) "Cavity expansion theory and its application to the analysis of pressuremeters." DPhil Thesis, Oxford University, UK.

Yu, H.S. (1993) A new procedure for obtaining design parameters from pressuremeter tests. *Trans the Institution of Engineers, Australia, Civil Engineering,* (4). *Methods Geomech.*, 22, 621–653.

Yu, H.S. (2000) *Cavity expansion methods in Geomechanics*. Springer Science & Business Media.

Yu, H.S. and Collins, I.F. (1998) Analysis of self-boring pressuremeter tests in overconsolidated clays. *Geotechnique*, *48*(5), pp. 689–693.

Yu, H.S. and Houlsby, G.T. (1995) A large strain analytical solution for cavity contraction in dilatant soils. *International Journal for Numerical and Analytical Methods in Geomechanics*, *19*(11), pp. 793–811.

Yu, H.S., Charles, M.T. and Khong, C.D. (2005) Analysis of pressuremeter geometry effects in clay using critical state models. *International Journal for Numerical and Analytical Methods in Geomechanics*, *29*(8), pp. 845–859.

Zhou, H., Liu, H., Kong, G. and Huang, X. (2014) Analytical solution of undrained cylindrical cavity expansion in saturated soil under anisotropic initial stress. *Computers and Geotechnics*, *55*, pp. 232–239.

Chapter 6

Interpretation of pressuremeter tests

6.1 INTRODUCTION

This chapter deals with the practical interpretation of pressuremeter tests since the cavity-expansion theories do not necessarily apply, as explained in Chapter 5. The engineering parameters derived from a pressuremeter test are a function of the type of pressuremeter, the method of installation, the test procedure, the ground type and the method of analysis. The various assumptions and techniques used to assist with the interpretation of tests are described below.

There are three approaches in use:

(a) Semi-empirical method in which key data points are selected using criteria that relate to an ideal pressuremeter test. The data points either define a ground property or are used to define the limits of a best-fit line which relates to a ground property.
(b) Semi-empirical method in which a constitutive model is used to develop a pressuremeter curve, and then the parameters that define the constitutive model are adjusted until the predicted curve aligns with the measured curve based on a statistical approach, such as the method of least squares.
(c) Numerical method using finite elements, for example, and a constitutive model that are defined by a range of parameters derived from various types of tests. The parameters are adjusted until the predicted curve aligns with the measured curve.

6.2 DATA QUALITY AND GROUND TYPE

6.2.1 Introduction

Inspection of a test curve in the field allows an operator to make a preliminary assessment of the quality of the data and installation technique and gain some useful information about the ground type.

6.2.2 Quality of installation

Figure 6.1 shows several tests in which the calibrated data are plotted in terms of applied stress and cavity strain.

Table 6.1 gives further details of the reasons for the shape of the curves.

DOI: 10.1201/9781003028925-6

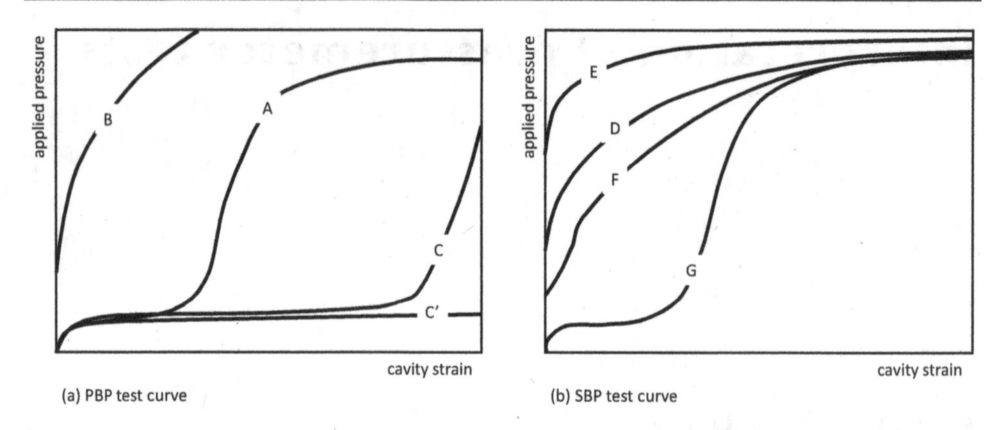

Figure 6.1 Effect of installation on the shape of a test curve: (a) PBP test (A, classical curve – ideal installation; B, probe pushed in undersize pocket; C, pocket too large – limited information; C′, pocket too large – no information); (b) SBP test (D, classical curve – ideal installation; E, probe partially pushed in; F, some over-drilling; G, wash boring – pocket too large).

Table 6.1 Interpretation of the quality of tests (see Figure 6.1)

Test	*General description*	*Probe*	*Comments*	*Effect on results*
A	Ideal	PBP	1. Typical curve, relaxation of pocket walls during installation	1. Estimate of horizontal stress 2. Initial modulus of disturbed ground 3. Ménard modulus 4. Unload/reload stiffness 5. Estimate of strength
B	Pushed in	PBP	1. Pocket smaller diameter than probe, expansion of cavity during installation	1. Initial modulus of disturbed ground 2. Unload/reload stiffness 3. Estimate of strength
C	Oversize	PBP (volume)	1. Oversize pocket 2. Pocket with hard and soft layers 3. Pocket with fissures	1. Limited results, possibly only initial modulus which is a function of disturbance and ground type 2. Test curve together with drilling log used to judge quality of rock 3. Unload/reload stiffness of disturbed ground 4. No results if pocket too large
C′	Oversize	PBP (displacement)	1. Oversize pocket if all arms are moving together 2. Pocket with fissure if some arms are moving significantly more than others	1. As C 1. Interpretation of some transducers may give full information as A

Table 6.1 Cont.

Test	*General description*	*Probe*	*Comments*	*Effect on results*
D	Ideal	SBP	1. Pocket diameter unchanged during installation	1. Horizontal stress 2. Initial modulus 3. Unload/reload stiffness 4. Strength
E	Undersize	SBP	1. Probe pushed in, expansion of cavity during installation	1. Estimate of horizontal stress 2. Unload/reload stiffness 3. Estimate of strength
F	Oversize	SBP	1. Pocket larger diameter than probe	1. Estimate of horizontal stress 2. Initial modulus of undisturbed ground 3. Unload/reload stiffness 4. Estimate of strength
G	Oversize	SBP	1. Pocket larger diameter than probe	1. Estimate of strength and stiffness
E	Ideal	FDP	1. Typical curve, expansion of cavity during installation	1. Unload/reload stiffness 2. Estimate of strength

It has been usual to plot volume change on the vertical axis and applied pressure on the horizontal axis for MPM tests but, to be consistent with all other tests, applied pressure is plotted against cavity strain. Any disturbance will alter the properties of the ground adjacent to the probe, but the magnitude of that effect will depend on the amount of disturbance and the parameters being interpreted. An operator can inspect a test curve and make a preliminary assessment of the quality of a test and adjust the drilling technique if necessary. There are a number of empirical and semi-empirical methods to account for disturbance.

6.2.3 Ground type

The shape of ideal pressuremeter curves can be used to identify the major ground type.

Figure 6.2 shows the type of curves obtained from ideal pressuremeter tests in different ground conditions. After yield, the slope of a test curve for clay (AB in Figure 6.2) is significantly less than that for a sand (CD). The shape of a test curve for rock can vary between AB and CD, depending on the stiffness and permeability of the rock.

A further qualitative indication of soil type can be obtained from the unloading curve of the test provided that there was no over-drilling during installation. The final unloading curve of a test in sand shows a distinct plateau at the ambient pore pressure (EF). The cavity may collapse in loose sands, giving an unloading curve EG. In clays,

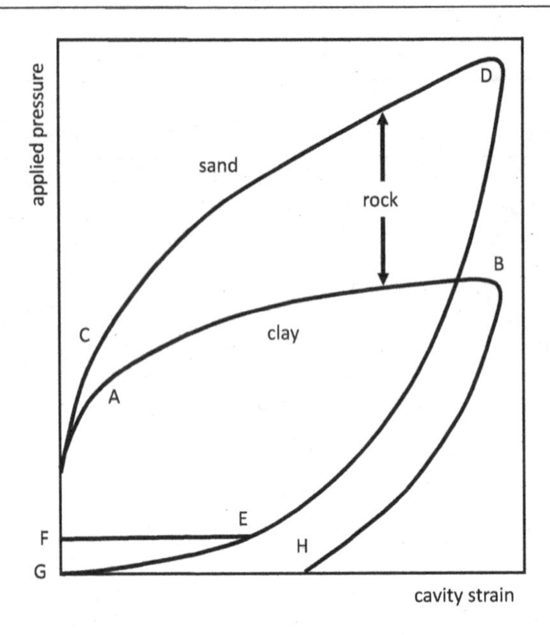

Figure 6.2 The effect of ground type on the shape of a test curve showing that the form depends on the permeability of the ground.

the membrane may return to the original diameter but, due to suction pressures in the clay, it often remains expanded as shown at H.

Baguelin (1982) proposed that the shape of an SBP test could be used to give a soil identification coefficient, β, such that

$$\beta = \left(\frac{p_{20} - p_5}{p_{20} - \sigma_h} \right) \times 100 \tag{6.1}$$

where p_{20} is the applied pressure at 20% strain and p_5 is the applied pressure at 5% strain. Figure 6.3 shows a summary of the soil classification developed by Baguelin (1982) and Becue et al. (1986) for SBP tests.

In theory it is possible to imply that the shape of an SBP test could be of the variability of the pocket diameter, there is much greater scatter in the equivalent β parameter. An alternative method is to correlate the stiffness, representing the initial portion of the test curve, and strength, representing the latter part of the curve, with ground type as shown in Table 6.2.

Baud (2005) and Baud and Gambin (2014) developed a spectral diagram (Figure 6.4) linking the soil and rock type to the limit pressure, p^*_{lm}, and pressuremeter modulus, E_m.

6.3 INTERPRETATION OF AN MPM TEST

6.3.1 The pressuremeter modulus and modified limit pressure

The Ménard pressuremeter modulus, E_m, and limit pressure, p_{lm}, are method-specific parameters taken directly from an MPM test (Figure 6.5a).

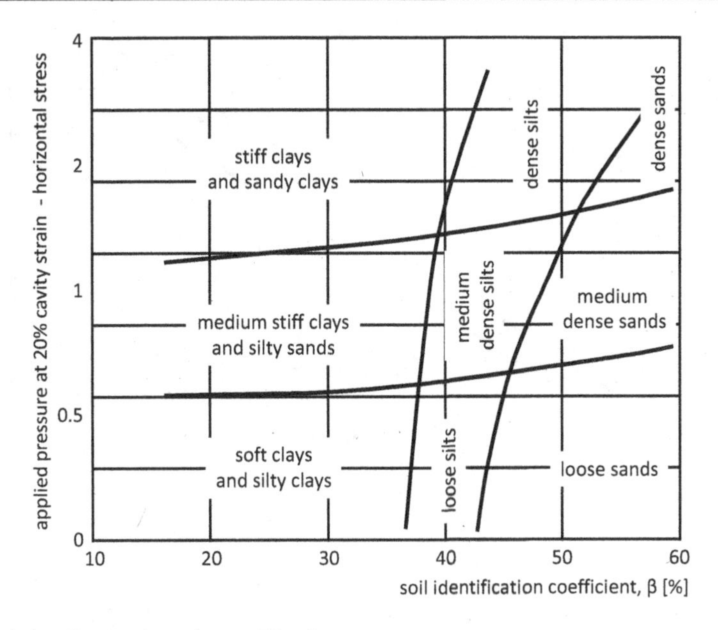

Figure 6.3 Soil classification based on self-boring pressuremeter tests.

Source: After Becue et al., 1986.

Table 6.2 Soil type from MPM tests

Ground type	E_m/p_{lm}
Very loose to loose sand	4–7
Medium dense to dense sand	7–10
Peat	8–10
Soft to firm clay	8–10
Stiff to very stiff clay	10–20
Loess	12–15
Weathered Rock (depends on degree of weathering)	8–40

The initial volume of the cavity is identified as the point A in Figure 6.5b, at which the pressure increases linearly with strain. This is not the same as the in situ horizontal stress, since the pocket wall is unloaded during preboring to either the mud pressure in a mud-filled hole or zero pressure in a dry hole. The pocket wall will unload elastically and, in some cases, even fail in extension during preboring. The pocket wall will respond elastically on reloading once the membrane is in contact with the wall. This will continue until the ground adjacent to the probe yields at point B.

Points A and B are difficult to identify, so use is made of the "creep" curve. This curve, shown in Figure 6.5c, is the rate of change of volume during any pressure increment plotted against that increment. The rate of volume change is at a minimum during the elastic response of the cavity wall. The pressure, p_o, at point A is at the start of the minimum creep, and the pressure, p_f, at point B, is at the end of the minimum creep.

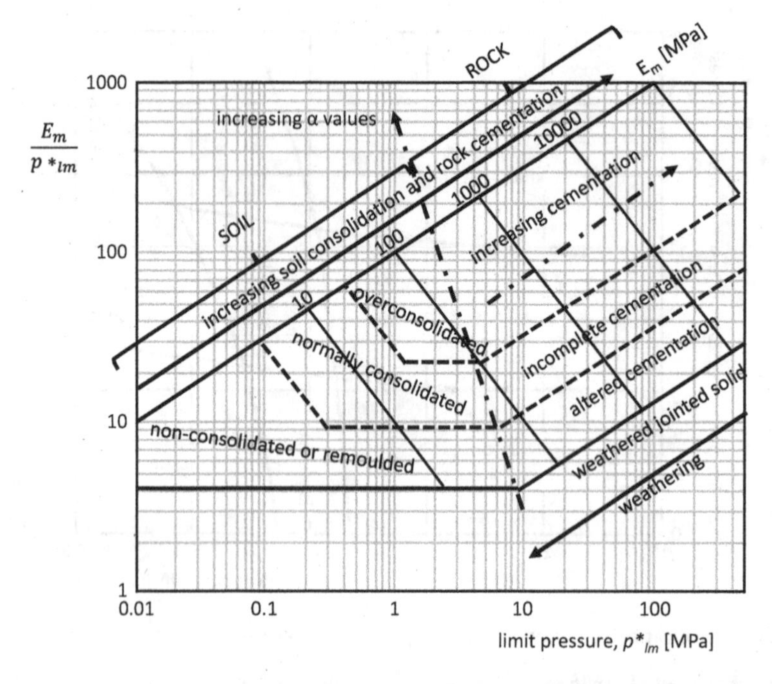

Figure 6.4 The Pressiorama diagram developed by Baud (2005) and Baud and Gambin (2014) linking the MPM parameters, Em and p*lm, with soil and rock type.

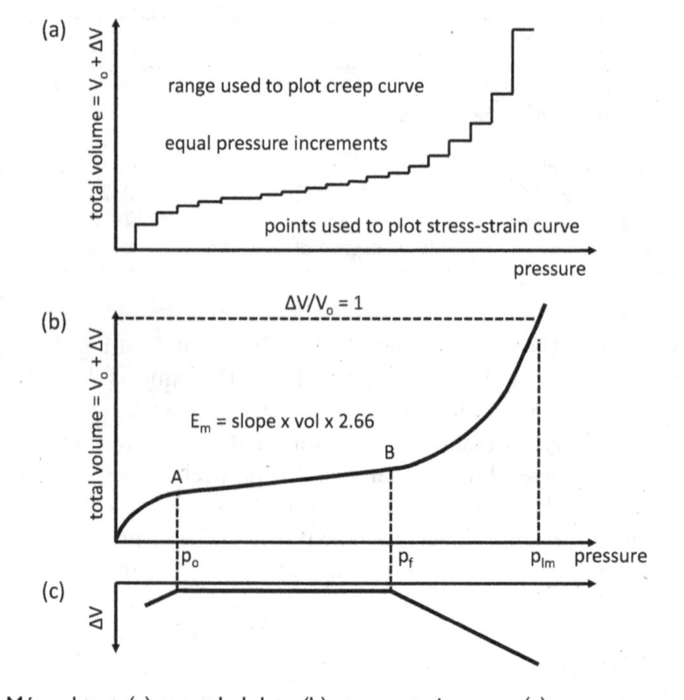

Figure 6.5 The Ménard test: (a) recorded data; (b) stress–strain curve; (c) creep curve.

The Ménard modulus or pressuremeter modulus is an initial elastic modulus taken from the slope AB, which is identified from the creep curve as the limits of the elastic response. The slope AB is a function of the shear modulus of the disturbed annulus and gives the Ménard modulus, E_m, defined as

$$E_m = 2.66\left[V_o + 0.5(V_B - V_A)\right]\left(\frac{p_A - p_B}{V_A - V_B}\right) \tag{6.2}$$

where V_o is the volume of the probe, V_A is the volume at pressure p_A and V_B is the volume at pressure p_B. The factor 2.66 is based on the assumption that Poisson's ratio for soils is 0.33. The ASTM standard quotes the same formula but permits other values of Poisson's ratio to be used.

In Russia, the modulus is derived from the same portion of the curve but is defined as

$$E = K_r a_A \left(\frac{p_A - p_B}{a_A - a_B}\right) \tag{6.3}$$

where K_r is a coefficient depending on soil type shown in Table 6.3, a_A is the radius of the cavity at A and a_B is the radius of the cavity at B.

In Sweden, where the MPM is often either pushed into the ground in a slotted tube or placed in a predrilled hole inside a slotted tube, the value of E_m is given by

$$E_m = 2.66\left(\frac{p_A - p_B}{V_A - V_B}\right)\left\{\left[V_o + 0.5(V_B - V_A)\right]\left[V_T + 0.5(V_B - V_A)\right]\right\}^{0.5} \tag{6.4}$$

where V_T is the volume of the slotted tube.

Ménard defined the limit pressure for an MPM test as the pressure required to double the initial volume of the cavity. This is equivalent to 41% cavity strain. This limit pressure, p_{lm}, is a function of the amount of disturbance during drilling and installation

Table 6.3 Values for the coefficient, K_r, for use in PBP tests

Soil type	*Properties*	*Test depth (m)*	*Test type*	K_r
Sands and gravels			Slow	1.3
Clayey and silty sands				1.35
Clays				1.42
Sands	e <0.5	0–10	Quick	2.5
	0.5 <e <0.8			2.25
	e > 0.8			2.0
Clays	LL <0.25	0–10	Quick	2
	0.25 <LL <0.5			3
	LL > 0.5			4
Clays	LL <0.25	10–20	Quick	1.75
	0.25 <LL <0.5			2.5
	LL > 0.5			3.5

Source: After GOST 20276-12 (2012).

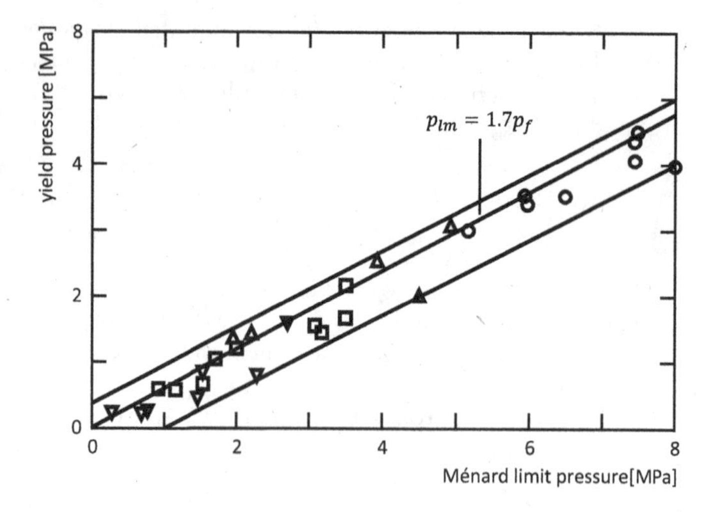

Figure 6.6 A correlation between yield pressure and limit pressure from MPM tests in rock.

Source: After Walker and Jewell, 1979.

and the properties of the ground. It is a method specific-parameter used directly in design rules based on this method of testing. The limit pressure is little affected by the use of a slotted tube (Hansbo and Pramborg, 1990) provided that corrections for the stiffness of the tube are applied.

It is often impossible to reach 41% cavity strain in tests in rock; however, Walker and Jewell (1979) showed that there is, for several rock types, a correlation between p_f and p_{lm} (Figure 6.6).

The Ménard parameters can only be obtained from PBP tests carried out in a specified manner. Baguelin et al. (1978) suggested that an equivalent limit pressure can be obtained from SBP tests. They found, from observing many tests, that the pressure needed to reach 20% cavity strain was similar to the limit pressure from an MPM test. Briaud et al. (1983) showed that the Ménard modulus, E_m, was equivalent to the unloading secant stiffness at radial strains of between 5 and 10% for tests at shallow depths in a variety of soils. A correlation between the maximum modulus from the unloading curve, E_{uo}, and E_m was found to be

$$E_{uo} = 0.27E_m^{1.45} \tag{6.5}$$

Thus, it is possible to apply the Ménard design rules using results from other tests, but this must be done with care.

6.3.2 Fitting a curve to an MPM test

Installing a pressuremeter disturbs the soil to some extent. The theoretical interpretation of a pressuremeter test depends on the correct selection of the reference datum. Baud and Gambin (2008) used a double hyperbolic model (Baud et al., 1992) to model an MPM test. The first hyperbola models the initial part of the curve, which is affected by the installation disturbance. The second hyperbola models the continuing expansion, which allows a stress strain curve to be derived from an MPM test. The volumetric strain due to the net cavity pressure, p^*, is

$$\frac{V - V_1}{V_p} = e_o + \frac{p^*}{E_o} + \frac{Rp^*_{lm}}{Kp^*_{lm} - p} \tag{6.6}$$

where V is the corrected volume reading, V_1 the volume at the start of the elastic phase, V_p the volume of the probe when in contact with the borehole wall, e_o is the cavity dimension at the start of a test $\left(= -\frac{R}{K}\right)$, E_o the slope of the initial curve, R a function of the minimum radius of curvature of the hyperbola, K the ratio between the Ménard limit pressure, p^*_{lm}, and the true limit pressure, p^*_l.

$$K = \frac{p^*_{lm}}{p^*_l} \tag{6.7}$$

$$R = K(K-1)\left(1 - \frac{p^*_{lm}}{E_o}\right) \tag{6.8}$$

Oztoprak et al. (2018) proposed a method to model pressuremeter test curve allowing for disturbance due to installation and degradation of stiffness with strain. They applied this to SBP tests in sand, which often show signs of disturbance, but it could be applied to MPM tests. They used the modulus-reduction curves developed by Oztoprak and Bolton (2013) and a strain-hardening/softening Mohr-Coulomb constitutive model. Thirteen parameters are needed to complete the analysis, but several of these are taken from laboratory tests and published information. By adjusting the parameters they were able to predict the field curve provided there were at least two unload/reload cycles.

6.4 ESTIMATING HORIZONTAL STRESS FROM A PRESSUREMETER TEST

It is assumed in the analyses presented in Chapter 4 that the reference datum (or origin) is a_o, the initial cavity radius, such that the pressure on the cavity wall at that radius is equal to the in situ total horizontal stress, σ_h. Analysing a pressuremeter test as if it were the expansion of a cavity from the in situ conditions requires that the radius of the cavity at the horizontal stress be clearly identified. The relative positions of the reference datum for the three types of probe are shown in Figure 6.7.

The SBP test, if the probe is drilled in correctly, can be approximated to an ideal test since a_o is equal to the radius of the probe and the pressure, after allowing for the membrane compression and stiffness as the membrane lifts off from the body of the probe, is the horizontal stress. This point is often not clearly identified because of the compliance and characteristics of the probe and the stiffness of the ground. Hence, its selection is subjective.

It is difficult to identify the datum on a PBP test curve. Figure 6.8 shows a typical PBP test.

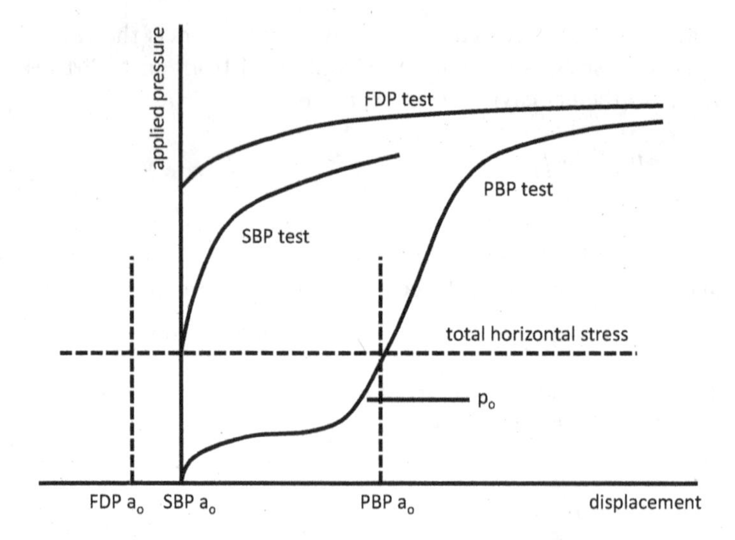

Figure 6.7 The position of the reference datum, a_o, on PBP, SBP and FDP test curves.

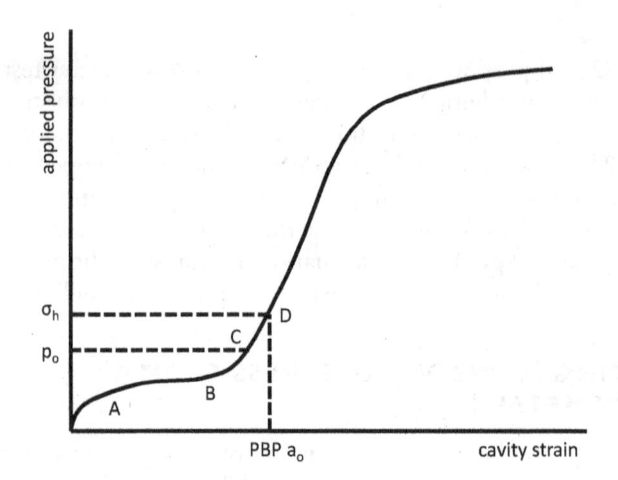

Figure 6.8 A typical PBP test.

From the start of the test the pressure is increased until, at A, it balances the membrane stiffness plus mud pressure. It then expands with a slight increase in pressure to overcome the membrane stiffness until B where, to continue expanding, the pressure has to be increased to compress the softened layer formed during drilling. The point C, where the slope of the curve becomes linear, is taken to be the point at which the membrane reaches the pocket wall. p_o is the pressure at this point, but it is not equal to σ_h since the pocket wall was unloaded during drilling. There is some point D ($p_d > p_o$) at which the pressure is equal to σ_h, and that is the reference datum. If the elastic behaviour of the ground is the same on unloading as on loading then there will be no point of inflexion at p_d. It is very difficult to determine the horizontal stress from PBP tests.

The horizontal stress cannot be identified directly from an FDP test curve. There will be an increase in stress acting on the membrane during installation of the FDP since creating its test pocket is similar to the expansion of a cavity. Hence, the applied pressure during loading always exceeds the horizontal stress. Further, since there is no identifiable datum point that can be used in the interpretation of a test, these tests are analysed using semi-empirical correlations with the unloading curve.

Several methods have been developed to assist in the selection of horizontal stress from pressuremeter test curves since, at best, the selection is subjective. These include

(a) the lift-off method;
(b) methods based on shear strength;
(c) methods based on test procedure;
(d) fitting functions to the test curve;
(e) empirical correlations with other data.

Many of these methods have been developed specifically for either PBPs or FDPs but they can be used to check the determined value of σ_h from SBP tests if there is any doubt about the quality of drilling.

6.4.1 Lift-off method

Self-boring probes were originally developed to enable direct measurements of total horizontal stress to be made. The French probe, the PAF, is a pressuremeter with the membrane supported on water, hence it is effectively a pressure transducer. The first UK probe was a total stress cell (Wroth and Hughes, 1973). Direct readings were made by observing the pressure on the side of the rigid probe using strain-gauged total pressure transducers. The lift-off method applies to all radial displacement type self-boring probes and volume-displacement-type self-boring probes in which the membrane is supported by the body of the probe at the start of the test.

The pressure at which the membrane lifts off from the body of the probe is theoretically equal to the total horizontal stress. The difference between a self-boring load cell and a radial displacement type SBP is that, with the former probe, the pressure is measured directly whereas, with the latter probe, the pressure is indirectly assessed from a change in displacement. Thus, at best, the lift pressure is only an indication of total horizontal stress.

The lift-off method is based on the premise that the probe is drilled in without changing the magnitude of the horizontal stress. Thus, prior to lift off there should be no movement of the inside of the membrane. After lift off the membrane should expand.

Figure 6.9 shows a typical self-boring pressuremeter test in clay in which the complete field curve for one of the arms has been plotted in the inset.

Identifying the lift-off point is very nearly impossible since the stiffness of the soil and the compliance of the probe appear to be the same at the scale used. It is necessary to inspect only the initial portion of the test, perhaps from 0 to 0.5% cavity strain, as shown in Figure 6.9. The membrane appears to move at about 150kPa, which is equal to the membrane stiffness. The initial slope of the curve above that pressure is a

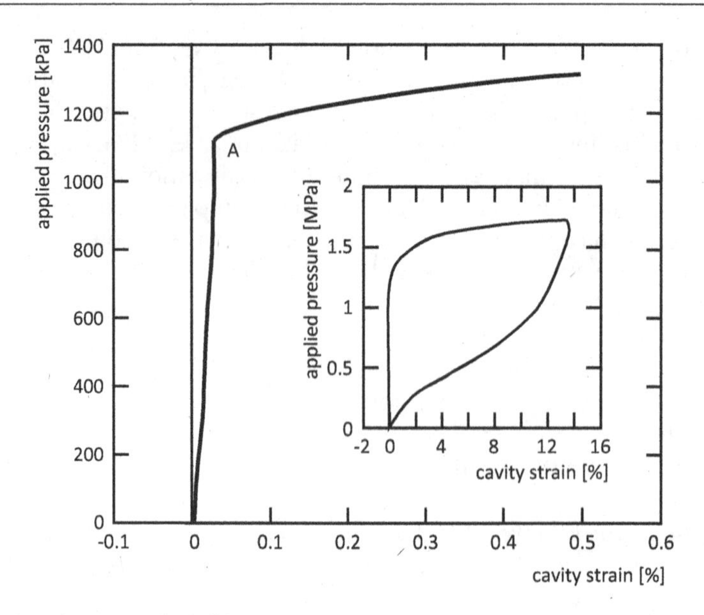

Figure 6.9 The initial portion of a self-boring pressuremeter test showing the displacement prior to lift off at A. The inset is the full test curve.

function of the system compliance though it can include the effects of drilling and the inclination of the probe. The lift-off point is the point at which the curve deviates from the initial slope, point A. This point can be clearly identified in normally consolidated and lightly overconsolidated soils where the stiffness of the soil is less than the compliance of the probe. As the overconsolidation ratio increases, it becomes more difficult to identify point A since the stiffness of the soil increases. Further, as σ_h increases, the initial compliance of the probe, represented by OA and referred to as the signature of the transducer, may no longer be linear which, together with installation disturbance, gives possible curves as shown in Figure 6.10.

The signature of the transducer can be affected by its design which could result in similar curves to those shown in the figure.

Improvements in the design of displacement transducers, which include rollers (Fahey and Jewell, 1990) and buttons (Clarke and Allan, 1990), have reduced some of these errors. A comparison, Figure 6.11, between the initial portion of the pressuremeter curve for tests in stiff clay using the SBP with and without 'Fahey' rollers shows a significant improvement in the initial shape of the test curve though the curves still do not conform to the ideal shape.

There will be some disturbance due to either probe deviation, under-drilling or over-drilling. The effects of under-drilling and over-drilling may not be noticed since the clay will relax onto the probe resulting in an apparently ideal curve. It is for this reason that values of horizontal stress taken from pressuremeter tests should be compared with other assessments to produce a degree of confidence in the results.

Three field curves are obtained from the CSBP test since measurements of displacement are taken from three displacement transducers. Figure 6.12 shows the data from three arms plotted separately.

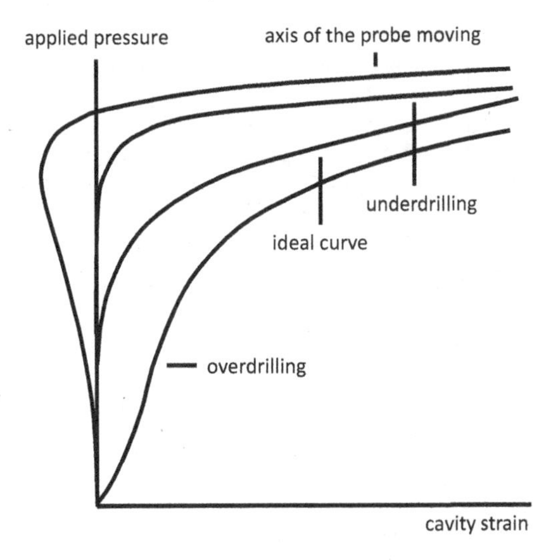

Figure 6.10 Possible shapes of the initial portion of an SBP test due to drilling techniques and system compliance.

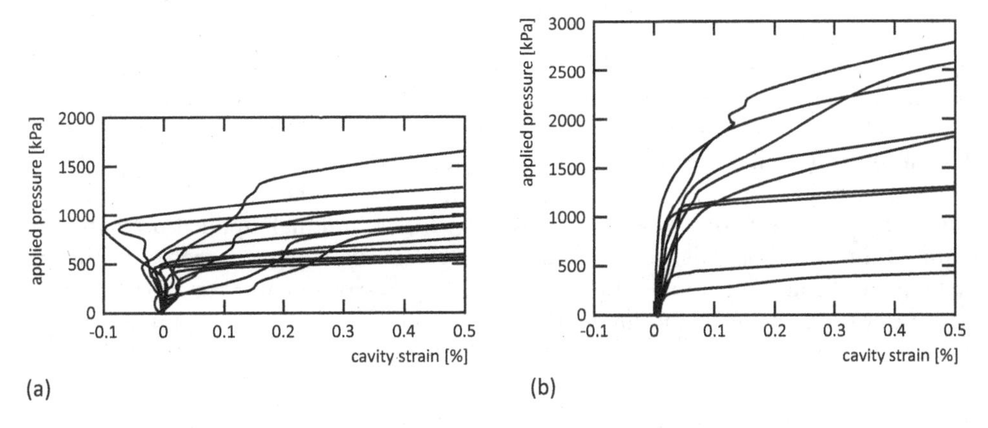

Figure 6.11 A comparison of the initial portion of CSBP curves in stiff clay using spring-loaded feeler arms (a) without and (b) with "Fahey" rollers showing that the rollers reduce the fluctuation in the output.

Source: after Clarke, 1993

They show lift off at different pressures. This may be interpreted to show that the horizontal stress varies but, before making this assumption, it is necessary to consider the method of insertion and the mechanics of the probe.

The probe may deviate, as shown in Figure 5.3. Hence, there may be an increase in stress on one side and a reduction in stress on the other. Furthermore, the arms may be indicating a component of vertical stress. The calibration for membrane stiffness shows that the lift-off pressure in air for each arm is different. Tests in 50 m deep boreholes full of water have shown that the lift-off pressures are still different even

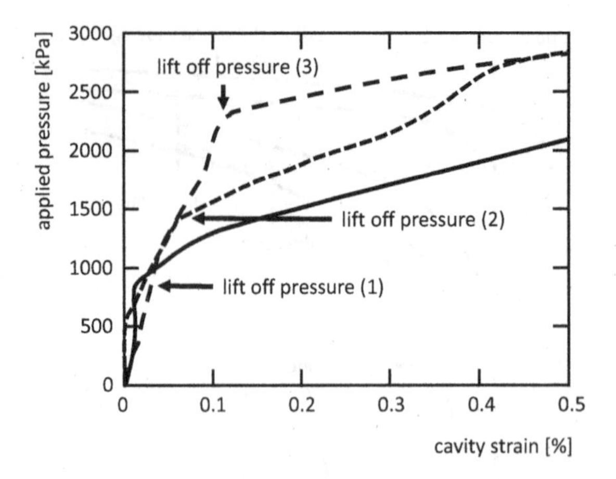

Figure 6.12 The variation in initial ground response around a self-boring pressuremeter showing the lift off pressure can vary between the three arms.

though the external pressure is hydrostatic. Indeed, for one probe, the difference in pressure between the average of the three arms and one of those arms was found to be 14% of the average. This is a consequence of the variation in membrane stiffness.

It has been noticed that, in some tests, especially those at high pressures, one or more of the arms may be moving inwards and the others moving outwards at a greater rate than the compliance of the system. It is suggested that the body of the probe is moving in the pocket or that the arms are not seated correctly. If this error is consistently found, then the probe may have to be thoroughly cleaned and the transducers possibly replaced.

The possibilities of an inclined probe and movement of the body of the probe are usually found in stiff soils where large insertion forces and high applied pressures are required. In those soils, the soil stiffness and the compliance of the probe are similar, making it difficult to identify lift off. The initial response or signature of a displacement transducer is often similar for different tests in the same borehole. Thus, if it is possible to identify, with confidence, the total horizontal stress at one depth and relate that to the signature of that arm, then this signature can be used to determine the horizontal stress at other test depths. Unless there is a significant difference between the arms or the horizontal stresses are expected to differ – for example, in the clay core of an earth dam – the average of the lift-off pressures should be quoted as the total horizontal stress.

It is unlikely that the horizontal stress can ever be measured directly in sands. Drilling the probe into the ground inevitably disturbs the sand. Clarke (1993) suggested that, even with a SBP, the disturbance during installation could exceed 0.5% cavity strain. Newman (1991) suggested that good quality tests in sands are those for which the cavity strain at the true horizontal stress is less than 0.2%.

Soils exhibit hysteretic behaviour, therefore unloading stiffness is likely to be different from the reloading stiffness. This implies that there will be a change in the shape of the curve at about the horizontal stress when the membrane is expanded if the lift-off pressure is less than the total horizontal stress. This can be used to estimate the horizontal stress in sands where it is known that unloading is likely to occur during drilling. This is similar to the techniques used to evaluate σ_h directly from PBP tests.

It is impossible to install a pressuremeter in rock without creating a pocket first. The pocket diameter in excess of the probe diameter varies from about 0.25% cavity strain for self-boring probes designed to drill in rock to 10% for a PBP. It is impossible to push a FDP into rock. The wall of the pocket in rock will not relax onto the probe, therefore it is only possible to deduce the horizontal stress by inspection of the curve – that is, to look for changes in slope. This is more likely to be successful with the SBP because the probe is installed at the time of drilling. There is no flow of drilling fluid past the pocket wall and there is little time for any softening to take place.

Often an SBP probe is installed in an apparently correct manner and the shape of the curve confirms this. However, the lift-off pressure may not be equal to the in situ horizontal stress because there is some under- or over-drilling, which produces either an expansion or a contraction of the pocket during drilling. In both cases the shape of the curve will appear to be satisfactory, since the amount of disturbance is still small. It has no effect on a modulus taken from an unload/reload cycle provided that the cycle is well beyond the zone of disturbance. Further, for a perfectly plastic soil, the amount of disturbance has little effect on the strength. It is for this reason that the first cycle is commonly specified at 1% cavity strain for SBP tests, and strengths are taken from the latter part of the curve.

The lift-off pressure is likely to be an overestimate of the horizontal stress if the probe has been pushed in (or under-drilled) and an underestimate if the probe has been over-drilled. An extreme example of over-drilling is shown by a PBP test curve. The start of the linear portion of the curve is taken as the point at which the membrane starts to pressurise the pocket wall. It is a lower bound value of horizontal stress, since the pocket wall has been unloaded during preboring or drilling.

It is for the reasons given above that several methods (Table 6.4) have been proposed to assist in the assessment of horizontal stress.

Many of these methods are based on fitting a soil model to the pressuremeter curve.

Table 6.4 Techniques to estimate horizontal stress from pressuremeter tests

Technique	*Reference*	*Soil type*	*Probe*	*Advantages*	*Disadvantages*
Lift off	Wroth (1982)	All soils	SBP	1. Direct measurement	1. Difficult to correct for system compliance and drilling
Strength	Marsland and Randolph (1977)	Stiff clays	SBP PBP		1. Iterative procedure 2. Difficult to assess yield point
	Hawkins et al. (1990)	Stiff clays	SBP	1. Mutual consistency between test and shear stress curves	1. Iterative procedure 2. Difficult to assess yield point
	Newman (1991)	Sands	SBP		1. Only applies to quality tests 2. Difficult to assess yield point 3. Yield probably occurs during drilling

(continued)

Table 6.4 Cont.

Technique	*Reference*	*Soil type*	*Probe*	*Advantages*	*Disadvantages*
	Denby and Hughes (1980)	Clays	SBP		1. Only perfectly plastic soils 2. Curve little affected by small changes in strain yet small changes in strain produce large changes in horizontal stress
	Fahey and Randolph (1984)	Sands	SBP		1. Only perfectly plastic soils 2. Curve little affected by small changes in strain yet small changes in strain produce large changes in horizontal stress
Test procedure	Lacasse and Lunne (1982)	All soils	SBP	1. Simple	1. Affected by stress rate and soil stiffness
	Wroth (1982)	Soft clays	SBP	1. Simple	1. Only applies to normally consolidated clays 2. Pore pressure readings have to be reliable
	Jefferies et al. (1987)	All clays	SBP		1. Pore pressure readings have to be reliable 2. Perfectly plastic soil
Curve fitting	Arnold (1981)	Clays	SBP PBP		1. Arbitrarily chosen parameters
	Denby and Clough (1980)	Clays	SBP		1. Only perfectly plastic soils 2. Curve little affected by small changes in strain yet small changes in strain produce large changes in horizontal stress
	Jefferies (1988)	Clays	SBP	1. Complete loading and unloading curve	1. Perfectly plastic soils
	Ferreira and Robertson (1992)	Clays	SBP	1. Complete loading and unloading curve	1. Perfectly plastic soils
	Withers et al. (1989)	Sands	SBP		1. Insensitive to parameters chosen 2. Not a unique set of parameters
Correlations				1. Consistency between derived parameters	

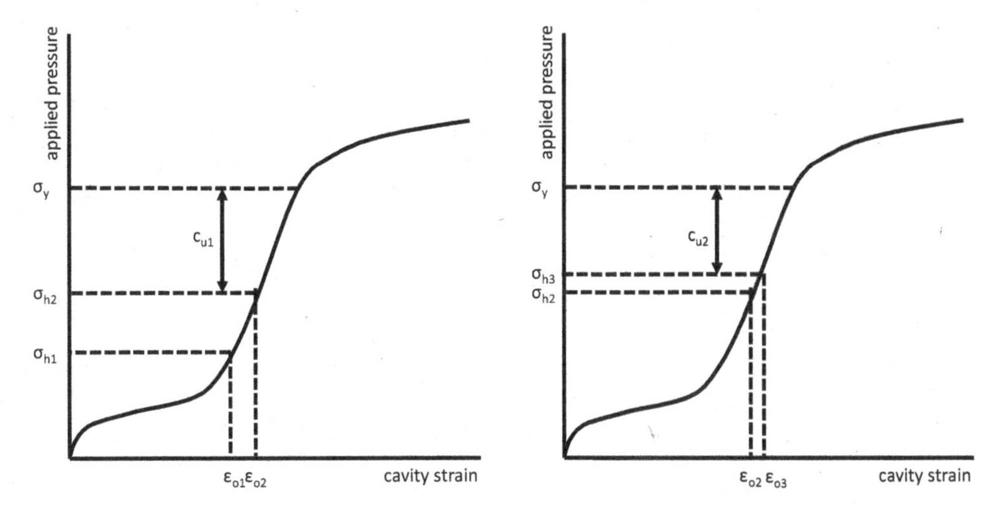

Figure 6.13 The assessment of horizontal stress from a PBP test in clay which uses the yield point and undrained shear strength in an iterative manner to determine the horizontal stress.

Source: After Marsland and Randolph, 1977.

6.4.2 Methods based on shear strength

There are two methods based on the shear strength of a clay. The Marsland and Randolph (1977) approach, based on peak strength and yield stress, was developed for PBP tests in stiff clays. It can be used for SBP tests. The other approach, Hawkins et al. (1990), also developed for stiff clays, is known as the modified Marsland and Randolph approach. In this method the test curve and derived shear stress–strain curve are forced to be mutually consistent during the initial portion of the test.

The yield point for tests in stiff clays is equal to the sum of σ_h and c_u and is identified as a significant change in slope on the test curve. The peak shear strength is determined from equation (6.9):

$$\tau = \frac{dp}{d[\ln \Delta V / V]} \tag{6.9}$$

The value of peak strength will depend on the chosen reference datum which depends on the selected value of total horizontal stress. An iterative approach was proposed by Marsland and Randolph (1977). This method applies to clays which have a significant linear elastic behaviour and show a marked yield point. As an example, consider the test shown in Figure 6.13.

1. An initial estimate of horizontal stress is made, point A, perhaps from an estimate of the coefficient of earth pressure at rest.
2. The cavity diameter, ε_{o1}, at that stress is found from the test curve.
3. The new strains, ε_{corr}, are calculated to correct for the new cavity diameter using the formula

$$\varepsilon_{corr} = \frac{\varepsilon_c - \varepsilon_o}{1 + \varepsilon_o} \tag{6.10}$$

where ε_o is the measured cavity strain (reference datum) at the selected value of σ_h and ε_c is the measured cavity strain.

4. The undrained strength, c_{ul}, of the clay is taken from the peak slope of the pressure against In (volumetric strain) plot.
5. The limit of the elastic zone, p_y, or onset of yield, is identified from the test curve. If the clay is truly linear elastic then there will be a significant change in the slope of the curve at yield.
6. A new value of horizontal stress is found using the equation

$$\sigma_{h2} = p_y - c_{u1} \tag{6.11}$$

7. If this coincides with the first value chosen then the horizontal stress has been correctly identified.
8. If it is not, then the steps from item 2 are repeated using ε_{o2} to determine c_{u2}.

Hawkins et al. (1990) proposed the test curve and shear strength plot be mutually consistent prior to yield. They recommend that, in practice, this is achieved by identifying and using the yield point. An initial estimate of horizontal stress is made, perhaps from knowledge of the stress history of the soil, and this is used to determine the cavity strain, ε_{o1}, at the in situ stress. The strains are corrected using equation (6.6). The reference datum, ε_{o1}, is adjusted until equation (6.8) is satisfied, as shown in Figure 6.14:

$$p - \sigma_h = \frac{dp}{d[\ln \Delta V / V]} \tag{6.12}$$

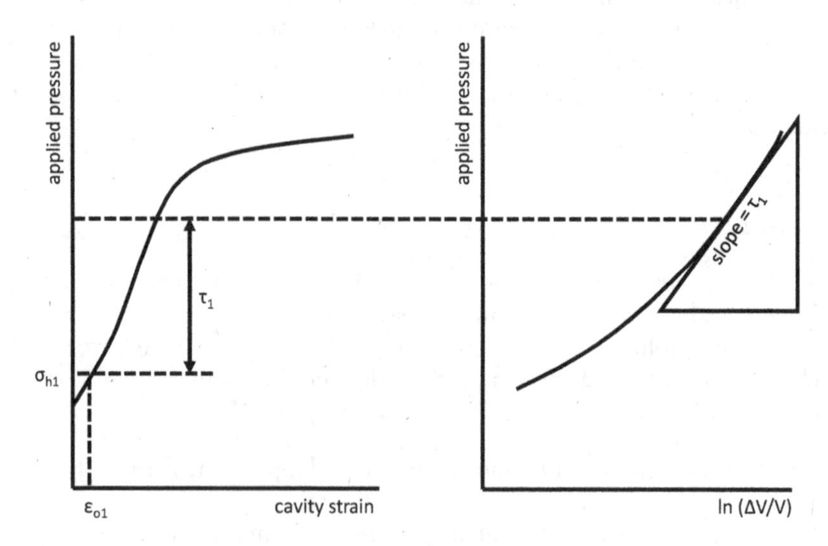

Figure 6.14 The assessment of horizontal stress from an SBP test in clay which uses the shear stress at any point post yield in an iterative manner to determine the horizontal stress.

Source: After Hawkins et al., 1990.

The two methods will give the same answer if the clay is truly linear elastic perfectly plastic.

Baguelin et al. (1978) and others have shown that the interpreted value of peak strength is significantly affected by any installation disturbance. Under-drilling causes a probe to be partly pushed in, which results in a cavity expansion bringing the soil adjacent to the probe to post-yield conditions. Over-drilling results in a reduction in stress, perhaps causing the soil to fail in extension. In either case an analysis could result in an apparent false peak strength unless the correct reference datum is chosen. It is possible to satisfy the criteria for both methods, yet choose the wrong value of in situ stress.

If an SBP is drilled in correctly then, for a linear elastic soil, there will be an easily identified change in slope at the yield stress. Stiff clays are not linear elastic, therefore there is a gradual change in the slope of the test curve and, in that case, it is difficult to identify the yield stress. This is a common problem with these methods.

An apparent yield stress can be identified from PBP test curves and test curves from SBPs, which are over-drilled. In both cases the pocket wall is unloaded during drilling and reloaded during the test. The effect of unloading produces, in practice, an annulus of disturbed ground which exhibits different properties from those in situ. Hence, on reloading there is a definite change in the slope at the point at which yielding of the undisturbed soil occurs, but this will be greater than the true yield stress for undisturbed soil at the pocket wall. This is similar to the change in slope at the end of an unload/reload cycle.

Newman (1991) proposed a method for SBP tests in sands not too dissimilar to the Marsland and Randolph approach. This method is based on the assumption (see equation (4.62)) that the sand is linear elastic perfectly plastic and the sand begins to yield at the pocket wall when

$$p' - \sigma_h' = (1 + \sin \phi') \tag{6.13}$$

The angle of friction is determined from the slope of the graph of effective applied pressure against current cavity strain plotted to logarithmic scales. If the sand is linear elastic perfectly plastic then the start of the linear portion of the curve coincides with the yield stress. It is more likely, for natural dense sands, that the slope of the logarithmic plot will be S-shaped as a result of post-rupture behaviour. Thus, the steepest slope will coincide with the peak failure of the sand and not the yield stress.

As an example, consider the test shown in Figure 6.15.

1. An initial estimate of effective horizontal stress, σ_{h1}' is made, point A, perhaps from an estimate of the coefficient of earth pressure at rest.
2. The cavity strain, ε_{01}, at that stress is found from the pressuremeter curve.
3. The strains are adjusted to correct for the new cavity radius using equation (6.6).
4. The average angle of friction is determined using the method proposed by Hughes et al. (1977) (see section 6.7).
5. The limit of the elastic zone, p_y', or onset of yield is the point at which the log plot deviates from the theoretical curve.
6. A new value of horizontal stress is found using equation (6.13).

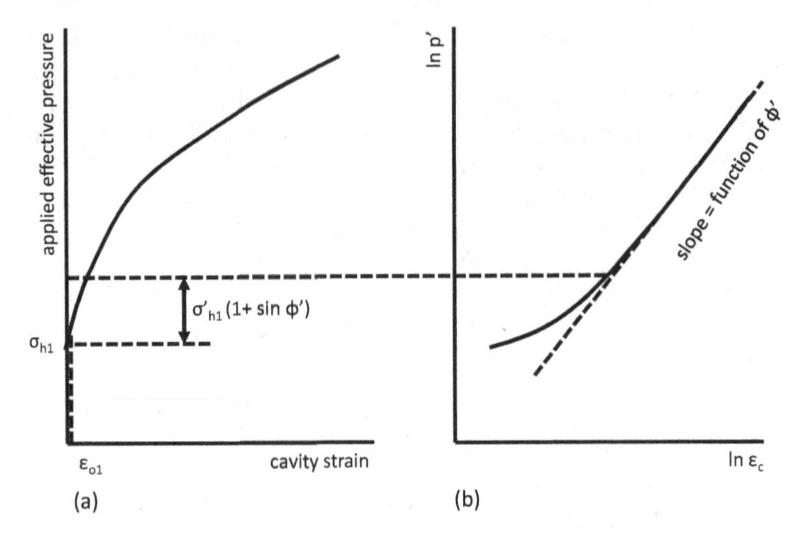

Figure 6.15 Estimating the horizontal stress from a test in sand. The pressuremeter test curve (a) is used to select the reference datum and the angle of friction (b) from the slope of the curve post yield.

7. If this coincides with the first value chosen, then the horizontal stress has been correctly identified.
8. If it does not coincide, then the steps from item 2 are repeated.

It is inevitable during drilling that sand particles in line with the edge of the probe will be moved, either into the probe or pushed out. This movement corresponds to 0.15% cavity strain if an 80 mm probe is drilled into fine sand. If the sand is loose and normally consolidated then, for a test at 5 m, yield will occur about 0.13% cavity strain; that is, the sand is brought to failure during drilling. In practice, the stress on the probe is not the total horizontal stress. This method can be used to estimate a lower bound value of horizontal stress if the peak angle of friction is used.

Denby (1978) and Fahey and Randolph (1984) suggested a method based on the assumption that soil is perfectly plastic. As an example, consider the tests shown in Figure 6.16.

1. Several reference data are chosen.
2. The strains are adjusted to correct for each reference datum using equation (6.10).
3. A plot of applied pressure against ln (volumetric strain) for clays or applied pressure against current cavity strain, both to a log scale, for sands is produced showing several curves representing the various chosen reference data (see Figure 6.16).
4. The reference datum that gives the longest straight line is chosen to represent the initial diameter of the pocket wall.
5. The pressure at that cavity diameter taken from the pressuremeter curve is the total horizontal stress.

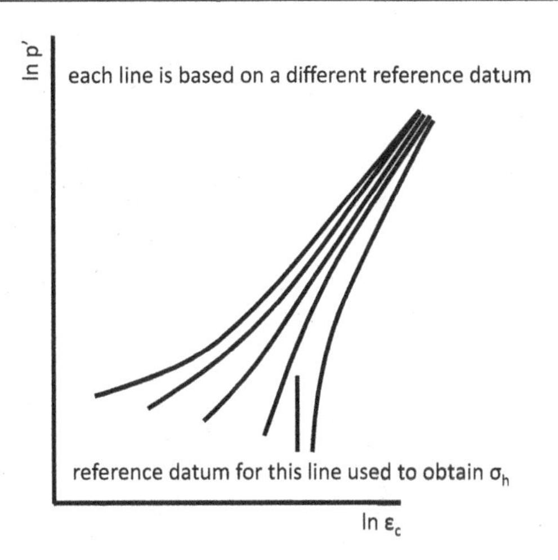

Figure 6.16 The selection of horizontal stress by adjusting the reference datum.

This method produces realistic values of horizontal stress if a soil is perfectly plastic. Most soils are strain-hardening or strain-softening, therefore, the plots are non-linear. Further, for stiff soils a small change in reference datum produces very little change in strength but has a significant effect on the horizontal stress.

6.4.3 Methods based on test procedure

Various authors have suggested that the SBP strain-controlled test procedure can be used to determine horizontal stress from SBP tests. Lacasse and Lunne (1981) suggest that a plot of strain against time will show a change in strain rate at the horizontal stress if a strain-control unit with an overriding stress-rate control is used to control a test. Initially the stress is increased in equal increments, as shown in Figure 6.17.

The strain rate is initially constant and a function of the system compliance. At the horizontal stress the strain rate begins to change and continues to change until the test becomes a strain-controlled test, possibly beyond yield, though this will depend on the strain- and stress-rates set.

Law and Eden (1980) proposed an alternative version of this method in which the applied pressure is plotted against the cavity strain to a logarithmic scale. A curve is produced which should amplify the change in slope at lift off. The strain rate varies between lift off and yield. If the soil stiffness is the same order of magnitude as that of the probe there will be no clearly defined point of inflexion at lift off. In that case these techniques are no more helpful than the direct lift-off method. They were developed for tests in soft clays where p_y coincides with σ_h.

Wroth (1982) suggested that a change in pore pressure in tests in clays is an indication of lift off. This method can only be used with probes fitted with pore-pressure transducers. Initially, the change in cavity pressure must equal the change in effective pressure. Once the soil begins to yield, pore pressures are generated. This is the horizontal stress in normally consolidated clays.

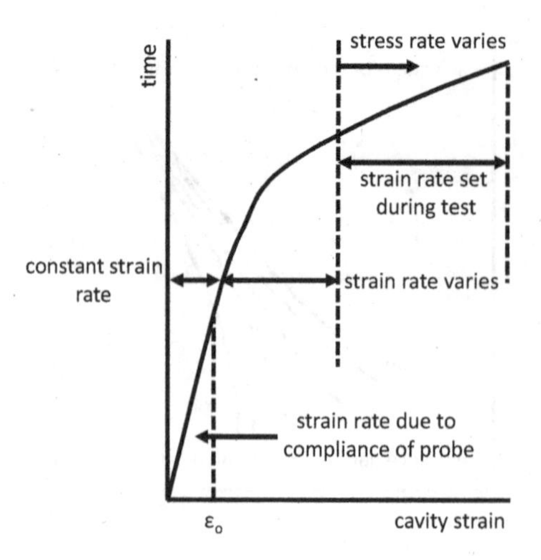

Figure 6.17 The variation in strain rate during a strain-controlled test with an overriding stress-rate control and the selection of the reference datum.

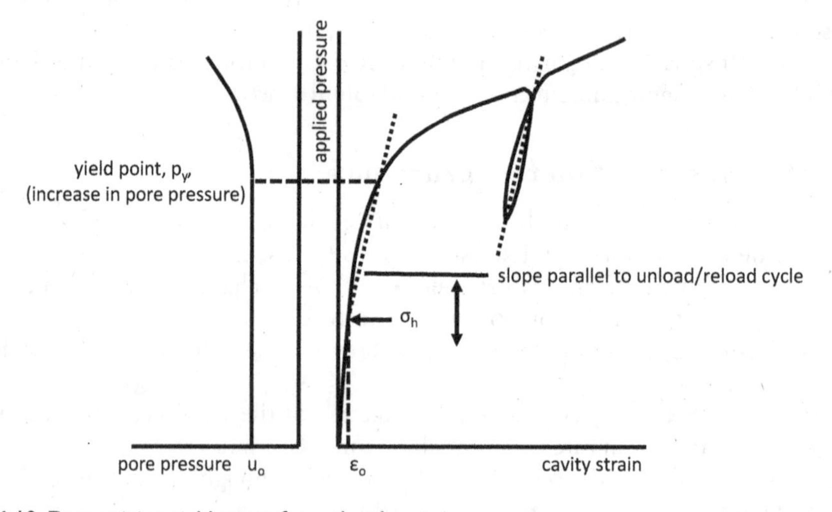

Figure 6.18 Determining yield stress from the change in pore pressure.

Jefferies et al. (1987) extended this method to cover all clays by using the critical state framework to show that a change in pore pressure occurs when the clay adjacent to the membrane yields. The horizontal stress is then estimated by constructing a line from p_y parallel to the slope of an unload/reload cycle. The intersection with the pressuremeter curve is σ_h, as shown in Figure 6.18.

6.4.4 Curve-fitting methods

Most interpretations of pressuremeter tests involve fitting curves to the test data. If a single continuous function is used to describe those data (for example Arnold, 1981) then the conventional least squares method can be used to fit the curve. If a multi-function is used (for example, those that separate elastic from plastic behaviour such as that proposed by Gibson and Anderson, 1961), it is usual to separate the functions and fit each function to the relevant part of the curve or use some form of graphical procedure. Numerical methods include optimisation techniques with a defined constitutive model and artificial neural networks. The most appropriate set of variables that define the constitutive model are found by adjusting the variables until the predicted curve and measured curve are similar. More-advanced models require additional information, often from triaxial tests. Note that more than one combination of variables will produce a fit to the data. This means a subjective assessment may have to be made to decide the most appropriate set.

A single continuous function can be applied to pressuremeter data. The ordinate given by that function at the reference datum is the horizontal stress. Arnold (1981) proposed that a hyperbolic relationship, used to fit a curve to the pressuremeter test data, takes the form

$$p = Q + \left(\frac{\varepsilon_c}{a + b\varepsilon_c} \right) \tag{6.14}$$

where Q is the intersection of the curve with the vertical axis and a and b are constants. Q is the total horizontal stress in an ideal test. In practice, Q will not be equal to σ_h, since there is some disturbance during installation. The general form of equation (6.14) is

$$p = Q + \left(\frac{\varepsilon_c - \varepsilon_{01}}{a + b(\varepsilon_c - \varepsilon_{01})} \right) \tag{6.15}$$

where ε_{01} is the offset on the strain axis to allow for disturbance.

Figure 6.19 shows the ideal case and the more realistic case.

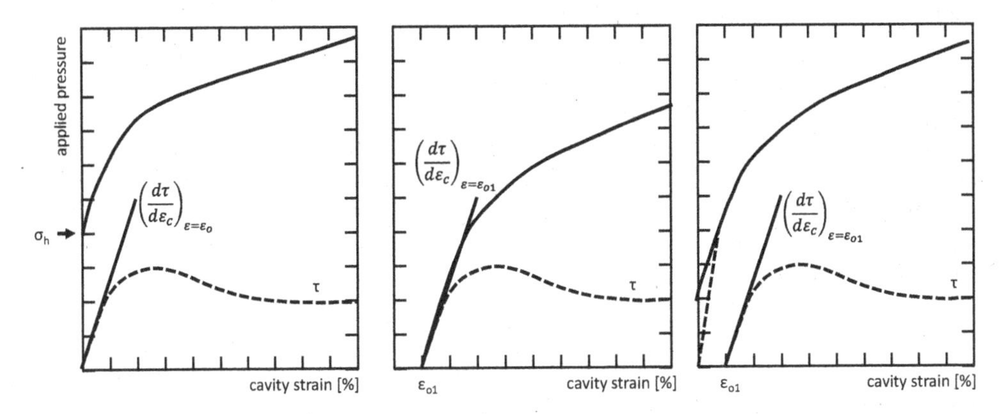

Figure 6.19 The definition of disturbance used by Arnold (1981): (a) ideal test, p at Q = σh; (b) total release of in situ stress; (c) partial release of in situ stress.

The unknown parameters, Q, a, b and ε_{o1} are adjusted until a reasonable fit is obtained. Arnold suggests that this is done in two stages to overcome fluctuations in data because of the recording system. The first is to assume that the reference datum is zero and calculate modified parameters Q', a' and b' using equation (6.9). The second stage is based on equation (6.10) using the modified parameter b'. The maximum shear stress (equivalent to the shear stress at the maximum strain for a hyperbolic model) is given by $(1/4b)$ and the initial modulus by $(1/a)$.

The main drawback of this method is assigning values to the constants – the values are not necessarily related to the fundamental soil parameters and therefore may be chosen arbitrarily. The hyperbolic model proposed by Denby and Clough (1980) for clays can be used to estimate horizontal stress in the same way as that proposed by Arnold (1981). An expression, equation (6.11), relating the total horizontal stress to the maximum shear stress is derived from the pressuremeter curve using the hyperbolic model. A plot of σ_h with strain can be obtained by altering the reference datum from which the most likely value is selected:

$$\sigma_h = p - \left(\frac{c_u}{R_f}\right) \ln\left[1 + \left[\frac{R_f}{c_u}\right] 2G_i \varepsilon_c\right] \tag{6.16}$$

Huang et al. (1986) proposed that a simplex algorithm could be used for multi-functions as well as single functions. This has the advantage that the optimisation routine is only dependent on the stress–strain relationship assumed. Full details of the method are given by Caceci and Cacheris (1984) and Huang et al. (1986), and the flow chart is shown in Figure 6.20.

Huang et al. suggested that the deviator stress should be expressed in terms of the cavity strain and the parameters optimised to satisfy those functions. Table 6.5 gives examples of functions and the parameters to be optimised.

Jefferies (1988) proposed that the complete loading and unloading curve could be used to obtain a value of horizontal stress. Values of G, c_u and σ_h are selected and a test curve produced using equations (4.27) and (4.30). The shear modulus controls the curvature, the strength controls the size of the complete cycle and the horizontal stress controls the average stress level, as shown in Figure 6.21.

Thus it is possible to find, very rapidly, suitable values that give a reasonable fit to the data. G is not taken from an unload/reload cycle but is a consequence of curve fitting. It is a secant modulus.

The membrane continues to expand as the pressure is reduced during unloading. This is a consequence of the control system. Jefferies suggests that the fit could be improved by selecting a maximum displacement greater than a_{max} to force the shape of the initial unloading phase to be the ideal shape (similar to a mirror image of the initial loading curve).

The value of horizontal stress is insensitive to the variations in G and c_u and therefore is insensitive to a small amount of disturbance. This point is important since it overcomes the criticism of the selection of σ_h from SBP tests where there will be some disturbance during installation.

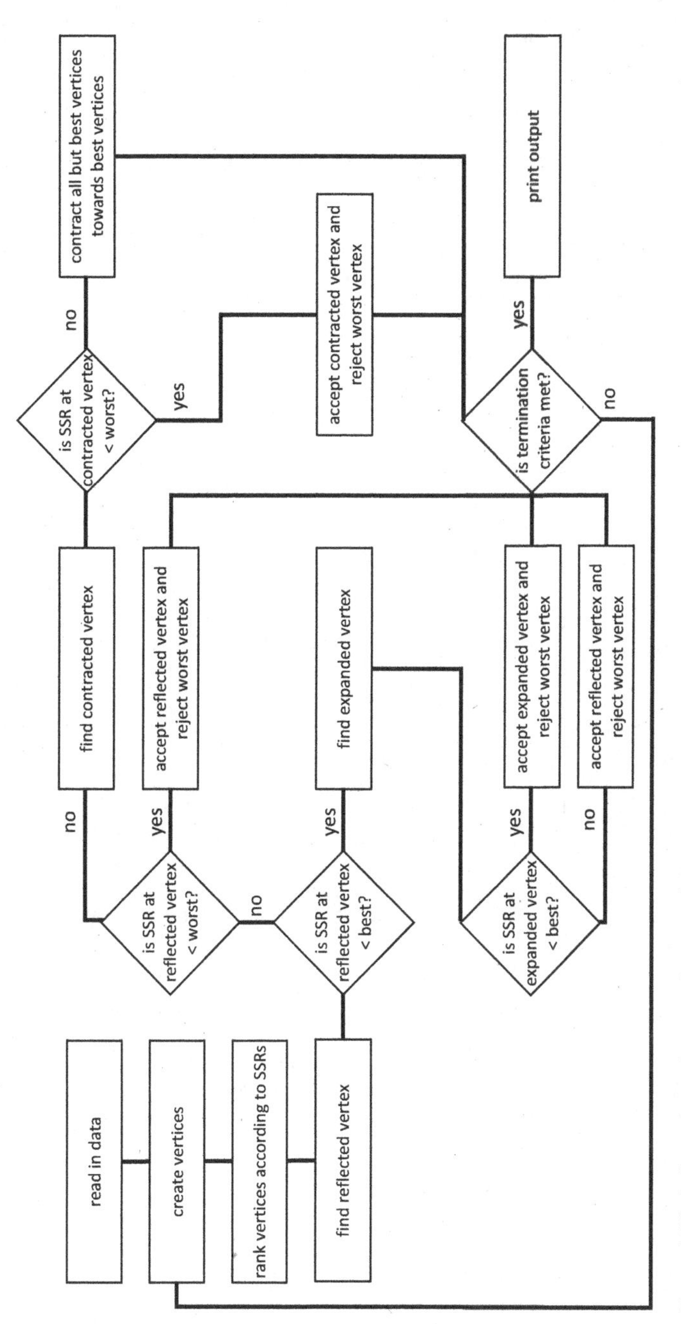

Figure 6.20 A flow chart showing the use of the simplex method to fit curves to pressuremeter data: SSR = sum of squares residuals.

Source: After Caceci and Cacheris, 1984.

Table 6.5 Functions and parameters to be optimised for pressuremeter tests in cohesive soils

Reference	*Test curve*	*Shear stress–strain curve*	*Parameters to be optimised*
Prévost and Hoeg (1975)	$\sigma_h + \frac{\tau_{ult}}{\sqrt{3}} \ln\left(\frac{\alpha+\gamma}{\gamma}\right)$	$[\gamma/(\alpha + \gamma)]\tau_{ult}$	$\alpha, \tau_{ult}, \sigma_h$
	$(A/2\sqrt{3})[B \ln [(1+\gamma^2)+2 \tan^{-1}(\gamma)]+\sigma_h$	$A[(B\gamma^2 + \gamma)/(1 + \gamma^2)]$	A, B, σ_h
Windle and Wroth (1977)	$c_u [1+\ln (G/c_u)+\ln(\Delta V/V)]+\sigma_h$	c_u	G, c_u, σ_h
Denby and Clough (1980)	$(c_u/R_f) \ln [1+2R_f G_i\varepsilon_c/c_u]+\sigma_h$	$\varepsilon_c/\{[1/2G_i]+[R_f/c_u]\varepsilon_c\}$	G, R_f, c_u, σ_h
Arnold (1981)	$Q+[(\varepsilon_c - \varepsilon_{o1})/(a + b(\varepsilon_c - \varepsilon_{o1}))]$	$ab\varepsilon_c/(a+\varepsilon_c)^2$	Q, a, b

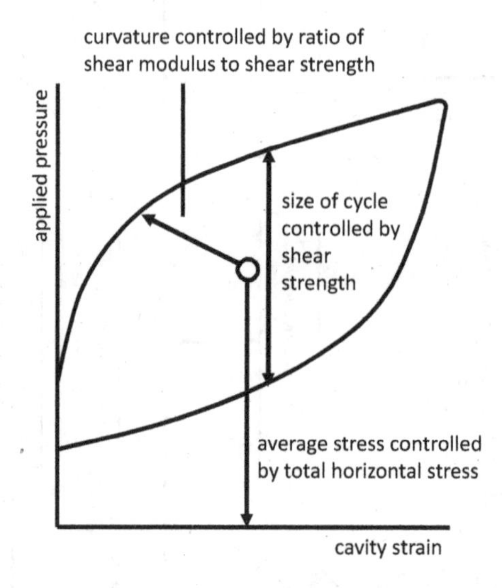

Figure 6.21 The influence of horizontal stress, stiffness and strength on the shape of a pressuremeter curve.

Ferreira and Robertson (1992) suggested that the horizontal stress, initial tangent modulus and shear strength could be obtained by fitting a hyperbolic model to the field data using the following equations for loading and unloading:

$$p = \sigma_h + \tau_{ult} \ln\left[1+\left(2G_i \frac{\varepsilon_c}{\tau_{ult}}\right)\right] \tag{6.17}$$

$$p = p_{max} + \tau^*_{ult} \ln\left[\frac{1}{1-\left[2G_i\left(\varepsilon_c - \varepsilon_{max}\right)\right]/\left[\left(1+\varepsilon_{max}\right)\tau^*_{ult}\right]}\right] \tag{6.18}$$

Equation (6.17) contains three unknowns, σ_h, G_i and τ_{ult} of which two, σ_h, and G_i, are affected by installation disturbance. This suggests that it is better to use the unloading data to estimate G_i and τ^*_{ult}. In order to complete the interpretation it is necessary to assume a value of the ratio of τ_{ult} to τ^*_{ult}. This could be assessed from laboratory stress path tests on undisturbed specimens (Jefferies, 1988) but in practice it is assumed that $\tau_{ult} = 0.5\tau^*_{ult}$ (Jefferies, 1988, and Houlsby and Withers, 1988). G_i, which is a function of the curve-fitting procedure, must be within certain limits if it is to be a reasonable assessment of modulus. An initial value could be found by assuming G_i to be the stiffness over 0.05% cavity strain (Ferreira, 1992).

The values of G_i and τ_{ult} are then substituted into equation (4.41). Ferreira and Robertson (1992) suggest that the latter part of the loading curve should be used, since the initial portion is affected by installation disturbance. This also eliminates the effects of strain softening. In theory, the ultimate shear strength cannot be reached, since a hyperbolic model is used. It is suggested that τ_{ult} is equivalent to the shear stress at the maximum cavity strain reached during a test. This implies that different results would be obtained for different tests at the same depth if the maximum cavity expansion differs. Clarke (1994) has shown, using a simple perfectly plastic model, that the shear stress is very nearly constant beyond 6% cavity strain for SBP tests and 16% cavity strain for PBP tests. Therefore, provided there is sufficient expansion, the assumption that the ultimate shear strength is the strength at the maximum cavity strain is reasonable. Ferreira and Robertson (1992) have shown that the choice of G_i has little effect on the best-fit curve but the choice of σ_h has a significant effect.

Withers et al. (1989) derived an expression for the expansion of a cavity in sand. A pressuremeter curve is generated using assumed values of G, ϕ', ψ and σ'_h and compared to the actual curve. The parameters are adjusted until a good fit is obtained. Some guidance to the choice of parameters is obtained from the modulus from an unload/reload cycle and strength from the unloading portion of the test. However, the generated curve is not too sensitive to the parameters chosen and, hence, it is unlikely that a unique set of parameters will be produced.

The principles of these approaches can be applied with other models, including the power law (Bolton and Whittle, 1999), Cam Clay (Cao et al., 2001) and a state parameter model (Mo and Yu, 2017).

6.4.5 Correlations

Values of σ_h, G and c_u are often independently derived from a test curve, but they are dependent on each other. Therefore, it should be possible to develop a framework in which the values obtained are assessed using correlations with other data. Examples of correlations are given in Chapter 7. Alternatively, the independent values obtained could be used to predict the test curve using a theoretical relationship. As an example, consider equation (4.24), which relates the limit pressure to the stiffness and strength of a linear elastic perfectly plastic soil. The horizontal stress is given by

$$\sigma_h = p_1 - c_u\left[1 + \ln\left(\frac{G}{c_u}\right)\right] \quad (6.19)$$

The limit pressure, strength and modulus are derived separately from the test curve using methods described below and substituted into equation (6.19) to obtain σ_h. They are not independent parameters but are dependent on each other and the reference

Table 6.6 Examples of limits to the chosen values of horizontal stress

$(\sigma_h - \sigma_v) < 2c_u$	for all clays
$\sigma_h > u_o$	for all soils
$c_u / \sigma_h' > 0.3$	(based on Skempton's relationship for normally consolidated clays)
$\sigma_h' / \sigma_v' > [(1 - \sin\phi') / (1 + \sin\phi')]$	for sands
$K_a < (\sigma_h' / \sigma_v') < K_p$	for all soils

Table 6.7 Terms used to define moduli taken from pressuremeter tests

G_i	Initial secant shear modulus
E_m	Ménard modulus
G_{ur}	Secant shear modulus from an unload/reload cycle
G_u	Secant shear modulus from an unloading curve
G_r	Secant shear modulus from a reloading curve
E_{m^-}	Secant elastic modulus from an unloading curve
E_{m^+}	Secant elastic modulus from a reloading curve
E_{mo}	Maximum elastic modulus from an unloading curve
E_{ro}	Maximum elastic modulus from a reloading curve
G_n	Secant shear modulus measured over strain range n%
G_o	Maximum shear modulus
G_s	Equivalent element modulus
G_{uro}	Equivalent shear modulus at the in situ effective stress

datum. An iterative procedure could be used to adjust the reference datum to produce a better estimate of σ_h. The shear modulus could be taken from an unload/reload cycle over a stress range equal to the strength.

6.4.6 The subjectivity of the selection of horizontal stress

The methods described above were developed because of the subjective nature of selecting the lift-off point. In theory, they offer a neat mathematical solution but, in practice, the models used do not describe the soil adequately. More importantly, the authors of several of the methods state that they only apply to quality tests. A quality test could be described as a test on soil that has suffered minimum disturbance during drilling. In many quality tests, the lift-off method is acceptable provided an experienced engineer reviews the data and, in that case, the other methods are unnecessary.

The selection of the lift-off pressure is subjective but, using the other parameters obtained from the pressuremeter curve, it is possible to develop a consistent framework which can be used to give confidence in the point chosen. Table 6.6 gives some examples.

6.5 MODULUS

There are several values of modulus quoted in the literature, as shown in Table 6.7.

The shear moduli, G_i, G_u, etc., are given by

$$G = 0.5\frac{a}{a_o}\frac{dp}{d\varepsilon_c} \quad (6.20)$$

$$G = V\frac{dp}{dV} \quad (6.21)$$

These are not the moduli of an element of the ground but represent the average stiffness response. The elastic pressuremeter moduli, E_m, E_{m^-} and, E_{m^+} are derived from MPM tests in which Poisson's ratio is assumed to be 0.33. The ratio E_{m^-} / E_{m^+} is known as a rheological coefficient, α.

6.5.1 Initial modulus

The initial modulus, G_i, is taken from the initial slope of the pressuremeter curve. At small strains (a_1/a_0) is small and $V \doteqdot V_o$, therefore, the initial stiffness is given by

$$G = 0.5(p - \sigma_h) / \varepsilon_c \quad (6.22)$$

$$G = V_o(p - \sigma_h) / \Delta V \quad (6.23)$$

In practice, there must be some disturbance, no matter how small, when any of the pressuremeters are inserted into the ground. This disturbance can cause a change in the properties of the annulus of soil surrounding the probe such that the initial modulus reflects the properties of that ground, not the undisturbed ground. Therefore, the parameter G_i should be treated with caution.

6.5.2 Unload/reload modulus

If the modulus of intact ground is required, it is necessary to expand the membrane far enough to test undisturbed ground. This will differ according to the probe used since the installation disturbance is different for different pressuremeters. Once the membrane has been expanded sufficiently, it is then unloaded, and this unloading will be elastic. The moduli, G_u and E_{m^-}, can be found from this unloading curve (Figure 6.22).

If the soil is unloaded too far it will fail in extension, so the membrane should be expanded before this occurs. The reloading will be elastic until yield occurs at about the point at which the membrane was unloaded. G_r and E_{m^+} can be determined from the reloading curve as shown in Figure 6.22.

It is clear from Figure 6.22 that a pressuremeter modulus varies with the strain range over which it is measured. It is for this reason that a modulus should be quoted together with the initial stress and strain at the start of unloading or loading and stress and strain magnitude of cycle. Alternatively, a modulus can be calculated over a common strain range, such as 0.1%.

G_{ur} is usually determined from a best-fit line to the data, as shown in Figure 6.23.

The actual cycle and best-fit line should be inspected, since the best-fit line may not be correct. Consider a stress-controlled test in which the membrane is unloaded from

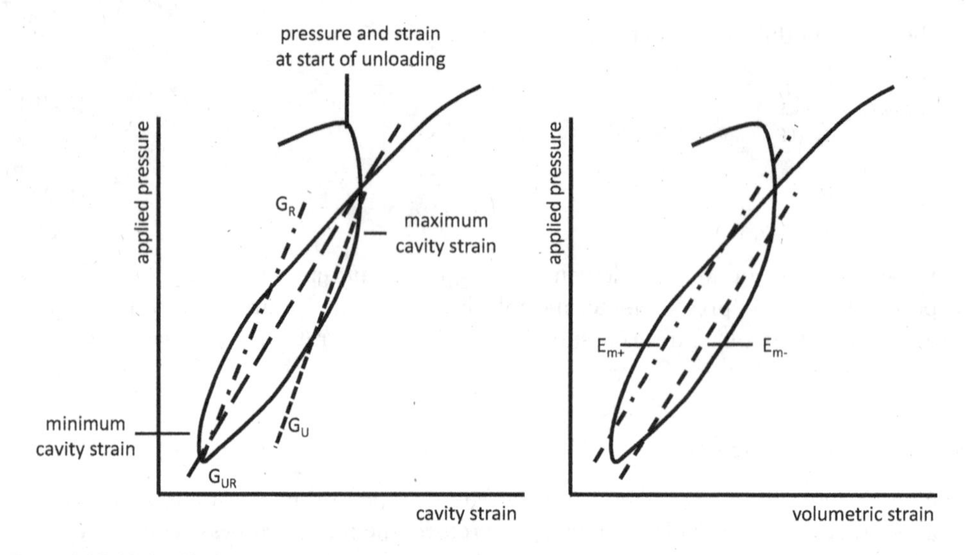

Figure 6.22 Unload/reload cycles showing the slopes used to assess G_u, G_r, E_{m-}, E_{m+}, and G_{ur} (note the slope is twice the shear modulus): (a) shear moduli; (b) elastic moduli.

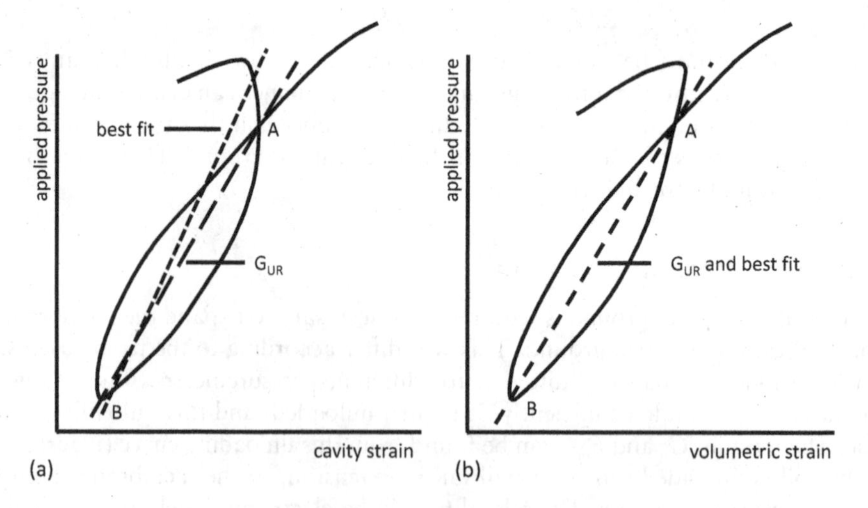

Figure 6.23 The difference between modulus cycles from (a) strain- and (b) stress-controlled tests showing that the best fit to the cycle is different from the slope between AB for strain-controlled tests but the same for stress-controlled tests.

point A (Figure 6.23b). The membrane will retract. At point B the applied pressure is increased, and the membrane will expand. In that case the data will lie about a straight line which is likely to be coincident with the best-fit line. In strain-controlled tests the membrane continues to expand even though the pressure is reduced (Figure 6.23a) because of creep. In that case the data used should be between the intersection of the unloading and reloading curve and the lowest point; that is CB.

This may still result in a visually poor fit. It is prudent to inspect the unload/reload cycle at a large scale and make manual adjustments, if necessary, or fit a line such that the areas between the best-fit line and the unloading and reloading curves are equal.

The modulus from an unload/reload cycle is the mean of the unload and reload secant moduli, that is

$$G_{ur} = 0.5\left[0.5\frac{a_u}{a_r}\left(\frac{dp_u}{d\varepsilon_c}\right)_u + 0.5\frac{a_e}{a_r}\left(\frac{dp_r}{d\varepsilon_c}\right)_r\right] \tag{6.24}$$

where 'u' refers to the unloading portion, 'r' refers to the loading portion, $\left(\frac{dp_u}{d\varepsilon_c}\right)_u$ is the radius at the start of unloading, $\left(\frac{dp_u}{d\varepsilon_c}\right)_r$ is the radius at the end of unloading and at the start of reloading, and $\left(\frac{dp_u}{d\varepsilon_c}\right)_e$ is the radius at the end of reloading. The ratios (a_u/a_r) and (a_e/a_r) are small and can be ignored. Therefore, for practical purposes

$$G_{ur} = 0.5\left(\frac{\Delta p}{\Delta\varepsilon_c}\right) \tag{6.25}$$

Similarly, for volume-displacement systems

$$G_{ur} = V\left(\frac{\Delta p}{\Delta V}\right) \tag{6.26}$$

where V is either the current volume at the start of unloading or, more correctly, the average volume over the whole cycle.

There is a limit to the elastic behaviour in clays. Consider the ideal test in a linear elastic perfectly plastic material shown in Figure 6.24.

The initial elastic phase is denoted by AB. At B the soil adjacent to the membrane fails. If the membrane were unloaded at C it would unload elastically until failure in extension at D. The distance CD is twice the undrained shear strength and this is the limit of the elastic unloading behaviour. On expansion from D the soil will behave elastically until it fails in shear at E; that is approximately the same as C. The unload/reload modulus, G_{ur}, will be the best-fit to the data included in CDE, and it will be an elastic modulus provided the magnitude of the cycle is less than the compression strength of the ground. In practice, the compression strength is unknown, therefore, it is usual to use an estimated value of the shear strength to ensure that the cycle remains in the theoretical elastic zone.

If the unloading cycle is conducted at a different stage of the test, such as F, the mean total stress is greater than that at C. However, once at failure, the clay, while the test remains undrained, will experience little change in effective stress. The deduced shear modulus should remain the same and be independent of the stage of the test at which it is measured provided the magnitude of the cycle remains the same. In reality, some drainage will take place during the test, giving rise to consolidation. This means the effective stress will increase near the pressuremeter, resulting in an increase in stiffness.

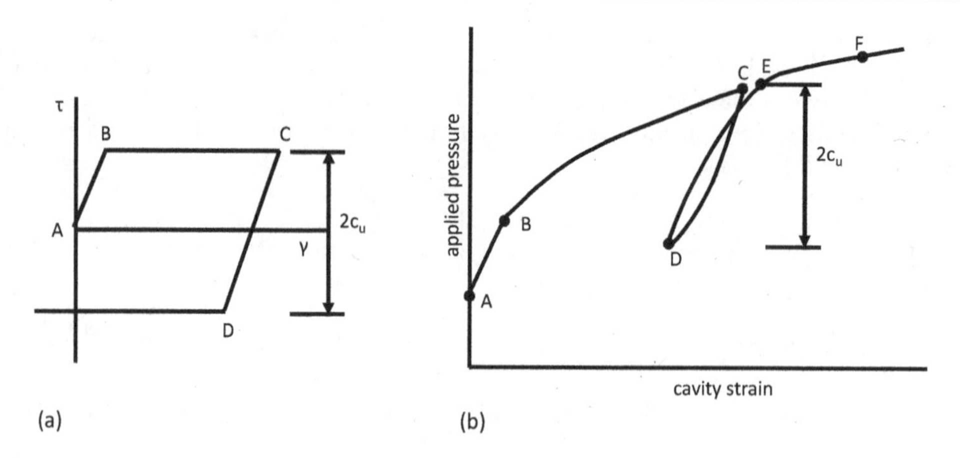

Figure 6.24 The limit of elastic behaviour on unloading in clays (a) shear stress–strain curve; (b) pressuremeter test curve.

Source: After Wroth, 1982.

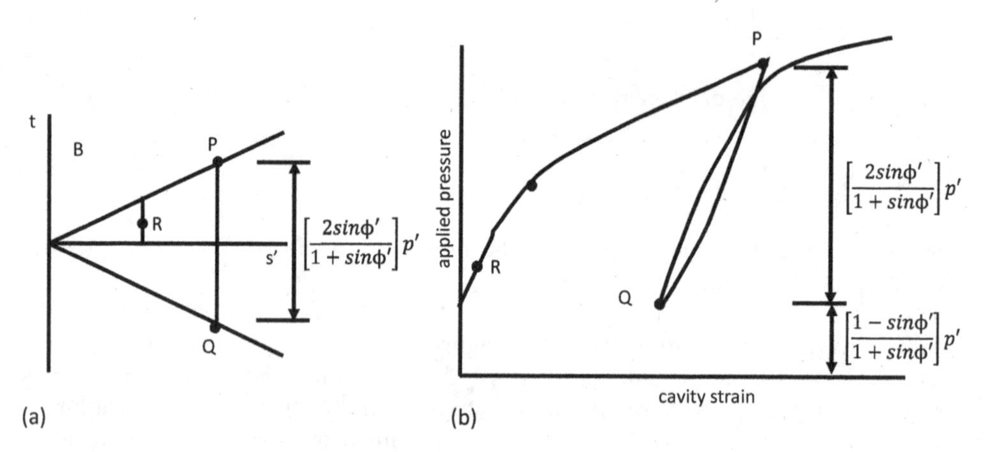

Figure 6.25 The limit of elastic behaviour on unloading in sands (a) stress path; (b) pressuremeter test curve.

Source: After Wroth, 1982.

The limit to the elastic range of behaviour for unloading in sands is shown in Figure 6.25.

An element of sand at the membrane/soil interface follows the stress path, ORP. At P the cavity pressure is reduced, the mean effective stress remains constant and the stress path followed is PQ with failure occurring in extension at Q. The distance PQ, which is the limit of elastic behaviour, is given by

$$\Delta p = (p - u_o)_{max} \frac{2\sin\phi'}{1+\sin\phi'} \tag{6.27}$$

where $(p - u_o)_{max}$ is the effective cavity pressure at the start of unloading and u_o is the ambient pore pressure.

Tests in sand are drained, therefore no excess pore pressures are generated. The mean effective stress must increase as the applied pressure increases provided the sand is at failure. Therefore, G_{ur} increases as the cavity strain increases.

The angle of friction is unlikely to be known at the time of testing, and it is not possible to predict the amplitude of the unload/reload cycle. Since most sand deposits are medium-dense to dense, the angle of friction will be greater than that at constant volume. The use of the angle of friction at constant volume will give a lower bound to the maximum amplitude to prevent failure in extension. For most quartz sands this angle is 35°, therefore the amplitude is 0.72 $(p - u_o)_{max}$. Fahey (1992) suggested that it should be limited to half that value to limit the hysteresis.

6.5.3 Non-linear stiffness profile

An unload/reload cycle in any ground shows some hysteresis even if it is within the elastic limits. This is expected, since ground does have a non-linear response. It is possible to carry out an unload/reload cycle with all pressuremeters, though the accuracy of the data obtained will depend on the type of probe used.

Briaud et al. (1983) showed that, as a first approximation, the reciprocal of the stiffness taken from an unload/reload cycle is linearly related to the volumetric strain (Figure 6.26).

This suggests that the stiffness could be represented as a hyperbolic function of strain. The tests carried out by Briaud et al. (1983) were part of a study of road bases and sub-bases which consisted of gravels, sands and clays. All the tests were above the groundwater level and were considered to be drained tests. The maximum unloading

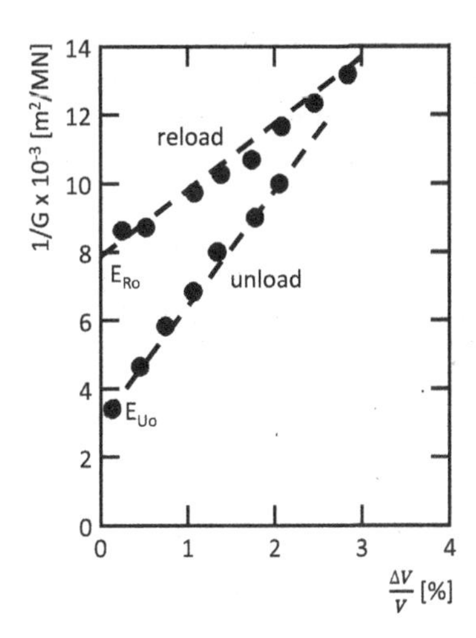

Figure 6.26 The variation in secant modulus with strain from unload/reload cycles.

Source: After Briaud et al., 1983.

and reloading moduli, E_{uo} and E_{ro}, were related to the applied pressures at the start of unloading, p_u, and the start of reloading, p_r, by

$$\frac{E_{uo}}{E_{ro}} = \left(\frac{p_u}{p_r}\right)^n \tag{6.28}$$

n was found to vary with soil type between 0.38 and 1.45. Briaud et al. (1983) suggested that the ratio (p_r/p_u) should be greater than 0.3 to prevent failure on unloading. The modulus will vary with stress level since the tests were drained. Briaud et al. also suggested that the maximum modulus is given by

$$E_{uo} = K(p_u)^n \tag{6.29}$$

where K is a constant.

Bolton and Whittle (1999) suggested a power law could be used to show the variation in shear stress with strain for reducing stiffness

$$\tau = \alpha\gamma^{\beta} \tag{6.30}$$

where α and β are constants to be derived from the pressuremeter test. Thus, the secant shear modulus, G_s, is

$$G_s = \alpha(\gamma_c)^{\beta-1} \tag{6.31}$$

where γ_c is the shear strain at the cavity wall. A plot of unload/reload cycles gives a linear plot (Figure 6.27) with a gradient $1/\beta$ and an intercept $ln(\alpha/\beta)$.

Figure 6.28 shows the variation of secant shear modulus normalised with respect to the effective horizontal stress against the current cavity strain for several tests in London Clay, which included two cycles per test.

The effective stress is taken as the in situ effective horizontal stress since this represents the stress to which the soil is consolidated prior to undertaking a test. The mean effective stress changes very little once yield has occurred (Wroth, 1982) hence, the modulus should not vary significantly as the membrane expands. It is independent of the strain level at which the cycle is carried out. The current cavity strain, ε_{curr}, is based on the maximum strain for the unloading portion of the cycle and the minimum cavity strain for the reloading portion and is given by

$$\varepsilon_{curr} = \frac{\varepsilon_c - \varepsilon_{um}}{1 + \varepsilon_{um}} \tag{6.32}$$

where ε_{um} is the maximum cavity strain during unloading for G_u or cavity strain at the start of reloading for G_r.

The results show that the reloading stiffness responses are independent of the stress level and strain level at which they are carried out. The unloading stiffness responses vary between cycles and are different from the reloading stiffness responses for the same cycles. This may be due to consolidation taking place during unloading. Clarke (1993) concluded that the reload modulus is more consistent for tests in clays.

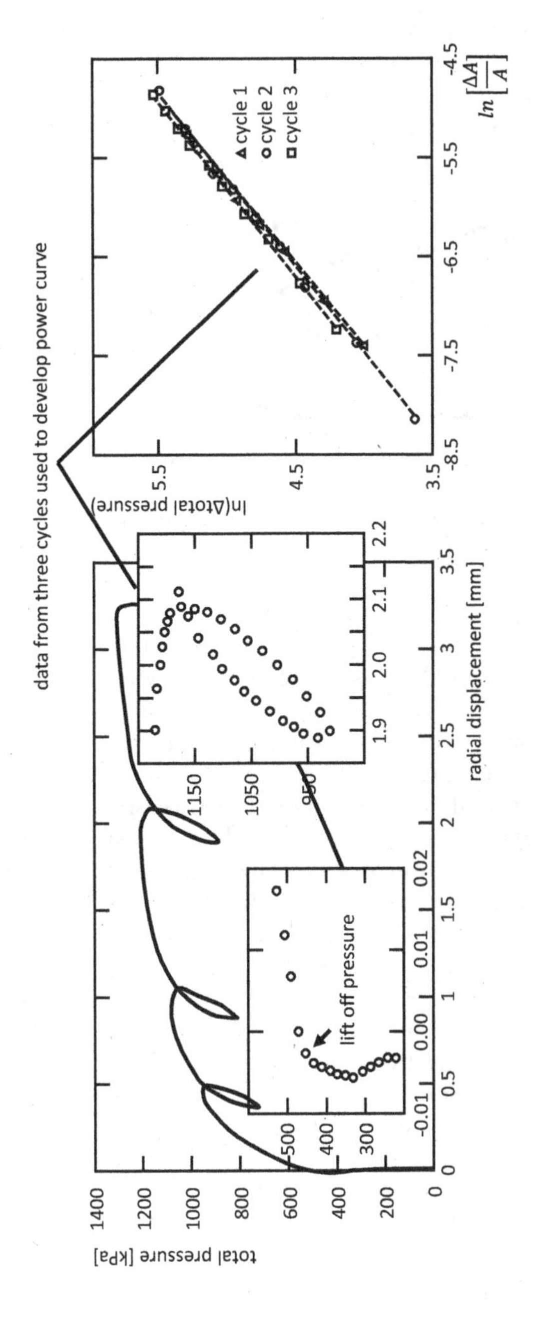

Figure 6.27 Deriving the power law relationship for shear modulus from a self-boring pressuremeter test in London Clay.

Source: After Bolton and Whittle, 1999.

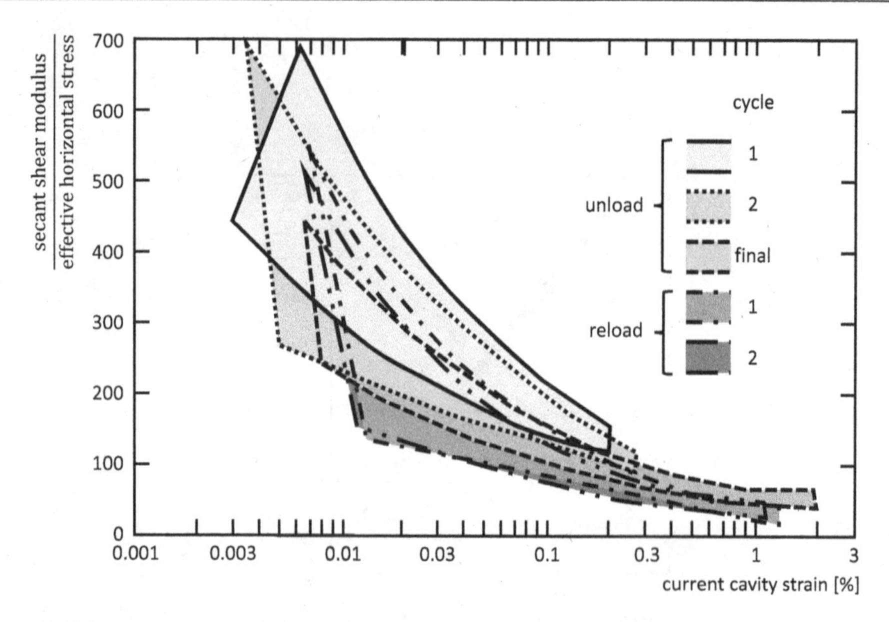

Figure 6.28 The variation in modulus with strain for tests in London Clay.

Source: After Clarke, 1993.

The potential of using an unload/reload cycle to determine an incremental modulus and a non-linear stiffness profile was recognised by Muir-Wood (1990). Jardine (1991) compared results from triaxial tests and pressuremeter tests carried out at the Cannons Park research site in North London using the equi-stiffness curves described in chapter 4. The pressuremeter results in Figure 6.29 lie below those from the triaxial tests, but this could be due to differences in initial stresses, stress rates and strain rates.

The pressuremeter data are taken from unloading curves, with the strain origin chosen as the maximum strain on unloading. The analysis shows a tendency towards infinite stiffness rather than the maximum modulus, G_o.

Robertson and Ferreira (1993) suggested that it is impossible to measure the true maximum shear modulus, G_o, because of the lack of sensitivity of the measuring systems over the small strain range. The value of G_i derived from fitting a hyperbolic stress–strain relationship to the test curve is not equal to G_o. G_o could be taken from a seismic test, for example, but this produces unacceptably high values of modulus from a pressuremeter curve. Robertson and Ferreira (1993) suggest that the true variation of stiffness with strain is a function of an initial modulus, which is less than G_o, but greater than G_i as shown in Figure 6.30.

Non-linear stiffness profiles can also be developed from tests in sand using the method suggested by Bellotti et al. (1989). The stiffness will vary with the stress-and-strain level at the start of unloading since the test is drained as the cavity expands. It is necessary to evaluate the stiffness profile at a common stress for comparisons to be made.

Figure 6.31 shows the variation in stiffness with strain for several tests in Thanet Sands between 25 and 40 m below ground level.

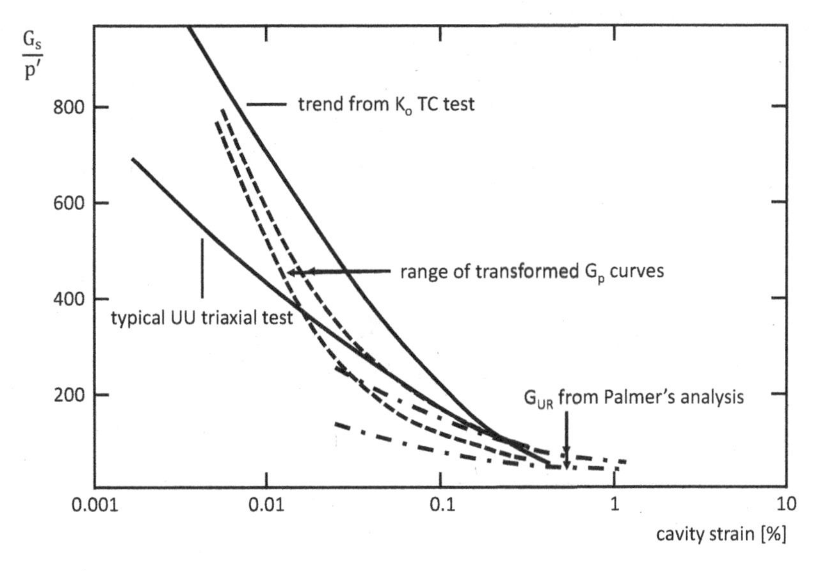

Figure 6.29 A comparison between triaxial and pressuremeter derived secant moduli.

Source: After Jardine, 1991.

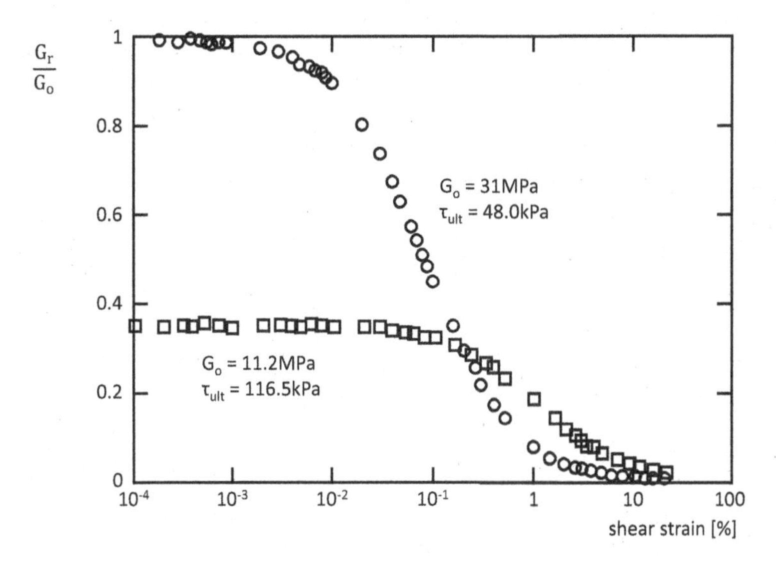

Figure 6.30 An example of the variation in shear modulus with strain assuming an initial modulus, Gi, derived from the pressuremeter test and an initial modulus, Go, derived from the seismic cone.

Source: After Robertson and Ferreira, 1993.

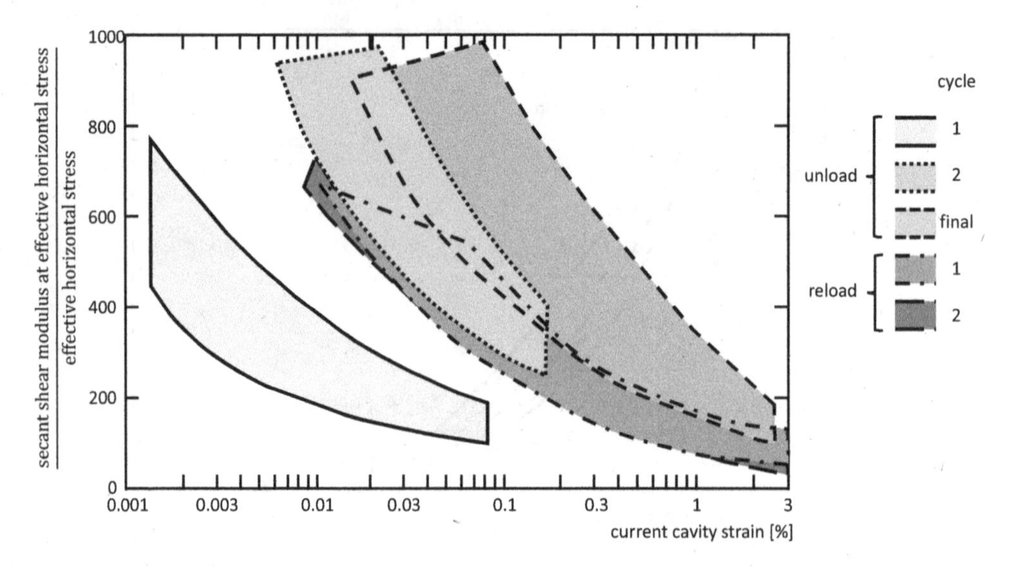

Figure 6.31 The variation in modulus with strain for tests in Thanet Sands.

Source: After Clarke, 1993.

The profiles of the two reloading curves are similar. Clarke (1993) suggests that the unloading curves are affected by creep, which could account for the differences in unloading curves. The strain is expressed in terms of either the radius at the start of loading or the maximum radius on unloading using equation (6.33). The secant moduli, G_u and G_r, have been converted to the stiffnesses, G_{uo} and G_{ro}, at the in situ effective horizontal stress by

$$G_{uro} = G_{ur} = \left(\frac{\sigma'_h}{\sigma'_{av}}\right)^n \tag{6.33}$$

n has been taken as 0.4 (Bellotti et al., 1989). The average effective horizontal stress, σ'_{av} at the start of unloading for these tests is equal to

$$\sigma'_{av} = \sigma'_{ho} + 0.2\left(p'_u - \sigma'_{ho}\right) \tag{6.34}$$

where p'_u is the maximum applied effective stress at the start of unloading. The constant 0.2 is an average for all of these tests in dense sands. Bellotti et al. found that 0.2 was valid for their chamber tests on medium dense sand, suggesting that 0.2 applies to all sands in practice.

The current cavity strain is restricted to the range 0.001–1%, which covers medium-to-large shear strains (Bellotti et al. 1989). The resolution of the displacement measuring systems is equivalent to ± 0.0005% cavity strain, but the resolution of the transducers in the pressuremeter is probably 0.01% because of system compliance. It is noted that for cavity strains less than 0.01% the data are unreliable.

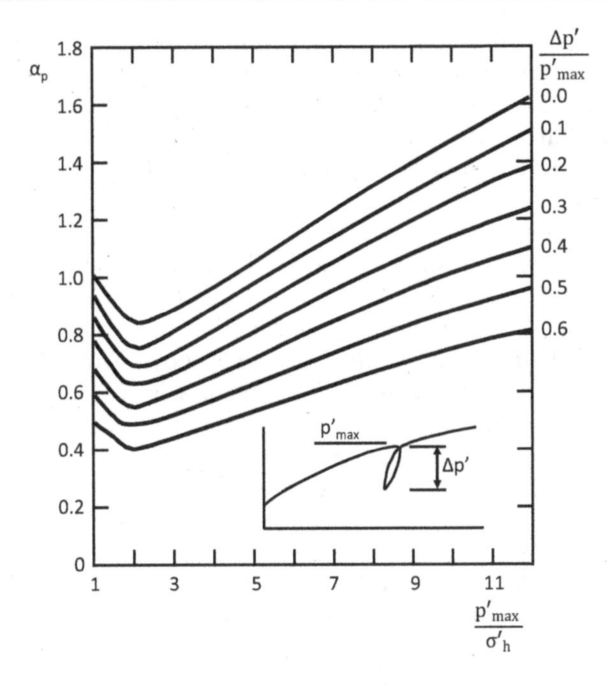

Figure 6.32 Estimating Go from an unload/reload cycle from an SBP test in sand.

Source: After Byrne et al., 1990.

Byrne et al. (1990) suggested that G_o for sands could be obtained from the unload/reload cycle using the relationship

$$G_o = G_{ur}\alpha_p\alpha_D \tag{6.35}$$

where α_p is a factor taken from Figure 6.32 and α_D is a factor to correct for disturbance.

Figure 6.32 was produced from a numerical analysis assuming non-homogeneous elastic conditions in the sand. Comparisons with data produced by Bellotti et al. (1989) showed that the disturbance factor, α_D, was 1.6 for the CSBP and 1.4 for the PAF-76.

6.6 UNDRAINED SHEAR STRENGTH

The parameters, horizontal stress and shear modulus are, theoretically, both independent of the type of test, since they are independent of the drainage conditions. Other engineering parameters derived from the test data are dependent upon the drainage conditions.

An undrained test is used in fine-grained soils where permeability is such that, for practical purposes, excess pore pressures do not dissipate during a test. Baguelin et al. (1986) undertook a numerical study of the generation of pore pressure around an expanding cavity using modified Cam Clay and an elastic perfectly plastic models. They established that the permeability of the soil had little effect on the shape of the curve, and therefore on the derived strength. Tests in some rocks are also undrained, but this depends on the stiffness of the rock as well as the permeability and potential to fracture (Haberfield and Johnston, 1993). A stiff rock with a low permeability will behave as a drained material.

A precise shear stress–strain curve can be derived from an ideal pressuremeter test in clay (Palmer, 1972). The curve produced generally shows a peak strength but Baguelin et al. (1978) have shown that this is significantly affected by the installation technique, which inevitably disturbs the ground. The interpreted value of undrained shear strength from pressuremeter tests is dependent on selecting the correct reference datum. It is for this reason that semi-empirical methods are used for PBP tests and the unloading curve for FDP tests and SBP tests, which show disturbance. A quality SBP test can be interpreted using the analyses described in Chapter 4.

There are a number of methods used to determine undrained shear strength of clays including modelling the soil as linear elastic perfectly plastic material, curve-fitting routines, correlations with other tests and empirical factors relating a measured parameter to strength. The two most widely used techniques in practice are those developed for prebored pressuremeter (PBP) tests in which the shear strength is taken as a factor of a modified limit pressure (Ménard, 1975), and for self-boring pressuremeter (SBP) tests, in which the soil is modelled as an elastic perfectly plastic material (Windle and Wroth, 1977).

Powell (1990) suggested that the shear strength should be determined from the true limit pressure – that is, the pressure required for infinite expansion. This should be independent of the installation technique and therefore should give similar values of strength for all pressuremeters. Ghionna et al. (1982) showed that the limit pressure depended on the L/D ratio of the probe, therefore strength would be a factor of probe type. Clarke (1994) showed that the limit pressure, which is found by extrapolating the test curve, is a function of the strain range to which the extrapolated curve is fitted, therefore the shear strength will be a function of the amount of expansion during a test.

There are other factors which affect the strength obtained. These include installation and testing procedures such as the rate of loading, the inclusion of unload/reload cycles, the time between drilling and testing, drainage conditions, end effects and length of test section. Thus, if the pressuremeter is to be used to determine strength there is a need to give a clear definition of the parameter being measured rather than quote average values over varying strain ranges or applying arbitrary factors related to other tests.

6.6.1 General analysis

The analysis developed by Palmer (1972) and others is a theoretically sound approach to determine the strength properties of clays. The shear stress–strain curve is derived directly from the test data without imposing any rheological restrictions on the analysis using equation (4.18). An example is shown in Figure 6.33.

The peak stress is a function of not only the properties of the clay but also of effects due to installation, lack of homogeneity of the soil and the variation in strain rate within the soil.

Figure 6.34 shows two ways that equation (4.19) can be applied.

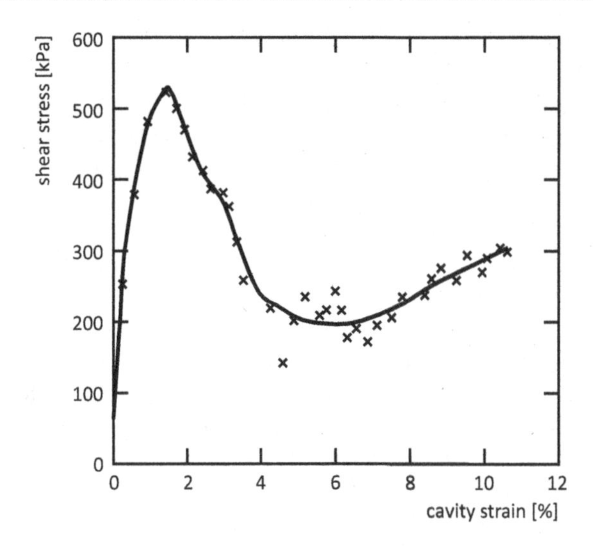

Figure 6.33 Using Palmer's analysis to determine the peak and post-peak strength from an SBP test.

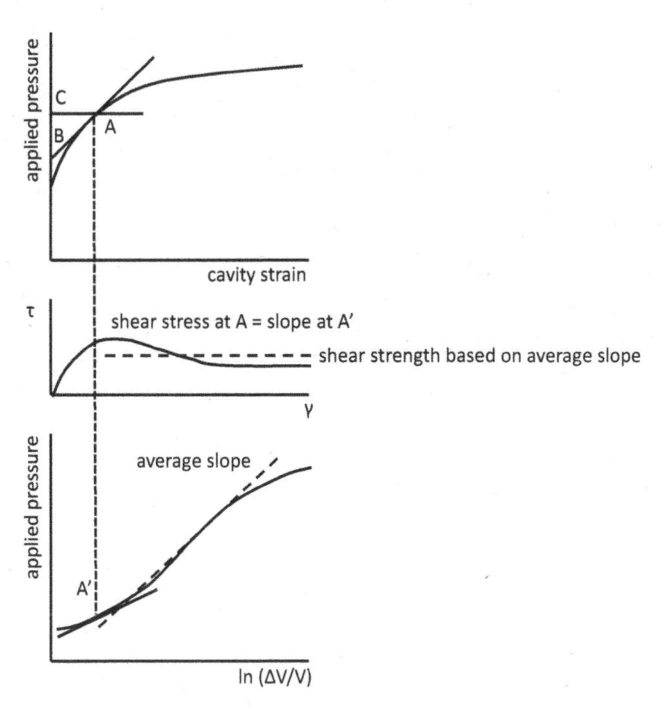

Figure 6.34 Stress–strain curve and average shear strength derived from an ideal pressuremeter test in clay.

The tangent at point A on the test curve is projected back to meet the vertical axis at B. A line parallel to the horizontal axis is drawn through point A to meet the vertical axis at C. The distance BC is the shear stress at A and the shear strain is twice the cavity strain. This is repeated at several points on the test curve and a smooth curve is drawn through the values to obtain the shear stress–strain curve. An alternative method is to use equation (4.20) and plot the cavity pressure against the volumetric strain. The slope at A′ is equal to the shear stress at the cavity strain A′. Note, if the soil is perfectly plastic, this will give the same result as the Windle and Wroth analysis since the shear stress will be constant.

Installation of all types of pressuremeters gives rise to a disturbed annulus of ground around the pressuremeter. This disturbed annulus is, in turn, surrounded by undisturbed ground, which is likely to be stiffer and stronger, and therefore the soil will no longer be homogeneous. Analysing the test as if the soil were homogeneous may give rise to an initial strength, which may be greater than the actual strength of the soil. Baguelin et al. (1979), Prévost (1979), Sayed and Hamed (1988), Prapaharan et al. (1990) and others have studied this problem. The authors claim that the peak strength can be overestimated by between 10% to 100%, depending on the soil model chosen. An example of the effect that disturbance has upon the derived stress–strain curve is shown in Figure 6.35.

The strain rate around an expanding cavity varies inversely with the square of the radius. A cavity strain rate of 1%/min becomes a circumferential strain rate of 0.01%/min at a radius ten times the cavity radius. If the stiffness response is strain dependent, then ground adjacent to the membrane, which governs the initial yield, will appear stiffer than that further away. Thus, the ground may exhibit a peak shear strength, which could be higher than the actual strength.

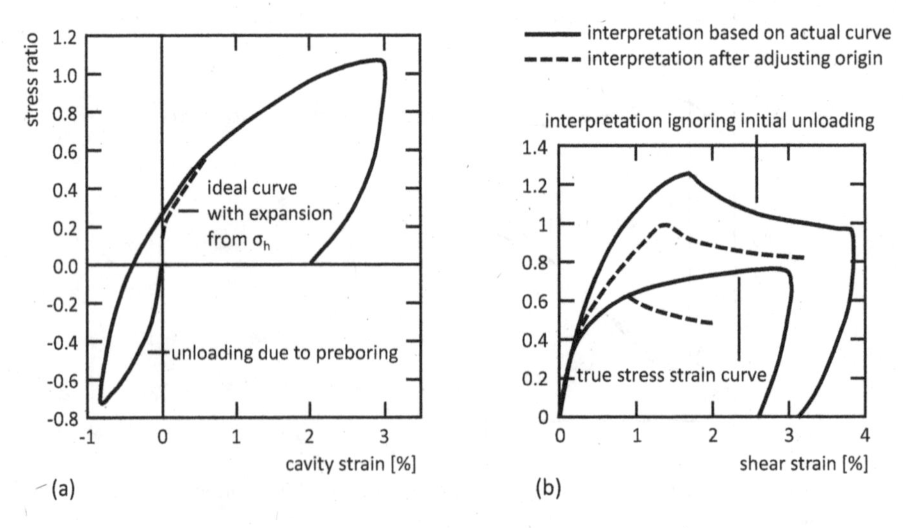

Figure 6.35 The effect of disturbance on the derived stress–strain curves: (a) test curve; (b) stress–strain curve.

Source: After Prévost, 1979.

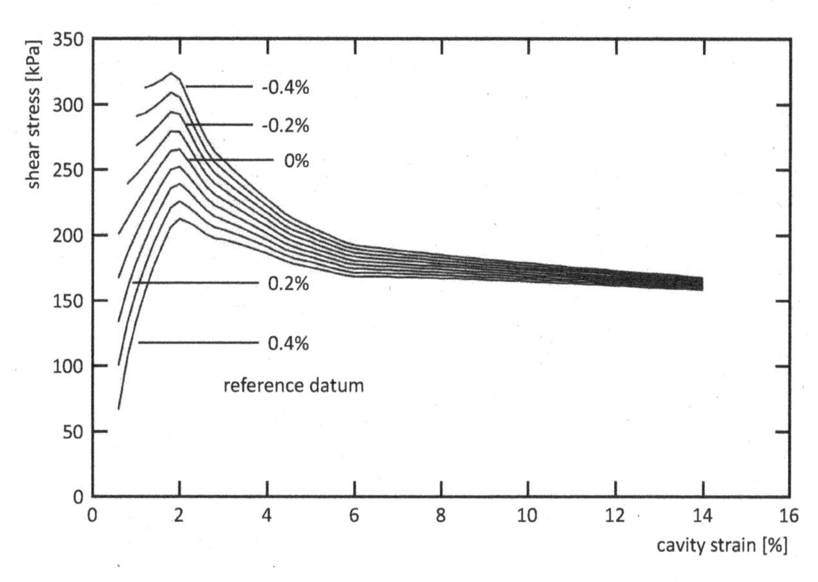

Figure 6.36 The effect of the choice of reference datum on peak strength and post-peak strength based on the Palmer (1972) analysis.

The selection of the reference datum affects the peak strength as well as the average strength. A strain correction of –0.3% in Figure 6.36 represents the creation of a pocket during installation, which has a diameter 0.3% greater than the SBP.

A strain correction of 0.5% is equivalent to a cavity expansion of 0.5% during installation. Unloading of the cavity produces higher peak strengths; loading reduces the peak strengths. In practice, it is not known whether a pocket wall has been unloaded or loaded. Beyond 4% strain the strength obtained, for practical purposes, is independent of the correction for the reference datum. Evidence from a significant number of SBP tests in stiff clays shows that there is a post-peak plateau developed at corrected strains of greater than 6%, provided the correction is less than 1%.

Mair and Wood (2013) consider the peak strength is unreliable and suggest that the shear strength should be deduced from large strains. An average strength will be obtained which is not too dissimilar from that given by the more simple interpretation based on an elastic perfectly plastic soil. For that reason the simpler method is more commonly used.

Disturbance during installation of a prebored pressuremeter is inevitable, but it may be possible to estimate the average shear strength if the expansion is sufficient.

The inset in Figure 6.37 shows a disturbed SBP test in stiff clay, which has a similar form to a PBP test. Note there were two unload/reload cycles in the early part of the test, and these have been removed for clarity. The true cavity diameter is probably within the strain range 2–4% cavity strain. Figure 6.37 shows the derived shear stress–strain curves for a range of strain corrections between 1.5 and 3.5%. The first peak stress is a function of the amount of assumed movement of the cavity wall. Other peaks are a function of the unload/reload cycles. The average shear strength is affected by any unload/ reload cycles.

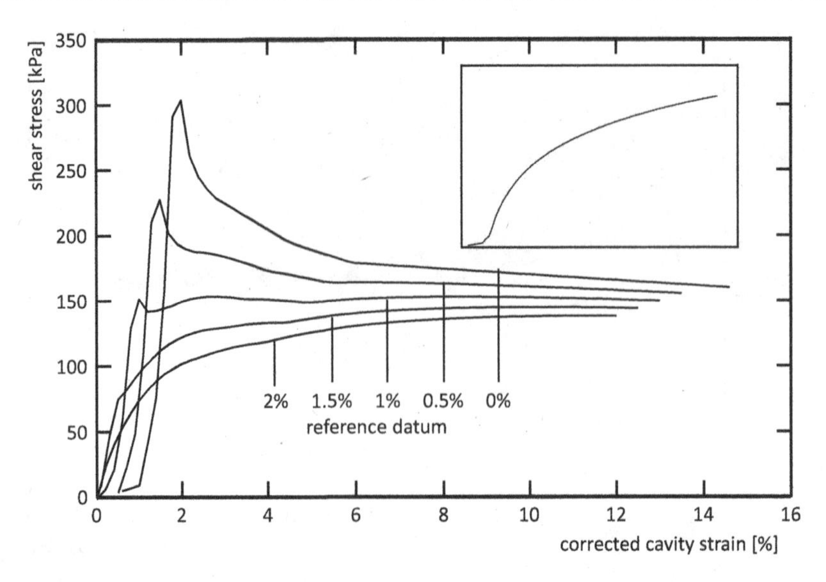

Figure 6.37 The determination of strength from tests in which disturbance is large using a PBP test as an example. The figure shows the effect of the reference datum on shear stress.

It is noted that, at large cavity strains the shear stress tends towards a common value independent of the chosen reference datum. This is expected, since the effects of installation reduce as the cavity expands. It is for this reason that this method could be used to determine a post-peak strength from PBP tests provided that there is sufficient expansion of the membrane, rather than the method based on limit pressure described below.

6.6.2 Elastic perfectly plastic soil

The soil is assumed to fail when the applied pressure is equal to the sum of the undrained shear strength and the horizontal stress and, thereafter, the undrained shear strength is assumed to be a constant. This is sometimes known as the average strength. A SBP test is plotted in Figure 6.38 with the applied pressure plotted against the natural logarithm of the volumetric strain.

Inspection of equation (4.23) shows that the gradient of the best-fit line should be equal to the undrained shear strength. Note that the data are only plotted for strains greater than 3.5% because of the effects of unload/reload cycles and installation.

Gibson and Anderson (1961) considered the situation where the pocket wall was completely unloaded, as is the case with the PBP test in a dry hole. They assumed that the wall unloaded elastically during preboring and proposed that the applied pressure is given by

$$p = \sigma_h + c_u\left[1 + ln\left(\frac{G}{c_u}\right) + ln\left\{\frac{\Delta V}{V} - \left(1 - \frac{\Delta V}{V}\right)\frac{\sigma_n}{G}\right)\right\}\right] \tag{6.36}$$

This is the same as equation (4.24) if no unloading occurs during installation.

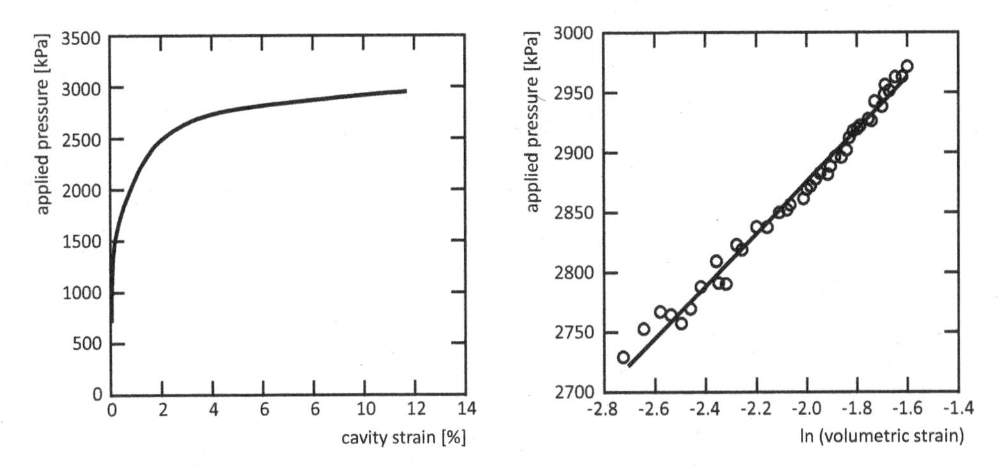

Figure 6.38 Interpretation of an SBP test to determine the undrained shear strength showing (a) the test curve and (b) the slope of the test curve when plotted in terms of ln (volumetric strain).

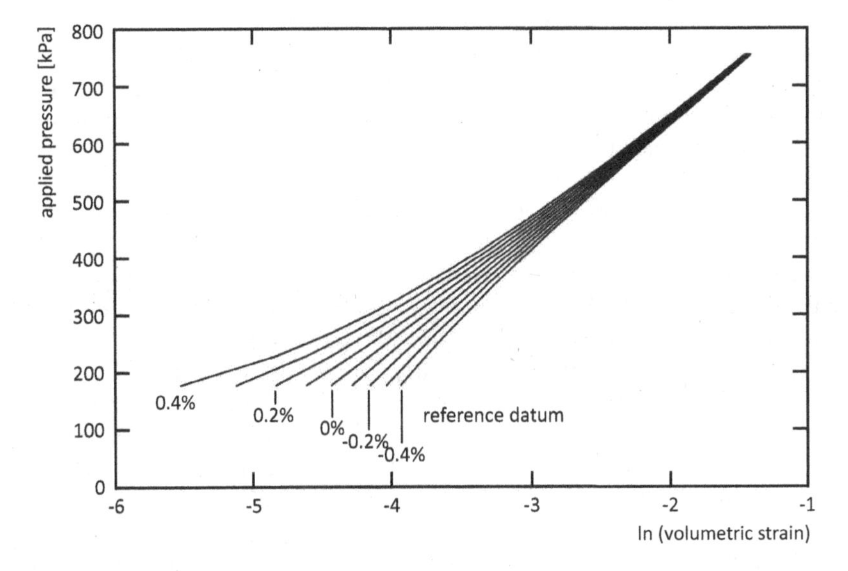

Figure 6.39 The effect of the reference datum on the assessment of shear strength using Gibson and Anderson's method from SBP tests in clay.

All the curve-fitting methods used to determine shear strength require a correct assessment of the reference datum. Figure 6.39 shows the effect a change in reference datum has upon the derived shear strength.

Clearly, an increase in disturbance due to installation has a greater effect on the average shear strength. As the membrane expands, the effect of the disturbance on derived strength reduces since an increasing amount of soil being loaded is unaffected by installation. Hence, in Figure 6.37, the data are only plotted for cavity strains

Table 6.8 The effect the strain range has upon the average strength and limit pressure from a number of tests in London Clay

	c_u / σ'_h			p_l / c_u		
Strain range (%)	*Min.*	*Av.*	*Max.*	*Min.*	*Av.*	*Max.*
1–2	0.35	**0.52**	0.69	7.83	**9.57**	11.98
2–3.5	0.48	**0.70**	0.92	6.70	**7.80**	9.40
4–6.5	0.65	**0.79**	1.10	5.85	**7.81**	8.48
6.5–10.5	0.39	**0.55**	0.86	7.15	**9.94**	12.02

Source: After Clarke, 1994.

greater than 3.5%. This method can be used to determine the shear strength from PBP tests provided there is sufficient expansion of the membrane and a reasonable estimate of the reference datum is made.

Most soils are not perfectly plastic and therefore the gradient of the line will vary depending upon the strain level. Table 6.8 shows that the strain range over which the strength is measured affects the quoted average shear strength for SBP tests in stiff clay. There appears to be a peak strength between 4% and 6.5% cavity strain. Clarke (1994) suggests that a post-peak strength can be obtained from above 6% cavity strain. A further implication of this is that the limit pressure, which is determined by extrapolation of the curve used to determine shear strength, will depend on the strain range over which the extrapolated line is fitted. It is necessary to quote the strain range over which the strength is measured.

It should be possible to obtain post-peak strengths from PBP tests provided there is sufficient cavity expansion to ensure that undisturbed clay is tested. For a test in which the strain correction to obtain the reference datum is about 6%, the probe would have to be expanded to at least 16% and the post-peak strength taken from between 15% and 16% cavity strain after correcting for installation effects.

Figure 6.33 shows a shear stress–strain curve derived from a typical SBP test curve. Two features are noted: the peak stress, which may be the true strength of the clay, and/or a function of installation or test procedure; and an apparent increase in strength post-peak, which may be due to partial drainage (Pyrah et al., 1988) or end effects (Yeung and Carter, 1990) or due to non-cylindrical expansion (Houlsby and Carter, 1993). Yeung and Carter (1990) suggested that, for a range of rigidity index of between 7.5 and 375, the correct strength, c_{uc} is a function of the measured strength, c_{um} given by equation (6.37). This was developed from a parametric study using a numerical analysis with an elastic perfectly plastic soil. The ratio, c_{uc}/c_{um}, varies between 0.7 and 0.8 and, for practical purposes, the quoted shear strength is three-quarters of the derived value:

$$\frac{c_{uc}}{c_{um}} = 0.65 + 0.06\log\left(\frac{G}{c_{uc}}\right) \tag{6.37}$$

Houlsby and Carter (1993), using finite element analysis, demonstrated that, for an L/D ratio of 6 and a rigidity index of between 50 and 1000, the derived strength over the strain range 2–5% could be 25–43% greater than the actual strength due to non-cylindrical expansion. This difference between measured and actual shear strength continues beyond 5% strain. Battaglio et al. (1981) concluded, after analysing over 200 pressuremeter tests using L/D ratios of 2, 4 and 6, that the ratio had a significant effect on the undrained shear strength and limit pressure.

Clarke (1994) used the work of Houlsby et al. and Windle and Wroth to produce a semi-empirical equation for the applied pressure in terms of volumetric strain:

$$p = \sigma_h + c_u\left\{1 + \ln\left[\left(\frac{G\Delta V}{c_u V}\right) + 0.025\left(\frac{G}{c_u}\right)^2\left(\frac{\Delta V}{V}\right)^2\right]\right\} \tag{6.38}$$

The second term in the bracket, $\{0.025((G\Delta V)/(c_u V)^2)\}$, is the correction for non-cylindrical expansion. Table 6.9 shows the effect that this has upon the average shear strength derived from a pressuremeter curve.

Clarke (1994) suggested that the average strength derived from a SBP test curve can be corrected for the effects of non-cylindrical expansion by following the procedure described below:

1. The minimum post-peak plateau strength is derived from the uncorrected curve by assessing the minimum slope of the data when plotted as applied pressure against logarithm of volumetric strain. Evidence from a significant number of tests shows that this occurs between 2 and 6% cavity strain.
2. A shear modulus is taken from an unload/reload cycle. It is recommended that the stress range should be equal to c_u since this conforms to the limit of elastic behaviour during the initial loading.
3. The function given below is calculated for each data point using the minimum post-peak strength and the secant shear modulus:

$$\frac{G\Delta V}{c_u V} = 0.025\left(\frac{G\Delta V}{c_u V}\right)^2 \tag{6.39}$$

   Note that this function is not too sensitive to the values of stiffness and strength.
4. The peak and post-peak strengths are derived from the corrected curve by assessing the slope of the data when plotted as applied pressure against volumetric strain.
5. The limit pressure is found from equation (6.28) using the post-peak strength and a $\Delta V/V$ value of 1.

Table 6.9 Effect of non-cylindrical expansion on ratio of actual strength to measured strength for London Clay (L/D = 6)

Strain Range (%)	*50*	*100*	*200*	*500*	*1000*	*Rigidity index*
2–5	0.92	0.87	0.80	0.70	0.62	Houlsby et al..
2–5	0.93	0.88	0.81	0.70	0.62	Equation (6.26)
6–10	0.87	0.79	0.71	0.61	0.56	Equation (6.26)
6–15	0.85	0.77	0.69	0.60	0.55	Equation (6.26)

It is more common to determine shear strength from PBP tests using empirical methods. Cassan (1972) and others suggested that strength could be derived from the net limit pressure using equation (6.40). It is only a constant for linear elastic perfectly plastic clays but, in practical terms, this is a reasonable assumption:

$$p_1 = \sigma_h = c_u\left[1 + ln\left(\frac{G}{c_u}\right)\right] \tag{6.40}$$

The rigidity index (G/c_u) varies between 200 and 2000. Thus, the parameter $(1 + \ln(G/c_u))$, known as the pressuremeter constant, β^*, would range between 6.3 and 8.6. The horizontal stress can either be determined from the test curve using one of the methods given in section 6.4 or derived from an assumed value of K_0. For normally consolidated clays this is approximately 0.3; for stiff clays, where the limit pressure is much greater than σ_h, the chosen value of σ_h is not critical. In practice, β^* typically ranges between 3.3 for soft clays and 12 for stiff clays, as shown in Figure 6.40.

The discrepancy arises because of the difficulty in measuring σ_h, the difference between the laboratory determined values of shear strength – on which the rigidity index is based – and the in situ pressuremeter strengths, the influence of the disturbed zone and anisotropy. Table 6.10 gives a summary of the relations between c_u and p_{lm} developed for PBP tests.

Table 6.8 shows that the shear strength to limit pressure ratio is dependent on the minimum strain and the range over which it is measured. Therefore, using the limit pressure to determine strength can only be an approximate method.

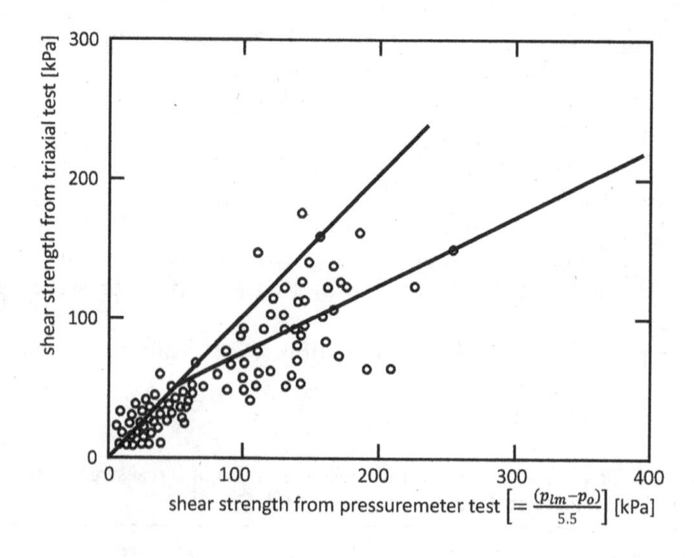

Figure 6.40 The variation in the relation between undrained shear strengths derived from pressuremeter and triaxial tests.

Source: After Amar et al., 1975.

Table 6.10 Empirical relations between undrained shear strength and net limit pressure

	Clay type	*Reference*
$(p_{lm} - \sigma_h)/k$	*K = 2 to 5*	Ménard (1957d)
$(p_{lm} - \sigma_h)/5.5$	Soft to firm clays	Cassan (1972), Amar and Jézéquel (1972)
$(p_{lm} - \sigma_h)/8$	Firm to stiff clays	
$(p_{lm} - \sigma_h)/15$	Stiff to very stiff clays	
$(p_{lm} - \sigma_h)/6.8$	Stiff clay	Marsland and Randolph (1977)
$(p_{lm} - \sigma_h)/5\text{-}1$	All clays	Lukas and LeClerc de Bussy (1976)
$(p_{lm} - \sigma_h)/10 + 25$		Amar and Jézéquel (1975)
$(p_{lm} - \sigma_h)/10$	Stiff clays	Martin and Drahos (1986)
$p_{lm}/10 + 25$	Soft and stiff clay	Johnson (1986)

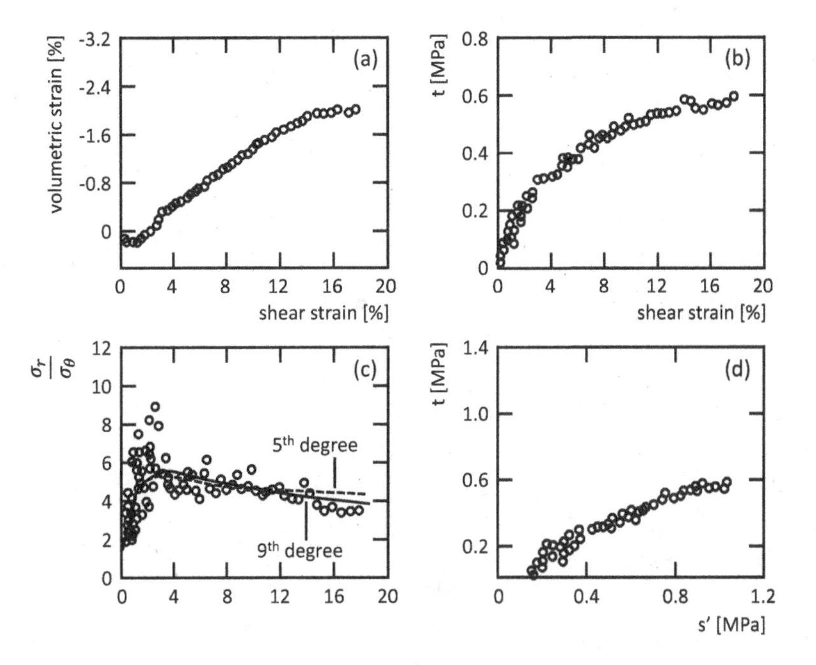

Figure 6.41 The interpretation of an SBP test in sand showing (a) the volumetric strain against shear strain, (b) the shear stress against shear strain, (c) stress ratio against shear strain and (d) shear stress against mean normal stress.

Source: After Manassero, 1989.

6.7 ANGLES OF FRICTION AND DILATION

An angle of friction can be obtained from tests in sands. The ideal test can be analysed using the method proposed by Manassero (1989), since it applies to all sands independent of density. Figure 6.41 shows the results of an analysis of an SBP test performed in a calibration chamber in Ticino sand.

The scatter in the derived shear stress–strain curve is removed by fitting a polynomial equation to the data. Manassero suggested that polynomial degrees of between

4 and 7 gave an acceptable fit. In practice, there will be some disturbance during installation and this method, like all other methods, is sensitive to the reference datum chosen.

Inspection of Figure 6.41 shows that there is a reduction in volume followed by an increase such that the ratio of volumetric to shear strain becomes constant. This is the assumption made by Hughes et al. (1977) for their widely used method for dense sands. If data from an ideal pressuremeter test in dense sand are plotted on logarithmic scales with the vertical axis being the effective applied pressure and the horizontal axis being the current strain, then the points will lie on a straight line. The slope of this line, s, is

$$s = \left(\frac{1 - N}{1 + n} \right) \tag{6.41}$$

The current strain is given by

$$\varepsilon_{curr} = \frac{a_i - a_o}{a_i} \tag{6.42}$$

$$\varepsilon_{curr} = \frac{1}{1 + \varepsilon_c} \tag{6.43}$$

The angles of dilation and friction are given by

$$sin\phi' = \frac{s}{1 + (s - 1) sin\phi'_{cv}} \tag{6.44}$$

$$sin\,\psi = s + (s - 1) sin\phi'_{cv} \tag{6.45}$$

Figure 6.42 shows an SBP test in Thanet Sands, a dense sand.

The data, replotted on a logarithmic scale, do lie about a straight line, which gives an angle of friction of 41°. Note that the test includes two unload/reload cycles, and it is for this reason, as well as inevitable disturbance during installation, that only the latter part (> 3.5% cavity strain) of the curve is plotted. This gives a post-peak angle of friction.

ϕ'_{cv} can be taken from the results of drained triaxial or direct shear tests. Houlsby et al. (1986) have suggested that ϕ'_{cv} could be found by interpreting the unloading portion of the curve (see chapter 4). Alternatively, a typical value can be taken from Table 6.11.

The choice of value of ϕ'_{cv} is not critical, as shown in Figure 6.43.

This solution is particularly sensitive to the reference datum chosen. Fahey and Randolph (1984) undertook a series of tests in which the SBP was intentionally drilled to disturb the sand using oversize shoes. They showed that adjusting the reference datum to take into account disturbance produced reasonable values of ϕ', though this was no substitute for quality drilling. Several reference data are chosen, and the most linear curve is taken to be that representing the true sand behaviour, which produce a series of curves similar to those shown in Figure 6.44.

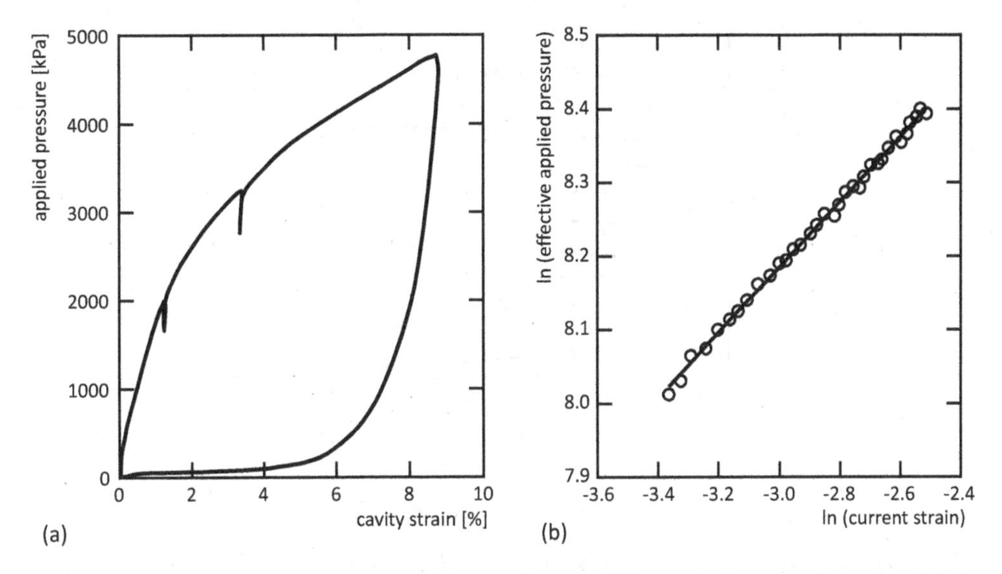

Figure 6.42 The assessment of the angle of friction from an SBP test in dense sand showing (a) the recorded data and (b) the relationship between effective applied pressure and strain.

Table 6.11 Typical values of ϕ'_{cv}

Soil type	ϕ'_{cv}
Well-graded gravel-sand-silt	40
Uniform coarse sand	37
Well-graded medium sand	37
Uniform medium sand	34
Well-graded fine sand	34
Uniform fine sand	30

Source: After Robertson and Hughes, 1986.

The most linear curve is taken to represent the strength of the sand.

The analysis proposed by Hughes et al. assumes that the relation between volumetric strain and shear strain is very nearly linear, and elastic strains are negligible. This is reasonable for dense sands, but not for loose sands, where shear-strain levels of between 20 and 30% are necessary for the assumption to be valid, as shown in Figure 6.45.

Interpreting tests in medium dense and loose sands will result in low angles of friction if the Hughes et al. method is used.

Robertson and Hughes (1986) proposed a semi-empirical method based on laboratory tests to correct the angles obtained from the Hughes et al. analysis. The shear strain in the annulus of sand under stress due to the expansion of the cavity varies, as shown in Figure 6.46.

The estimated average shear-strain level in the sand around a probe expanded to 10% cavity strain is approximately 10% with the maximum value, at the cavity wall, being about 20%.

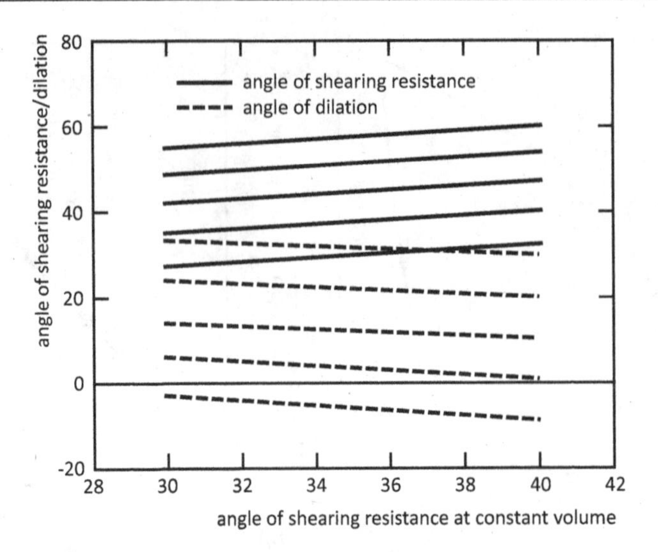

Figure 6.43 The effect of varying the angle of friction at constant volume with the angles of friction and dilation.

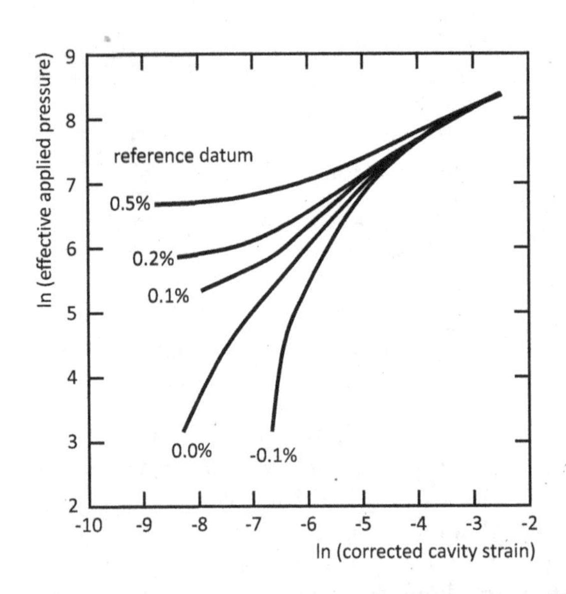

Figure 6.44 Adjusting the reference datum to obtain the angle of friction from SBP tests in sands.

Charts (Figure 6.47) used to select the angles of friction, are based on tests on Ottawa sand (Vaid et al. 1981), assuming that the cavity is expanded by 10% from the original diameter of the cavity at the in situ horizontal stress.

These charts have limited use because they are based on one type of sand. The methods proposed by Robertson and Hughes and Hughes et al. converge, as shown in Figure 6.48, as the cavity is expanded further and more sand reaches failure.

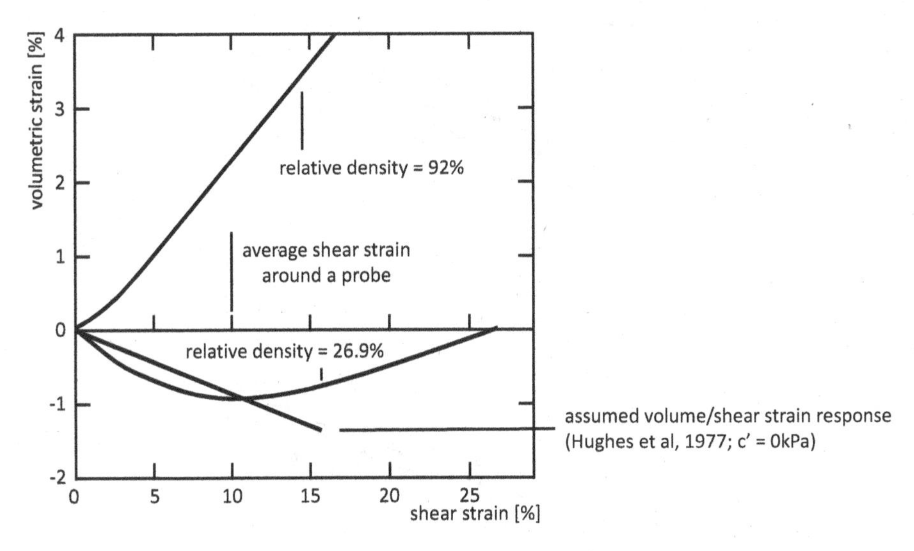

Figure 6.45 The effect relative density has upon the relation between volumetric and shear strain for sands.

Source: After Robertson and Hughes, 1986.

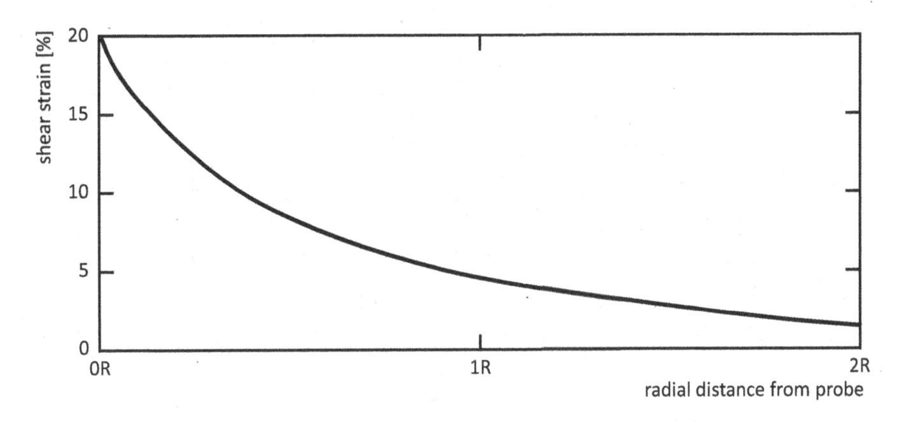

Figure 6.46 The approximate variation in average shear strain in the sand around an expanding cavity.

Withers et al. (1989) produced a series of nomograms, shown in Figure 6.49, to determine the angle of friction from SBP and FDP tests using both the expansion and contraction curves. The interpretation was based on Hughes et al. (1977) and Houlsby et al. (1986).

The small-strain analysis proposed by Hughes et al. (1977) does not strictly apply to FDP tests because of the large strains induced during installation. The interpretation of FDP tests in sands is based on empirical relationships developed from tests in calibration chambers (for example, Nutt and Houlsby, 1991; Schnaid and Houlsby, 1990). Results of chamber tests in silica and carbonate sands shown in Figure 6.50 suggest the following relationships:

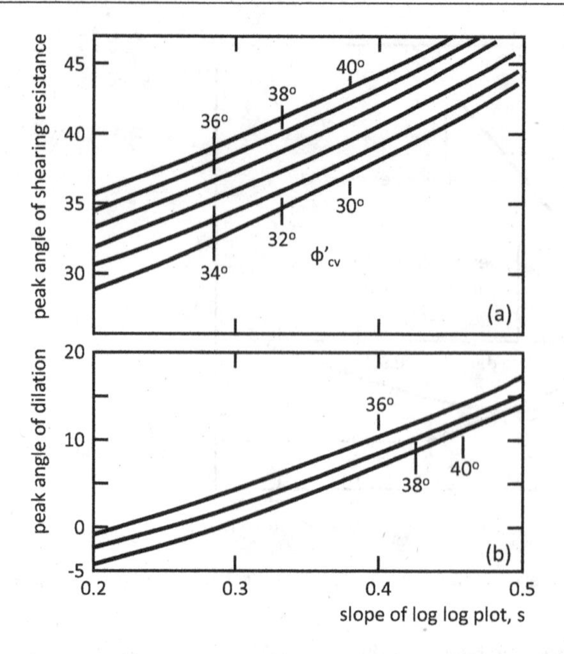

Figure 6.47 Correlation between (a) peak angle of friction and (b) peak angle of dilation corrected for strain level for different values of angle of friction at constant volume.

Source: After Robertson and Hughes, 1986.

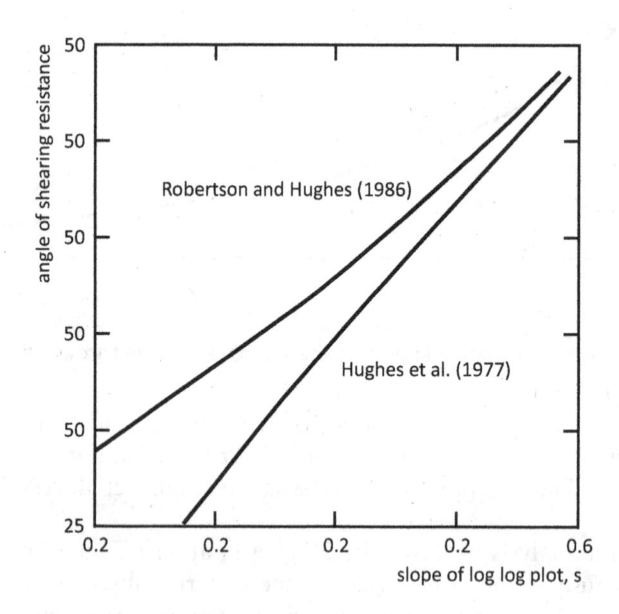

Figure 6.48 A comparison between the Robertson and Hughes (1986) empirical method and the Hughes et al. (1977) theoretical method of interpreting tests in sands to obtain the angle of friction.

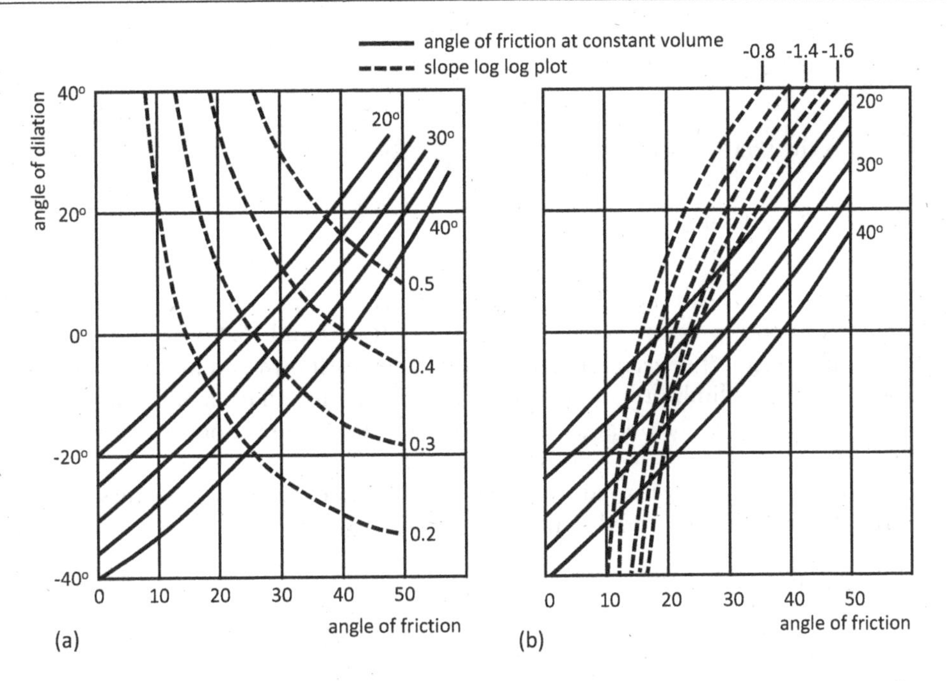

Figure 6.49 Nomograms to derive friction and dilation angles from pressuremeter expansion and contraction curves: (a) loading curves; (b) unloading curves.

Source: After Withers et al., 1989.

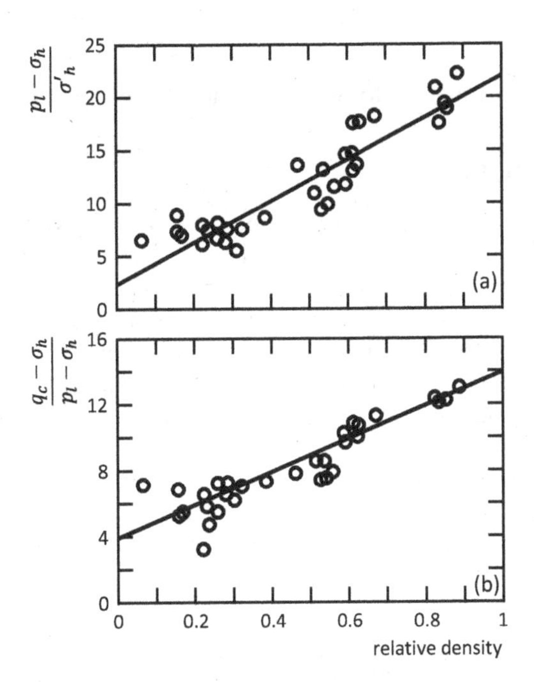

Figure 6.50 Correlation between relative density and (a) limit pressure and (b) cone resistance.

Source: After Houlsby and Nutt, 1993.

$$\frac{p_l - \sigma_h}{\sigma_h'} = 2.21 + 19.35D_r \tag{6.46}$$

where D_r is the relative density, and

$$\frac{q_c - \sigma_h}{\sigma_h'} = 3.80 + 9.84D_r \tag{6.47}$$

where q_c is the cone resistance. The excess pore pressure can be assumed to be zero and, if the ambient pore pressure is known, equations (6.46) and (6.47) can be combined to give a relationship between relative density, cone resistance and limit pressure. The cone resistance and limit pressure are measured during installation and testing. Thus, the relative density can be obtained and, hence, the angle of friction from correlations between D_r and ϕ'.

Van Wieringen (1982) showed, from theoretical and experimental studies, that the cone resistance could be related to the limit pressure by

$$q_c = 3p_1 \text{ for clays} \tag{6.48}$$

$$q_c = 15(\tan\phi)^{1.75}\, p_l \textit{ for sands} \tag{6.49}$$

Thus, correlations of q_c with ϕ' could be used to give an indirect measurement of ϕ'.

Mair and Wood (2013) suggest that PBP tests should not be used to evaluate ϕ' since the disturbance is so great, but Ménard suggested that

$$p_{lm} = b2^{(\phi'-24)/4} \tag{6.50}$$

where b is 1.8 for wet sand, 3.5 for dry sand and 2.5 on average.

6.8 LIMIT PRESSURE

The limit pressure, p_l is not an intrinsic property of soil but is used in design to determine other parameters from the test curve and to compare results from different tests. p_l is defined as the maximum pressure reached in a pressuremeter test at which the cavity will continue to expand indefinitely. In practice, it is not possible to reach this pressure since the expansion of the membrane is limited. The limit pressure can be obtained by extrapolating the test curve to infinity. Ménard redefined the limit pressure as the pressure required to double the cavity diameter so that it can be directly measured (see section 6.3).

Equation (6.51) gives a relationship between the applied pressure and volumetric strain for clays which, if plotted to a log scale (Figure 6.38), produces a straight line, which has a slope equal to c_u and an intercept equal to p_x. p_x is the pressure when the volumetric strain is 1; that is, the change in volume is equal to the current value $[\Delta V / V = \Delta V / V_o + \Delta V]$.

$$p - \sigma_u = c_u\left[1 + \ln\left(\frac{G}{c_u}\right) + \ln\left(\frac{\Delta V}{V}\right)\right] \tag{6.51}$$

A graph of pressure against logarithm of volumetric strain for a test in clay produces an S-shaped curve (see Figure 6.34) reflecting the reduction in strength with strain; that is, post-rupture behaviour. The slope of the tangent at any point on the curve is the shear stress. The intercept of that tangent with the pressure axis is the limit pressure. Thus, the limit pressure will reduce as the strain increases, since the slope of the tangent reduces (note that the curve lies to the left of the vertical pressure axis). This implies that the limit pressure should be determined by extrapolating the latter part of the test, assuming that a critical state condition is reached. Clarke (1993) showed, for CSBP tests in London Clay that the slope of the curve was constant above 6% cavity strain, suggesting that the limit pressure for tests in stiff clays could be taken from the extrapolated fit to the latter part of the test curve.

Hughes et al. (1977) developed a relationship between applied pressure and strain for tests in dense sands of the form

$$\ln\left(\varepsilon_{curr} + \frac{c}{2}\right) = \frac{1}{s}\ln(p') + C2 \tag{6.52}$$

where c, s and C2 are constants, c is negligible for dense sands and at an infinite expansion ε_{curr} is 1, therefore, using the definition of limit pressure given, ln (p_1) is equal to C2, the intercept given by the best-fit line. For the test shown in Figure 6.42 the intercept is 13.6 MPa. This limit pressure is an overestimate of the true limit pressure unless constant volume conditions are achieved. This method cannot be applied to medium dense sand or loose sands, since critical state conditions are not reached during a test.

Ghionna et al. (1989, 1990) developed a method based on Manassero's analysis that allows the limit pressure to be derived from

$$p_1' = p_{cv}'\left(\frac{\varepsilon_{vcv} + 2}{\varepsilon_{vcv} - 2\varepsilon_{cv}}\right)^{(1-1/K_{pcv})/2} \tag{6.53}$$

where p'_{cv} is the applied effective pressure at the onset of constant volume behaviour, ε_{vcv} is the volumetric strain of an element adjacent to the membrane at the onset of constant volume behaviour and ε_{cv} is the cavity strain at the onset of constant volume behaviour. This is based on the assumption that the volumetric strain is the sum of the radial and circumferential strains. At large strains the volumetric strain is given by

$$\varepsilon_v = \varepsilon_r + \varepsilon_\theta + \varepsilon_r\varepsilon_\theta \tag{6.54}$$

hence

$$p_1' = p_{cv}'\left(\frac{\varepsilon_{vcv} + 1}{\varepsilon_{vcv} - (\Delta V / V)_{cv}}\right)^{(1-1/K_{pcv})/2} \tag{6.55}$$

where $\Delta V / V_{cv}$ is the volumetric strain at the onset of constant volume. The onset of constant volume behaviour is selected from the curve of effective applied pressure against ln $(0.5\varepsilon_{vcv} + \varepsilon_{cv})$ $\left(or\left(\varepsilon_{vcv} - (\Delta V / V)_{cv}\right)\right)$ shown in Figure 6.51.

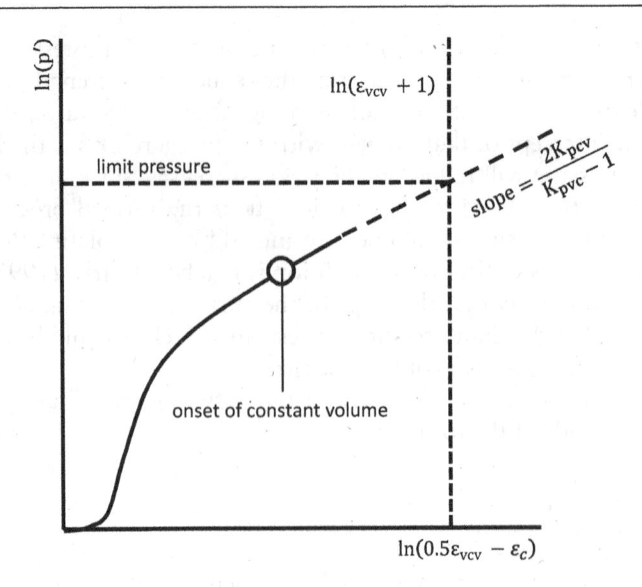

Figure 6.51 The assessment of limit pressure from tests in dense and loose sands.

Source: After Ghionna et al. 1989.

It is taken as the point at which the slope of the line becomes constant and equal to $\left(1-1/K_{pcv}\right)/2$. This only applies if there is sufficient strain for the sand at the cavity wall to reach critical state.

Plots of p' against $(0.5\varepsilon_{vcv} + \varepsilon_{cv})$ or p' against $(\varepsilon_{vcv} - \Delta V/V)$ to log scales will be linear but, if the sand has reached a state of constant volume, the slope will be $[0.5(1 - K_{pcv})]$. The line is extrapolated until ε_{curr} =1 or $\Delta V/V$ = 1 to obtain the limit pressure.

Ghionna et al. (1989) demonstrated that the magnitude of the volumetric strain at the onset of constant volume behaviour, ε_{vcv}, is small if the sand is dense, and ignoring it produced an underestimate of p_1' which is within 8% of the actual value. The ratio of limit pressure using the small-strain formulation (equation (6.37)) to that of the large-strain formulation (equation (6.40)) is about 1.2.

Constant volume conditions may not be reached in a test. In such cases, Ghionna et al. (1989) suggest that a line is drawn at a slope of $2/(1 - 1/K_{pcv})$ from the last data point on a plot of applied pressure against current cavity strain plotted to natural log scales, as shown in Figure 6.51. The pressure at a current cavity strain of 1 is the effective limit pressure. This will overestimate p_1'.

6.9 CONSOLIDATION AND CREEP

Pressuremeters can be used to measure time-dependent parameters by observing changes at the cavity wall while maintaining either a constant pressure or a constant displacement. Figure 6.52 shows the results of a constant displacement holding test carried out in soft clay with a CSBP.

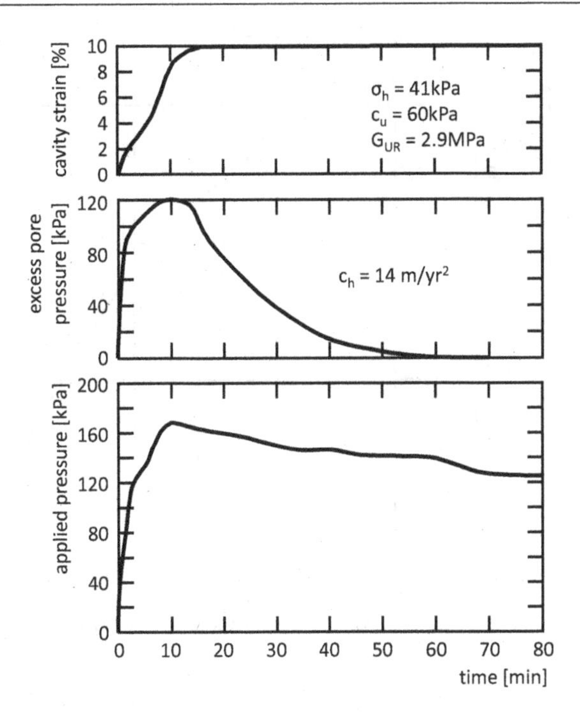

Figure 6.52 A typical holding test result.

Source: After Clarke et al., 1979.

There are several points to note. The predicted maximum pore pressure, 128kPa, is greater than the measured pore pressure since dissipation occurs throughout the test. The control system used for these tests cannot change instantly from a strain rate of 1%/min to 0%/min. It has to be done in stages, or the cavity wall overshoots and then retracts. There is further dissipation of pore pressure during this change. Huang et al. (1990) showed, from experimental studies, that the initial distribution of pore pressure has little effect on the time for dissipation. This implies that the effect of drainage during the loading stage is not critical.

The interpretation of this test is based on the time for 50% of the theoretical maximum pore pressure to dissipate. A single point is chosen because the test curve differs from the predicted curve. This is no different from other tests and since the curve is steepest at 50% the interpreted value of coefficient of consolidation is not greatly affected by the choice of time.

Parametric studies of the effect of permeability on a pressuremeter loading curve and the decay in total and pore pressure during a holding test by Baguelin et al. (1986) and Clarke (1990) showed that the test is only applicable to soils with a permeability less than 10^{-9} m/s (Figure 6.53).

Almost full drainage occurs in soils with coefficients of permeability greater than this, even though the shape of the total pressure curve changes by a small amount. The decay in total pressure differs from the decay in pore pressure, but at 50% of dissipation of pore pressure, the decay in total pressure is about 50% for practical purposes, as shown in Figure 6.54.

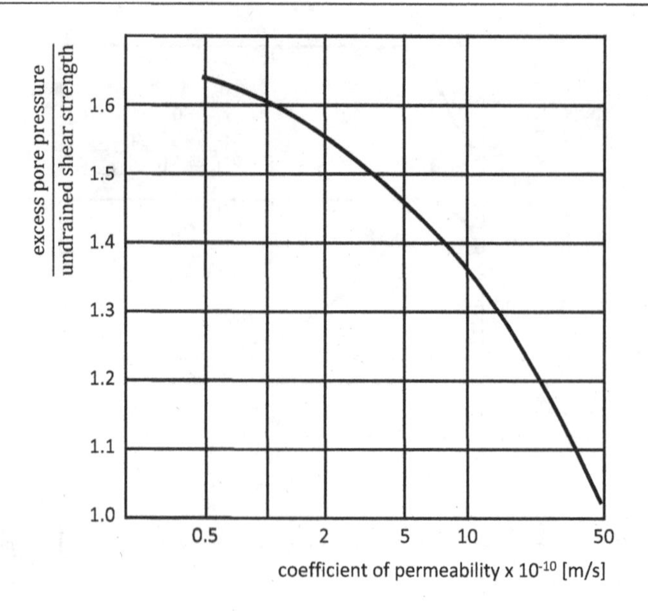

Figure 6.53 The influence of permeability on the initial excess pore pressure distribution at the start of a holding test for $\left[2\varepsilon_c \frac{G}{c_u} - 6.77\right]$.

Source: After Baguelin et al., 1986.

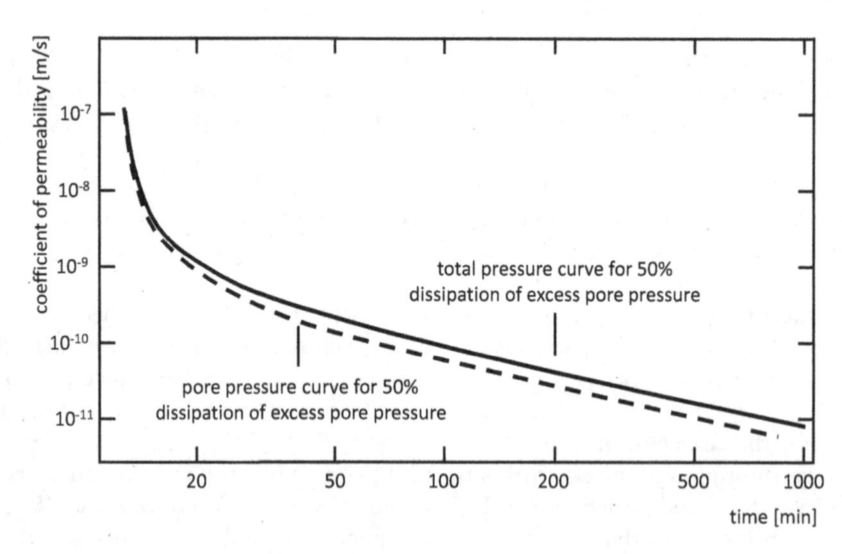

Figure 6.54 A comparison between the decay of pore pressure and total pressure using the 50% dissipation point.

Source: After Clarke, 1990.

Thus, the total pressure can be used to measure consolidation time. The advantage of this is that the total pressure is easily measured, whereas there are difficulties with the pore pressure measurement.

Clarke et al. (1979) found that CSBP holding tests gave values of c_h, an order of magnitude greater than laboratory-measured values. McKinlay and Shwaik (1983) found that this was the case when they used a PBP to carry out pressure- and displacement-controlled holding tests in till.

Creep in sands has been reported by Withers et al. (1989) and in rocks by Clarke and Smith (1992b), but they did not use the test to measure creep. Shields et al. (1989) used constant pressure holding tests lasting up to 119 hours to measure the creep of ice. Figure 6.55 shows that there is a correlation between the minimum strain rate and applied stress, which could be used to predict long-term creep settlements of ice.

It is normal to carry out quick undrained tests in clays. The Russian Standard recommends that slow tests are carried out to obtain drained moduli of elasticity. Denisov et al. (1982) described constant pressure tests in stiff clay using a 108 mm diameter PBP (Denisov et al., 1980). A typical test included several stages of loading with each increment being maintained for up to 28 hours. They concluded that the drained modulus of elasticity was about one-third of the undrained value of between 28 and 56 MPa.

Liu et al. (2018) suggested that the overall compression in the soil medium is more representative of consolidation progress than excess pore-water pressure decay at the cavity boundary. Thus, consolidation only occurs within the soil annulus that has yielded. The coefficient of horizontal permeability, k, is

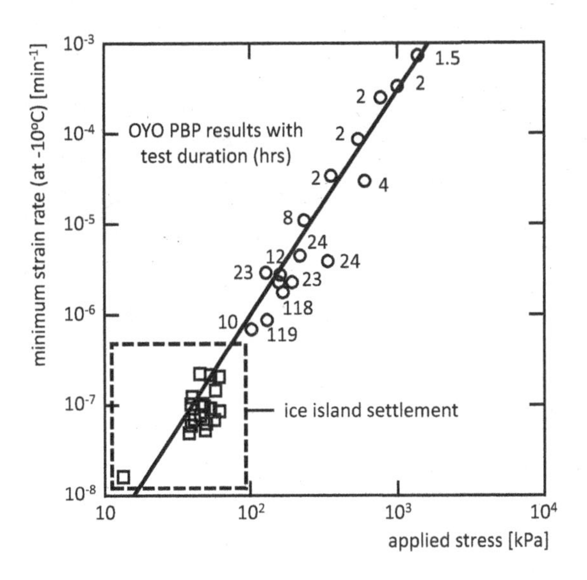

Figure 6.55 Creep rates determined from stress-controlled holding tests compared with those predicted from settlement of ice islands.

Source: After Shields et al. 1989.

$$k = \frac{m_h^e \gamma_w r_0^2}{t_{17}} \left(\frac{\Delta u_{max}}{a c_u} \right)^{1/b} \tag{6.56}$$

where m_h^e is the coefficient of volume compressibility

$$m_h^e = \frac{\kappa}{(1+e)\sigma'_r} = \frac{1}{M} = \frac{1-2\upsilon}{2G(1-\upsilon)} \tag{6.57}$$

where σ'_r is the effective radial stress at yield, e the void ratio, r_o the initial cavity radius, Δu_{max}, the maximum excess pore pressure generated during expansion. t_{17} is the time to 17% consolidation.

The cavity strain increment from constant-pressure pressuremeter tests is plotted against the root time (Figure 6.56) to obtain an S-shaped curve. Figure 6.57 shows the predicted time to reach 17% consolidation in a range of soils. The maximum slope on each test curve represents 17% consolidation. a and b are constants used to predict the time factor, T^e

$$\frac{\Delta u_{max}}{c_u} = ae^{(bT^e)} \tag{6.58}$$

They are selected from Figure 6.57
in which the values are plotted against the plastic volumetric strain ratio, Λ

$$\Lambda = 1 - \frac{\kappa}{\lambda} = 1 - \frac{C_r}{C_c} \tag{6.59}$$

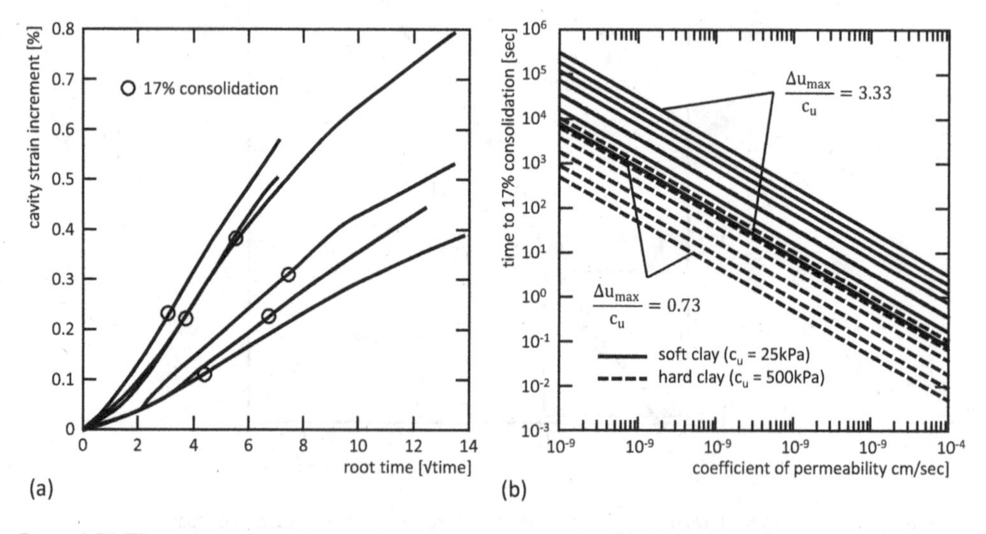

Figure 6.56 The root time against cavity strain increment for (a) a number of tests in Seattle Clay showing the time to 17% consolidation and (b) the time to 17% consolidation for a range of soils with $\Lambda = 0.75$.

Source: After Liu et al., 2018.

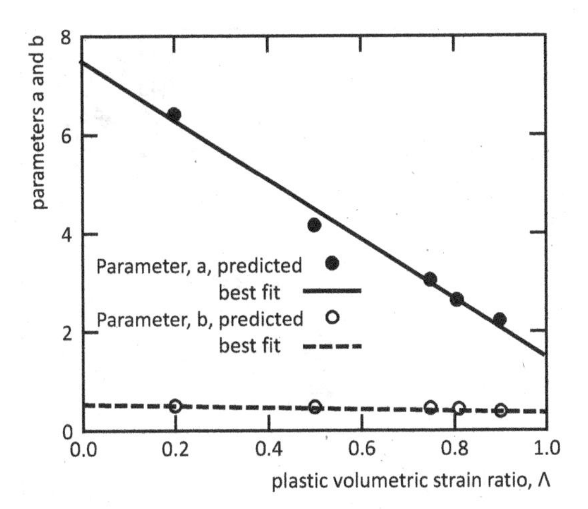

Figure 6.57 Values of the parameters a and b predicted from a numerical study for a range of plastic volumetric strain ratios.

Source: After Liu et al., 2018.

The transient response of the soil was analysed using a fluid–stress coupled model under fixed boundary loadings with the soil modelled as a modified Cam clay material, but with perfectly plastic behaviour for undrained expansion following the onset of yielding.

6.10 OVERCONSOLIDATION RATIO

The excess pore pressure generated during a pressuremeter test in clay is a function of the overconsolidation ratio. Mayne and Bachus (1989) undertook a study of excess pore pressures created during the installation of a piezocone and flat dilatometer (DMT) to develop empirical relationships between the pre-consolidation pressure and excess pore pressure. These were based on a theory combining critical state concepts with cavity expansion (Mayne and Bachus, 1988), which gives a relationship between the effective pre-consolidation pressure, p_c', and the excess pore pressure, Δu, of the form

$$p_c' = \frac{4\Delta u}{M \ln(G / c_u)} \tag{6.60}$$

$$p_c' = \frac{4c_u}{M} \tag{6.61}$$

Figure 6.58, showing the results of 42 pressuremeter tests, including CSBP, Oyometer and PAF tests, suggests that there is a correlation between p_c' and c_u and G which is a function of the fabric of the clay (intact or fissured).

Mori and Tajima (1964), Lukas and LeClerc de Bussy (1976) and Martin and Drahos (1986) suggested that the creep pressure, p_f, obtained from MPM tests is related to the pre-consolidation pressure, as shown in Table 6.12.

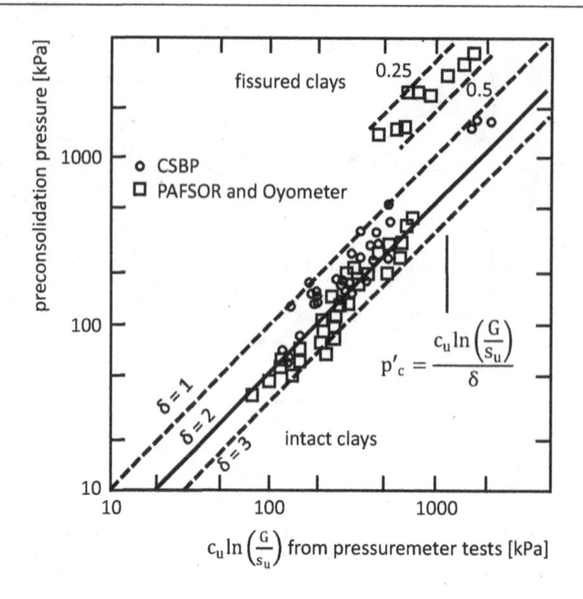

Figure 6.58 A relationship between pre-consolidation pressure, shear strength and rigidity index for clays.

Source: After Mayne and Bachus, 1989.

Table 6.12 Correlations between p_f and p_c

Author	*Soil type*	p_c
Mori and Tajima (1964)	All soils	p_f
Lukas and LeClerc de Bussy (1976)	Till	p_f
Martin and Drahos (1986)	Overconsolidated clay	$0.6p_f$

The correlations vary and Baguelin et al. (1978) suggested that correlations are site specific.

6.11 NUMERICAL ANALYSES

Predictions of ground deformation are increasingly dependent on the use of advanced constitutive models which, together with numerical studies, can model the complex stress paths associated with construction and operation. Therefore, to have confidence in the chosen model it is necessary to calibrate the model against field data. These advanced models are developed from theoretical studies and results of laboratory and field tests. The purpose of the calibration is to determine a set of parameters that best-fit a set of real soil test data to a pre-defined set of mathematical equations. The best-fit is assessed using an optimisation technique such as genetic algorithms. Examples of this approach include

- Maranha and Neves (2000) predicted surface settlements with some confidence due to an EPBS tunneling machine boring through alluvial clays using 3D analysis with linear elastic perfectly plastic soil model based on a variety of tests including pressuremeter tests;
- Monnet and Allagnat (2006) using results from MPM tests carried out in fill overlying gravel to develop a linear elastic plastic Mohr-Coulomb model to assist in the design of a diaphragm wall;
- Ho and Hu (2014) used a hardening model to predict the deformation of the diaphragm walls of a cut-and-cover station box constructed in fill, silty sands and varved glacial silts. K_o was assessed from a range of tests, including SBP tests; shear strength from triaxial tests; and stiffness from SPT tests using correlations with pressuremeter tests. A hardening soil model incorporating non-linear stress–strain behaviour and time-dependent properties of soils was used to predict the wall depths for under-seepage cutoff;
- Janin et al. (2015), using the Analysis of Controlled Deformation in Rocks and Soils technique, undertook a 3D back analysis of the observations of a tunnel bored through phyllitic bedrock. A prediction is made of the tunnel face and stability measures installed where necessary in advance of the excavation. A non-linear elasto-plastic hardening soil constitutive model was used to predict the ground deformation. The ground stiffness was obtained from MPM tests and shear strength parameters from a back analysis of the construction of an existing tunnel. They observed that the prediction using a 3D analysis was similar to the observations.
- Pedro et al. (2017) describe a hierarchal approach using a combination of a non-linear elastic non-linear elastic small-strain stiffness model stiffness model and a linear/non-linear Mohr-Coulomb plastic model. They used data from field and laboratory tests, including pressuremeter tests and, using engineering judgement, arrived at a set of properties for the ground profile consisting of a clay and silty sand. These were validated against observations of historical excavation in the ground profile.
- Paternesi et al. (2017) used a hardening soil model with small-strain stiffness in a 3D analysis of the deformation associated with the construction of twin tunnels in over consolidated fine-grained deposits. Parameters were derived from triaxial and MPM tests. The model was calibrated against triaxial and pressuremeter tests. This led to engineering judgement to finely tune the parameters. The predictions of ground deformation were compared with those actually observed showing general agreement.
- Conte et al. (2018) investigating a landslide in layers of silty sand and clayey silt underlain by evaporitic limestone. They concluded that the landslide was triggered by a significant increase in groundwater level that occurred after a long rainy period characterised by rainfall, not particularly intense, but distributed with time. A linear elastic perfectly plastic Mohr-Coulomb model with angle of dilatancy set to zero was used to model the behaviour of all soils with the exception of the gneiss and the marly clay. Linear elastic behaviour was assumed for the gneiss since it was not involved in the landslide. The strain softening behaviour of the marly clay was modelled using a Mohr-Coulomb model with

strain-softening behaviour modelled by reducing the strength parameters with the accumulated deviatoric plastic strain. The pressuremeter modus, E_m, limit pressure, p_{lm}, were used to develop the constitutive models.

- Miliziano and Lillis (2019) used a 3D finite element numerical model to simulate the main features of tunnel excavation and construction processes influencing the surface settlements, such as front pressure, geometry (including cutter head overcut and conicity) and weight of the shield, tail void grouting and grout hardening over time. They found that a simple elastic perfectly plastic soil model, calibrated at small-strain, led to reasonably accurate predictions of settlements due to a tunnel boring machine (TBM) used in man-made ground, medium-dense to loose coarse-grained soil, clayey silt and sandy silt, very dense silty sand and clayey silt, clayey silt and silty clay, and sandy gravel. The soil was modelled as an elastic perfectly plastic material with the stiffnesses derived from PMT, CPT and SPT tests. They also investigated the use of a more-advanced model, the hardening soil model to show that the simple model adequately predicted the settlement but suggested that more advanced models would be appropriate for detailed analysis of the behaviour of the ground.
- Cardoso Bernardes et al. (2021) used a hardening soil model with parameters derived from MPM tests to predict the settlement of axially loaded piles in a sandy clay and lateritic residual soils. They were able to predict the settlement at working load with some confidence but underestimated the ultimate load.

These examples suggest that it is possible to use the results for a range of tests, including pressuremeter tests, to develop an appropriate constitutive model. The model can be calibrated against pressuremeter data and validated against field observations. However, it is necessary to apply engineering judgement to refine the parameters. These examples are unique in that only one interpretation of each case study is available. Doherty et al. (2018) described an international exercise aimed at assessing the geotechnical engineering profession's ability to predict the response of shallow foundations on soft clay subjected to undrained loading. The ground profile consisted of a crust of alluvial clayey silty sand over soft estuarine clay, which is underlain by a transition zone of clay, silt and sand, then sand of varying thickness. The site was characterised using laboratory, SBP and CPT tests. They found that the predictions of bearing capacity varied by more than an order of magnitude and settlement by more than two orders of magnitude. This indicates that engineers with the same data and problems judge the data in different ways and ultimately produce very different results. It was demonstrated that the foundation performance could be reasonably predicted using simple models with parameters derived from the in situ and laboratory tests. A key issue is the fact that the predictive models are disconnected from the data that informs them. This is evident in the case studies presented above. Therefore, engineering judgement has to be used to refine the models, and case studies have to be assumed to validate the model.

Another aspect of these examples is that the more complex models are developed from a range of sources, including pressuremeter tests, yet the factors that affect the results vary between the tests, suggesting that soil properties are test-dependent. Richards et al. (2014) suggest that the interpretation of in situ tests requires an appreciation of the effects of installation procedures, loading direction, stress paths, strains and drainage conditions during testing. Further, the relative volume of ground tested means that the interpretation may yield either intact or mass properties. Thus, there is no reason to expect agreement between values measured from pressuremeter tests and those measured in other ways.

6.12 SUMMARY

A Ménard test is interpreted to give an elastic modulus and limit pressure, which are selected from specific parts of the test curve. These do not represent fundamental properties of the ground but are test-specific parameters used in design rules, as explained in Chapter 7.

Horizontal stress can be estimated directly from quality SBP tests or estimated from any pressuremeter test using a variety of methods. The selection of the correct value of horizontal stress is essential if peak strengths are required, since the cavity radius at that stress is the reference datum for further analysis of a test curve. Post-peak strengths and stiffnesses from unload/reload cycles are not too dependent on the reference datum chosen, provided that there is sufficient strain to overcome the effects of installation.

The post-peak undrained shear strength is simply the slope of applied pressure against natural logarithm of the volumetric strain for a test in clay. This should be corrected for non-cylindrical expansion. Soil is likely to exhibit strain softening or hardening, therefore the line will be curved. The strength should be taken from the latter part of the test curve, and the reference datum and strain range over which the strength is measured should be quoted.

The peak angle of friction derived from tests in sands is more sensitive to the reference datum than is the peak strength of clays. It is recommended that a post-peak value is quoted, since any installation will produce disturbance in bringing the sand adjacent to the membrane to yield, which implies it is unlikely that a peak angle of friction can be determined. The post-peak value is simply a linear function between the effective applied pressure and current cavity strain, both plotted to natural logarithm scales provided that the correct reference datum is chosen and the sand is dense. Tests in loose to medium dense sand should be interpreted using the method proposed by Manassero, unless constant volume conditions are achieved in the test. In that case the simple method is valid.

Angles of friction cannot be found directly from PBP and FDP tests. Empirical correlations do exist, and those for FDP tests are supported by quality chamber tests though only on a limited number of types of sand.

A secant modulus is simply half the slope of an unload/reload cycle, and this is an elastic modulus provided the amplitude of the cycle is limited to prevent failure in extension. This modulus is an average stiffness representing the integrated effect of the ground around the cavity. The applied pressure and strain at the start of unloading and the strain amplitude of the cycle should be quoted, since the shear modulus varies with effective stress and strain. Non-linear average and incremental stiffness profiles can be evaluated from an unload/reload cycle.

Constant displacement and pressure tests can be interpreted to evaluate time-dependent parameters such as the coefficient of consolidation and creep.

Pressuremeter tests are used to validate numerical studies of geotechnical structures incorporating more complex constitutive models.

REFERENCES

Amar, S., Baguelin, F., Jézéquel, J.F. and Le Méhauté, A. (1975) In situ shear resistance of clays, *Proc. ASCE Spec. Conf. In Situ Measurement of Soil Properties, Raleigh,* Vol. 1, pp. 22–45.

Arnold, M. (1981) Empirical evaluation of pressuremeter test data, *Can. Geotech. J., 18*(3), 455–459.

Baguelin, F. (1982) Rules of foundation design using self boring pressuremeter test results, *Proc. Int. Symp. Pressuremeter Marine Appl.,* Paris, pp. 347–360.

Baguelin, F., Jézéquel, J.F. and Shields, D.H. (1978) *The Pressuremeter and Foundation Engineering,* Trans. Tech. Publication.

Baguelin, F., Jézéquel, J.F. and Le Méhauté, A. (1979) Le pressiomètre autoforeur et le calcul des fondations, *Proc. 7th Eur. Conf. SMFE,* Brighton, Vol. 2, pp. 185–190.

Baguelin, F., Frank, R.A. and Nahra, R. (1986) A theoretical study of pore pressure generation and dissipation around the pressuremeter, *Proc. 2nd Int. Symp. Pressuremeter Marine Appl., Texam,* United States, ASTM STP950, pp. 169–186.

Battaglio, M., Ghionna, V., Jamiolkowski, M. and Lancellotta, R. (1981) Interpretation of selfboring pressuremeter tests in clays, *Proc. 10th Int. Conf. SMFE, Stockholm,* Vol. 2, pp. 433–438.

Baud J-P (2005) Analyse des résultats pressiométriques Ménard dans un diagramme spectral [Log(p_{LM}), Log (E_M/p_{LM})] et utilisation des regroupements statistiques dans la modélisation d'un site. In: Gambin, Magnan, Mestat (eds) *ISP5 – PRESSIO* 2005, vol 1. Presses ENPC, Paris, pp 167–174

Baud, J.P. and Gambin, M. (2008) Homogenising MPM test curves by using a hyperbolic model. *Huang, A.-B., & Mayne, PW* (eds) *Geotechnical an Geophysical Site Characterization, Proc. ISC'3* Taipei, Taiwan, 1–4 April 2008.

Baud, J.P. and Gambin, M. (2014) Soil and Rock classification from high pressure borehole expansion tests. *Geotechnical and Geological Engineering, 32*(6), pp. 1397–1403.

Baud, J.P., Gambin, M. and Uprichard, S. (1992) Modeling and automatic analysis of a Ménard pressuremeter test. *Géotechnique et informatique.*

Becue, J., Brucy, F. and Le Tirant, P. (1986) Proposed methods for application of pressuremeter, test results to designing of offshore foundations, *Proc. 2nd Int. Symp. Pressuremeter Marine Appl. Texam,* United States, ASTM STP 950, pp. 357–375.

Bellotti, R., Ghionna, V., Jamiolkowski, M., Robertson, P.K. and Peterson, R.W. (1989) Interpretation of moduli from self-boring pressuremeter tests in sand, *Geotechnique, 39*(2), 269–292.

Bolton, M.D. and Whittle, R.W. (1999) A non-linear elastic/perfectly plastic analysis for plane strain undrained expansion tests. *Géotechnique, 49*(1), pp. 133–141.

Briaud, J.-L., Lytton, R.L. and Hung, J.T. (1983) Obtaining moduli from cyclic pressuremeter tests, *J. Geotech. Engng Div., ASCE,* 109 (NGT5), 657–665.

Byrne, P.M. Salgado, F.M. and Howie, J.A. (1990) Relationship between the unload shear modulus from pressuremeter tests and the maximum shear moduli for sand, *Proc. 3rd Int. Symp. Pressuremeters,* Oxford, pp. 231–242.

Caceci, M.S. and Cacheris, W.P. (1984) Fitting curves to data, the simplex algorithm is the answer, *Byte,* Vol. 5, 339–360.

Cao, L.F., Teh, C.I. and Chang, M.F. (2001) Undrained cavity expansion in modified Cam clay I: Theoretical analysis. *Geotechnique, 51*(4), pp. 323–334.

Cardoso Bernardes, H., Martines Sales, M., Rodrigues Machado, R., José da Cruz Junior, A., Pinto da Cunha, R., Resende Angelim, R. and Félix Rodríguez Rebolledo, J. (2021) Coupling hardening soil model and Ménard pressuremeter tests to predict pile behavior. *European Journal of Environmental and Civil Engineering*, pp. 1–20.

Cassan, M. (1972) Corrélation entre essais in situ en Mécanique des soil. *Internal Report, Fondasol,* Avignon.

Clarke, B.G. (1990) Consolidation characteristics of clays from self-boring pressuremeter tests, *Proc. 24th Ann. Conf. of the Engng Group of the Geological Soc.: Field Testing in Engineering Geology*, Sunderland, pp. 19–35.

Clarke, B.G. (1993) The interpretation of pressuremeter tests to produce design parameters, *Predictive Soil Mechanics, Proc. Wroth Memorial Symp.*, Oxford, pp. 75–88.

Clarke, B.G. (1994) Peak and post rupture strengths from pressuremeter tests, *Proc. 13th Int. Conf. SMFE,* Delhi, India, Vol. 1, pp. 125–128.

Clarke, B.G. and Allan, P.G. (1990) Self-boring pressuremeter tests from a gallery at 220 m below ground, *Proc. 3rd Int. Symp. Pressuremeters,* Oxford, pp. 73–84.

Clarke, B.G. and Smith, A. (1992b) Self-boring pressuremeter tests in weak rocks, *Construction and Building Materials,* 6(2), 91–96.

Clarke, B.G., Carter, J.P. and Wroth, C.P. (1979) In-situ determination of the consolidation characteristics of saturated clays, *Proc. 7th Eur. Conf. SMFE,* Brighton, Vol. 2, pp. 207–211.

Conte, E., Donato, A., Pugliese, L. and Troncone, A. (2018) Analysis of the Maierato landslide (Calabria, Southern Italy). Landslides, *15*(10), pp. 1935–1950.

Denby, G.M. (1978) "Self-boring Pressuremeter Study of San Francisco Bay Mud," PhD Thesis, University of Stanford.

Denby, G.M. and Clough, G.W. (1980) Self-boring pressuremeter tests in clay, *J. Geotech. Engng Div.,* ASCE, 106 (GT12), 1369–1387.

Denisov, V.N., Chetyrkin, N.S., Rumiantsev, B.N. and Semenov, V.I. (1980) Device for soil deformation determination in boreholes, *Bull. Goscomizobreteniy,* No. 27.

Denisov, V.N., Chetyrkin, N.S. and Golubev, A.V. (1982) Long term investigation of soil deformability by automatic pressuremeter, *Proc. Int. Symp. Pressuremeter and its Marine Appl.,* Paris, pp. 91–102.

Doherty, J.P., Gourvenec, S. and Gaone, F.M. (2018) Insights from a shallow foundation load-settlement prediction exercise. *Computers and Geotechnics*, *93*, pp. 269–279.

Fahey, M. (1992) Shear modulus of cohesionless soil: Variation with stress and strain level, *Can. Geotech. J., 29*, 157–161.

Fahey, M. and Jewell, R. (1990) Effect of pressuremeter compliance on measurement of shear modulus, *Proc. 3rd Int. Symp. Pressuremeters,* Oxford, pp. 115–124.

Fahey, M. and Randolph, M.F. (1984) Effect of disturbance on parameters derived from self-boring pressuremeter tests in sand, *Geotechnique, 34*(1), 81–97.

Ferreira, R.S. (1992) "Interpretation of pressuremeter tests using a curve fitting technique," PhD Thesis, University of Alberta.

Ferreira, R.S. and Robertson, P.K. (1992) Interpretation of undrained self-boring pressuremeter test results incorporating unloading, *Can. Geotech. J., 29,* 918–928.

Ghionna, V.N., Jamiolkowski, M. and Lancellotta, R. (1982) Characteristics of saturated clays as obtained from SBP tests, *Proc. Int. Symp. Pressuremeter and its Marine Appl.,* Paris, pp. 165–186.

Ghionna, V.N., Jamiolkowski, M., Lancellotta, R. and Manassero, M. (1989) Limit pressure of pressuremeter tests, *Proc. 12th Int. Conf SMFE,* Rio de Janeiro , Vol. 1, pp. 223–226.

Ghionna, V.N., Jamiolkowski, M. and Manassero, M. (1990) Limit pressure in expansion of cylindrical cavity in sand, *Proc. 3rd Int. Symp. Pressuremeters,* Oxford, pp. 149–158.

Gibson, R.E. and Anderson, W.F. (1961) In situ measurements of soil properties with the pressuremeter, *Civ. Engng Public Wks. Rev., 56*, 615–618.

GOST 20276–12 (2012) updated version of GOST 20276–85 (translated by Foque, J.B. and Sousa Coutinho, G.F.) *Soils Methods for Determining Deformation Characteristics.*

Haberfield, C.M. and Johnston, I.W. (1993) Factors influencing the interpretation of pressuremeters tests in soft rocks, *Proc. Conf. Geotechnical Engng Hard Soils-Soft Rocks,* Athens, Vol. 1, pp. 525–531.

Hansbo, S. and Pramborg, B. (1990) Experience of the Ménard pressuremeter in foundation design, *Proc. 3rd Int. Symp. Pressuremeters,* Oxford, pp. 361–370.

Hawkins, P.G., Mair, R.J., Mathieson, W.G. and Muir Wood, D. (1990) Pressuremeter measurement of total horizontal stress in stiff clay, *Proc. 3rd Int. Symp. Pressuremeters,* Oxford, pp. 321–330.

Ho, C.E. and Hu, S. (2014) Design optimization of underground subway station diaphragm walls using numerical modeling. In *Geo-Congress 2014: Geo-characterization and Modeling for Sustainability* (pp. 3122–3132).

Houlsby, G.T. and Nutt, N.R.F. (1993) Development of the cone pressuremeter, *Predictive Soil Mechanics, Proc. Wroth Memorial Symp.,* Oxford, pp. 254–271.

Houlsby, G.T. and Withers, N.J. (1988) Analysis of the cone pressuremeter test in clay, *Geotechnique, 38*(4), 575–587.

Houlsby, G.T., Clarke, B.G. and Wroth, C.P. (1986) Analysis of unloading of a pressuremeter in sand, *Proc. 2nd Int. Symp. Pressuremeter Marine Appl., Texam,* United States, ASTM STP 950, pp. 245–264.

Houlsby, G.T. and Carter, J.P. (1993) The effects of pressuremeter geometry on the results of tests in clay. *Geotechnique, 43*(4), pp. 567–576.

Huang, A.B., Chameau, J.L. and Holtz, R.D. (1986) Interpretation of pressuremeter data in cohesive soils by simple algorithm, *Geotechnique, 36*(4), 599–603.

Huang, A.B., Holtz, R.D. and Chameau, J.L. (1990) Pressuremeter holding tests in a calibration chamber, *Proc. 3rd Int. Symp. Pressuremeters,* Oxford, pp. 253–262.

Hughes, J.M.O., Wroth, C.P. and Windle, D. (1977) Pressuremeter tests in sands, *Geotechnique, 27*(4), 455–477.

Janin, J.P., Dias, D.F.R.H., Emeriault, F., Kastner, R., Le Bissonnais, H. and Guilloux, A. (2015) Numerical back-analysis of the southern Toulon tunnel measurements: A comparison of 3D and 2D approaches. *Engineering Geology, 195*, pp. 42–52.

Jardine, R.J. (1991) Discussion on "Strain dependent moduli and pressuremeter tests," *Geotechnique, 41*(4), 621–626.

Jefferies, M.G. (1988) Determination of horizontal geostatic stress in clay with self-bored pressuremeter, *Can. Geotech. J.,* 25 (3), 559–573.

Jefferies, M.G., Crooks, J.H.A., Becker, D.E. and Hill, P.R. (1987) Independence of geostatic stress from overconsolidation in some Beaufort Sea clays, *Can. Geotech. J.,* 24(3), 342–356.

Johnson, L.D. (1986) Correlation of soil parameters from in situ and laboratory tests for building, Proc. In situ'86: Use of In Situ Tests in Geot. Engng, Blacksburg VA, pp. 635–648.

Lacasse, S., Jamiolkowski, M., Lancellotta, R. and Lunne, T. (1981) In-situ characteristics of two Norwegian clays, *Proc. 10th Int. Conf. SMFE, Stockholm,* Vol. 2, pp. 507–511.

Lacasse, S. and Lunne, T. (1982) In situ horizontal stress from pressuremeter tests, Proc. Int. Symp. Pressuremeter Marine Appl., Paris, pp. 187–208.

Law, K.T. and Eden, W.J. (1980) Influence of cutting shoe size in self-boring pressuremeter tests in sensitive clays, *Can. Geotech. J., 17*(2), 165–173.

Liu, L., Elwood, D., Martin, D. and Chalaturnyk, R. (2018) Determination of permeability of overconsolidated clay from pressuremeter pressure hold tests. *Canadian Geotechnical Journal, 55*(4), pp. 514–527.

Lukas, G.L. and Ledere de Bussy, B. (1976) Pressuremeter and laboratory test correlations for clays, *J. Geotech. Engng Div., ASCE, 102* (GT9), 954–963.

Mair, R.J. and Wood, D.M. (2013) *Pressuremeter Testing: Methods and Interpretation*. Elsevier.

Manassero, M. (1989) Stress-strain relationships from drained self-boring pressuremeter tests in sands, *Geotechnique, 39*(2), 293–307.

Maranha, J.R. and das Neves, E.M., 2000, November. 3D analysis of ground displacements due to the construction of Lisbon underground. In *ISRM International Symposium*. OnePetro.

Marsland, A. and Randolph, M.F. (1977) Comparisons on the results from pressuremeter tests and large in-situ plate tests in London clay, *Geotechnique, 27*(2), 217–243.

Martin, R.E. and Drahos, E.G. (1986) Pressuremeter correlations for preconsolidated clay, *Proc. In situ '86: Use of In Situ Tests in Geot. Engng*, Blacksburg, VA, pp. 206–220.

Mayne, P.W. and Bachus, R.C. (1988) Profiling OCR in clays by piezocone, *Proc. ISOPT-1 Penetration Testing*, Orlando, Vol. 2, pp. 659–667.

Mayne, P.W. and Bachus, R.C. (1989) Penetration pore pressures in clay by CPTU, DMT, and SMP, *Proc. 12th Int. Conf. SMFE*, Rio de Janeiro, Vol. 1, pp. 291–294.

McKinlay, D.G. and Shwaik, R. (1983) Pressuremeter measurement of consolidation rate in a glacial till, *Proc. Int. Symp. Soil and Rock Investigations by In-situ Testing*, Paris, Vol. 2, pp. 341–346.

Ménard, L. (1957) Mesures in situ des propriétés physiques des sols, *Annales des Ponts et Chaussées*, Paris, No. *14*, 357–377.

Ménard, L. (1975) *The Ménard pressuremeter: interpretation and application of the pressuremeter test results to foundations design*, Sols Soils, No. 26, pp. 5–44.

Miliziano, S. and de Lillis, A. (2019) Predicted and observed settlements induced by the mechanized tunnel excavation of metro line C near S. Giovanni station in Rome. *Tunnelling and Underground Space Technology, 86*, pp. 236–246.

Mo, P.Q. and Yu, H.S. (2017) Undrained cavity expansion analysis with a unified state parameter model for clay and sand. *Géotechnique, 67*(6), pp. 503–515.

Monnet, J. and Allagnat, D. (2006) Interpretation of pressuremeter results for design of a diaphragm wall. *Geotechnical Testing Journal, 29*(2), pp. 126–132.

Mori, H. and Tajima, S. (1964) The applications of pressiometre method to design of deep foundations, *Soil and Foundations, Tokyo, 4*(2), 34–44.

Muir-Wood, D. (1990) Strain dependent moduli and pressuremeter tests, *Geotechnique, 40*(26), 509–512.

Newman, R.L. (1991) Interpretation of data from self-boring pressuremeter tests for the assessment of design parameters in sand, *Tech. Sem. Pressuremeters for Design in Geotechics*, Soil Mechanics, UK, No. 3.

Nutt, N.R.F. and Houlsby, G.T. (1991) Calibration tests on the cone pressuremeter in sand, *Proc. 1st Int. Conf. on Calibration Testing*, New York, pp. 265–276.

Oztoprak, S. and Bolton, M.D. (2013) Stiffness of sands through a laboratory test database. *Géotechnique, 63*(1), pp. 54–70.

Oztoprak, S., Sargin, S., Uyar, H.K. and Bozbey, I. (2018) Modeling of pressuremeter tests to characterize the sands. *Geomechanics and Engineering, 14*(6), pp. 509–517.

Palmer, A.C. (1972) Undrained plane-strain expansion of a cylindrical cavity in clay: a simple interpretation of the pressuremeter test, *Geotechnique*, 22 (3), 451–457.

Paternesi, A., Schweiger, H.F. and Scarpelli, G. (2017) Parameter calibration and numerical analysis of twin shallow tunnels. *Rock Mechanics and Rock Engineering, 50*(5), pp. 1243–1262.

Pedro, A.M., Zdravković, L., Potts, D. and e Sousa, J.A. (2017) Derivation of model parameters for numerical analysis of the Ivens shaft excavation. *Engineering Geology, 217*, pp. 49–60.

Powell, J.J.M. (1990) A comparison of four different pressuremeters and their methods of interpretation in a stiff heavily overconsolidated clay, *Proc. 3rd Int. Symp. Pressuremeters,* Oxford, pp. 287–298.

Prapaharan, S., Chameau, J.L., Altschaeffl, A.G. and Holtz, R.D. (1990) Effect of disturbance on pressuremeter results in clays, *J. Geotech. Engng Div.,* ASCE, *116*(1), 35–53.

Prévost, J.-H. (1979) Undrained shear tests on clays, *J. Geotech. Engng Div.,* ASCE, 105 (NGT1), 49–64.

Prévost, J.-H. and Hoeg, K. (1975) Analysis of pressuremeter in strain softening soil, J. Geotech. Engng Div., ASCE, 101 (GT8), 717–732.

Pyrah, I.C., Anderson, W.F. and Pang, L.S. (1988) Effects of test procedure on constant rate of strain pressuremeter tests in clay, *Proc. 6th Conf. Numerical Methods in Geomechanics,* Innsbruck, pp. 647–652.

Richards, D.J., Clayton, C.R.I., Powrie, W. and Hayward, T. (2014) Geotechnical analysis of a retaining wall in weak rock. *Proceedings of the Institution of Civil Engineers-Geotechnical Engineering, 157*(1), pp. 13–26.

Robertson, P.K. and Ferreira, R.S. (1993) Seismic and pressuremeter testing to determine soil modulus, *Predictive Soil Mechanics, Proc. Wroth Memorial Symp.,* Oxford, pp. 434–448.

Robertson, P.K. and Hughes, J.M.O. (1986) Determination of properties of sand from self-boring pressuremeter tests, *Proc. 2nd Int. Symp. Pressuremeter Marine Appl., Texam,* United States, ASTM STP 950, pp. 283–302.

Sayed, S.M. and Hamed, M.A. (1988) Pressuremeter test and disturbance effects, *J. Geotech. Engng Div.,* ASCE, *114*(5), 631–637.

Schnaid, F. and Houlsby, G.T. (1990) Calibration chamber tests of the cone pressuremeter in sand, *Proc. 3rd Int. Symp. Pressuremeters,* Oxford, pp. 263–272.

Shields, D.H., Domaschuk, L., Funegard, E.G. and Azizi, F. (1989) Pressuremeter creep tests in spray ice, *Proc. 12th Int Conf. SMFE,* Rio de Janeiro, Vol. 1, pp. 313–317.

Vaid, Y.P., Byrne, P.M. and Hughes, J.M.O. (1981) Dilation angle and liquefaction potential, *Proc. Int. Conf. Recent Advances in Geotechnical Earthquake Engineering and Soil Dynamics,* St Louis, Vol. 1, pp. 161–165.

Van Wieringen, J.B.M. (1982) Relating cone resistance and pressuremeter test results, *Proc. 2nd Eur. Symp. Penetrating Testing,* Amsterdam, pp. 951–955.

Walker, L.K. and Jewell, R.L. (1979) The selection of design parameters in weathered rocks, *Proc. 7th Eur. Conf. SMFE,* Brighton, Vol. 2, pp. 287–294.

Windle, D. and Wroth, C.P. (1977) Use of self-boring pressuremeter to determine the undrained properties of clays, *Ground Engng, 10*(6), 37–46.

Withers, N.J., Howie, J., Hughes, J.M.O. and Robertson, P.K. (1989) Performance and analysis of cone pressuremeter tests in sands, *Geotechnique, 39*(3), 433–454.

Wroth, C.P. (1982) British experience with the self-boring pressuremeter, *Proc. Int. Symp. Pressuremeter and its Marine Appl.,* Paris, pp. 143–164.

Wroth, C.P. and Hughes, J.M.O. (1973) An instrument for the in-situ measurement of the properties of soft clays, *Proc. 8th Int. Conf. SMFE,* Moscow, Vol. 1.2, pp. 487–494.

Yeung, S.K. and Carter, J.P. (1990) Interpretation of the pressuremeter test in clay allowing for membrane end effects and material non-homogeneity, *Proc. 3rd Int. Symp. Pressuremeters,* Oxford, pp. 199–208.

Chapter 7

Design rules and applications

7.1 INTRODUCTION

The majority of the technical risk of a project is in the ground because of the spatial variation in type of ground and its properties, the limited investigation imposed by time and cost, and limited understanding of the response of the ground to loading and unloading. Further, many design methods are semi-empirical developments of theoretical models based on laboratory tests. Test results are stress-path dependent, which means that the parameters derived from pressuremeter tests may be different from those derived from other tests. Thus, the impact of ground hazards are often reduced by applying relatively high factors of safety compared to those applied to manufactured materials.

A range of techniques is used to reduce the risk, including the following:

a. Databases of ground response linked to ground investigation data; the basis of the Ménard approach.
b. A regional database of ground properties to provide a robust means of validating site-specific ground investigation data (e.g., Jamiolkowski et al., 2009).
c. Multi-stage approach to ground investigation to produce a robust ground model.
d. Use of site-specific correlations to link simpler tests that can be used extensively within a project with more sophisticated tests. An example of this approach is the use of correlations between standard penetration blow counts and the stiffness of soil derived from pressuremeter tests.
e. Development of realistic constitutive models to improve the prediction of the behaviour of ground subject to unloading/loading using numerical methods. In this case, pressuremeter tests are used to validate the numerical methods.

However, when presented with the same data engineers will apply that data in different ways to predict ground behaviour (e.g., Briaud and Gibbens, 1999; Lehane, 2003; Lehane et al., 2008; Doherty et al., 2018). This is in part due to the fact that the constitutive models are often test-specific (e.g., many models are developed from the results of triaxial tests) and engineering judgement is used to select the appropriate parameters. Therefore, to reduce this uncertainty it is important to apply a consistent approach in undertaking pressuremeter tests and interpreting the data to produce the properties required for design.

DOI: 10.1201/9781003028925-7

Ground investigations are designed to determine the ground model, hydrogeological properties and geotechnical properties. Methods used to determine geotechnical properties include empirical and semi-empirical correlations, laboratory tests and in situ tests; and the investigation findings depend on the ground type, the availability of testing equipment and cost. This has led to numerous debates as to which are the best methods. In practice, there are advantages to all methods of testing and, provided test procedures are followed, it is possible to produce consistent results for any test. Different tests will give different properties because the boundary conditions and stress paths vary. Site-specific correlations are often developed to compare results from different tests. Ideally, a holistic approach should be taken when creating the ground model. This means the design methods and methods of analysis should be compatible with the selected tests. This is often not possible. Hence, the use of correlations. However, regardless of the test method, provided the tests are carried out correctly, it should be possible to derive parameters for design and analysis.

This chapter provides examples of the use of pressuremeter in design (Figure 7.1).

The pressuremeter is used to produce design parameters (e.g., the Ménard method), or determine intrinsic properties of the ground using theoretical and numerical methods. The former method relies on a systematic collection of subsoil information and the performance of structures. It was developed by Ménard for prebored pressuremeter tests and is the basis of French design codes which are now incorporated in Eurocode 7 (ENV 1997-1., 2004). This approach, the direct method, is based on the philosophy that the behaviour of full-scale foundations can be related to parameters obtained from empirical correlations supported by theory and is used with other in situ tests including standard penetration, cone penetration and plate tests. It is important that the design parameters are obtained in a standard manner and it is for this reason that the Ménard method is separately identified. In this method the probe, the installation and test procedure, and interpretation are all specified as discussed in previous chapters. The direct method has also been developed for use with SBP and FDP tests.

The latter, the indirect method, includes theoretical and close formed solutions and numerical methods using appropriate constitutive models, and site specific correlations. The parameters obtained are a function of the installation and test procedure and

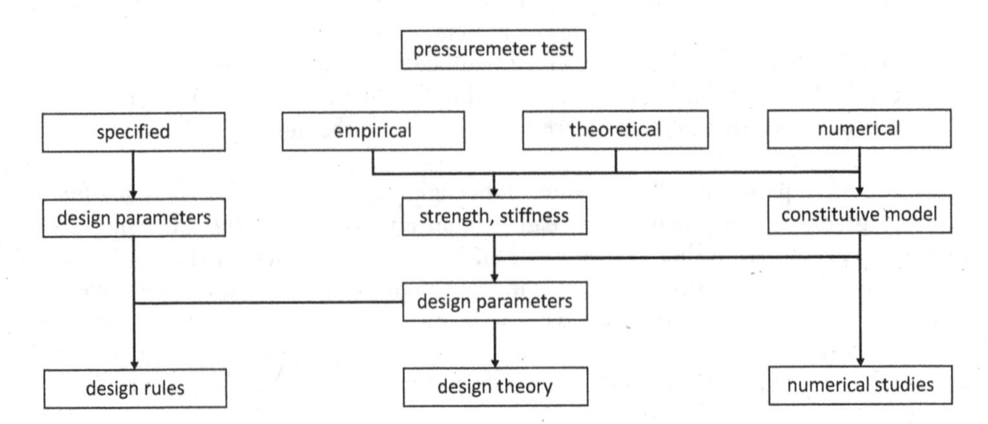

Figure 7.1 The pressuremeter in geotechnical design and analysis.

method of interpretation. It is, therefore, important to follow standard procedures whenever possible to standardise the effects of the site operations. The indirect method still requires a degree of empiricism in the application of the properties derived from a pressuremeter test since the test does not truly model the behaviour of the ground subject to loading or unloading. This is no different to other in situ and laboratory tests.

The theoretical and numerical methods use constitutive models which range from the simple linear elastic model based on a single parameter used to determine a secant shear modulus to sophisticated models that take into account non-linear elastic, strain hardening/softening, and structured soils. Typically simpler models (e.g. Gibson and Anderson, 1961; Palmer, 1972; Hughes et al., 1977; Clarke et al., 1979; Wroth, 1982; Jamiolkowski et al., 1985; Manessero, 1989; Houlsby and Withers, 1988; Schnaid, 1990; Byrne et al., 1990; Jardine, 1992; Fahey and Carter, 1993; Yu, 1994, 1996, 2000; and Bolton and Whittle, 1999) are used to interpret tests to produce the characteristic mechanical properties of the ground and the sophisticated models are used to gain an insight using numerical methods into the response of the ground to a pressuremeter test.

7.2 THE DIRECT METHOD: THE MÉNARD MÉTHOD

In the direct method developed for the Menard pressuremeter (MPM) it is assumed that the ultimate bearing capacity is related to the limit pressure, p_{lm}, and the settlement to the pressuremeter modulus, E_m. These relationships developed from theoretical studies of foundation behaviour and observations of full-scale structures which, over the years, has led to a substantial database that is used to refine and improve the design methods.

A pressuremeter bearing capacity factor, k, is defined by

$$k = \frac{q_u - \sigma_v}{p_{lm} - \sigma_h} \tag{7.1}$$

where q_u is the ultimate bearing capacity, σ_v is the total vertical stress at the formation level and a_h is the total horizontal stress at the pressuremeter test level. The factor k depends on

(a) the type of ground;
(b) the depth of the foundation;
(c) the shape of the foundation;
(d) the method of construction.

The minimum value of k (= 0.8) corresponds to a foundation at the surface, k increases with depth and becomes constant below a critical depth, which is a function of the equivalent dimension of the foundation, B_e, in homogeneous ground conditions. Critical depths are given in terms of B_e (equation (7.2)) in Table 7.1 where

$$B_e = \frac{4 \times area\ of\ foundation}{perimeter\ of\ foundation} \tag{7.2}$$

(B_e is twice the breadth of a strip footing).

Table 7.1 Critical depth below which k is constant when using the MPM design rules

Ground type	p_{lm} (kPa)	Category	Critical depth as a function of the equivalent dimension	
			Pad foundation	Strip foundation
Clay	0–1200	I	$2B_e$	$3B_e$
Silt	0–700			
Firm clay or marl	1800–4000	II	$5B_e$	$6B_e$
Compact silt	1200–3000			
Compressible sand	400–800			
Soft or weathered rock	1000–3000			
Sand and gravel	1000–2000	III	$8B_e$	$9B_e$
Rock	4000–10000			
Very compact sand and gravel	3000–6000	IV	$10B_e$	$11B_e$

Deep foundations have an additional load-carrying capacity due to the adhesion between the ground and the side of the foundation. The adhesion is a function of the limit pressure and depends on

(a) the shape of the foundation;
(b) the surface of the foundation;
(c) the method of construction.

The Ménard method for settlement is based on a modulus of elasticity, which is expressed in terms of the pressuremeter modulus, E_m. The settlement, s, is given by

$$s = \left(\frac{q - \sigma_v}{9E_m}\right) \times \text{a shape factor} \tag{7.3}$$

where q is the total bearing pressure.

The shape factor depends on

(a) the size of the foundation;
(b) the type of soil.

Details are given below of the application of equations (7.1) and (7.3) to determine the performance of shallow foundations, axially loaded single piles, horizontally loaded piles and grouted anchors.

7.2.1 Shallow foundations

The pressuremeter limit pressure and ultimate bearing capacity of a shallow foundation are both functions of the expansion of cavities. The pressuremeter is the expansion of a cylindrical cavity; a shallow foundation is analogous to the expansion of a spherical cavity. Figure 7.2 shows the variation in pressure with distance from the load and contours of equal stress for a shallow foundation and a pressuremeter test.

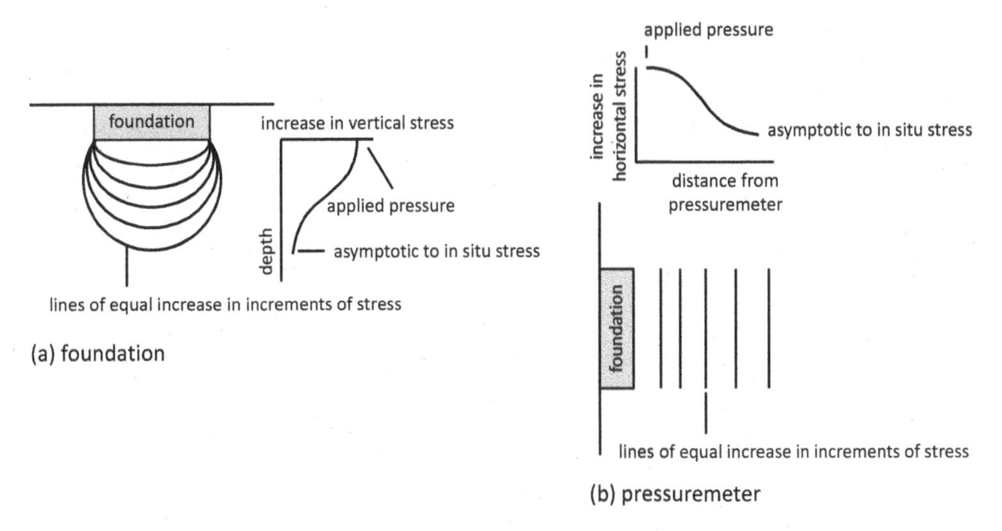

Figure 7.2 An analogy between (a) the ultimate capacity of a shallow foundation and (b) the pressuremeter limit pressure.

Source: After Briaud, 1992.

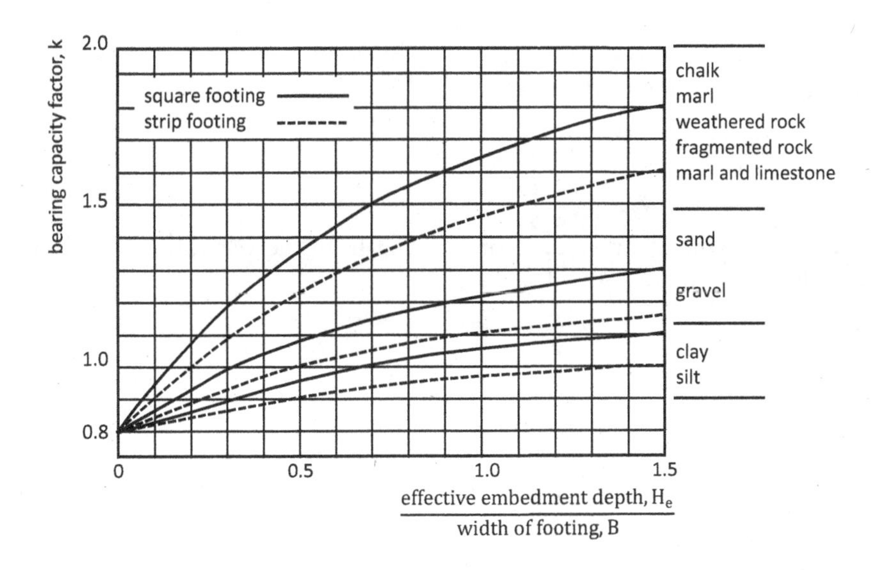

Figure 7.3 Pressuremeter bearing capacity factors for shallow foundations.

It can be shown, using these analogies, that k will vary between 1.3 and 4 depending on ground type.

The values of k (Figure 7.3) for shallow foundations are adjusted to take into account the depth of foundation and all ground conditions within the zone of influence of the foundation.

The ultimate bearing capacity is calculated in the following manner.

1. An average net limit pressure $(p_{1m} - \sigma_v)_e$ is calculated using the results of tests from within 1.5*B* of the formation level, both above and below. This is given by

$$(p_{1m} - \sigma_v)_e = \left[(p_{1m} - \sigma_v)_1 \times \ldots \times (p_{1m} - \sigma_v)_n\right]^{1/n} \quad (7.4)$$

   where *n* is the number of tests within 1.5*B* of the formation level and *B* is the width of the foundation.
2. A relative depth of foundation, H_e (or depth of embedment) is

$$H_e = \frac{1}{(p_{1m} - \sigma_v)_e} £\left[(p_{1m} - \sigma_v)_i\, z_i\right] \quad (7.5)$$

   where z_i is the thickness of a layer *i* for which the net limit pressure is $(p_{1m} - \sigma_v)_i$.
3. The value of *k* is taken from Figure 7.3 using the ground type, shape of footing and ratio of relative depth of foundation to width of foundation. The ground type is selected from Table 7.1.
4. The value of *k* can be adjusted for rectangular foundations using the factors given in Table 7.2.
5. The ultimate bearing capacity is calculated using equation (7.1).

The average ultimate bearing capacity for a foundation subject to an inclined load or adjacent to an excavation or slope is reduced by a factor, i_δ, given by

$$i_\delta = \left(1 - \frac{A'}{90}\right)^2 (1 - \lambda) + \left(1 - \frac{A'}{20}\right)\lambda \quad (7.6)$$

where A' is either the inclination, δ, of the load from the vertical, or for slopes the angle, β', shown in Figure 7.4 in degrees

$$\lambda = \lambda_D \lambda_M \quad (7.7)$$

where
$$\begin{aligned} \lambda_D &= 1(1 - D/B) && \text{for } 0 < D/B < 1 \\ &= 0 && \text{for } D/B > 1 \\ \lambda_M &= (1 - M) && \text{for } 0 < M < 1 \\ &= 0 && \text{for } M > 1 \end{aligned}$$

D is the depth of the foundation

Table 7.2 Bearing capacity factors for shallow foundations as a function of foundation shape

Footing	k value
Square or circular	K
Strip	k/1.2
Rectangle	(k/1.2)+(k/0.6) B/L

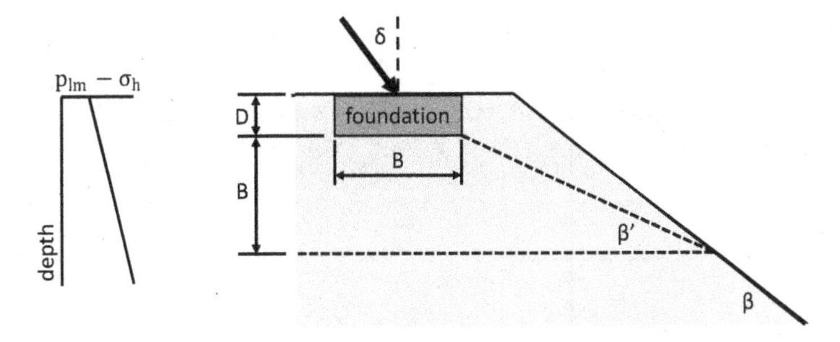

Figure 7.4 The reduction of bearing capacity to take into account an inclined load or an adjacent excavation or slope.

$$M = \frac{(p_{1m} - \sigma_h)_{[z=D]}}{(p_{1m} - \sigma_h)_{[z=(D+B)]}} \tag{7.8}$$

If the load is inclined and adjacent to a slope, the factor is adjusted such that

$$\begin{aligned} A' &= (\delta + \beta')\, if\ the\ load\ is\ directed\ towards\ the\ slope \\ &= the\ lowest\ value\ of\ \delta\ or\ (\beta' - \delta)\ if\ the\ load\ is\ directed\ away\ from\ the\ slope \end{aligned}$$

Eccentric loads are allowed for by assuming that the revised dimension is $B' = B - 2e$, where e is the eccentricity of the load. The pressure distribution beneath a foundation which supports an eccentric load or inclined load is calculated using a trapezoidal distribution. Stability is ensured if the average pressure is less than the average ultimate bearing capacity and the maximum pressure is less than 1.5 of the ultimate bearing capacity.

An indirect method of calculating bearing capacity could be based on the shear strength obtained from the limit pressure using data from Table 6.10 together with the formulae proposed by Brinch Hansen. Briaud (1992), using data produced by Amar et al. (1984), showed that the Ménard method and the indirect method gave a similar range of results for foundations on clay. The Ménard method shows less scatter than the indirect method for foundations in sand though this may be due to the difficulty in assessing an angle of friction from PBP tests.

The settlement of a shallow foundation is due to an increase in both the isotropic stress and the deviatoric stress. Equation (7.3) includes these two components of settlement. The isotropic component is a maximum immediately beneath the foundation, as shown in Figure 7.5, and the deviatoric component is a maximum at $(D + 0.5B)$.

Ménard and Rousseau (1962) undertook a series of full-scale tests on foundations up to 1 m wide and 1 m deep to develop the semi-empirical formula (equation (7.9)) based on the isotropic (E_v) and deviatoric (E_d) stiffness, to determine the settlement, s:

$$s = (q - \sigma_v)\left[\frac{2B_o}{9E_d}\left(\lambda_d \frac{B}{B_o}\right)^{\alpha} + \left(\frac{\alpha\lambda_v}{9E_v}\right)B\right] \tag{7.9}$$

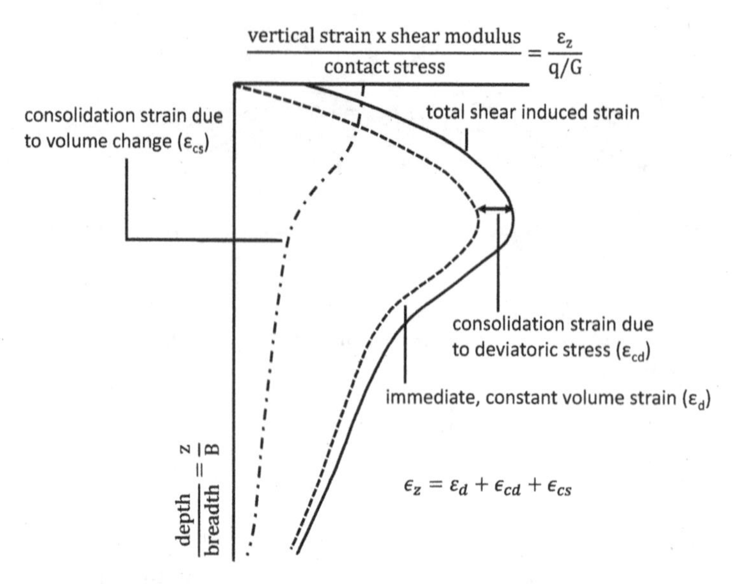

Figure 7.5 The components of settlement.

Source: After Baguelin et al., 1978.

Table 7.3 The shape coefficients λ_v and λ_d for shallow foundations

L/2B	*Circle*	*Square*	*2*	*3*	*5*	*20*
λ_v	1	1.12	1.53	1.78	2.14	2.65
λ_d	1	1.1	1.2	1.3	1.4	1.5

where B_0 is a reference width (60 cm), λ_v, λ_d are shape coefficients given in Table 7.3, α is a function of the ground type and ratio E_m/p_{lm} given in Table 7.4 and E_v, E_d are related to E_m.

This formula applies to foundations at a depth which is greater than twice the breadth. The settlement is increased by up to 20% for foundations above that. E_v and E_d are calculated as follows.

1. The ground below the foundation is divided into sixteen layers each $B/2$ deep, as shown in Figure 7.6.
2. E_v is equal to E_{ml}, the average value of E_m for the layer immediately beneath the foundation.
3. The average value of E_m for all layers is calculated.
4. The maximum difference between E_m for any layer and the average of all the layers is calculated.
5. If the maximum difference is less than 30% of the average value then $E_d = E_m$.
6. If the maximum difference exceeds 30% of the average, then E_d is calculated as follows.

Table 7.4 The coefficient α for shallow foundations

Ground type	*Description*	E_m/p_{lm}	α
Peat			1
Clay	Overconsolidated	>16	1
	Normally consolidated	9–16	0.67
	Remoulded	7–9	0.5
Silt	Overconsolidated	>14	0.67
	Normally consolidated	8–14	0.5
Sand		>12	0.5
		7–12	0.33
Sand and gravel		>10	0.33
		6–10	0.25
Rock	Extensively fractured		0.33
	Unaltered		0.5
	Weathered		0.67

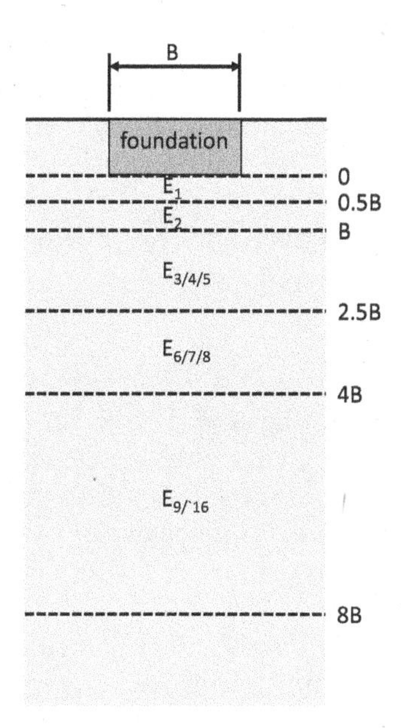

Figure 7.6 Calculating E_d and E_v in order to predict settlement.

(a) The harmonic mean for layers (3, 4, 5), (6, 7, 8) and (9 to 16) are calculated using as an example

$$\frac{3}{\left(E_{m(3,4,5)}\right)} = \frac{1}{E_{m3}} + \frac{1}{E_{m4}} + \frac{1}{E_{m5}} \tag{7.10}$$

where E_{m3} is the average value of E_m for layer 3.

(b) E_d is given by

$$\frac{1}{E_d} = 0.25\left[\frac{1}{E_{m1}} + \frac{1}{0.85E_{m2}} + \frac{1}{E_{m(3,4,5)}} + \frac{1}{2.5E_{m(6,7,8)}} + \frac{1}{2.5E_{m(9\ to16)}}\right] \tag{7.11}$$

This method only applies to relatively homogeneous ground in which there are no major changes in stiffness, such as a soft layer within a stiff deposit. Further adjustments are necessary if this does occur. If a shallow foundation rests on a soft layer and the thickness of that soft layer below the foundation is less than $B/2$, then only the settlement of the soft layer is considered. The increase in stress due to the foundation loading is calculated using elasticity theory. The settlement is

$$s = \beta\Sigma\left[\frac{\alpha_i \Delta\sigma_{vi} z_i}{E_{mi}}\right] \tag{7.12}$$

where β is 0.67 $[F/(F-1)]$ if $F = 3$, $\beta = 1$; $\Delta\sigma_{vi}$ is the increase in vertical stress due to the foundation in layer i, z_t is the thickness of the layer i, n is the number of layers within the soft soil, F is the factor of safety against general shear failure (usually 3).

If the soft layer is at depth, within or below a stiff layer and within $8B$ of the formation level, the settlement is a combination of the compression of the stiff layer and the soft layer. Equation (7.9) is used to calculate the settlement, assuming the soft layer has the same stiffness as the stiffer layer above. This settlement is added to the settlement of the soft layer to give the total settlement. The settlement of the soft layer is given by

$$s = \alpha h\,\Delta\sigma_v\left[\frac{1}{E_{ms}} - \frac{1}{E_{ma}}\right] \tag{7.13}$$

where E_{ms} is the stiffness of the soft layer and E_{ma} is the stiffness of the ground above.

Baguelin et al. (1978) quoted 26 case studies giving 45 measurements of settlement for strip footings, pad foundations, rafts and embankments covering a full range of soils from clays to weathered rock. Settlements up to 1 m were measured. In general, the predicted settlements were within 30% of the actual settlements as shown in Figure 7.7.

They concluded that the Ménard method is adequate for the majority of shallow foundations on a variety of natural and artificial ground but, where consolidation settlement predominates, an improved estimate may be made using moduli determined from oedometer tests, for example.

7.2.2 Axially loaded piles

The calculation for the end-bearing capacity of axially loaded piles is based on equation (7.14). The original proposals of Ménard (1963) were based on ground type, type of pile and relative pile base depth. These have been modified as further data became available (for example, see Bustamante and Gianeselli, 1981, Briaud, 1986a, Bustamante et al., 2009, Burlon et al., 2014). A recommended method (LCPC-SETRA,

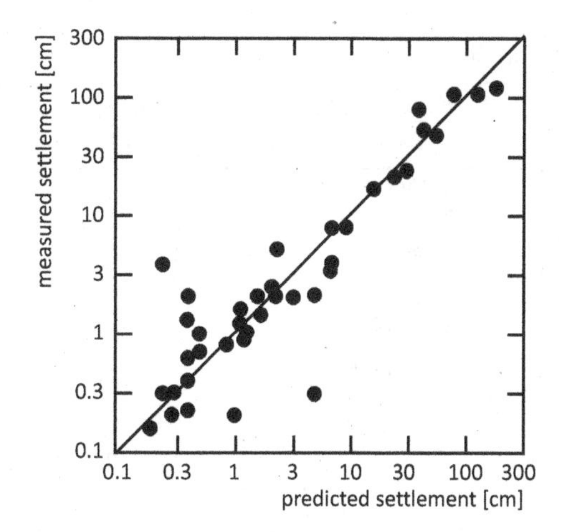

Figure 7.7 A comparison between measured and predicted settlements.

Source: After Baguelin et al., 1978.

1985) was developed from a study of over 200 instrumented piles and is based on ground type and net limit pressure.

The ultimate end-bearing capacity, Qp, for closed-ended piles is given by equation (7.14). The capacity for open-ended piles is half that:

$$Q_p / A = k[p_{1me} - \sigma_h] + \sigma_v \tag{7.14}$$

where A is the pile base area, p_{1me} is the equivalent limit pressure, σ_h is the total horizontal pressure at the base level, σ_v is the total vertical pressure at the base level and k is the bearing capacity factor from Table 7.5.

The equivalent limit pressure depends on the distance the pile penetrates the bearing layer and the degree of homogeneity of that layer. A homogeneous layer is defined as one in which the maximum value of p_{lm} is less than 1.5 of the minimum value of p_{lm} ($p_{1m\ min}$). The maximum value of p_{lm} is taken as $1.5p_{1m\ min}$ for a non-homogenous layer. The equivalent limit pressure, p_{lme}, is taken as the average limit pressure within a distance a below and a distance d above the pile-base level, that is,

$$p_{1me} = \frac{1}{a+b}\Sigma[p_{1mi}z_i] \tag{7.15}$$

where p_{lmi} is the limit pressure over depth z_i, which is the thickness of a layer at which p_{lm} is measured such that

$$z_1 + \ldots + z_n = a + d \tag{7.16}$$

a and d are distances that depend on the pile diameter and embedment length, d is equal to a or the distance between the pile base and the top of the bearing layer, whichever is the smallest, a is given by

Table 7.5 The bearing capacity factor, k, for axially loaded piles

Ground type	p_{lm} *(kPa)*	*Category*	*Bored piles and small displacement piles*	*Full displacement piles*
Clay	0–1200	I	1.2	1.8
Silt	0–700			
Firm clay or marl	1800–4000	II	1.1	3.2–4.2†
Compact silt	1200–3000			
Compressible sand	400–800			
Soft or weathered rock	1000–3000			
Sand and gravel	1000–2000			
Rock	4000–10000	III	1.8	2.6
Very compact sand and gravel	3000–6000	IV	1.1–1.8*	1.8–3.2†

Source: After LCPC-SETRA, 1985.

* 3.2 for dense sand or gravel; 4.2 for loose sand or gravel.

† limited database.

$$\begin{aligned} a &= 0.5 \quad \text{if } B_e < 1\text{m} \\ &= B_e/2 \quad \text{if } B_e < 1\text{m} \end{aligned} \tag{7.17}$$

where B_e = 4(base area of pile)/(base perimeter of pile). Note that in many cases $(a + d)$ is likely to be less than 2 metres and, since pressuremeter tests are carried out at a minimum spacing of 1 metre, p_{lme} will equal p_{lm}.

Values of k are given in Table 7.5. It is assumed that the pile penetrates the bearing layer such that the equivalent embedment depth, d_e, is greater than $5B$, where d_e is given by

$$d_e = \frac{1}{p_{1me}} \Sigma p_{1mi} z_i \tag{7.18}$$

k is reduced to k_e if $d_e < 5B$, where k_e is given by

$$k_e = 0.8 + \left(\frac{k - 0.8}{25}\right)\frac{d_e}{B}\left(\frac{10 - d_e}{B}\right) \tag{7.19}$$

The ultimate friction capacity, Q_f, is given by

$$Q_f = \Sigma[q_{si} z_i] \tag{7.20}$$

where q_{si} is the unit skin friction for soil layer i and z_t is the thickness of soil layer i. The unit friction capacity is obtained from Table 7.6 read in conjunction with Figure 7.8.

A comparison between the predicted and measured ultimate capacity of compression and tension pile tests is shown in Figure 7.9.

The measured load is the load required to cause a settlement of $(B/10 + Q_m L/AE)$ where $Q_m L/AE$ is the compression of the pile due to the ultimate load, Q_m. In general, the Ménard method over-predicts the compression capacity by 20% and the tension

Table 7.6 The selection of design curves for unit shaft friction

			Bored and lined		*Driven*		*Grouted*	
Soil type	p_{lm} *(MN/m²)*	*Bored concrete*	*Concrete*	*Steel*	*Concrete*	*Steel*	*Low pressure*	*High pressure*
Soft clay	0–0.7	A	A	A	A	A	B	
Stiff clay	1.2–2	A, (B)	A, (B)	A	A, (B)	A	B	E*
Very stiff clay	>2	A, (B)	A, (B)	A	A, (B)	A, B	E*	
Loose sand	0–0.7	A	A	A	A	A	B	
Medium dense sand	1–2	B, (C)	A, (B)	A	B, (C)	B	C	E
Very dense sand	>2.5	C, (D)	B, (C)	B	C, (D)	C	D	E
Completely weathered chalk	0–0.7	A	A	A	A	A	B	
Partially weathered chalk	>1	C, (D)	B, (C)	B	C, (D)	C	E	E
Marl	1.5–4	D, (F)	C, (D)	C	F	F	F	G
Stiff marl	>4.5	F					G	G
Weathered rock	2.5–4	G	G		G	G	G	G
Fractured rock	>4.5	G					G	G

Curves in parentheses only apply for well-constructed piles.

* If p_{lm}<1.5 MN/m².

Source: After LCPC SETRA, 1985.

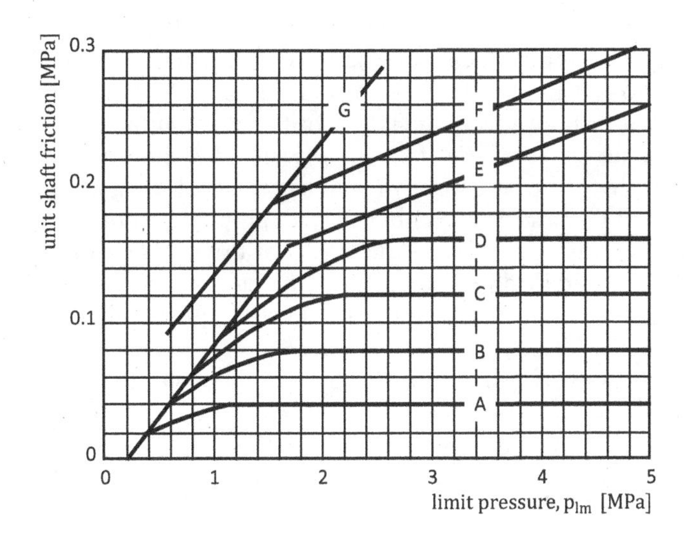

Figure 7.8 Unit skin friction for axially loaded piles.

Source: After LCPC-SETRA, 1985.

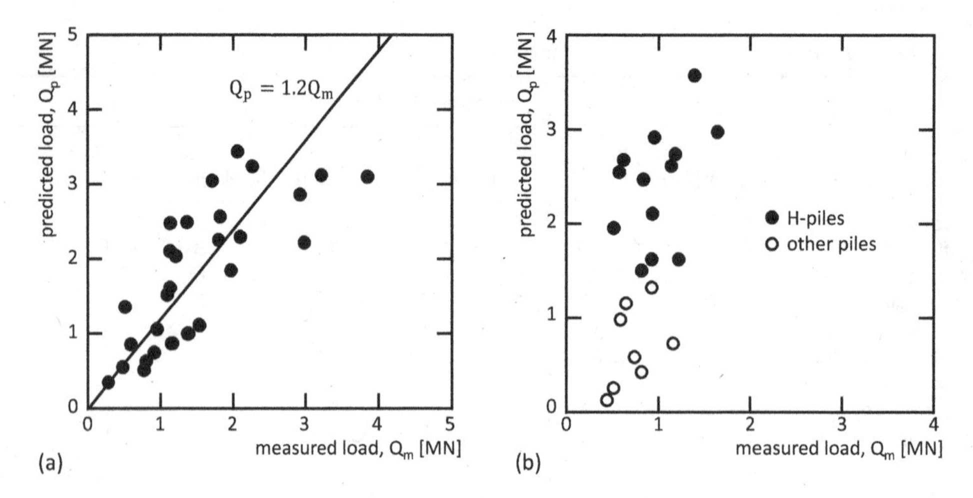

Figure 7.9 A comparison between the predicted and measured ultimate capacity of axially loaded piles: (a) compression tests; (b) tension tests.

Source: After Briaud, 1992.

capacity by 250%. The large discrepancy between the measured and predicted tension capacity may be due to the difference in compressive and tensile friction for H piles (Briaud et al., 1986).

The factors given above are only truly valid for calculating the capacity of piles installed in the same soils for which those factors were determined. The Ménard method is often adapted to suit local soils. Kruizinga (1989) gave an example of a 25 m long steel tubular pile driven into soft alluvial deposits underlain by sand. Five static load tests were carried out with the base of the pile driven to five different levels. The conclusion reached was that the base capacity could be predicted using the factors proposed by Bustamante and Gianeselli (1981) but that all proposed factors overestimated the shaft friction. This was attributed to the effects of driving, repeated testing and the time between testing and loading. Note that the Ménard method is based on long-term capacity therefore any short-term load test is likely to give a different capacity than that predicted.

The ultimate capacity of a shallow foundation is based on experiments in which the foundation was loaded to produce a settlement of $B/10$. This is the basis of predicting settlement of piles. The settlement of a pile is taken to be 0.6% of the pile diameter for bored piles and 0.9% of the pile diameter for driven piles together with, for long piles, the elastic shortening of the pile.

Gambin (1963) proposed a load transfer method to calculate settlement which was subsequently modified by Marchai (1971). The pile is divided into a series of elements of length h. The settlement, s_b, of the tip of the pile is given by

$$s_b = \frac{\lambda B Q_p}{4AE_{m^+}} \tag{7.21}$$

where λ is 1 for cylindrical piles and 1.13 for square piles. The settlement, s_i, of each of the other elements (1, 2, 3…,i) is given by

$$s_{i+1} = s_i + \left[\frac{0.5(\sigma_{i+1} + \sigma_i)_h}{E_p} \right] \tag{7.22}$$

where σ_i is the stress in the pile at the base of element i and E_p is the pile modulus.

The stress at the base of an element is given by

$$\sigma_{i+1} = \frac{\sigma_i \left[1 + \left(\frac{2h^2 E_m}{E_p r_o^2 C} \right) + \left(\frac{8 s_i E_m h}{r_o^2 C} \right) \right]}{1 - \left(\frac{2h^2 E_m}{E_p r_o^2 C} \right)} \tag{7.23}$$

where C is a displacement coefficient and E_m is the pressuremeter modulus at pile element i.

Figure 7.10 shows a comparison between a prediction using equation (7.21) and the settlement of piles in a variety of materials.

Frank and Zhao (1982) developed a semi-empirical method from elasticity theory using MPM results to fit experimental data to predict the load settlement curve at the

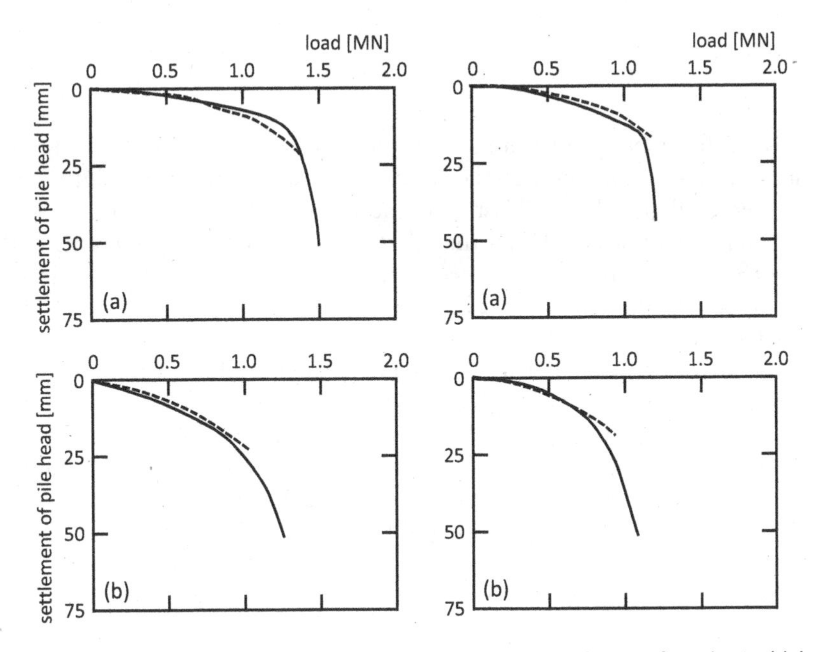

Figure 7.10 A comparison between the measured and predicted settlement for piles in (a) Lulea Site (loose silt over dense sand) and (b) Goteburg Site (soft clay overlying sand).

Source: After Sellgren, 1981.

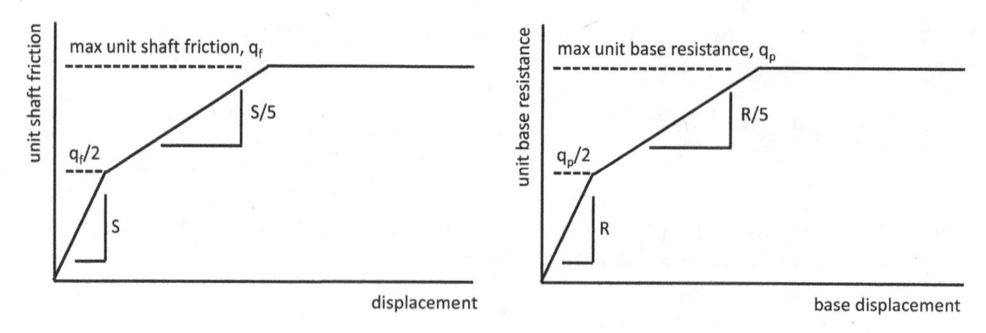

Figure 7.11 A model to determine the settlement of an axially loaded pile using a load transfer method.

Source: After Frank and Zhao, 1982.

top of the pile. It is a form of the load transfer method based on the model shown in Figure 7.11.

The slope, S, used to determine the mobilised skin friction, is given by

$$S = \frac{2CE_m}{B} \tag{7.24}$$

where C is 1 for cohesive soils and weak rocks and 0.4 for sandy soils. The slope R used to determine the base settlement is given by

$$R = \frac{2C'E_m}{B} \tag{7.25}$$

where C' is 5.5 for cohesive soils and weak rocks and 2.4 for sandy soils. Frank et al. (1991) showed that the method applies to complex soils including residual soils, soft rock, engineered fill and clays. Christoulas and Frank (1991) suggest that this method predicts the settlement of piles better than that based on the theory of elasticity (Poulos and Davis, 1980) using stiffness parameters derived from correlations with in situ tests including the Cone penetrometer (CPT), Standard penetration test (SPT), prebored pressuremeter test (PBP) as shown in Figure 7.12.

Note that the method of elasticity is based on a single stiffness value.

Boumedi et al. (2009) describe predictions of the settlement of 40 m high 180 m diameter LNG tanks supported on 316, 47 m long 1.2 m diameter piled foundations using the Ménard method. The design was validated against axial and laterally loaded pile tests. The ground profile consisted of 13 m of loose to medium dense sand, 17 m of soft clay and 7 m of stiff clay overlying dense sand. They found that the Ménard rules were accurate.

Eurocode 7 (ENV 1997-1., 2004) requires that a design for piled foundations is validated against the results of static load tests to ensure that there is an adequate factor of safety. A model factor is applied to ensure that factor of safety. Burlon et al. (2014) developed a model factor for the ultimate limit state of axial compression piles using Ménard pressuremeter test results. They used a database of 174 full-scale static

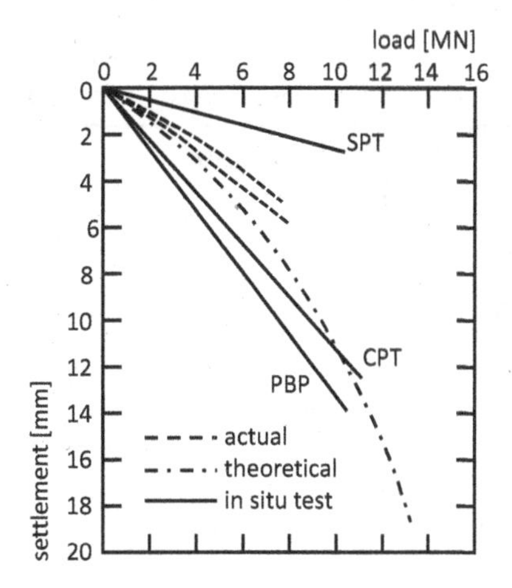

Figure 7.12 A comparison of predicted settlements using the load-transfer method, method of elasticity and observed settlements.

Source: After Christoulas and Frank, 1991.

pile load tests carried out by the Laboratoire Central des Ponts et Chausées, over the last 40 years. The database covers 20 categories of piles distributed into eight classes and two groups. The design resistance, $R_{c,d}$, of bearing capacity of an axially loaded compression pile is

$$R_{c,d} = \frac{R_{b,k} + R_{s,k}}{\gamma_t} or = \frac{R_{b,k}}{\gamma_b} + \frac{R_{s,k}}{\gamma_s} \quad (7.26)$$

where the subscripts b and s refer to the base and shaft capacity, k the characteristic value and γ is the resistance factor. The characteristic design resistance takes into account the spatial variation of soil properties and errors attributed to the method of calculation. Thus, the model factor, $\gamma_{R,d}$, is

$$R_k = \frac{R_{cal:sp}}{\gamma_{R,d}} \quad (7.27)$$

where $R_{cal:sp}$ is the calculated value of the ultimate bearing capacity. Eurocode suggests two procedures to calculate $R_{cal:sp}$ based on correlation factors to take into account the spatial variability or a statistical analysis of the variation of pile capacity across the site. Burlon et al. (2014) introduced a third method, which is compatible with Eurocode 7, to produce a model factor based on Ménard pressuremeter parameters, E_m and p_{lm}. This is an update of the model developed by Bustamante and Gianeselli (1981) where

$$unit\ base\ capacity, q_b = k_p p_{lm}^* \quad (7.28)$$

$$unit\ shaft\ capacity, q_s = \kappa g_i p_{lm}^* \quad (7.29)$$

Table 7.7 Description and characteristics of instrumented piles used by Bustamente et al. (2009) to update the pile design parameters for the French design codes

Group	*Type*	*Number Tested*	*Max and Min Diameter [mm]*	*Max and Min Embedment [m]*	*Description*
1	1	8	500 – 2000	11.5 - 23	pile or barrette in the dry
	2	64	270 – 1800	6 - 78	pile or barrette bored with slurry
	3	2	270 – 1200	20 - 56	bored and cased (permanent) pile
	4	28	420 – 1100	5.5 - 29	bored and cased (recoverable) pile
	5	4	520 – 880	19 - 27	dry bored piles
2	6	50	410 – 980	4.5 - 30	bored pile (cfa)
3	7	38	310 – 710	5 – 19.5	screwed pile cast in place
	8	1	650	13.5	screwed pile with casing
4	9	30	28 – 520	6.5 – 72.5	precast or prestressed concrete driven pile
	10	15	250 – 600	8.9 - 20	coated driven pile
	11	19	330 – 610	4 – 29.5	driven cast in place pile
	12	27	170 – 810	4.5 - 45	driven steel pile closed end
5	13	27	190 – 1220	8 - 70	driven steel pile open end
6	14	23	260 – 600	6 - 64	driven H-pile
	15	4	260 – 430	9 – 15.5	driven grouted H-pile
7	16	15	-	2.5 – 12.5	driven sheet pile
1	17	2	80 – 140	4 - 12	micropile type I
	18	8	120 – 810	8.5 - 37	micropile type II
8	19	23	100 – 1220	8.5 - 67	micropile type III
	20	20	130 – 660	7 - 39	micropile type IV

The measured capacity of the piles listed in the database were compared to the calculated capacity. This showed that the model factor is 1.15 except for piles in chalk ($\gamma_{R,d}$ = 1.4), micropiles ($\gamma_{R,d}$ = 2) and some types of H-piles ($\gamma_{R,d}$ = 2). Bustamante et al. (2009) updated design parameters for piles based on MPM tests by analysing load tests on 400 instrumented piles for which an overview is given in Table 7.7.

They identified seventeen pile types and produced revised values of the end bearing capacity factor, k_p (Table 7.8) and the unit skin friction factor q_s (Figure 7.13).

The Q_i line is selected according to Table 7.9.

The settlement of axially loaded compressive piles can be estimated from a *t-z* curve, which relates the displacement to the applied stress. A *t-z* curve is defined by the initial slope, the ultimate capacity and the shape of the curve. The initial slope is a function of E_m, the ultimate capacity on p_{lm}. The curvature depends on the model. Frank and Zhao (1982) suggested a trilinear model (Figure 7.11) defined by Equations 7.24 and 7.25. Abchir et al. (2016) proposed non-linear models (Figure 7.14).

In this case, the mobilised shaft friction, τ, for a displacement, s, is

$$\frac{d\tau}{ds} = \frac{q_s - \tau}{\lambda_s} \tag{7.30}$$

where λ_s is a resistance mobilisation factor depending on E_m, the soil type and the pile diameter, B.

Table 7.8 Revised values for the end bearing factor, k_p, to be used with the modified limit pressure, p^*_{lm}

Group	*Clay and Silt*	*Sand and Gravel*	*Chalk*	*Marl and Limestone*	*Weathered Rock*
1	1.25	1.2	1.6	1.6	1.6
2	1.3	1.65	2.0	2.0	2.0
3	1.7	3.9	2.6	2.3	2.3
4	1.4	3.1	2.4	2.4	2.4
5	1.1	2.0	1.1	1.1	1.1
6	1.4	3.1	2.4	1.4	1.4
7	1.1	1.1	1.1	1.1	1.1
8	1.4	1.6	1.8	1.8	1.5

Source: After Bustamante et al., 2009.

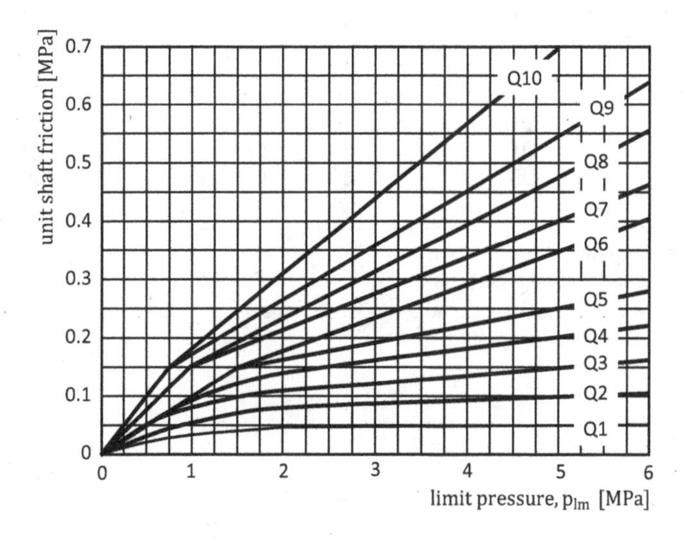

Figure 7.13 Values of unit shaft friction from the Ménard limit pressure for a range of piles.

Source: After Bustamente et al., 2009.

$$\lambda_s = \frac{q_s B}{\alpha_s E_m} \tag{7.31}$$

$$\lambda_b = \frac{q_b B}{\alpha_b E_m} \tag{7.32}$$

where α_s and α_b are factors dependent on soil type.

Abchir et al. (2016) introduced a further refinement (Figure 7.14) which is independent of the ultimate resistances. The initial stiffnesses, k_s and k_b, are

Table 7.9 The relevant Qi line for different pile types (see Table 7.7) and soil types to be used with Figure 7.13

Pile Type	*Clay and Loam*	*Sand and Gravel*	*Chalk*	*Marl and Limestone*	*Weathered Rock*
1	Q2	Q2[1]	Q5	Q4	Q6[2]
2	Q2	Q2	Q5	Q4	Q6[2]
3	Q1	Q1	Q1	Q2	Q1[2]
4	Q1	Q2	Q4	Q4	Q4[2]
5	Q3	Q3[1]	Q5	Q4	Q6
6	Q2	Q4	Q3	Q5	Q5[2]
7	Q3	Q5	Q4	Q4	Q4[2]
8	Q1	Q2	Q2	Q2	Q2[2]
9	Q3	Q3[2]	Q2	Q2[2]	[4]
10	Q6	Q8	Q7	Q7	[4]
11	Q2	Q3	Q6[2]	Q5[2]	[4]
12	Q2	Q2[2]	Q1	Q2[2]	[4]
13[3]	Q2	Q1	Q1	Q2	[4]
14[3]	Q2	Q2	Q1	Q2[2]	[4]
15[3]	Q6	Q8	Q7	Q7	[4]
16[3]	Q2	Q2	Q1	Q2[2]	[4]
17	Q1	Q1	Q1	Q2	Q6[2]
18	Q1	Q1	Q1	Q2	Q6[2]
19	Q6	Q8	Q7	Q7	Q9[2]
20	Q9	Q9	Q9	Q9	Q10[2]

[1] If ground properties permit;
[2] Use load tests to accept a higher value;
[3] Cross section for steel piles is area enclosed by the piles; the perimeter is the area in contact with the soil;
[4] Possible to use values for marl and limestone.

$$k_s = \frac{\alpha_s E_m}{B} \tag{7.33}$$

$$k_b = \frac{\alpha_b E_m}{B} \tag{7.34}$$

and the asymptotic values are

$$\frac{d\tau}{ds} = R_{fs} k_s \tag{7.35}$$

$$\frac{d\tau}{ds} = R_{fb} k_b \tag{7.36}$$

The soil pile stiffness relation is

$$\frac{d\tau}{ds} = R_{fs} k_s + \frac{\left(1 - R_{fs}\right) k_s}{1 + \left(s / \delta_s\right)^2} \tag{7.37}$$

where R_{fs} and R_{bs} are reduction factors for the soil–pile stiffness and δ_s and δ_b are

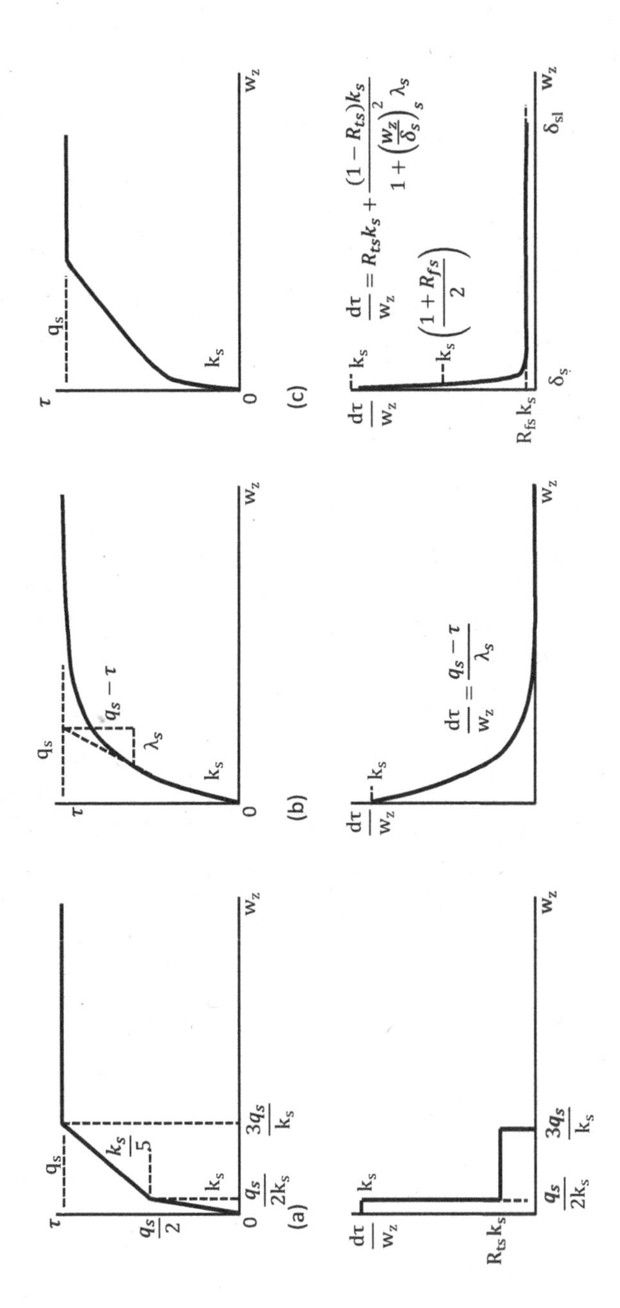

Figure 7.14 t-z models to be used with MPM parameters to predict the settlement of axially loaded compression piles (a) Frank and Zhao (1982); (b) AB1 model; and (c) AB2 model.

Source: After Abchir et al., 2016.

$$\delta_s = \frac{B}{1000(\alpha_s E_m + b_s)} \tag{7.38}$$

$$\delta_b = \frac{B}{1000 b_b} E_m^{-a_b} \textit{ for clays} \tag{7.39}$$

$$\delta_b = \frac{B}{1000(\alpha_b E_m + b_b)} \textit{ for other soils} \tag{7.40}$$

where a_s, b_s, a_b and b_b are model factors.

An analysis of ninety piles in the IFFSTARR database shows that none of the methods accurately predicted the actual settlement. However, at the working load all three methods give a reasonable prediction. A partial factor is introduced to ensure the predicted settlement is acceptable with the factor varying from 1.35 to 2 depending on the degree of risk.

7.2.3 Horizontally loaded piles

The finite length of a pressuremeter and the horizontal direction of loading are similar to the length and loading of an element within a horizontally loaded pile. Ménard et al. (1975) and Gambin (1969) developed a method to predict the deflected shape of a horizontally loaded pile based on the modulus of subgrade reaction, k_s, which is expressed as a function of the pressuremeter modulus, E_m. The pile is divided into a series of elements and the resistance to deflection, p_d, for each element is calculated from

$$p_d = k_s y \tag{7.41}$$

where y is the displacement. This only applies if $p_d < p_f$ and the element is below the critical depth. The maximum value of p_d, that is, the ultimate resistance of the horizontally loaded element, is p_{lm}. The resistance to displacement for $p_{1m} < p < p_{lm}$ is

$$p_d = 0.5 k_s y \tag{7.42}$$

The modulus of subgrade reaction for each element, for piles at least 0.6 m wide, is given by

$$\frac{1}{k_s} = \frac{2B_o}{9E_m}(2.65B / B_o)^{\alpha} + \frac{\alpha B}{6E_m} \tag{7.43}$$

where B is the pile diameter, B_o is a reference diameter (= 0.6 m) and α is a coefficient dependent on soil type (see Table 7.4).

The following equation applies to smaller piles.

$$\frac{1}{k_s} = B\left[\frac{4(2.65)^{\alpha} + 3\alpha}{18E_m}\right] \tag{7.44}$$

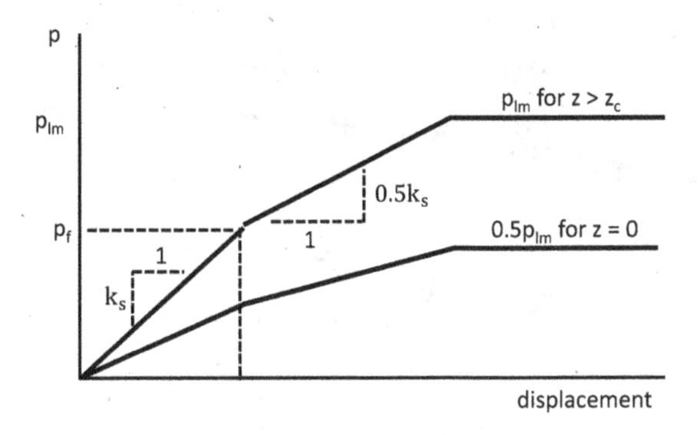

Figure 7.15 Reaction curves within and below the critical depth, z_c.

It is recommended that the ultimate resistance at the surface is reduced to $0.5p_{lm}$ since the method over-predicts the settlements near the surface. The effect that this has upon the reaction curves is shown in Figure 7.15.

This reduction only applies above a critical depth which, for clays, is $2B$ and for sands, $4B$.

7.2.4 Grouted anchors

The creation of a pressure-grouted anchor is similar to carrying out a PBP test since a pocket is drilled and filled with grout under pressure, which causes the pocket to expand, predominantly as a circular cavity. The friction between the ground and the grout when the anchor is loaded is similar to the friction on a pile. The load on the end of the grouted section is similar to the load on buried plate.

Bustamante and Doix (1985) undertook a study of experimental data from 213 anchors in a variety of ground conditions. A correlation between limit pressure, p_{lm}, and skin friction on the anchors for different ground conditions is shown in Figure 7.16. Two types of grouting systems are referred to. The IGU anchor is one in which the grout is injected under pressure once. The IRU anchor is one in which the grout is injected under pressure several times. The Ménard limit pressure is less than or equal to the injection pressure.

The ultimate load, T, is given by

$$T = \pi \Sigma D_{ai} q_{si} L_i \tag{7.45}$$

where D_{ai} is the diameter of the anchor in soil i, q_{si} is the unit skin friction in soil i and L_i is the length of anchor in soil i. The total length, L, required is

$$L = \Sigma L_i \tag{7.46}$$

The diameter of the anchor, D_a, is a function of the pocket diameter, a_p, such that

$$D_a = \alpha a_p \tag{7.47}$$

where α is a constant taken from Table 7.10.

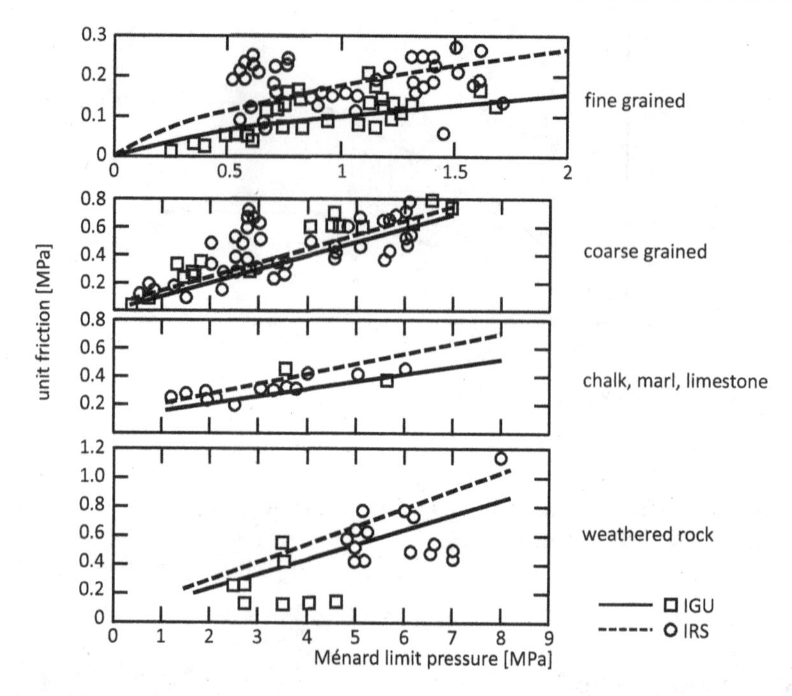

Figure 7.16 Design curves for ground anchors.

Table 7.10 Coefficients for design of ground anchors

Ground type	α *IRS*	*IGU*	*Minimum quantity of grout × Volume of grout bulb*
Weathered rock	1.2	1.1	1.1–1.5 for intact rock 2 for fissured rock
Marl	1.8	1.1–1.2	1.5–2 for intact rock
Marl and limestone	1.8	1.1–1.2	
Weathered chalk	1.8	1.1–1.2	2–6 for fissured rock
Gravel	1.8	1.3–1.4	1.5
Sandy gravel	1.6–1.8	1.2–1.4	1.5
Gravelly sand	1.5–1.6	1.2–1.3	1.5
Coarse sand	1.4–1.5	1.1–1.2	1.5
Medium sand	1.4–1.5	1.1–1.2	1.5
Fine sand	1.4–1.5	1.1–1.2	1.5
Silty sand	1.4–1.5	1.1–1.2	1.5–2 for IRS 1.5 for IGU
Silt	1.4–1.6	1.1–1.2	2 for IRS 1.5 for IGU
Clay	1.8–2	1.2	2.5–3 for IRS 1.5–2 for IGU

7.2.5 Ground improvement

The Ménard pressuremeter is used as a means of assessing the increase in stiffness of loose soils subject to dynamic compaction (e.g. Hamidi et al., 2010, 2011, 2012; Varaksin and Hamidi, 2013, 2015), preloading (Bo et al., 2005), jet grouting (Wen, 2005), and treatment of metastable loose soils (Jefferson et al., 2005).

Ménard proposed that creep settlement, s_c, of artificial ground subject to dynamic compaction after one year could be predicted by

$$s_{c(1)} = H\frac{1-\frac{\alpha p'_{lm}}{2}}{\frac{\alpha p'_{lm}}{2}} \tag{7.48}$$

where H is the fill thickness in metres, p'_{lm} the effective limit pressure and α, a factor depending on soil type. Varaksin et al. (2005) extended that for creep settlement after n years

$$s_{c(n)} = s_{c(1)}\frac{\ln\left(\frac{t+1}{n}\right)}{\ln(t+1)} \tag{7.49}$$

Ménard suggested an adequate bearing capacity in sands after t years could be achieved if the limit pressure after compaction is at least 600kPa. Ménard also suggested that the limit pressure doubled for every 3% of strain to estimate the amount of dynamic compaction induced subsidence to reach self-bearing.

$$\epsilon = a\frac{\log\left(\frac{(p_{lm})_j}{(p_{lm})_i}\right)}{\log 2} \tag{7.50}$$

where a, is the strain needed to double the limit pressure (= 3%), p_{lmi} and p_{lmj} the limit pressures before and after compaction. Thus, the settlement, s, of the artificial ground due to dynamic compaction is:

$$s = \frac{a}{\log 2}\sum_{k=1,m} h_k \log\left(\frac{(p_{lm})_j}{(p_{lm})_i}\right) \tag{7.51}$$

where m is the number of pressuremeter tests, and h_k the testing interval.

Arulrajah et al. (2011) described the ground improvement works for the Changi East Reclamation Project in the Republic of Singapore, which involved reclamation using preloading of 2000 ha of land underlain by marine clay. As part of the construction works a test site was investigated in detail using a variety of laboratory and in situ tests. Two areas were investigated – one with vertical drains installed at 1.5 m square spacing and one without drains. Both areas were preloaded to the same magnitude. In situ tests were carried out using self-boring pressuremeter tests to determine the

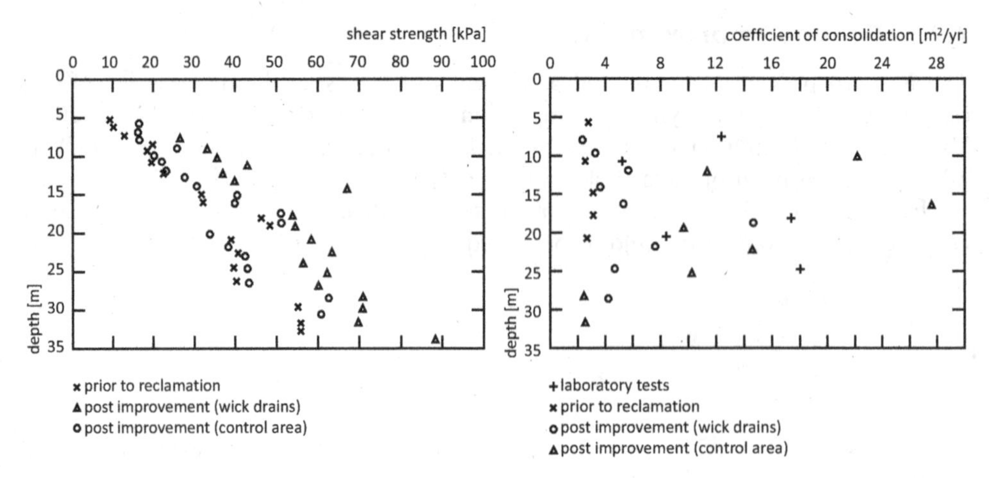

Figure 7.17 The use of the self-boring pressuremeter to determine (a) the undrained strength and (b) coefficient of consolidation of a marine clay before and twenty three months after preloading with prefabricated drains.

Source: After Arulrajah et al., 2011.

undrained shear strength and horizontal stress from which the over-consolidation ratio was calculated using the method proposed by Mayne and Mitchell (1988). In situ dissipation tests were also undertaken with the SBP to provide a means of evaluating the in situ coefficient of consolidation due to horizontal flow (c_h) and horizontal hydraulic conductivity (k_h) of Singapore marine clay at Changi. Figure 7.17 shows the effect of preloading and vertical drains on the underlying soils.

Kirstein et al. (2013) describe settlement predictions of embankments up to 15 m high built on mine waste up to 60 m deep. The mine waste was stiffened using either dynamic compaction or stone columns. MPM tests were undertaken before and after ground improvement to assess the increase in stiffness of the ground and the stiffness of the columns. The 15 m high embankment was built in 3 m stages. The settlements predicted using a FEM analysis with a hardening soil model using E_m taken from the MPM tests were acceptable.

Kassou et al. (2017) report on a comparison of settlements of four trial embankments in Morocco built on soft soils. Vertical drains were used to accelerate the consolidation. The embankments were modelled as strip footings applying a contact pressure due to the weight of the embankments, which varied from 6.5 m to 10.5 m in height. They found that settlements predicted using the recompression indices from oedometer tests overestimated the settlement whereas those based on the E_m value gave reasonable predictions.

Moon et al. (2019) undertook a series of trials to determine whether dynamic compaction could be used to improve the strength and stiffness of Sabkha soil, which is found throughout the Arabian Gulf Coast region. It is a problematic soil because of its loose density, soft consistency, high salinity and water content, and presence of fine sands and clays. It is generally highly compressible. Construction problems include excessive total and differential settlements due to the spatial variation in density; high collapse potential because of dissolution of sodium chloride, leaching of calcium ions, and the soil grain rearrangement. MPM tests before and after dynamic compaction showed that there was a significant improvement with a 20-ton tamper dropped 12 times at a height of 10–15m.

7.2.6 The Ménard method based on results of other pressuremeter tests

Baguelin (1982) describes a direct test method similar to that developed for the MPM but to be used with the PAF test. The net bearing pressure, q_n, is given by

$$q_n = \psi(p_{20} - \sigma_h) \tag{7.52}$$

where p_{20} is the pressure at 20% volumetric strain and ψ is the bearing capacity coefficient given in Figure 7.18.

The settlement, $_s$, is given by

$$s = \frac{q - \sigma_v}{10}\left[\frac{2B_o}{G_2}\left(\lambda_d \frac{B}{B_o}\right)^{1-\beta} + \lambda_v \frac{B}{G_5}\right] \tag{7.53}$$

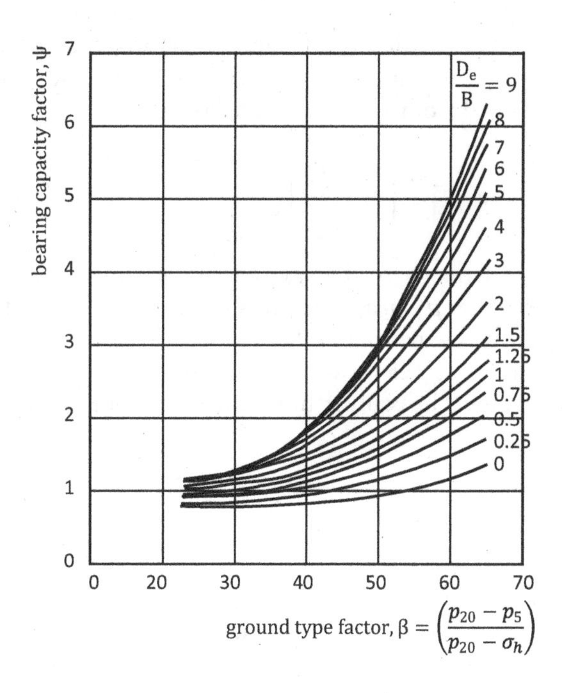

Figure 7.18 The coefficient of bearing capacity, ψ, for strip footings.

Source: After Baguelin, 1982.

where B_o is a reference length (60 cm), λ_v λ_d are shape coefficients given by Ménard in Table 7.3, β is a function of the ground type given by equation (6.1) and G_2 and G_5 are secant moduli between 0 and 2% and 0 and 5% strain.

Powell et al. (2001) suggested that the principles of the design of axially loaded compression piles based on MPM tests could be applied to result of FDP tests. The net limit pressure, p^*_{lm}, from a FDP test exceeds that from the MPM test. Therefore, the unit shaft friction, q_s, uses a reduction factor to allow for that:

$$q_s = \left(0.08\frac{p^*_{lm}}{1.5}\right)\left(2 - \frac{p^*_{lm}}{1.5}\right) if\ p^*_{lm} < 1.5 \tag{7.54}$$

$$q_s = 0.08 p^*_{lm}\ if\ p^*_{lm} > 1.5 \tag{7.55}$$

The unit end bearing, q_b, is

$$q_b = 1.4 p^*_{lm}\ for\ soft\ clays (p_{lm} < 0.7 MPa) \tag{7.56}$$

$$q_b = 1.5 p^*_{lm}\ for\ firm\ clays (1.2 MPa < p_{lm} < 2 MPa \tag{7.57}$$

7.3 OTHER DIRECT DESIGN METHODS FOR HORIZONTALLY LOADED PILES

Results of PBP and FDP tests are used directly in design. Suyama et al. (1982) describe a method used to predict the deflected shape of a horizontally loaded pile which is based on the subgrade reaction deduced from PBP tests. The modulus of subgrade reaction, k_s, is determined from a PBP test as follows:

$$k_s = 0.5\pi\left[2a_o(r_m - a_o)^2\right]^{0.25}\left(\frac{p_f - p_o}{a_f - a_o}\right) \tag{7.58}$$

where a_o is the initial diameter of the cavity (see section 6.3), a_f is the radius of the cavity at the yield pressure, p_o, p_f are the pressures at a_o and a_f, and $r_m = (a_f - a_o)/2$.

The pile is divided into elements and the weighted average value of k_s is determined for the layer adjacent to each element. More credit is given to the upper layers since deflections will be greater. The equivalent subgrade reaction, k_e, for a pile of diameter, B, is given by

$$k_e = \frac{k_s}{[B]^{0.25}} \tag{7.59}$$

The horizontal load for a given deflection, is calculated using the method proposed by Chang (1937). A design value of subgrade reaction, fk_d, for that deflection is given by

$$k_d = \frac{k_e}{[y_i]^{0.5}} \tag{7.60}$$

The load, H_i, for each element i of the pile is given by

$$H_i = y_i \left[\frac{12EI\beta_i^3}{(4-3f)(1+\beta_i h)^3 + 1} \right] \tag{7.61}$$

where EI is the pile stiffness, β_i is a function of the pile and soil stiffness, f is degree of fixity and h is the distance between the top of the pile and the ground surface.

$$\beta_i = \left[\frac{k_{di} 2r_o}{4EI} \right]^{0.25} \tag{7.62}$$

$$f = \frac{(1+\beta_i h)^3 + 0.5}{(\beta_i h)^3} \tag{7.63}$$

A comparison between the measured and estimated behaviour of a pile in layered sands and clays (Figure 7.19) shows good agreement.

Briaud et al. (1983) proposed that the deflected shape of a horizontally loaded pile is a function of the normal and shear stresses between the soil and the pile developed from a PBP or SBP test curve. The normal load, Q, for an element of a pile in a soft layer is given by

$$Q = (p - \sigma_h) B\alpha_n \tag{7.64}$$

where B is the width or diameter of the pile, p is the applied pressure measured in a pressuremeter test (prebored curve for bored piles; reload curve or SBP curve for

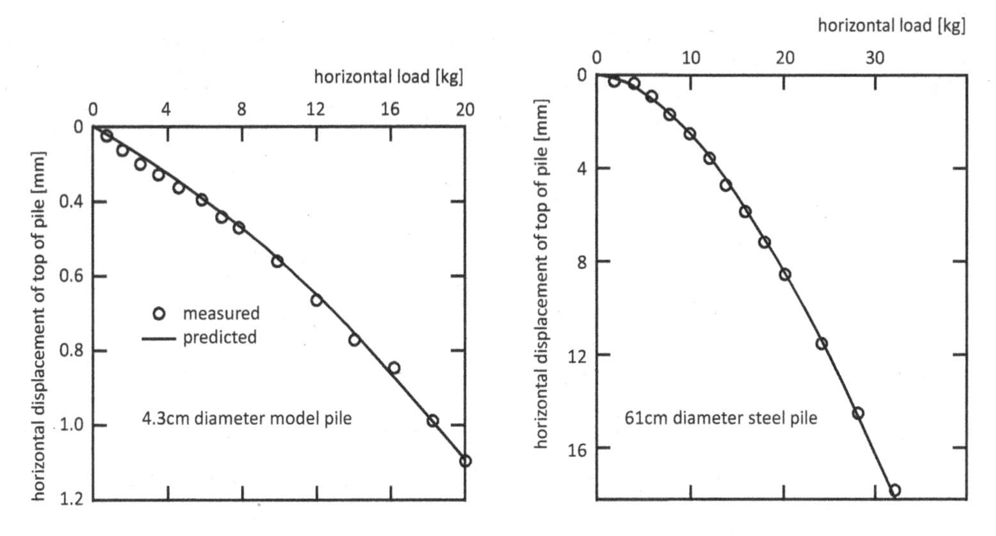

Figure 7.19 A comparison between the measured and predicted behaviour of horizontally loaded piles.

Source: After Suyama et al., 1983.

driven piles) and α_n is a shape factor (= 1 for square piles, 0.75 for circular piles). The shear load, F, on the side of the pile is given by

$$F = c_u B \alpha_n \tag{7.65}$$

where c_u is the average shear strength and a_s is a shape factor (= 2 for square piles, 1 for circular piles). There is an additional resistance, F_b, if the pile is short and the base of the pile moves. This resistance is equal to the shear stress multiplied by the base area and applies if the length of the pile is less than 3*B*.

The displacement of the pile, *y* is given by

$$y = \left(\frac{a_i - a_o}{2a_o}\right) B \tag{7.66}$$

where a_o, a_i are the cavity radii at the start and during a test. The normal and shear loads are determined for a given displacement. The *p–y* curve is the sum of the normal and shear load curves expressed in terms of the displacement, *y*.

The soil resistance is reduced above a critical depth which is a function of the pile rigidity and limit pressure. The relative rigidity, *RR*, is defined as

$$RR = \frac{1}{B}\left(\frac{EI}{p_1 - \sigma_h}\right)^{0.25} \tag{7.67}$$

where *E* is the pile material modulus and *I* is the moment of inertia. The critical depth is taken from Figure 7.20.

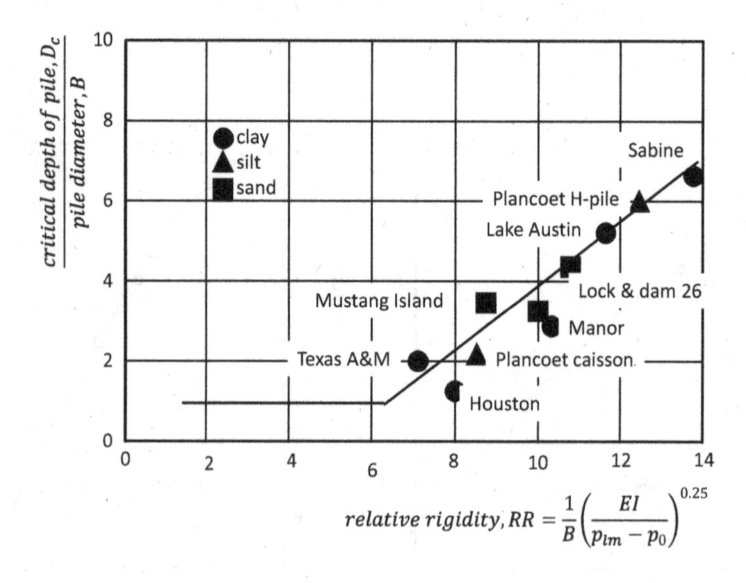

Figure 7.20 The critical depth of a laterally loaded pile as a function of the pile stiffness and size.

Source: After Briaud et al., 1983.

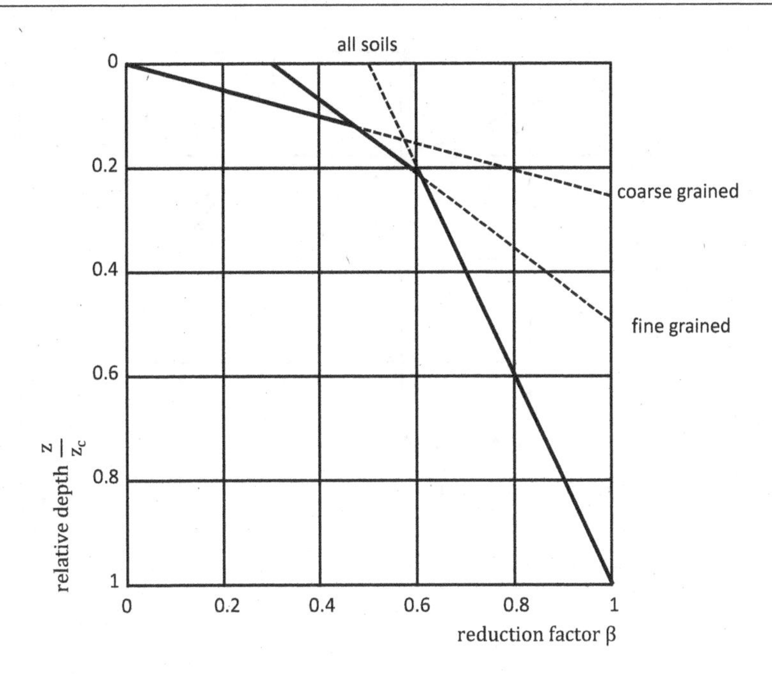

Figure 7.21 Reduction factors for pressuremeter test results at shallow depth.

Source: After Briaud et al., 1983.

The pressuremeter curve for tests above the critical depth is adjusted by a reduction factor, β, which is taken from Figure 7.21.

The ultimate resistance, Q, within the critical depth is reduced by a factor, α, given in Figure 7.22.

Robertson et al. (1986) suggested a method that could be used to determine the deflected shape of laterally loaded piles from the results of FDP tests. Figure 7.23 shows that the displacements in a soil around a pressuremeter and a laterally loaded pile are similar if the pressuremeter and pile are installed in the same way.

The curves obtained from these two horizontally loading devices are similar, as shown in Figure 7.24, except that the limiting soil reaction for the pile is greater than that for a pressuremeter.

Hughes et al. (1979) and Robertson et al. (1984) suggested that the applied pressure obtained from a pressuremeter test should be increased by χ to account for this, where $\chi = 2$ for cohesive soils and 1.5 for cohesionless soils. However, Anderson et al. (2009) found that the method only applied to fully drained conditions; that is, in sands and partially saturated clays. Predictions for saturated clays were found to be too stiff and, therefore, unsafe.

The χ values only apply at depth since, near the surface, the response is affected by the reduced vertical stress. A reduction factor, β, proposed by Briaud et al. (1983) is used to modify the pressuremeter test curves within the zone above the critical depth

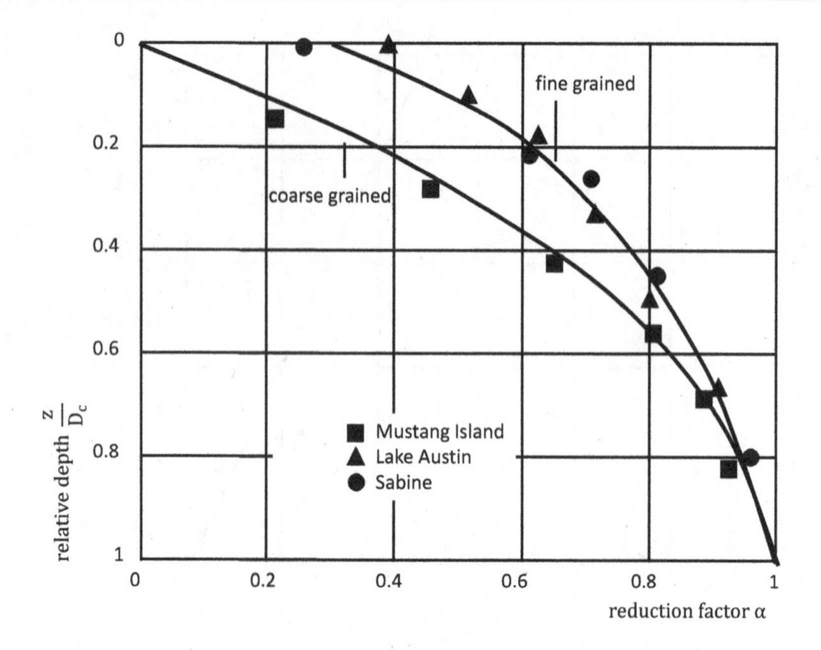

Figure 7.22 Reduction factors for horizontal pile capacity at shallow depth.

Source: After Briaud et al., 1983.

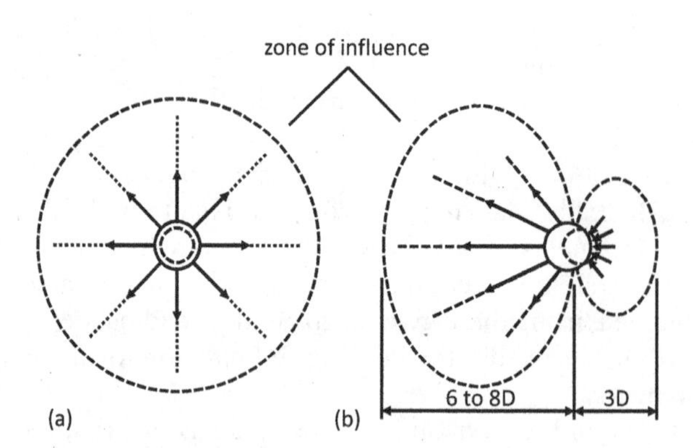

Figure 7.23 A comparison between the deformation around (a) a pressuremeter and (b) a horizontally loaded pile.

Source: After Robertson et al., 1986.

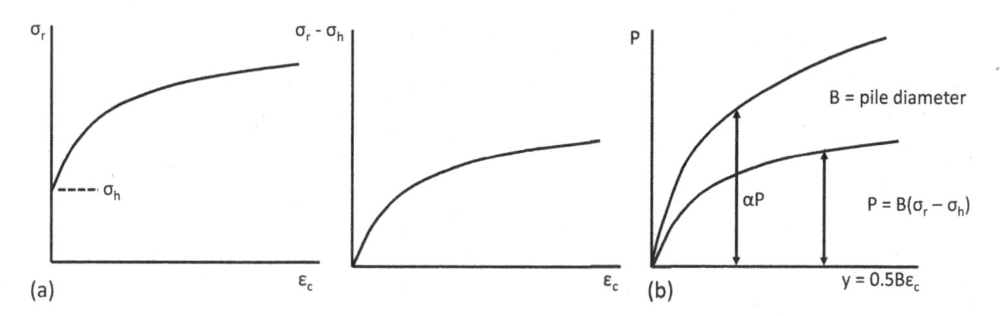

Figure 7.24 The development of a p-y curve (b) from a pressuremeter curve (a).

Source: After Robertson et al., 1986.

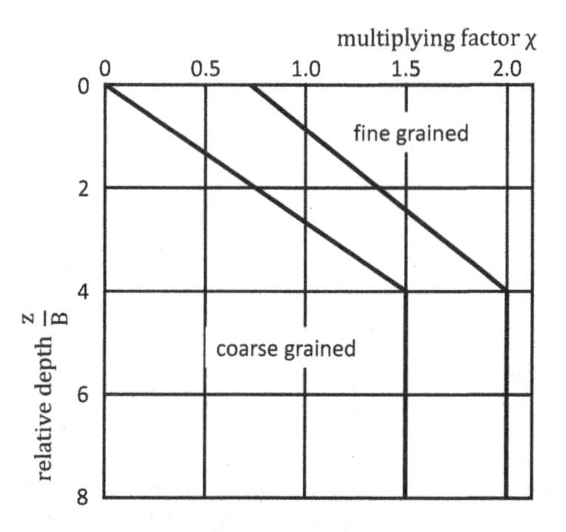

Figure 7.25 Correction factors to take into account the influence of the surface.

Source: After Robertson et al., 1986.

(see Figure 7.21). Robertson et al. (1986) proposed that χ should be reduced within the critical depth. The reduction is given in Figure 7.25.

The procedure to determine the *p-y* curve is as follows.

1. The critical depth and reduction factor, β, for the pressuremeter test are found from Figures 7.20 and 7.21.
2. The factor, χ, to adjust the pressuremeter test curve is taken from Figure 7.25.
3. The pressure, p, and strain, ε_c, at each data point on the pressuremeter curve are adjusted to give the *p–y* curve:

$$P = \left(\frac{p}{\beta}\right)B \quad (7.68)$$

where B is the pile width

$$y = \varepsilon_c \left(\frac{B}{2} \right) \tag{7.69}$$

Robertson et al. (1986) compared predicted deflected shapes using this method with measured deflections for a variety of pile types and sizes and found reasonable agreement for short-term loading.

Meyerhof and Sastry (1987) undertook a series of tests with instrumented rigid model piles under eccentric and inclined loads. The piles were jacked into three soil layers: a clay, a sand and a layered clay and sand. A horizontal load and moment were applied to the top of the piles and the deflection and rotation of the top of the pile and soil pressures on the pile were measured. A profile of test curves was obtained using a FDP. Meyerhof and Sastry defined the limit pressure (p_{lms}) as the applied pressure at 10% cavity strain which corresponds to the point at which there was little further increase in applied pressure. A stiffness, E_s, was taken as the secant modulus of deformation over the pressure increment $0.5(p_{lms} - p_o)$ where p_o is the pressure at the point the membrane lifts from the body of the probe. This is not equal to σ_h because of installation effects.

Meyerhof and Sastry (1987) proposed that the pile capacity in terms of horizontal load, H, and pure moment, M_o, can be estimated from

$$H = a\sigma_{hb}BD \tag{7.70}$$

$$M_o = b\sigma_{hb}BD^2 \tag{7.71}$$

where σ_{hb} is the horizontal stress at the base of the pile, a, b are constants depending on soil type, B is the pile diameter and D is the embedded length of pile. Meyerhof and Sastry found that a_h was simply equal to $\alpha(P_{1ms} - p_o)$ where a is α function of soil type. The value of α is reduced if the depth of pile is less than a critical depth equal to 20 probe diameters. The values of a, b and α deduced by Meyerhof and Sastry are given in Table 7.11.

The displacement and rotation of a free pile head were estimated from elasticity theory using derived parameters from FDP tests. It was assumed that a triangular distribution of stiffness exists for sands with the stiffness taken from a test carried out at about 13 times the probe diameter. The stiffness was assumed to be constant with depth for clays.

Alternatively the deflection, y, and rotation, θ, at the ground surface can be determined from

Table 7.11 Coefficients for design of laterally loaded rigid piles based on PIP tests

Soil type	*a*	b	α
Sand 0.125	0.125	0.09	2.5
Clay 0.4	0.4	0.2	1
Layered clay and sand			1.75

Source: After Meyerhof and Sastry, 1985.

Table 7.12 Pile displacement factors m, n

Soil type	*Load*	*m*	*N*
Sand	Horizontal load	$(I_{ph}K_bF_b)/1.3f_q$	$(I_{.h}K_bF_b)/1.3f_q$
	Pure moment	$(I_{pm}K_bF'_b)/1.3f_q$	$(I_{\theta h}K_bF'_b)/1.3f_q$
Clay	Horizontal load	$(I_{ph}K_cF_c)/1.5f_q$	$(I_{\theta h}K_cF_e)/1.5f_q$
	Pure moment	$(I_{ph}K_cF'_c)/1.5f_c$	$(I_{\theta h}K_cF'_c)/1.5f_c$

Source: After Meyerhof and Sastry, 1985.

Notes:

I = elastic influence factors given by Poulos and Davis (1980)

K = coefficients of net passive pressure for weight and cohesion (Meyerhof et al. (1981))

F = lateral resistance factors for weight and cohesion (Meyerhof et al. (1981))

f = limit pressure factors for weight and cohesion (Baguelin et al., 1978)

$$\frac{y}{B} = m\varepsilon_c \tag{7.72}$$

$$\theta = n\varepsilon_c \frac{B}{D} \tag{7.73}$$

where y is the horizontal displacement for a pure horizontal load, m, n are constants depending on soil type given in Table 7.12 and θ is the rotation for a pure moment.

Meyerhof and Sastry found reasonable agreement between the predictions and measurements but consider that without further studies the method has yet to be proved for full-scale piles.

7.4 COMPARISONS BETWEEN RESULTS OF PRESSUREMETER AND OTHER TESTS

It is common practice to compare results of pressuremeter tests with results from other in situ tests and laboratory tests. This may justify the method of interpretation of the pressuremeter tests or produce correlations so that pressuremeter test results can be used in design methods based on those other tests. A lack of understanding of the pressuremeter equipment, installation and test procedure and interpretation has led to criticism of the results if they do not agree with those expected.

There is no reason why pressuremeter test results should be similar to those from other tests. The rate of loading, the stress path followed during installation and testing and the boundary conditions are different. Prévost (1979) proposed a general analytical model to describe anisotropic, elastic plastic, path-dependent stress–strain properties of rate-dependent soils to show that the undrained strengths from pressuremeter tests are never equal to those from triaxial, plane strain and simple shear tests. Briaud (1986b) gave examples of correlations between PBP tests and cone, SPT and triaxial tests from a large database. There is very little agreement between the results of the different tests. If there is a poor correlation between two sets of results it does not imply that one set of results is wrong. Comparisons do allow an assessment to be made of the quality of the results, to compare predicted and actual behaviour and to formulate correlations with results from other tests.

7.4.1 Total horizontal stress

The selection of horizontal stress is subjective, therefore comparisons tend to be used to justify the selection. There are many published examples of total horizontal stress profiles. These include measurements in soft clays, stiff clays, sands and weak rocks. Generally, results are quoted for one site and comparisons made with other data from that site, including predictions based on assumed values of horizontal stress. An example of this is given in Table 7.13 in which results of SBP tests have been compared to the best estimates of total horizontal stress.

Graham and Jefferies (1986) concluded from eight programmes of SBP testing on six sites composed of hydraulic fill that the in situ horizontal stress gives values of K_o between 0.7 and 1.3, suggesting that Jaky's formula for normally consolidated soils does not apply to hydraulic fill.

Jefferies et al. (1987) used the results of Cambridge self-boring pressuremeter (CSBP) tests in clays at 20 sites in the Arctic to produce profiles of σ_h. These were compared to the results of oedometer tests to show (Figure 7.26) that there is a relation between K_o and OCR which tends to be site specific and not as clearly defined as is often assumed.

Benoit and Lutenegger (1993) presented initial results of a study into the assessment of laboratory tests and intrusive and non-intrusive in situ tests for determining soil parameters. Figure 7.27 is a comparison of the effective horizontal stress either assumed, measured directly or determined indirectly.

They concluded that the CSBP can be used to provide the reference values for other tests since at those sites the CSBP results give a lower bound profile, which is consistent with the predicted profile. They stressed the importance of the improvements that have taken place in the instrumentation of the CSBP making the lift-off method the best method to determine horizontal stress.

Table 7.13 Estimates of total horizontal stress from SBP tests

Site	*Reference*	*OCR*	*Probe*	*L/D*	*Source of σ_h*	p_o/σ_h	p_{oD}/σ_h	p_{oM}/σ_h
Porto Tolle	Ghionna et al. (1982)	1	PAF	2	Field/lab	1.05 ± 0.12	0.96 ± 0.16	1.04 ± 0.15
			PAF	4		1.12 ± 0.15	1.07 ± 0.14	1.15 ± 0.18
Boston	Ladd et al. (1979)	3–7	PAF	2	Field/lab	<0.75		0.99 ± 0.18
			CSBP	6		<0.85		0.98 ± 0.14
Bandar Abbas	Ghionna et al. (1982)	1.5–2	PAF	2	Lab	1.12		0.95
Trieste	Ghionna et al. (1982)	1	PAF	2	Lab	1.29 ± 0.18		1.29 ± 0.15
			PAF	4				
Onsoy	Lacasse et al. (1981)	1	CSBP	6	Field/lab		1.15 ± 0.08	
	Ghionna et al. (1982)							1.02 ± 0.09
Drammen	Lacasse et al. (1981)	1	CSBP	6	Field/lab		0.89 ± 0.09	
	Ghionna et al. (1982)							0.95 ± 0.05

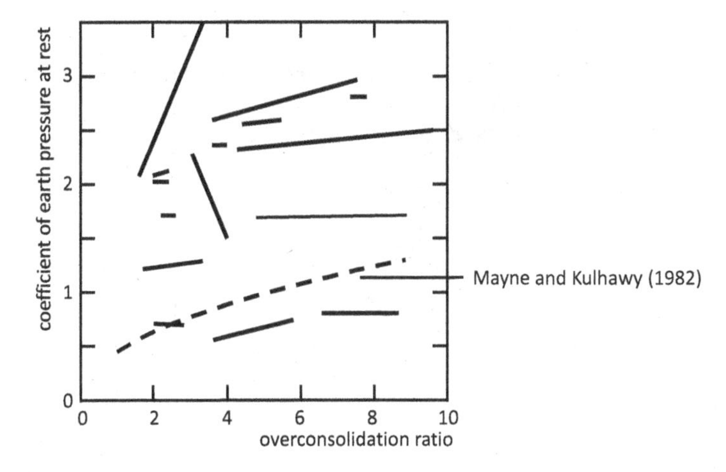

Figure 7.26 A comparison between Ko estimated from SBP tests and OCR estimated from oedometer tests from a number of sites.

Source: After Jefferies et al., 1987.

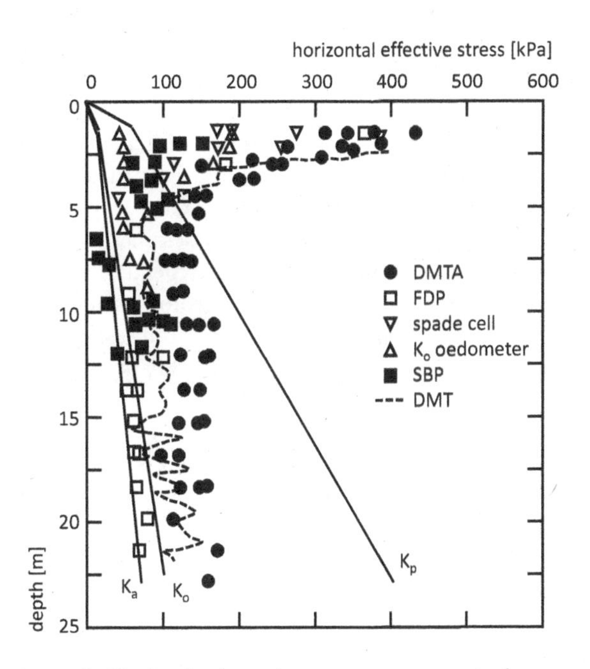

Figure 7.27 A comparison of effective horizontal stress measurements from a soft clay site at the University of Massachusetts.

Source: After Benoit and Lutenegger, 1993.

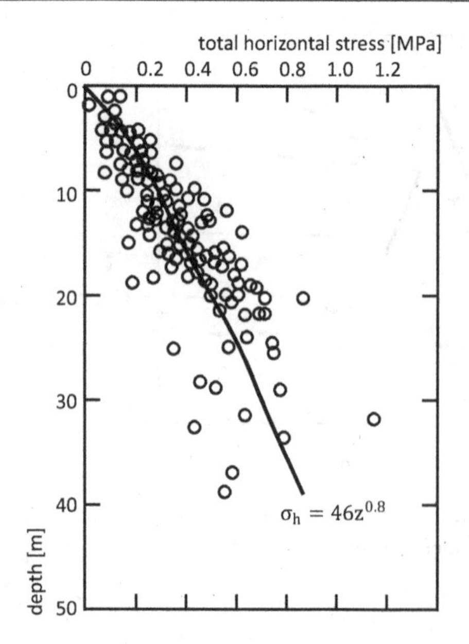

Figure 7.28 Values of total horizontal stress from fifteen sites in London Clay.

Source: Clarke, 1993.

The selection of horizontal stress becomes more difficult and more subjective the stiffer the ground. Clarke (1993) produced a summary of measurements of horizontal stress using the lift-off method from tests in London Clay (Figure 7.28).

An arbitrary fit to the data gives a preliminary assessment of the variation in horizontal stress, which is not too dissimilar from that expected. It does not represent the variation with depth at all sites because the local geology and pore pressure distribution at each site will be different, but does allow an initial estimate to be made.

The lift-off method was used to select Σ_h from rock self-boring pressuremeter (RSBP) tests in Keuper Marl by Clarke et al. (1990). The results, shown in Figure 7.29, vary with depth.

This is a result of the natural variation within the ground which is confirmed by other studies (for example, Chandler, 1969) and it is also a consequence of the installation system. The RSBP is drilled into the ground with a cutter that is slightly larger than the probe, hence a cavity will be created. At the start of a test the pressure on the probe is equal to the mud pressure or zero if the borehole is dry, therefore, the selection of horizontal stress is based on an assessment of the initial slope of the test curve.

Clarke and Allan (1990) describe the use of the RSBP from a gallery at 220 m below ground level. Two boreholes were drilled, one vertical and one horizontal, with the arms pointing in known directions. Individual measurements of stress were made using the three arms. It was possible to determine the variation in stress on horizontal and vertical planes using Mohr's circle of stress. The results, shown in Figure 7.30, suggest that the total vertical stress some distance from the tunnel axis was equal to the

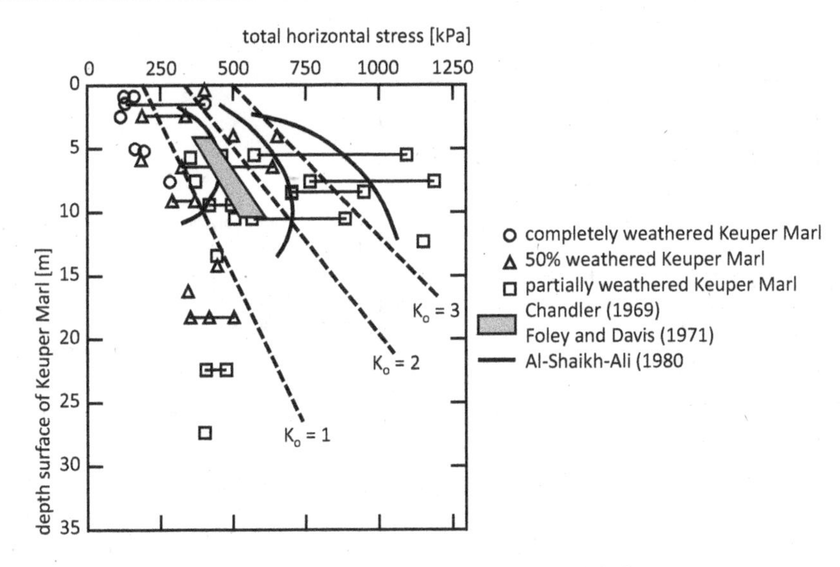

Figure 7.29 Profiles of horizontal stress taken from tests in Keuper Marl.

Source: After Clarke., 1993.

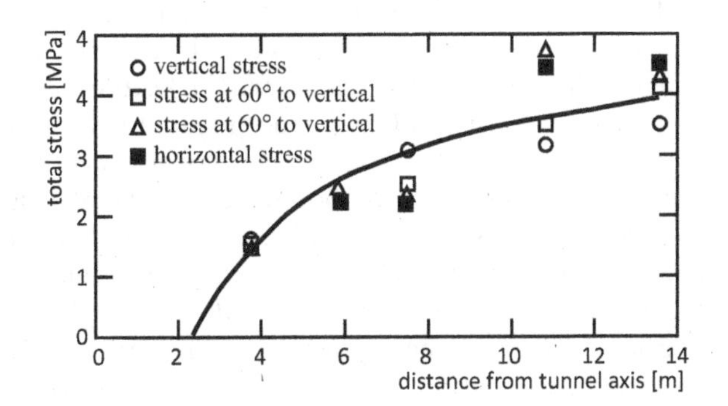

Figure 7.30 Variation in total horizontal stress with distance from tunnel axis as measured with the RSBP.

Source: After Clarke and Allan, 1990.

estimated vertical geostatic stress confirming the method of selecting the in situ stresses from the RSBP test.

Both the total vertical and horizontal stresses increased from zero with distance from the tunnel.

The comparisons given above were made using results of SBP tests. They show that, with good quality drilling and proper interpretation of the test data, it is possible to derive profiles of horizontal stress which compare favourably with other estimates from other sources. The selection of horizontal stress will always be subjective since any installation must disturb the ground, but it is possible to obtain realistic values especially when using a SBP.

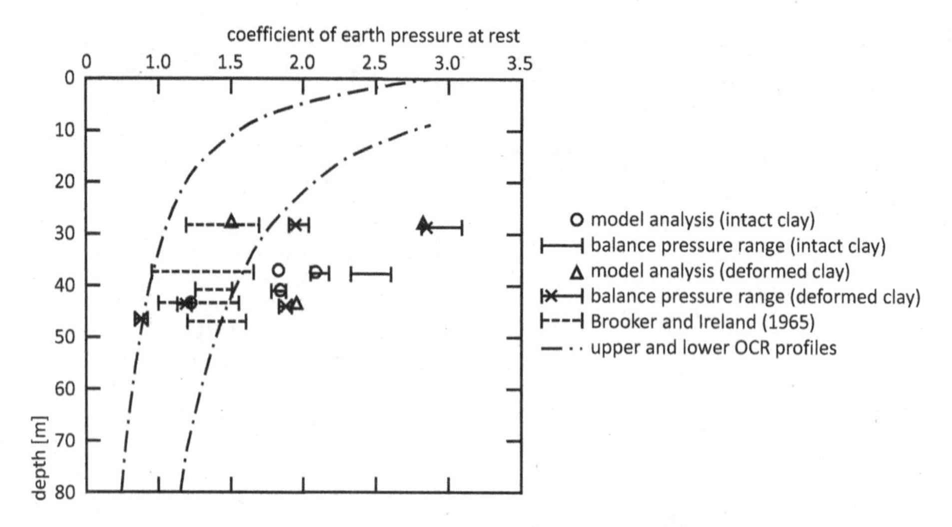

Figure 7.31 Profile of Ko for Seattle Clay using a range of techniques including those from SBP tests using the principle that the unloading curve is a function of the in situ horizontal stress (after).

Hoopes and Hughes (2014) conducted pressuremeter testing for the State Route (SR) 99 Bored Tunnel project in Seattle, Washington, to estimate in situ soil stress-deformation parameters. The tests were conducted in Seattle clay, a very stiff to hard glaciolacustrine clay, which is known for deep-seated slope failures associated with release of high, locked-in lateral stresses. It was not feasible to use a self-boring pressuremeter because of the presence of boulders and the strength of the clay. A prebored pressuremeter was used. The total horizontal stress was derived from the pressuremeter test using the Gibson and Anderson (1961) model and a balanced pressure curve and compared to results derived K_o-OCR and K_o-PI-OCR relationships. They found that the in situ measurements were generally greater than the predicted values and attributed this to shearing within the fabric of the clay influencing the in situ stress state and response to lateral unloading rather than a simple, laterally constrained, vertical loading and unloading stress path due to glaciation. The balanced pressure curve is based on the principle that the unloading of a pressuremeter test is a function of the total horizontal stress. During the unloading phase, a number of holding tests are carried out at different pressures to determine whether the membrane would expand (cavity pressure > total horizontal stress) or contract (cavity pressure <total horizontal stress). Hoopes and Hughes (2014) found that this method gave comparable results to the Gibson and Anderson (1961) method and exceeded those derived from the simple relationships by up to a factor of two (Figure 7.31).

7.4.2 Stiffness

The pressuremeter was developed to measure ground stiffness. It is often stated that the modulus measured with a pressuremeter represents the horizontal stiffness of the ground and therefore is not valid in assessing the performance of vertically loaded foundations. Baguelin et al. (1978) suggested that the stiffness mobilised during vertical loading is a function of the vertical and horizontal moduli since there is a change in isotropic and deviatoric stress. Lee and Rowe (1989) found that anisotropy has little effect on the settlement of vertically loaded foundations. Several authors, including Leischner (1966) and Shields and Bauer (1975), found that in situ tests in vertical and horizontal boreholes gave similar values of stiffness, suggesting that the vertical and horizontal stiffnesses are similar for practical purposes.

Many examples of modulus profiles have been published, but very rarely are there any references to the stress and strain range, so comparisons with other data may not be valid. The profiles are a useful database.

Figure 7.32 shows a profile of G_{ur} taken from CSBP tests in London Clay expressed in terms of the undrained modulus of elasticity.

Typically, the range of stress for a cycle was equal to the undrained shear strength. The scatter in the data arises from differences in local geology as well as differences in the strain range over which the moduli were measured. The results agree favourably with those inferred from observations of the deformation of vertically and horizontally loaded structures (St John, 1975). Further, the average profile is not too dissimilar to that proposed from observations of structures.

O'Brien and Newman (1990) reported a series of CSBP tests in London Clay. Secant shear modulus profiles were taken from unload/reload cycles over 0.1% and 0.25% cavity strain and compared to those back-figured from observations of full-scale structures, as shown in Figure 7.33.

The quoted modulus represents an average stiffness of the ground but, providing it is derived over values of cavity strain, which are similar in magnitude to the strains in the ground, the moduli can be used to predict deformation. Figure 7.34 shows the shear strains under different loading conditions.

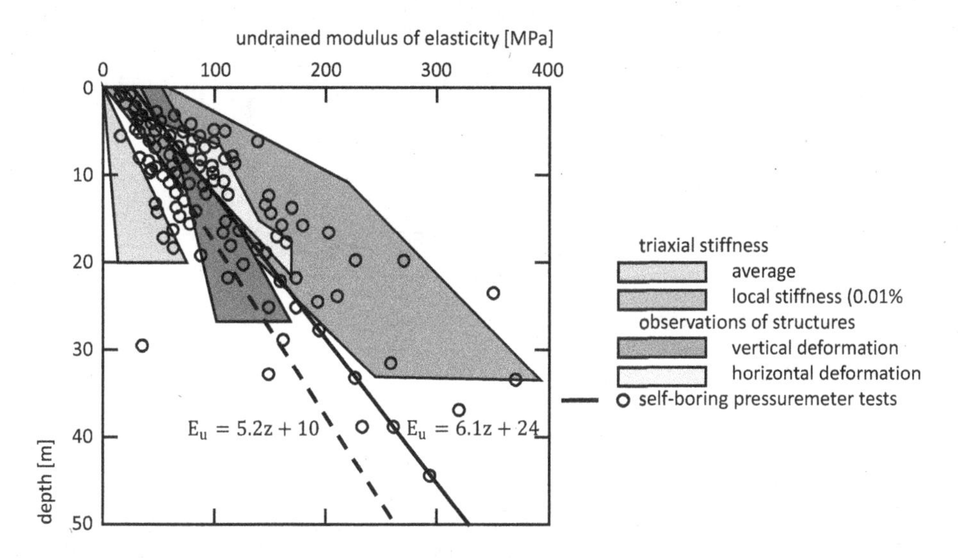

Figure 7.32 A comparison between G_{ur} from SBP tests and observations of the behaviour of structures.

Source: Clarke, 1993.

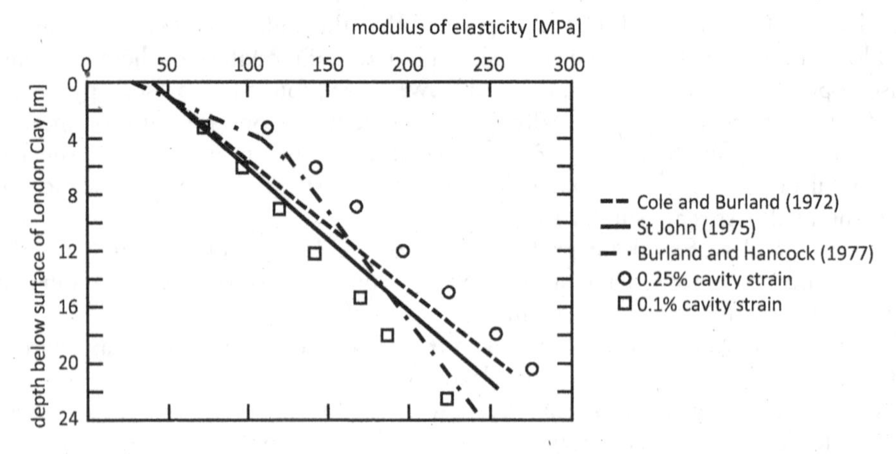

Figure 7.33 Profile of modulus of elasticity with depth from CSBP tests in London Clay compared to published data.

Source: After O'Brien and Newman, 1990.

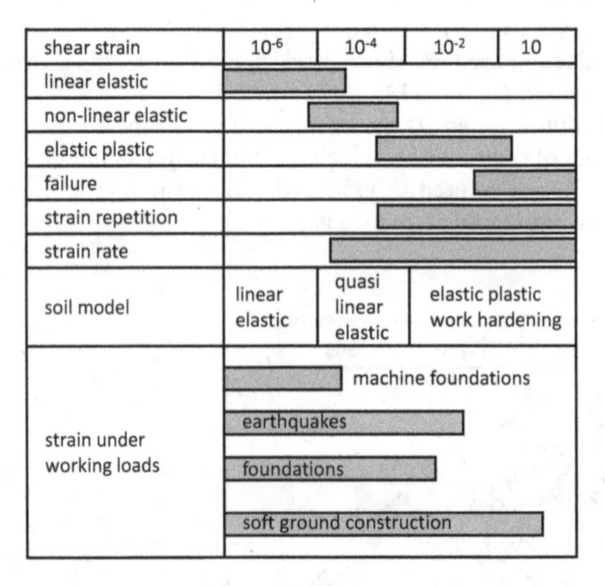

Figure 7.34 The amount of deformation under different loading conditions.

Source: After Bellotti et al., 1989.

Inspection of equation (6.15) shows that for small strains, the shear modulus is independent of the initial cavity diameter; therefore, provided a pressuremeter is expanded sufficiently to load ground unaffected by installation, the moduli derived from an unload/reload cycle should be independent of the pressuremeter type and installation. Powell (1990) showed (Figure 7.35) that results from CSBP, PBP and FDP tests are

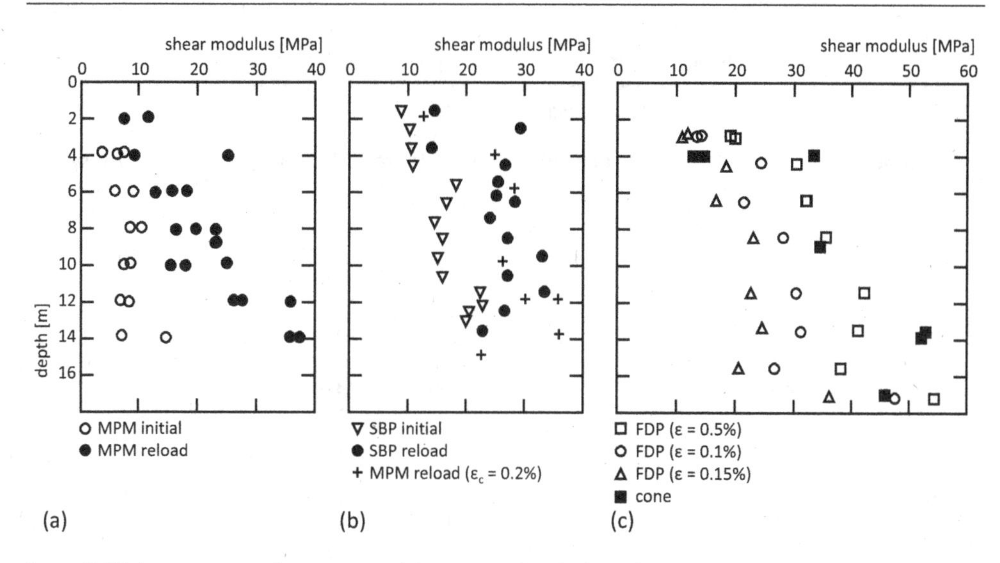

Figure 7.35 A comparison between moduli measured with (a) Ménard; (b) self-boring and Ménard; and (c) pushed-in pressuremeters.

Source: After Powell, 1990.

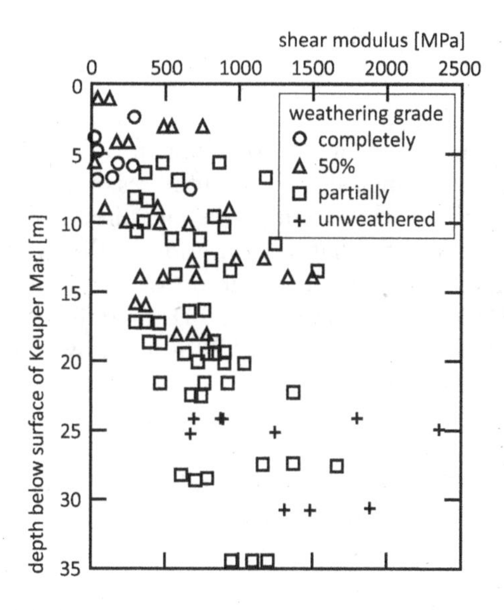

Figure 7.36 The shear moduli of completely weathered to unweathered Keuper Marl from SBP tests.

Source: After Clarke et al., 1990.

similar, though no conclusions can be drawn since the strain ranges over which they were measured were not necessarily the same.

Figure 7.36 shows a comparison of moduli determined at one site from tests in Keuper Marl, which has a variable weathering profile ranging from completely weathered to intact rock.

Table 7.14 Stiffnesses of Middle Coal Measures of the upper carboniferous series

	Shear modulus, G (MN/m^2)					
Rock profile (> 50 m)	*HPD G_i*	*HPD G_{ur}*	*Cross hole geophysics*	*Side wall sonic*	*Multichannel sonic*	*Modulus of elasticity, E (MN/m^2) Laboratory*
Sandstone	330	1294	2380		4760	7200
Coal			901		1103	1400
Seat earth			825		3100	
Mixed sst, mst				8000	4960	6960
Coal	93	630		6550	2000	
Mixed sst, mst				12310	8140	9970
Mudstone	420	1633	4 580	9290	5450	3 950
Mixed sst, mst					4890	5050
Coal			5 861	6890		
Seat earth					3310	
Mudstone	950	2766	3 690	10140	5 930	5070
Coal	180	426	4 720	12000	2550	2600
Seat earth				12760	5 860	20000
Siltstone	867	2160	6138	11540	6350	5 500
Siltstone	1204	2808	8 530	9 580	5 800	12600
Mixed sst, mst			10180	11030	6790	
Sandstone	1087	2720	10119	10930	6630	12 300

Source: After Wilson and Corke, 1990.

Notes: sst, sandstone; mst, mudstone.

The parameters determined from the MPM tests form the lower bound to the data, as expected, since the quoted moduli are the stiffnesses of a disturbed zone. The scatter in all the data arises because the moduli were measured over different strain ranges and at different stress levels; the natural variation in the ground is also a contributory factor. The results of the RSBP and high pressure dilatometer (HPD) are similar, supporting Powell's (1990) conclusion. The HPD can only be used in rock in which it is possible to form a stable pocket. The RSBP forms its own pocket and, hence, there is not the problem of pocket collapse.

A comparison of stiffnesses obtained from a variety of in situ and laboratory tests is shown in Table 7.14.

Wilson and Corke (1990) concluded that stiffnesses obtained from unload/reload cycles from HPD tests in rock represent the stiffness of the intact rock confirming the conclusions reached by Haberfield and Johnston (1993).

Muir-Wood (1990) and Jardine (1991) suggested that the cavity modulus from an unload/reload cycle in clay can be converted to a secant modulus similar to that measured in a triaxial test. The shear modulus is an average stiffness, not an incremental stiffness. Data at two sites from tests in London Clay are plotted in Figure 7.37.

The cavity strains have been converted to shear strains using the proposal of Jardine (1991). Jardine (1992) stated that reasonable agreement is found between stiffness strain curves derived from pressuremeter tests in two clays, including London Clay, and those measured in the triaxial tests, provided the shear strain is greater than 0.01%. Differences are mainly attributed to disturbance during installation and identification

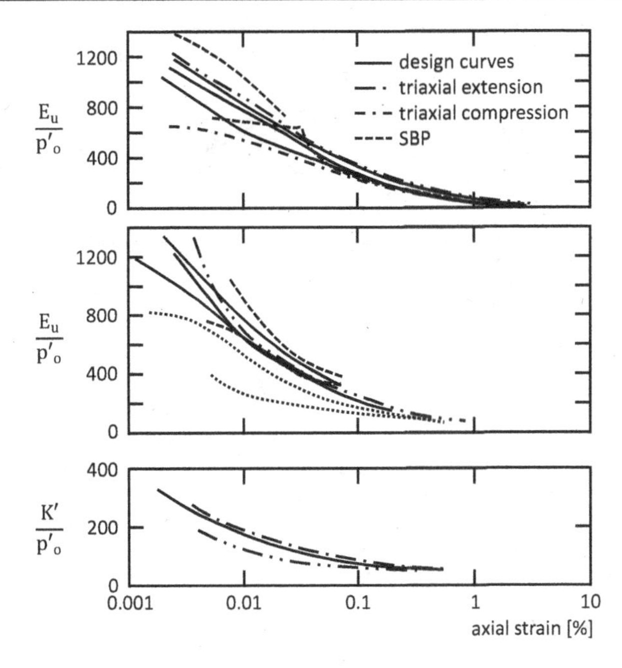

Figure 7.37 A comparison between the variation in modulus with strain from triaxial and pressuremeter tests: (a) weathered London clay; (b) and (c) unweathered London clay.

Source: After Hight et al., 1993.

of the reference datum for the unload/reload cycles because of the test procedure and creep phenomena.

Bellotti et al. (1989) compared the results of CSBP tests in sand in a chamber and in situ with those from resonant column tests as shown in Figure 7.38.

The values of G_{ur} obtained from complete unload/reload cycles were converted to the equivalent values at the effective horizontal stress using equations (4.88 to 4.90). The agreement may be fortuitous since the CSBP results are taken from one cycle, whereas the resonant column results represent the values after a number of cycles.

Data from the reloading curves shown in Figure 6.28 have been replotted in Figure 7.39.

The average shear strain, γ_{av}, is obtained from the cavity strain using equation (4.93). Bellotti et al. (1989) found that β was equal to 0.5 for medium dense sand but for these tests in a very dense sand β varied from about 0.6 to 0.2 as the applied pressure increased. The results from the second cycle give a better agreement than those from the first cycle perhaps because the first cycle is dominated by installation effects.

It is not possible to determine the moduli at strains less than 0.01% because of system compliance. Robertson and Ferreira (1993) suggested that the maximum value of modulus, G_o, could be obtained from a seismic test. Kalteziotis et al. (1990) undertook MPM tests on a variety of soils in Greece and compared the results with those from cross-hole seismic tests to show (Figure 7.40) that there is a practical correlation between G_o and E_m.

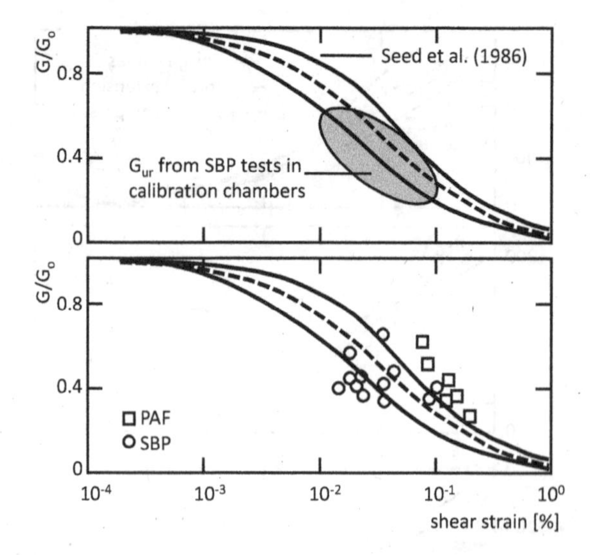

Figure 7.38 A comparison between the moduli from SBP tests and resonant column tests.

Source: After Bellotti et al., 1989.

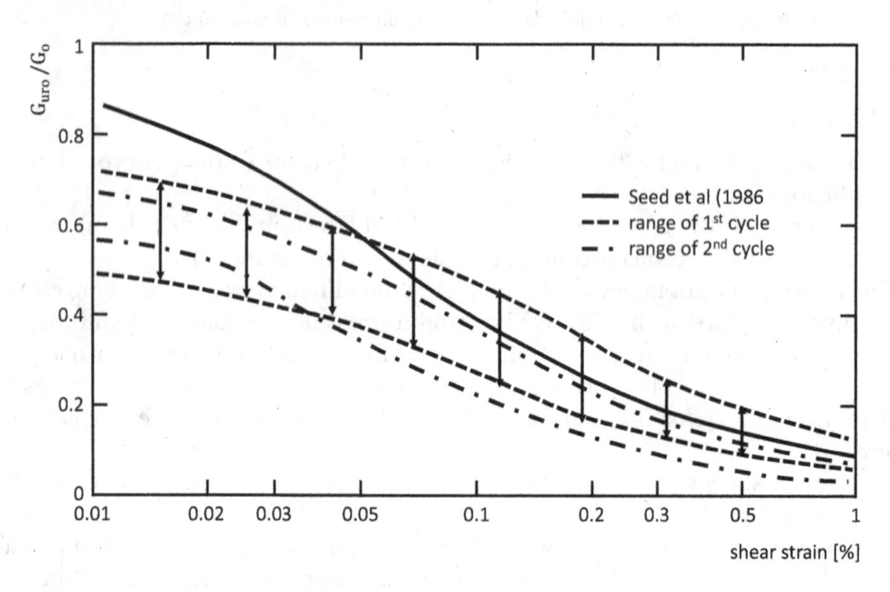

Figure 7.39 A comparison between reload stiffness and the theoretical proposal of Seed et al. (1986).

Source: After Clarke, 1993.

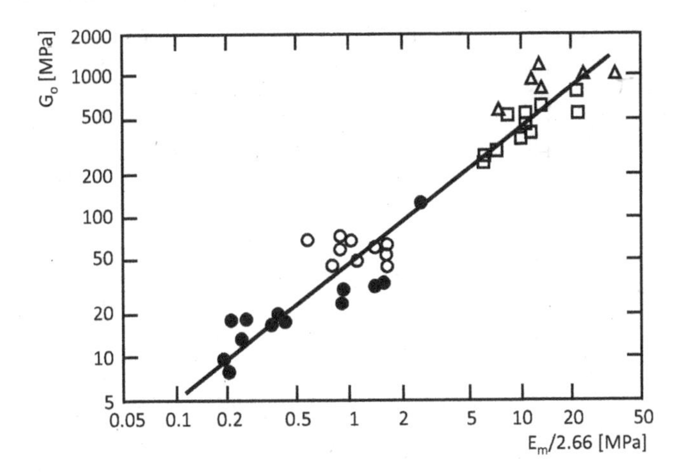

Figure 7.40 A correlation between Go and Em.

Source: After Kalteziotis et al., 1990.

Robertson and Hughes (1986) suggested that the shear modulus, G, from pressuremeter tests are related to the mean effective stress, p', at which it is measured:

$$\frac{G}{p_a} = K_G\left(\frac{p'}{p_a}\right)^n \tag{7.74}$$

where K_G is the modulus number, and, n, a modulus exponent, which is typically 0.5. Lashkaripoura and Ajalloeian (2003) suggested that K_G is 313, 422 and 516 for loose, medium and dense sand respectively based on chamber tests. Rashed et al. (2012) used a linear genetic programming (LGP) technique to develop empirical relationships for the pressuremeter secant modulus. The shear modulus is a function of the soil composition and density. The optimal relationship is:

$$E_m = 28 - 4FC + 4\gamma_d\left(\frac{\frac{288}{w\gamma_d^3}\left(\frac{12\sqrt{CU}+12FC}{\gamma_d}+7\right)^2+60}{FC}\right) \tag{7.75}$$

where E_m (kg/cm^2) is the Ménard pressuremeter modulus, FC the fines content, γ_d the dry density and CU the coefficient of uniformity.

Several authors have proposed that the pressuremeter can be used to determine the effect of cyclic loading and unloading on stiffness. Jézéquel and Le Méhauté (1982) demonstrated the change in stiffness at the in situ stress and at 10% strain with number of cycles (Figure 7.41).

The results given above indicate that stiffnesses from unload/reload cycles are similar to the average ground stiffness determined from observations of structures. Further, a

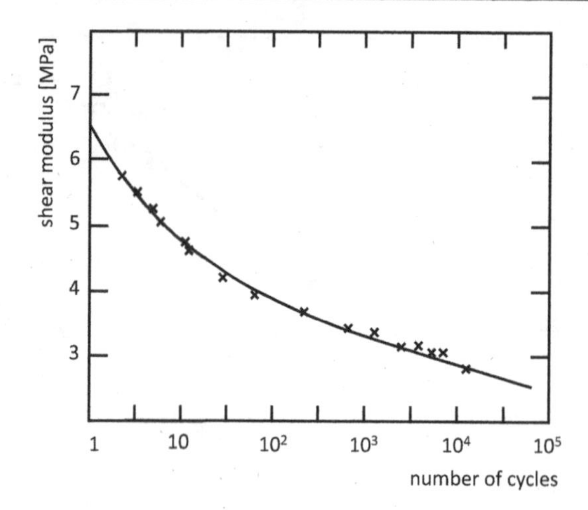

Figure 7.41 The variation in stiffness with number of cycles.

Source: After Jézéquel and Le Méhauté, 1982.

non-linear incremental stiffness profile can be determined from either an unloading or a reloading curve, though the reloading curve may give more consistent results. It is possible to carry out an unload/reload cycle with any form of pressuremeter in any ground condition provided that the probe has sufficient pressure and displacement capacity to ensure that ground undisturbed by installation is tested.

The stiffness from an unload reload cycle should represent the stiffness of the undisturbed ground no matter which pressure is used and the start of the cycle is at sufficient strain to overcome the effect of installation disturbance. However, the stiffness is also dependent on the magnitude of stress at which it is measured. Whittle et al. (2017) undertook a series of tests in highly weathered chalk using the CSBP, HPD and reaming pressuremeter (RPM). Most of the tests were carried out with an RPM fitted with a dummy cone, which was pushed into a pocket created by a 10cm² wireline CPT because of the presence of flints, making it difficult to form test pockets for the CSBP and HPD. It was assumed that the tests were fully drained and interpreted using a development of the Carter et al. (1986) analysis for c'-ϕ' material with non-linear elasticity based on the Bolton and Whittle (1999) power law. They applied a modified version of the method proposed by Bellotti et al. (1989) and by Whittle and Liu (2013), which take into account stress dependency of the moduli to show that the stiffness from the RMP and SBP were similar. In order to determine the various constants it was necessary to complete four good-quality unload/reload cycles per test and, importantly, to ensure a consistent approach to the installation of the probes to obtain repeatable disturbance.

The modulus degradation curve is defined by the variation in stiffness determined from an unload/reload cycle and the initial stiffness, G_o, which is related to the shear wave velocity. The shear wave velocity, V_s, is a function of the density of the ground as is the stiffness. Therefore, it is reasonable to assume that a relationship exists between the stiffness derived from pressuremeter tests and shear wave velocity determined

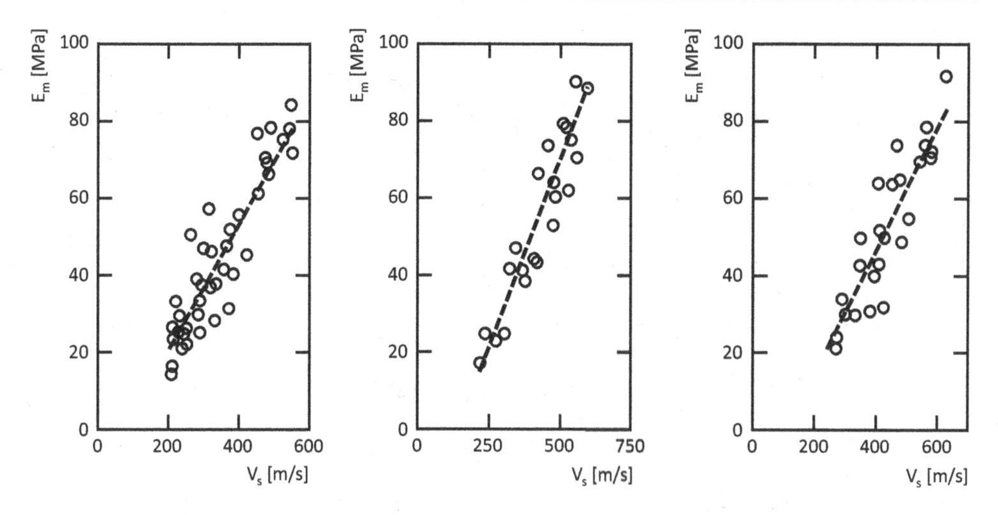

Figure 7.42 Relationships between the Ménard modulus and shear wave velocity.

Source: After Cheshomi and Khalili, 2021.

from geophysical tests. These will be site-specific relationships. Akkaya et al. (2019) suggested the following for CL and CH clay soils

$$E_m = 0.1819V_s - 26.94 \tag{7.76}$$

Cheshomi and Khalili (2021) investigated three sites in Iran with a range of sands, silts and clays to produce the correlations in Figure 7.42 which suggests that site specific correlations can be developed.

7.4.3 Undrained shear strength

Pressuremeter tests in clays and weak rocks are interpreted as undrained tests, even though some drainage will occur, as explained in chapter 5. The strength is a function of the chosen reference datum, the rate of testing, the type of test control and the method of analysis. The choice of reference datum is especially important for PBP and FDP tests unless empirical correlations are used.

Burland (1990) defines the post-peak plateau of shear stress from a triaxial test as the post-rupture strength. Leroueil et al. (1979) state that soil can exist in several forms: in situ it is in a structured form; once it has yielded it becomes destructured. The average shear strength taken from the latter part of a SBP curve is similar to a post-rupture strength – that is, a destructured strength. It is possible to obtain post-rupture (or average) strengths from PBP and SBP tests provided there is sufficient expansion of the cavity. This can be taken from the latter part of the derived shear stress curve (Palmer, 1972) or as an average value from the applied stress-strain curve (Windle and Wroth, 1977). The FDP will always give a post-rupture strength.

Generally, average strengths are quoted and often represent the average of all of the shear stress-strain curve, which must include a component of peak stress, a function

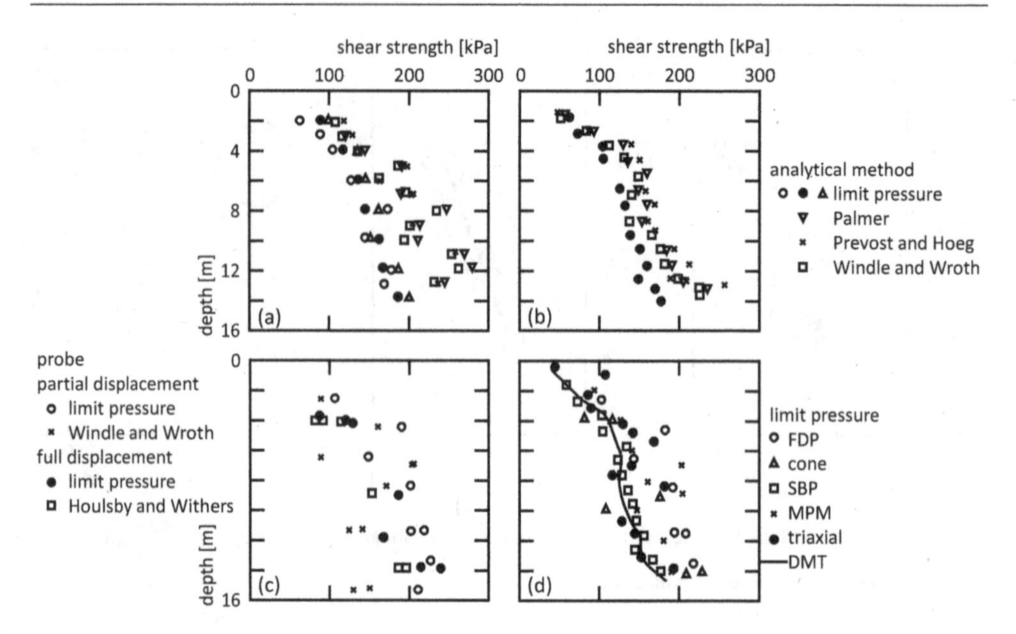

Figure 7.43 A comparison of shear strengths derived from (a) MPM; (b) SBP; and (c) full and partial displacement FDPs using different methods of analysis together with (d) a comparison of strengths from pressuremeter, triaxial and flat dilatometer tests.

Source: After Powell, 1990.

of installation disturbance (Clarke, 1994). The true average strength at large strains is unaffected by installation disturbance. The pressuremeter strengths shown in these figures are average strengths with no correction for strain range or non-cylindrical expansion. The data in Table 6.8 show that the pressuremeter strength could be overestimated by up to 50% if these effects are ignored.

Battaglio et al. (1981) showed, using Prévost and Hoeg's strain-hardening and strain-softening model, Denby and Clough's hyperbolic model and an elastic perfectly plastic model, that the true average strength obtained was independent of the analysis.

Powell (1990) reported that the method of analysis used has a significant effect on the strength (Figure 7.43) obtained, though the variation from the mean profile is less for SBP tests than for PBP tests.

This may be due to the difficulty in selecting the correct reference datum for PBP tests. Figure 7.43d shows the average strength values derived from limit pressures taken from FDP, SBP and PBP tests. This agrees reasonably well with triaxial and flat dilatometer (DMT) results. The strength was assumed to be equal to $(p_l - \sigma_h)/7.8$ (see Table 6.10).

Laçasse et al. (1981) compared strengths from CSBP with those from vane triaxial compression on two sites of soft clay. They concluded that the strength will depend on the reference datum chosen and the method of analysis, though in all cases the CSBP results were greater than the results of other tests by up to a factor of 2 (Figure 7.44).

Thus, the average strengths will depend on the method of analysis used but, if the average strength is taken from the latter part of the curve, it is more likely to be independent of the type of analysis.

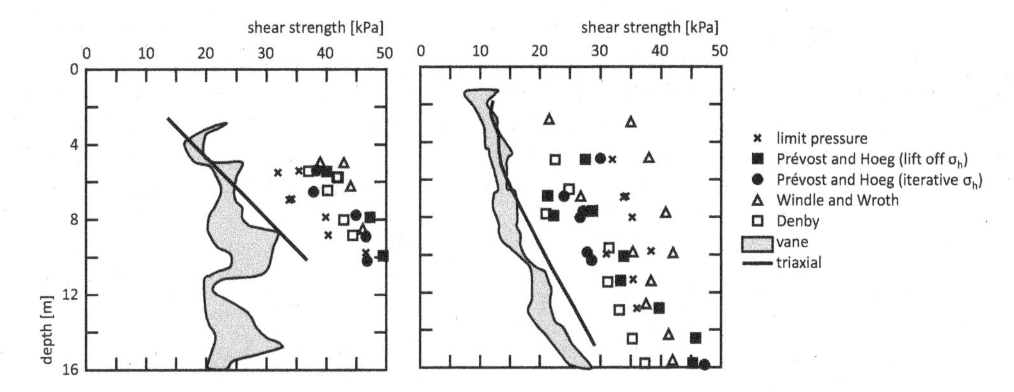

Figure 7.44 A comparison of shear strength measured with triaxial, vane and SBP tests showing the effect of varying the reference datum and method of analysis.

Source: After Lacasse et al., 1981.

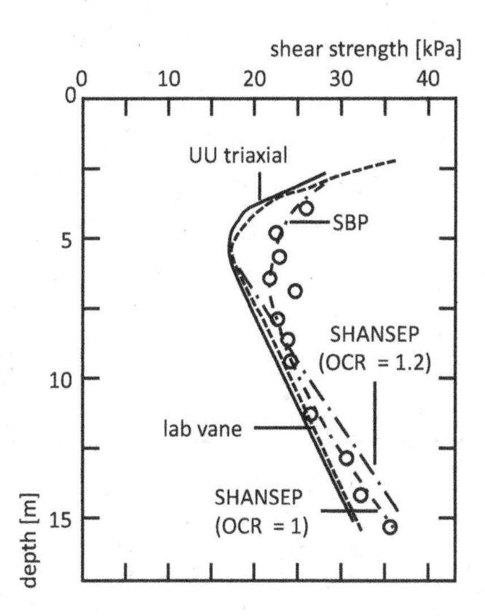

Figure 7.45 A comparison between strengths derived from CSBP, vane, quick undrained triaxial and consolidated triaxial tests on San Francisco Bay mud.

Source: After Clough and Denby, 1980.

Clough and Denby (1980) showed that the laboratory-measured values of strength are less than those obtained from CSBP tests in a soft clay (Figure 7.45).

They did find that laboratory tests on specimens reconsolidated to reduce the effects of sample disturbance gave similar results to the CSBP results. This applies to soft clays in which it is relatively straightforward to drill with minimum disturbance. The strength from the reconsolidated specimens of clays containing shell fragments was lower than the CSBP results. This discrepancy between the results arises from disturbance during installation causing changes in the soil properties, difficulties of identifying the reference datum as well as different test methods that will give different results, since the soil follows a different stress path.

Generally, it is observed that the pressuremeter gives higher values of strength than the vane. Konrad et al. (1985) showed that this depends on the OCR (Figure 7.46).

Houlsby and Nutt (1992) found that FDP tests in soft clay gave consistent values of strength, which were less than those obtained from the CSBP and triaxial tests (Figure 7.47).

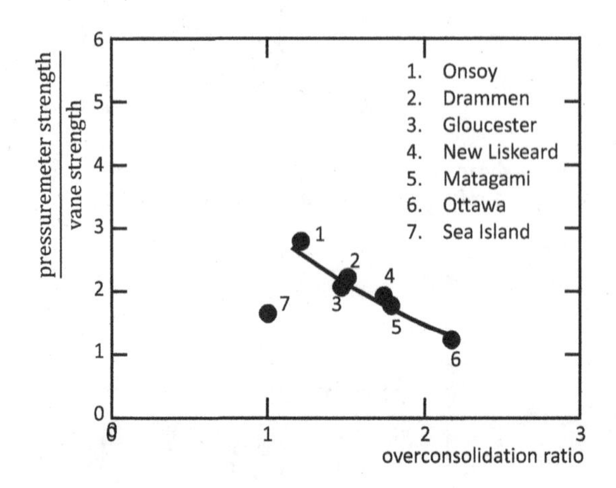

Figure 7.46 Correlation of CSBP/vane strength ratio and the OCR.

Source: After Konrad et al., 1985.

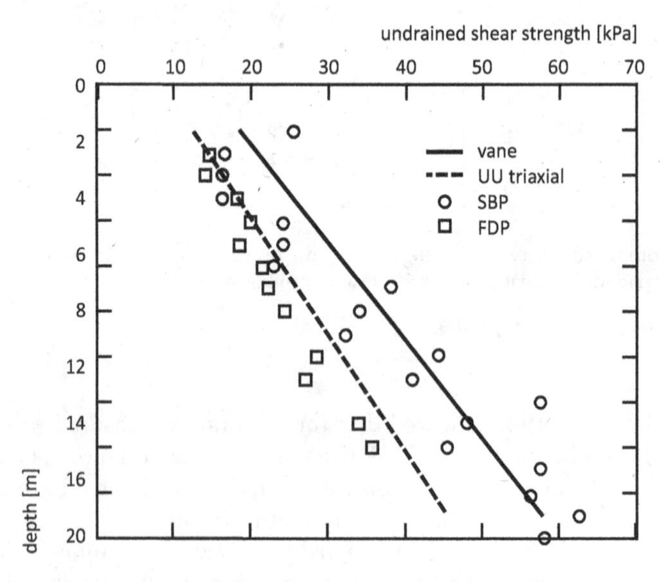

Figure 7.47 A comparison between the average strength from SBP and FDP tests and triaxial tests for a soft clay.

Source: After Clarke, 1990, and Houlsby and Nutt, 1992.

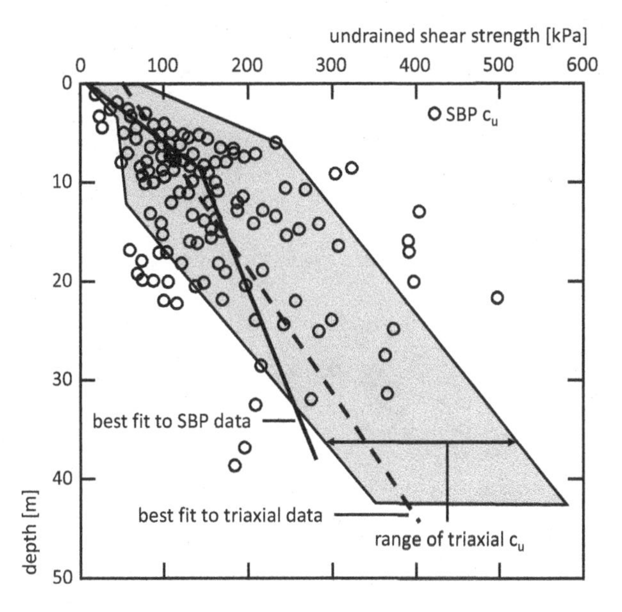

Figure 7.48 A comparison between the average strength from SBP tests and triaxial tests for London Clay.

Source: After Clarke, 1993.

Note that the strength from a FDP test is taken from the unloading curve whereas the strengths from the other tests are taken from loading curves and, therefore, should not necessarily be the same.

Eden and Law (1980) compared the strengths from in situ and laboratory tests on three types of clays ranging in stiffness from soft to firm. They concluded that the strength from pressuremeter tests would be greater than those from laboratory tests because of different stress paths and disturbance during installation. This conclusion was reached after undertaking a series of tests with a CSBP, in which the ground was intentionally disturbed by the use of over- and undersize shoes (Law and Eden, 1980). Clarke (1994) confirmed that the average strength quoted is greater if the strength is measured over the complete test curve and there is some installation disturbance.

Figure 7.48 shows average strengths obtained from CSBP and triaxial tests in London Clay.

The SBP results lie within the zone encompassing laboratory tests, suggesting that the pressuremeter gives the same strength for stiff clays as laboratory tests. Houlsby and Wroth (1989) confirmed that strength from CSBP, FDP, screw plate and triaxial tests on stiff clay give similar results.

PBPs and SBPs designed to drill into rock, have been used to determine the stiffness of weak rock. In many instances it is not possible to estimate the strength of the rock because the limit of the pressure capacity of the probe is less than that required to clearly define yield. A plot of applied pressure against cavity-strain to a log scale will produce a straight line even at small strains. The slope of this line is an index of rock strength if the rock has yielded. It is often quoted as the undrained strength but, as explained in Chapter 4, this is unlikely due to the stiffness of the rock. It can be used to

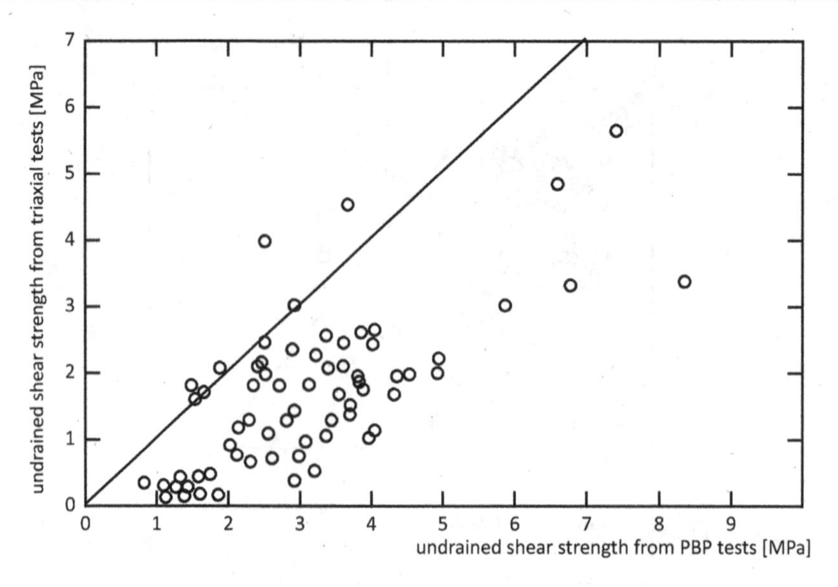

Figure 7.49 A comparison between strength from triaxial tests and HPD tests.

Source: After Hughes and Ervin, 1980.

determine the variability in the quality of the rock. Hughes and Ervin (1980) showed that this index of strength is less than that obtained from triaxial tests (Figure 7.49).

Jewell and Fahey (1984) were able to measure strengths of a siltstone up to 7 MPa. They reported that the pressuremeter strengths were greater than the strengths obtained from unconfined compression tests but do qualify this with the statement that the assumption of no volume change may not be valid.

The conclusions reached from the comparisons given above are in conflict. Generally, strengths obtained from pressuremeter tests are equal to those from triaxial tests on specimens in stiff clay, and greater than those from soft clay. The analysis used has an effect on the strength obtained if the whole of the loading curve is considered, and this may account for some of the discrepancies. A true average strength can be obtained from the latter part of an ideal SBP or PBP test, and it is likely that this strength will be more consistent.

Mayne (1995) presented a unified approach for profiling the effective yield stress of natural clays by in situ tests by using regression analyses of databases developed using in situ tests involving soft-to firm normally consolidated clays and stiff-to-hard over-consolidated clay deposits. The yield stress or pre-consolidation pressure, σ'_p, from SBP tests is

$$\sigma'_p = 0.454 p_l \tag{7.77}$$

$$\sigma'_p = 0.755 c_u ln\left(\frac{G}{c_u}\right) \tag{7.78}$$

A more generic form of these relationships is

$$\sigma'_p = \frac{1.6c_u ln\left(\frac{G}{c_u}\right)}{\delta} \tag{7.79}$$

where δ varies between 1.5 and 2.5 for intact clays and 0.4 and 0.8 for fissured clays. Theoretically, for intact clays,

$$\delta = \frac{4}{3}\frac{\phi'}{100} ln\left(\frac{G}{c_u}\right) \tag{7.80}$$

A criticism of field tests in fine-grained soil is that there is no control over the drainage conditions. Indeed, it could be argued that many tests are partially drained because of the time it takes to complete a test. This does not affect the shear modulus, as that is independent of drainage, and does not affect direct methods to determine horizontal stress. It does affect the undrained shear strength and indirect methods to determine horizontal stress based on strength. This is one reason why strengths determined directly from the field differ from those determined from laboratory tests. Jain and Nanda (2009) undertook laboratory tests on San Francisco Bay Mud to determine the parameters for the Modified Cam Clay model. They found that they could predict an SBP test provided the in situ permeability was used in the coupled Biot's theory of consolidation. The inference is the application of constitutive models in design for geotechnical structures in fine-grained soils is viable provided the in situ rather than the laboratory-determined value of permeability is used.

7.4.4 Angle of friction

It is difficult to obtain the strength of sand. Most forms of sampling from boreholes disturb the sand to such an extent that any form of laboratory test will only produce an estimate of strength. The installation of a PBP disturbs the sand, and therefore the evaluation of the angle of friction from PBP tests is discouraged, though empirical correlations do exist, such as those given in Table 7.15.

Table 7.15 A comparison of friction angles from MPM tests and triaxial tests

Ménard $p_{lm} = 2.5 \times 2^{(\phi' - 24)/4}$	*Triaxial*	*Modified Ménard* $p_{lm} = b \times 2^{(\phi' - 24)/4}$
46	35	42
50	35	45
77	35	63
44	35	41
47	42	43
53	42	47
34	34	34
32	30	33
34	30	34
29	28	31

Source: After Baguelin et al., 1978.

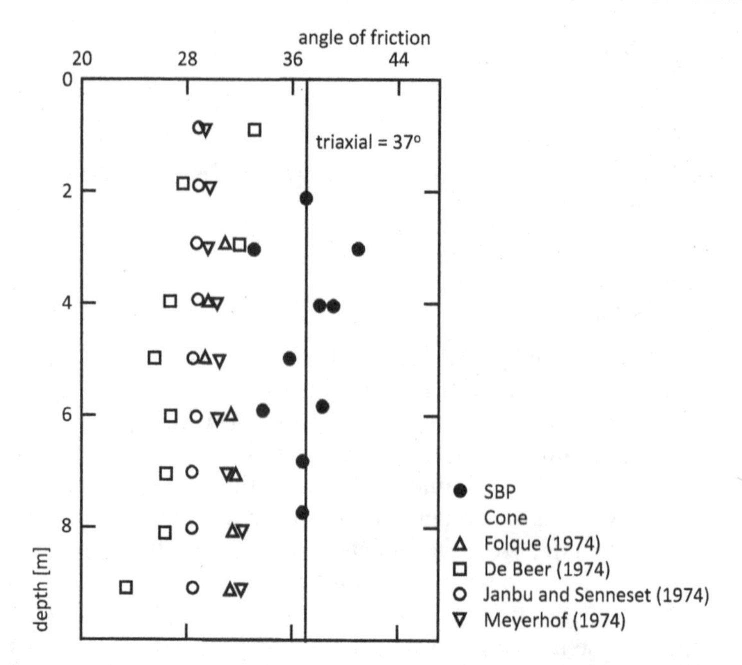

Figure 7.50 A comparison of derived values of ϕ' from cone, pressuremeter and triaxial tests.

Source: After Hughes et al., 1977.

The difference between the results is due to the effects of installation, choice of reference datum and loading conditions. Any disturbance, however small, will reduce the quality of the estimate of the angle of friction, since it is very dependent on the chosen reference datum from pressuremeter tests.

An in situ test on undisturbed sand will give a more realistic estimate of strength. Hence, it is possible to obtain an estimate of the angle of friction from good-quality SBP tests. Figure 7.50 shows a comparison between angles of friction taken from CSBP, cone and triaxial tests on a hydraulic micaceous fill.

Bruzzi et al. (1986) showed that, for dense sand, the angle of friction from pressuremeter tests was generally greater than that from cone tests (Table 7.16).

The cone data are obtained from correlations, which could account for the variation and possible difference.

In order to compare pressuremeter results with other test results, it is necessary to take into account the different stress paths followed during tests. Manassero (1989) proposed a procedure to produce unified results of secant angles of friction from triaxial and pressuremeter tests (as shown in Figure 7.51), using the relationship suggested by Lade and Lee (1976) that the angle of friction from a pressuremeter test, ϕ' – that is, the plane strain value – is a function of that for axially symmetric conditions, ϕ'_{tx}, such that

$$\phi'_{tx} = \frac{\phi' + 17}{15} \qquad (7.81)$$

Table 7.16 A comparison of derived values of ϕ' from cone and pressuremeter tests

Depth (m)	ϕ'*	ψ'†	ϕ'	ψ	ϕ'‡	*Reliability*
6.2	46.0	15.5	47.5	17.6	39	
7.7	48.6	19.2	50.5	21.0	38	
9.2	45.3	14.6	47.1	17.1	39	
10.7	39.8	7.3	42.7	11.2	40	
12.2	46.6	16.4	48.4	18.6	41	
15.2	41.2	9.1	43.7	12.6	–	
16.7	37.7	4.6	41.0	9.1	–	
18.2	38.4	5.5	41.5	9.7	40	
19.7	27.1	-7.9	34.5	0.9	37	NR
21.2	27.9	-7.0	34.9	1.2	39	NR
23.2	39.8	7.3	42.7	11.2	39	
7.4	30.3	-4.3	36.3	3.0	38	NR
8.9	39.1	6.4	42.1	10.2	40	
10.4	31.1	-3.4	36.7	3.6	38	NR
11.9	31.1	-3.4	36.7	3.6	39	NR
13.4	28.0	-7.0	34.9	1.2	39	NR
15.9	27.1	-7.9	34.5	0.9	37	NR
17.9	29.5	-5.2	35.8	2.5	38	NR
19.4	28.7	-6.1	35.5	2.0	36	NR
20.9	31.9	-2.6	37.1	4.1	–	NR
22.8	27.1	-7.9	34.5	0.9	36	NR
6.3	46.0	15.5	47.5	17.6	39	
7.3	49.3	20.2	51.2	21.6	38	
8.8	51.9	24.0	53.9	24.6	38	NR
10.3	48.0	18.3	49.7	20.3	36	
11.8	43.3	11.8	45.5	14.8	39	
13.3	49.3	20.2	51.2	21.6	37	
16.3	45.3	14.6	47.1	17.1	38	
17.8	53.9	27.0	55.5	27.2	38	NR
19.3	49.3	20.2	51.2	21.6	38	
20.8	62.8	41.1	62.8	41.1	41	NR
22.3	34.8	1.0	39.0	6.6	35	
23.8	23.0	-12.5	32.5	-1.5	36	NR
6.3	44.6	13.6	46.5	16.3	39	
7.3	32.6	-16.7	37.6	4.9	38	NR
8.8	40.5	8.2	43.3	11.7	38	
10.3	47.3	17.4	49.1	19.3	36	
11.8	46.6	16.4	48.4	18.6	39	

Source: After Bruzzi et al., 1986.

Notes:

* Hughes et al. (1977)

† Robertson and Hughes (1986)

‡ Cone

NR Not reliable.

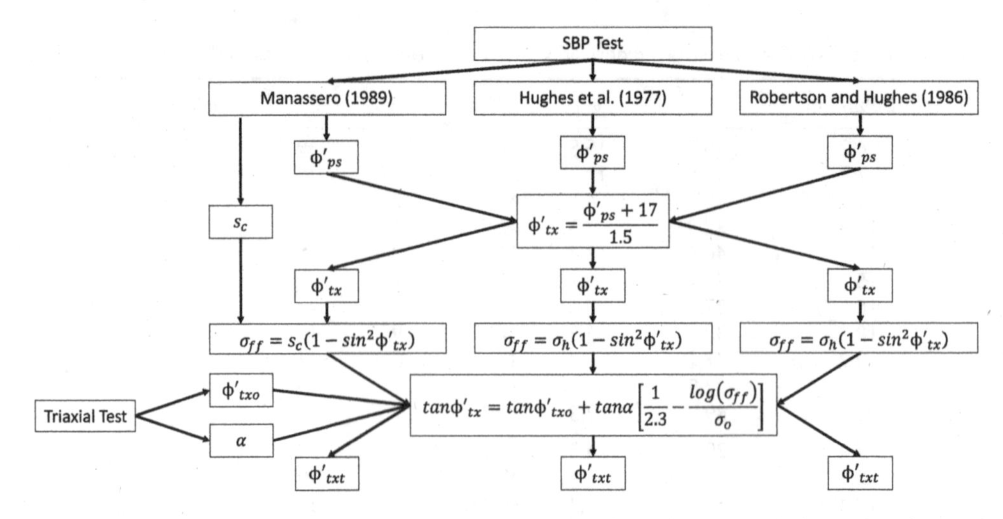

Figure 7.51 A procedure to evaluate pressuremeter test results and compare with results of triaxial tests: ϕ'_{ps} peak angle of friction in plane strain; ϕ'_{tx} peak angle of friction in axisymmetry; ϕ'_{txt} peak angle of friction from triaxial test; ϕ'_{txo} factor to describe the curved failure envelope; α factor to describe curved failure envelope; s_c isotropic stress in plane strain; σ_h horizontal stress; σ_{ff} normal stress on failure surface.

Source: After Manassero, 1989.

The secant angle of friction from a triaxial test is defined as the ratio between the shear stress and the normal stress. The failure envelope is not linear, but varies with the normal effective stress, σ'_n which, at failure, for a pressuremeter test is given by

$$\sigma'_n = \sigma'_h\left(1 - sin\phi'^2_{ax}\right) \quad (7.82)$$

A comparison of results from a test on Ticino sand is given in Figure 7.52 for a range of densities.

Potentially, the SBP is an ideal test for sand, since it should be possible to install it with minimum disturbance to the sand, and thus, the test could be analysed to give an angle of friction. SBPs have been used extensively to test sand using the method of Hughes et al. (1977) to determine an average angle of friction.

7.4.5 Limit pressure

The limit pressure is used to obtain other parameters from correlations with limit pressure (for example, shear strength) and represent the stiffness response of the ground. It should be independent of the type of pressuremeter and installation, since it is the pressure reached when the volumetric strain is 1, and at that strain undisturbed ground should be being tested. Figure 7.53 shows that the limit pressures derived from tests in clay using different pressuremeters follow a similar trend, but they are not the same.

Powell (1990) suggested that the difference is due to the time taken to carry out a test (Figure 5.8). The difference may also be due to the method used to derive p_1 since it is affected by the strain range over which it is determined (Clarke, 1994).

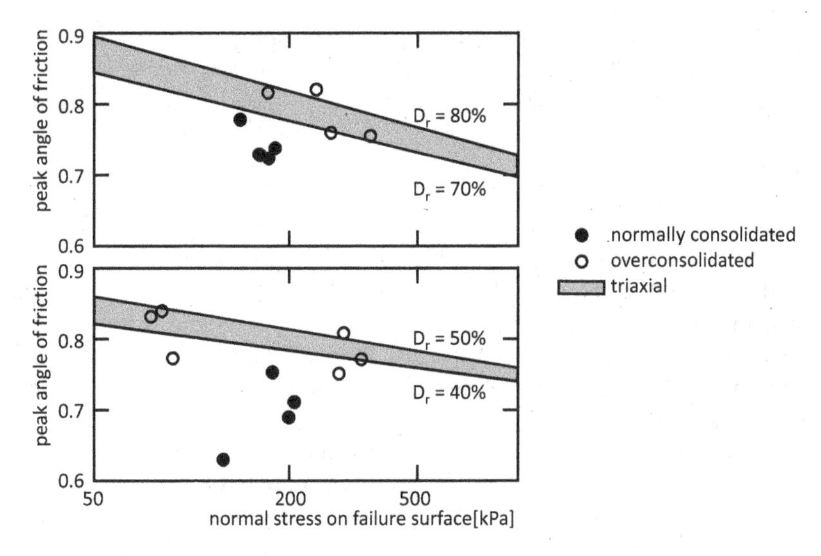

Figure 7.52 A comparison between results from pressuremeter and triaxial tests using the interpretative methods given in Figure 7.51.

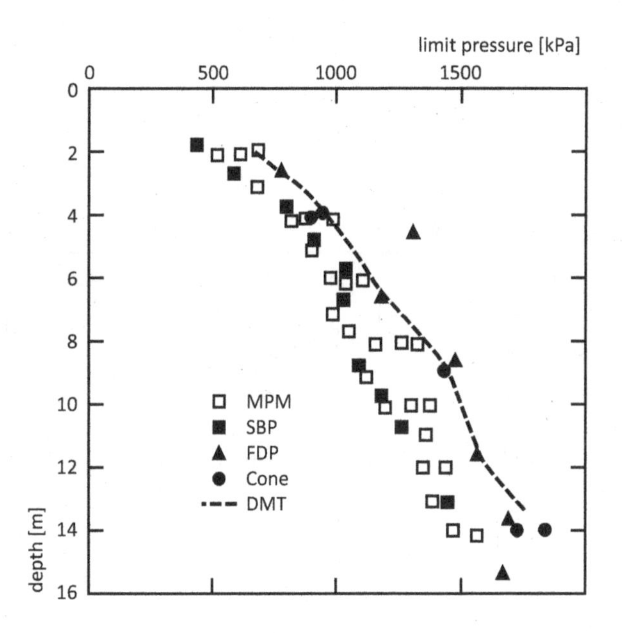

Figure 7.53 A profile of limit pressure obtained from FDP, SBP and PBP tests in stiff clay.

Source: After Powell, 1990.

The limit pressure is not an intrinsic property of the ground. It is a function of the in situ stress, stiffness and strength of the ground (σ_h, G, c_u or ϕ'). It is an indication of the variation in the quality of tests in a borehole. Profiles of limit pressure tend to be more consistent than profiles of other parameters, which may be a consequence of the greater volume of ground being tested to obtain the limit pressure. The limit pressure can be obtained from all pressuremeter tests in which yield in clay or constant volume conditions in sands are achieved.

7.4.6 Penetration tests

The use of site-specific correlations to link simpler tests to properties determined from more complex tests can be used to establish the spatial variability of a ground model for a specific site. Given that the SPT is universally used in ground investigations and is relatively inexpensive when compared to pressuremeter testing, a number of authors have developed correlations between the SPT blow count and E_m and p_{lm} (Table 7.17).

This may have value when assessing the characteristics of a site when the regional characteristics are known.

Goh et al. (2012) reported on a significant number of pressuremeter tests (Ménard, OYO and self-boring pressuremeters) carried out as part of the development of the Singapore infrastructure to understand the stiffness of local soils. They recommended that stiffness of a soil be derived from an unload/reload cycle carried out at sufficient strain such that the effects of installation disturbance are minimal. They suggested a nonlinear relationship between stiffness, N_{60} and cavity strain of the form:

$$\frac{E_p}{N_{60}} = A(\varepsilon_c)^B \tag{7.83}$$

where A and B are site-specific constants.

Table 7.17 Correlations between the Ménard modulus, limit pressure and N_{60}

Soil Type	*E_m [MPa]*	*p_l [MPa]*	*Authors*
Silty clay	$E_m = 388.67N_{60} + 4.554$	$p_l = 29.45N_{60} + 219.7$	Yagiz et al. (2008)
Silty clay	$E_m = 2.67N_{60}$	$p_l = 0.05N_{60} + 0.42$	Cheshomi and Ghodrati (2015)
Clayey soil	$E_m = 1.61(N_{60})^{0.71}$	$p_l = 0.26(N_{60})^{0.57}$	Bozbey and Togrol (2010)
Sandy soil	$E_m = 1.33(N_{60})^{0.77}$	$p_l = 0.33(N_{60})^{0.51}$	Ohya et al. (1982)
Clayey soil	$E_m = 1.93(N_{60})^{0.63}$	$p_l = 0.26(N_{60})^{0.57}$	Ohya et al. (1982)
Gravelly sand	$E_m = 0.908(N_{60})^{0.66}$		Ohya et al. (1982)
Clayey soil	$E_m = 0.29(N_{60})^{0.71}$	$p_l = 0.043(N_{60})^{1.2}$	Kayabasi (2012)
Clayey soil	$E_m = 2.611N_{60} - 26.03$	$p_l = 0.142N_{60} - 1.166$	Ozvan et al. (2018)
Sand	$E_m = 2.611N_{60} - 26.03$	$p_l = 0.12N_{60} + 0.1$	Narimani et al. (2018)
Clay	$E_m = 6.76e^{(0.05N_{60})}$	$p_l = 0.1(N_{60})^{1.06}$	Narimani et al. (2018)
		$p_l = 0.28N_{60} - 0.0021$	Cassan (1968, 1969a, 1969b)
		$p_l = 0.21N_{60} - 0.33$	Hobbs and Dixon (1969)
		$p_l = 0.056N_{60} - 0.092$	Wasachkowski (1976)
Silty clay to clayey silt	$E_m = N_{60} + 0.28$	$p_l = 0.86N_{60} - 2.21$	Liang et al. (2015)
Silty clay to gravelly silt	$E_m = 0.83N_{60} + 0.24$	$p_l = 0.92N_{60} + 1.97$	Liang et al. (2015)
Sand to gravelly sand	$E_m = 2.43N_{60} + 1.97$	$p_l = 2.28N_{60} - 0.27$	Liang et al. (2015)
Residual soil	$E_m = 0.4045N_{60} + 2.1056$		Zaki et al. (2020)
Residual soil	$E_m = 0.2961N_{60} + 3.0543$		Zaki et al. (2020)

The Chinese code for investigation of geotechnical engineering (Ministry of Housing and Urban-Rural Development of China (MOHURD), Code for Investigation of Geotechnical Engineering, Ministry of Housing and Urban-Rural Development of China, Beijing, China, GB 50021-2001, 2001) refers to the dynamic penetration test (DPT). This consists of repeatedly dropping a 120 kg hammer 100 cm onto an anvil that is connected by 60 mm diameter drill rods to a solid cone tip with a diameter of 74 mm and a cone angle of 60°. The DPT result (N_{120}) is the number of blows required to drive the cone 10 cm. Shan et al. (2021) produced correlations between the DPT and MPM results.

$$E_{pm} = 3.4437 + 0.5149N_{120} \tag{7.84}$$

$$p_{lm} = 0.4655 + 0.0845N_{120} \tag{7.85}$$

where E_{pm} is the pressuremeter modulus and p_{lm} is limit pressure.

Roy et al. (1999) undertook a statistical study of pressuremeter and cone results to determine how reliable the results were. Data from seven sand and silt sites in western Canada and one location in the United States were examined to show that the sensitivity of cone-tip resistance and the SBP data to the variability in the in situ state of packing is comparable. Laboratory chamber test results suggest that both CPT and SBP data are affected by the soil state, grain characteristics and environmental factors. They found that the measurements from carefully conducted SBP tests correlate well with the CPT cone tip resistances from the same site; and inherent soil variability affects the SBP and CPT measurements in a similar fashion. Procedural uncertainty associated with CPT results is due to dissimilar probe geometries since the installation method is prescribed in international standards and for the SBP results it is due to imperfect installation. Roy et al. (1999) found that these factors have a similar influence on CPT and SBP results. Table 7.18 are examples of published relationships between the Ménard limit pressure and modulus, and the cone resistance.

Table 7.18 Examples of correlations between cone resistances and limit pressure and pressuremeter modulus

	Correlation		
Soil type	*Net Limit Pressure* [p^*_{lm}]	*Pressuremeter Modulus* [E_m]	*Author*
Sand and gravel	0.08 to 0.2 q_c		Baguelin et al. (1978)
Compacted silt	0.25 to 0.33 q_c		Baguelin et al. (1978)
Very loose to loose sand and compressible silt	0.67 to q_c		Baguelin et al. (1978)
Very stiff to hard clay	0.25 to 0.33 q_c		Baguelin et al. (1978)
Firm to very stiff clay	0.29 to 0.4 q_c		Baguelin et al. (1978)
Very soft to soft clay	0.29 to 0.4 q_c		Baguelin et al. (1978)
Clay	0.2 q_c	2.5 q_c	Briaud (1992)
Sand	0.11 q_c	1.15 q_c	Briaud (1992)
Sand	0.22 q_c	1.35 q_c	Hamidi et al. (2015)

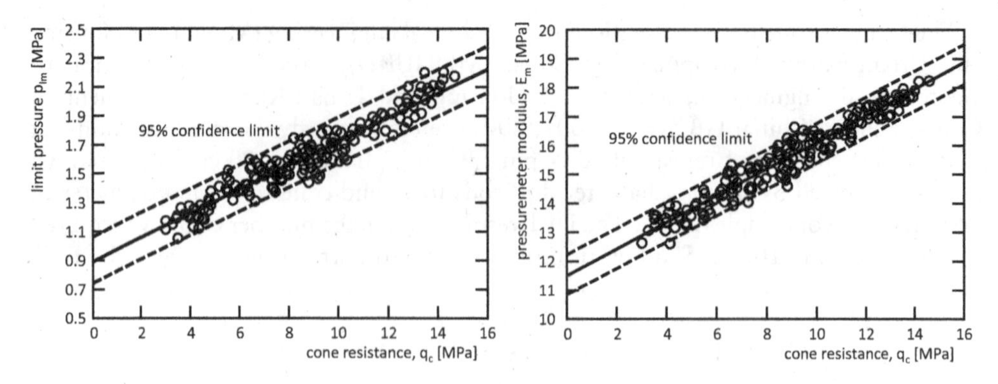

Figure 7.54 Correlations between the Ménard limit pressure and modulus and the cone resistance for desert sands.

Source: After Tarawneh et al., 2018.

Tarawneh et al. (2018) suggested relationships (Figure 7.54) between cone resistance (MPa) and limit pressure (MPa) for desert sands:

$$p_{lm} = 0.9 + 0.08q_c \quad (7.86)$$

$$E_m = 11.44 + 0.46q_c \quad (7.87)$$

These estimated pressuremeter parameters were used to predict the settlement of a 2.5 m square shallow foundation on silty sand showing good comparison with the actual behaviour.

7.5 APPLICATIONS

Pressuremeters have been widely used in site investigations for design, but the majority of reported case studies cover the use of the Ménard pressuremeter. These studies have been used to justify or amend the design method and update databases such as the IFFSTAR database. Increasingly the pressuremeter is being used to determine parameters for constitutive models used in designs based on numerical methods.

7.5.1 Use of pressuremeter results in design

Davidson and Bodine (1986) described the results of axial and tension pile tests using a variety of pile types up to 30 m long in 6 m of soft clay overlying stiff to hard clay. The capacities of the piles were predicted using four methods: the Ménard rules, the α method (Tomlinson, 1957), the λ method (Vijayvergiya and Focht, 1972) and an effective stress method, β method (Burland, 1973). The undrained strengths for the α and λ methods were derived from Ménard pressuremeter tests assuming that the strength was a function of the limit pressure (see Table 6.10). The drained parameters were obtained from triaxial tests. Realistic predictions which compared favourably with the test pile results were obtained, as shown in Table 7.19.

Table 7.19 A comparison between predicted and measured pile capacity using the Ménard, α, λ and β methods

	Prediction (tonnes)									
	A		*λ*		*β*		*MPM*		*Measurement: Load test*	
Pile type	*Friction*	*Total*	*Friction*	*Total*	*Friction*	*Total*	*Friction*	*Total*	*Compr.*	*Tension*
Steel pipe	201	217	279	294	223	319	210	226	230	230
Hel-cor	231	252	320	341	258	383	242	263	270	230
Step taper	215	239	326	350	222	356	220	244	270	290
Precast concrete	256	276	354	375	285	407	315	335	310	150*
Monotube	256	262	395	401	270	302	299	305	330	300

Source: After Davidson and Bodine, 1986.

Note: *Splice failed.

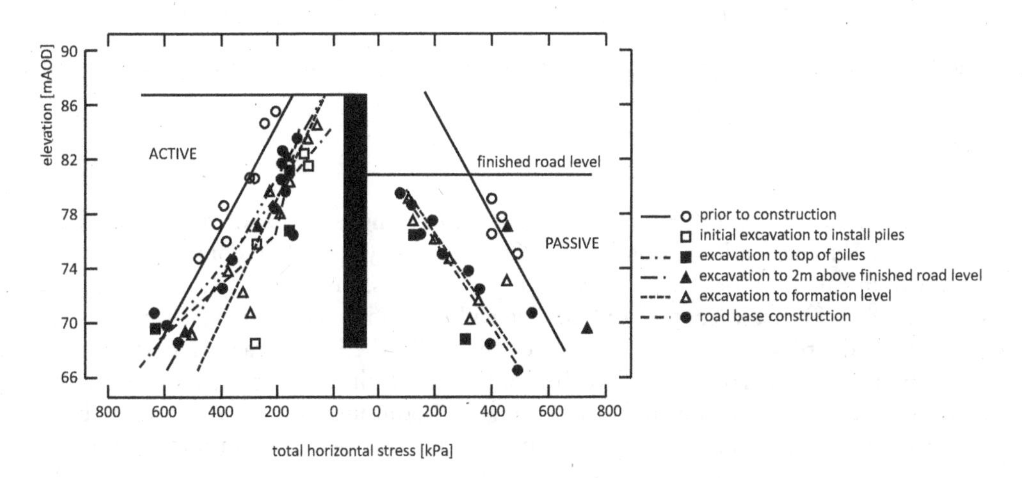

Figure 7.55 The variation in horizontal stress with construction of a contiguous bored pile wall.

Bergado et al. (1991) used E_m in an elastic method based on the drained stiffness to predict the settlement of trial embankments on soft clays. They found that $E' = 0.36E_m$ gave reasonable agreement with the observed settlements.

Clarke and Wroth (1984) used the CSBP to monitor stress changes during the construction of a contiguous bored pile wall within a stiff over-consolidated clay. The results, shown in Figure 7.55, suggest that the pressure in the soil on the active side reduced, but not to the active conditions.

The total horizontal stress measured with the CSBP on the passive side was possibly less than that in the ground due to the installation effects (see section 5.2.2). These data, together with the measured stiffness, were used to predict the deformation of the wall using a finite element analysis.

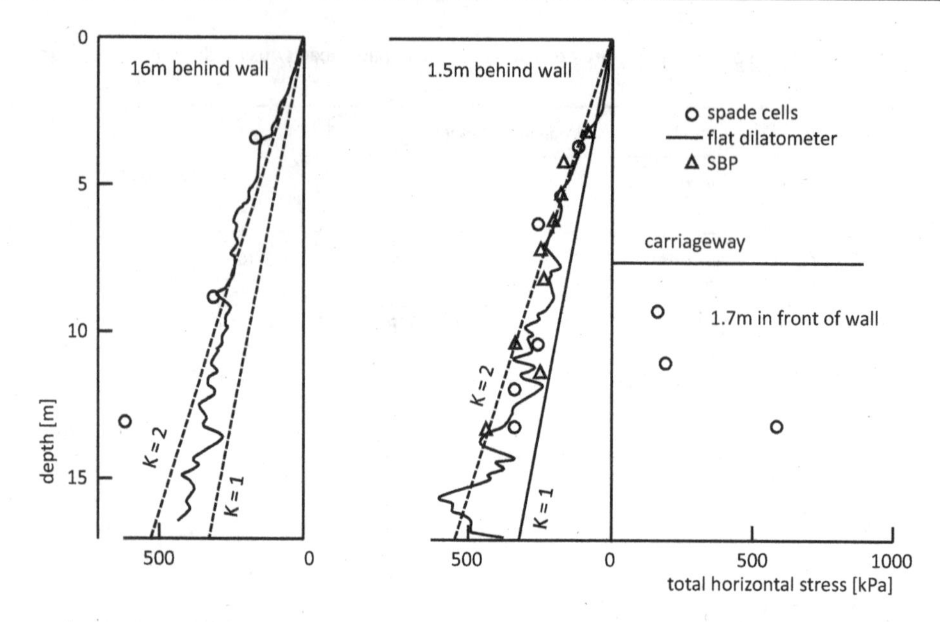

Figure 7.56 The variation in total horizontal stress ten years after constructing a contiguous bored pile wall in a stiff clay.

Source: After Symons and Carder, 1992.

Symons and Carder (1992) undertook a monitoring exercise of a propped contiguous bored pile wall in London Clay about ten years after construction. This included installation of spade cells and piezometers. In situ tests, including the CSBP and DMT, were carried out to determine profiles of total horizontal stress, modulus and strength on the active side of the wall. The DMT was inserted with the blade parallel and perpendicular to the wall to determine the stresses in orthogonal directions. The CSBP was installed with one arm pointing to the wall, thus enabling the stresses in that direction to be determined. The stresses on planes perpendicular to the wall were estimated from Mohr's circle of the stresses. Figure 7.56 shows the measured profiles of total horizontal stress remote from the wall and on a plane parallel and adjacent to the wall, that is, the in situ conditions.

They concluded that measured values of total horizontal stress on the active side were greater than those predicted from the inferred bending moments. Those predicted and measured on the passive side were similar.

Barksdale et al. (1986) proposed that the MPM could be used in residual soils with some success, especially as these soils are difficult to sample. Laboratory tests tend to over-predict settlements leading to costly foundation solutions. The MPM method was found to give satisfactory results for predicting settlements of large oil tanks and a 22-storey hotel. Barksdale et al. confirmed that the application of the MPM method requires experience of the local soils and for that reason it may not be universally applied without an alternative design method, such as that proposed by Terzaghi and Peck (1948), based on SPT tests. Barksdale et al. found a good correlation between E_m and N_{60} for Piedmont residual soils of the form

$$\ln(E_m) = 3.509 + 0.712\ \ln(N_{60}) \tag{7.88}$$

where N_{60} is the SPT blow count and E_m in ksf.

Hight et al. (1993) described the predicted and measured deformations associated with the construction of the Waterloo International Terminal. The site investigation in made ground, Thames gravel, London Clay and Woolwich and Reading Beds, included small-strain stiffness triaxial tests and pressuremeter tests. The results of those tests were used to predict a nonlinear stiffness profile, which was subsequently used in a plane-strain finite element analysis. They followed the construction of the original Waterloo terminal so that the kinematic nature of the soil stiffness could be accounted for. Figure 7.57 shows a comparison between predictions and measurements.

The difference is due, in part, to the difference in the modelled and construction sequence and the 3D effect. Note that the pressuremeter results were used to supplement other site investigation data.

St John et al. (1993) described predictions and measurements of movements associated with a deep excavation. The site investigation in made ground, alluvium, London Clay and Woolwich and Reading Beds include cone penetration tests, CSBP

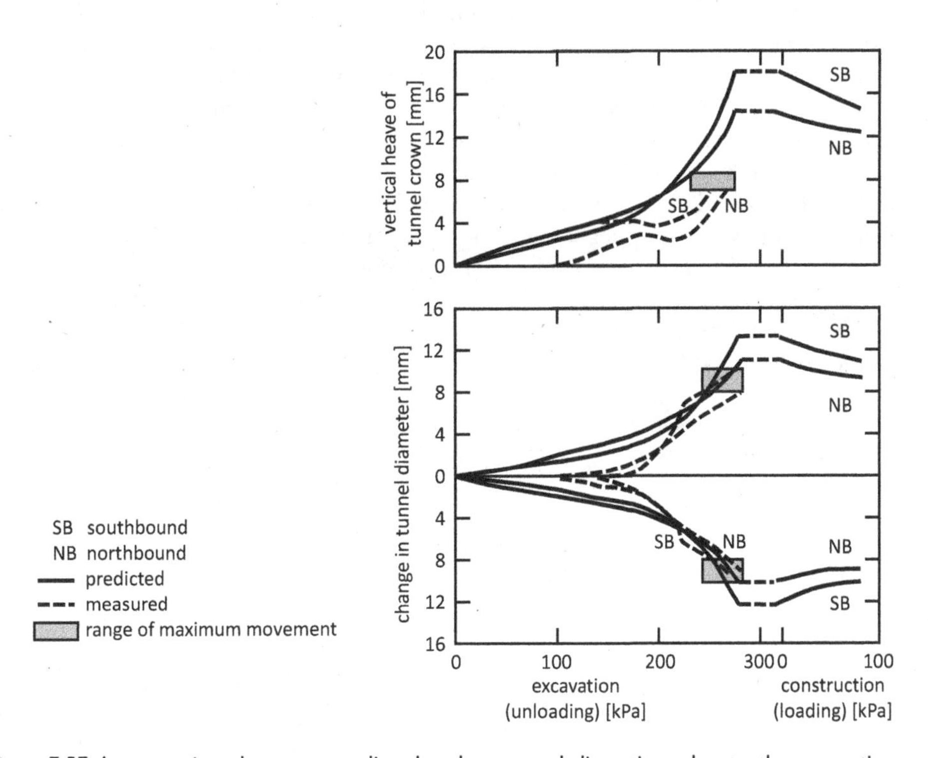

Figure 7.57 A comparison between predicted and measured distortions due to the excavation and construction of the northbound and southbound tunnels.

Source: After Hight et al., 1993.

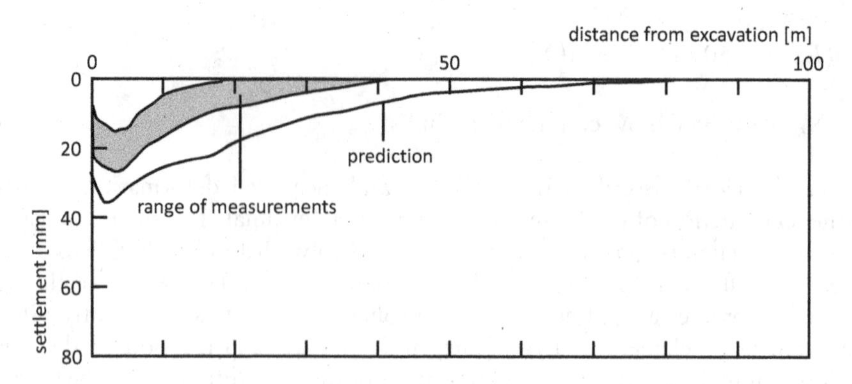

Figure 7.58 A comparison between predicted and measured settlements adjacent to the excavation.

Source: After St John et al., 1993.

tests and stress path tests. The CSBP was used primarily to obtain the initial stress conditions, and results from other tests were used to develop a profile for the site. They were able to predict the ground movements with reasonable accuracy, though this could only be achieved by a careful assessment of the soil properties from laboratory and in situ tests, other studies of foundation behaviour and a sophisticated finite element programme. St John et al. stressed the importance of the use of a finite element programme to investigate a range of construction options and to confirm the initial predictions with field observations. Figure 7.58 shows a summary of the predicted movements.

Thompson and Leach (1991) gave a comparison between predicted and measured settlements of several structures forming a nuclear power station founded on rock, the sequence of which is Sherwood Sandstone (Triassic), Permian Mudstones and Evaporites. Records were taken from the start of construction in 1979 for seven years, as shown in Figure 7.59.

A variety of in situ and laboratory tests were used to determine moduli profiles, which are given in Table 7.20.

The predicted settlements using elastic analyses shown in Figure 7.59 are in agreement with the measured settlement.

Campanella et al. (1989) report good agreement between the variation in surface deflection with lateral load and the deflected shape with depth for laterally loaded piles in normally consolidated clays, silts and sands using methods proposed by Robertson et al. (1984).

McCabe and Phillips (2008) found from a number of tests on shallow and piled foundations in soft estuarine silt (Belfast) that it was more appropriate to use a shear strength derived from a test that modelled the installation of the foundation. For example, for driven piles, piezocone and cone pressuremeter tests provide more appropriate strengths as they model the intense shearing associated with installing driven piles.

The increasing use of numerical methods in design means that there is a greater reliance on using ground investigation results to produce parameters that define

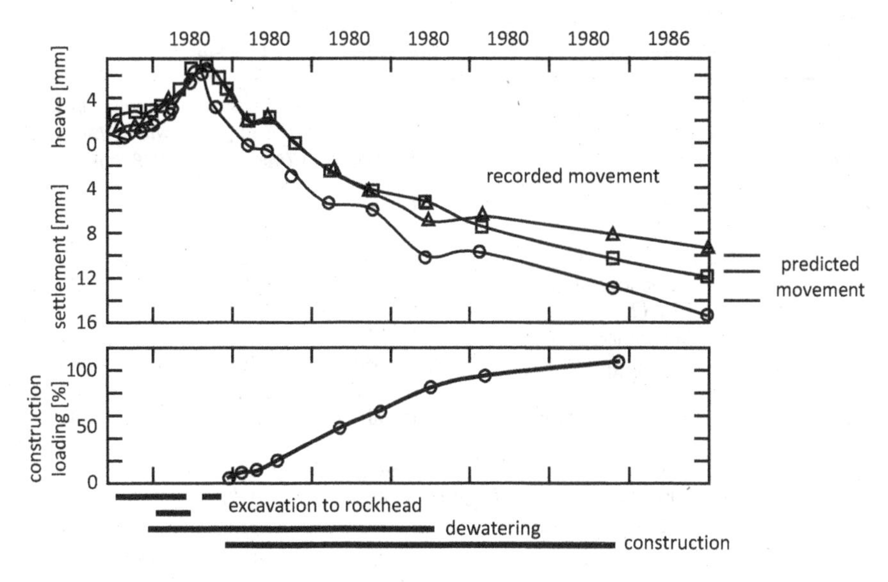

Figure 7.59 Records of the settlement of three structures founded on rock.

Source: After Thompson and Leach, 1991.

constitutive models. It is now possible to use more sophisticated models, but these require a number of parameters, which may have to be derived from a range of tests. There can still be a discrepancy between actual ground behaviour and that predicted from the models. This has been addressed by applying empirical factors based on field observations (e.g., Gambin et al., 1996; Fawaz et al., 2002), by developing correlations between pressuremeter test results and the results of other tests (e.g., Angelim et al., 2016), using optimisation procedures to adjust the parameters in order to model the pressuremeter test (e.g., Gaone et al., 2019; Rouainia et al., 2017) and by adjusting the parameters to validate the model against case studies (e.g., Fawaz et al., 2014; Angelim et al., 2016). Soil deformation is difficult to predict, yet it is a requirement of all designs. It is possible with pressuremeter tests to model the non-linear behaviour of soils (e.g., Robertson and Hughes, 1986; Briaud, 2013; Baud et al., 2012; Bolton and Whittle, 1999). The increasing sophistication of the constitutive models means that parameters may have to be derived from more than one test. Cardoso Bernandes et al. (2021) used this approach to calibrate a Hardening Soil model (Figure 7.60) to predict pile load-settlement behaviour.

The effective friction angle, effective cohesion and angle of dilatancy can be obtained from direct shear or triaxial tests, but Cardoso Bernandes et al. (2021) developed a curve-fitting approach based on cavity expansion theory using the cohesive-frictional model developed by Fontaine et al. (2005) with the hyperbolic extension proposed by Baud et al. (2012) to model the initial portion of the curve due to soil recompression. Cardoso Bernandes et al. (2021) were able to model the initial portion of the pile load settlement response (Figure 7.61) for bored and contiguous flight augur (CFA) piles using the hardening strain (HS) model.

Table 7.20 Average undrained stiffness profiles for the underlying rock at Heysham

Rock type	*Test*	*Modulus*	*Stiffness at the surface (MPa)*	*Variation in stiffness with depth (MPa/m)*
Sherwood sandstone	PBP	Initial	1200	19
		Reload	2 760	44
		Unload	1200	19
	Sonic logging (E_{max}/5)		1500	23.75
	SPT (2.3N)		1800	28.5
	Unconfined compression	Initial	1800	28.5
		Reload	2760	44
		Unload	2000	60
	Triaxial compression	Initial	2520	40
		Reload	3310	53
		Unload	2400	27
	Ultrasonic velocity (E_{max}/5)	1500	23.75	

Rock type	*Test*	*Modulus*	*Weak (MPa)*	*Medium (MPa)*	*Strong (MPa)*
Permian mudstones	PBP	Initial	70	470	2700
		Reload	390	1860	4600
		Unload	400	1860	
	Sonic logging (E_{max}/5)		300	1200	2000
	SPT (2.3N)		130	500	
	Triaxial compression	Undrained	265		5 500
		Drained	255		
	Oedometer		260		
Evaporites				30000	

Source: After Thompson and Leach, 1991.

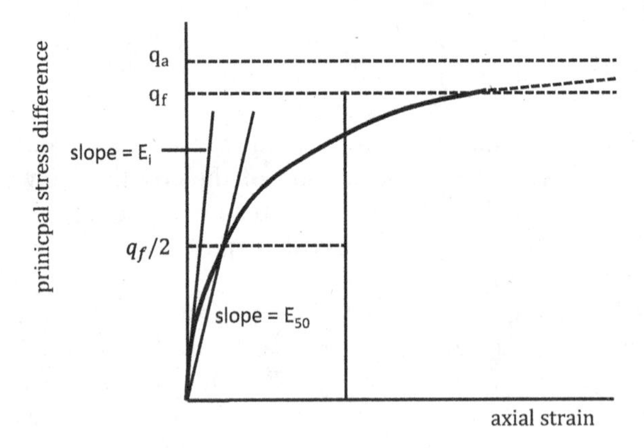

Figure 7.60 A hyperbolic stress strain relationship used to model pile-load settlement behaviour with the parameters derived from pressuremeter tests.

Source: After Cardoso Bernardes et al., 2021.

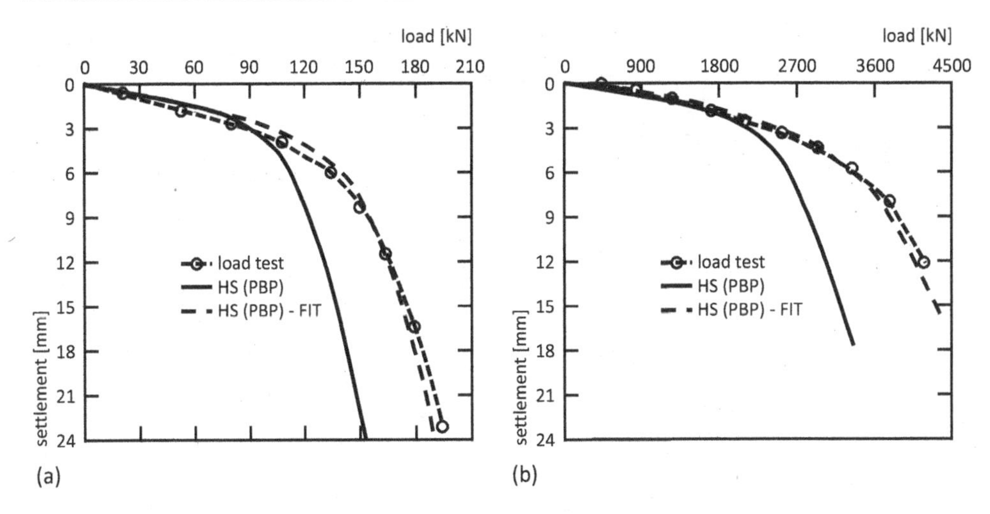

Figure 7.61 A comparison between the predicted pile settlement response to load using a hardening soil model for (a) a bored pile and (b) a cfa pile.

Source: After Cardoso Bernardes et al., 2021.

This represents the deformation at the working load. The method underestimated the ultimate pile capacity, but engineering the fit could be improved by increasing the cohesion (Figure 7.61). The fit to the initial portion of the pile load settlement curve is based on stiffnesses derived from pressuremeter tests. This has to be validated against pile tests. However, the prediction of the ultimate pile capacity is not proven.

Rouainia et al. (2017) examined the undrained behaviour of a deep excavation, which forms part of a 100 m wide 10 m deep basement excavation located in Boston, Massachusetts. The glacial marine clay was modelled within the framework of kinematic hardening plasticity (Rouainia and Muir Wood, 2000) using three constitutive models (a kinematic hardening model for structured soils (KHSM), its reduced bubble model version (KHM) and the Modified Cam Clay (MCC) model) and two elastic assumptions (traditional elasticity and a small-strain stiffness formulation). The pore pressure time histories beneath the centre of the excavation and the associated heave of its base were modelled with coupled finite element analyses. The chosen parameters were evaluated by means of numerical simulations of undrained triaxial, constant rate of strain and self-boring pressuremeter tests. The parameters were derived from triaxial, piezocone, seismic cone and self-boring pressuremeter tests. The finite element (FE) analyses were able to make a reasonable prediction of the wall deformation (Figure 7.62), base heave and pore pressure dissipation.

Critically, it was necessary to use the Viggiani and Atkinson (1995) formulation for the small-strain stiffness

$$\frac{G_o}{p_{ref}} = A\left(\frac{p}{p_{ref}}\right)^n R_o^m \tag{7.89}$$

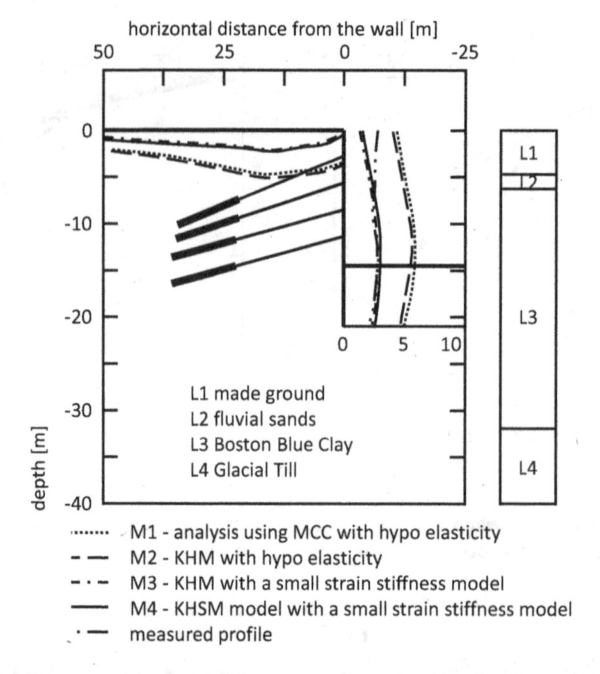

Figure 7.62 Predicted deformation of a tied back wall and surface settlements behind a wall in Boston Blue Clay using a kinetic hardening model for structured soils.

Source: After Rouainia et al., 2017.

where G_o is the small-strain shear modulus, p_{ref} a reference pressure, p the mean effective stress, R_o the isotropic over-consolidation ratio and A, m and n stiffness parameters. This analysis showed that predicting the lateral wall deflection is possible if the small-strain properties of the soil are included since they control the magnitude of the deformation.

The use of geostructures as heat exchangers is increasing because of the climate change target to attain net zero carbon by 2050. Bourne-Webb et al. (2013) suggested that the thermomechanical performance of energy pile foundations leads to negative friction due to the deformation of the pile. Eslami et al. (2017) suggested that the cyclic temperature evolution associated with energy piles would also affect the properties of the surrounding soils. Pressuremeter tests using the 380 mm long, 28 mm diameter APAGEO mini-pressuremeter were performed on an illitic soil in a thermo-regulated metre-scale container subjected to temperatures from 1 to 40°C. The results reveal a slight decrease in the pressuremeter modulus (E_m) and a significant decrease in the creep pressure (p_f) and limit pressure (p_l) with increasing temperature with the effect being most pronounced during the first cycle. The results also reveal the reversibility of this effect during a heating–cooling cycle throughout the investigated temperature range, whereas the effect of a cooling–heating cycle was only partially reversible.

These examples of the use of pressuremeters to obtain properties of the ground, which are subsequently used in design, demonstrate the potential of the pressuremeter.

However, they also indicate that the predictions have to be modified using engineering judgment to predict the measured deformation.

7.5.2 Use of pressuremeter tests in complex ground

Classic soil mechanics is based on fine-grained and coarse-grained soils; that is, they are distinguished by their composition, and this is the approach used in the interpretation of pressuremeter tests. In practice, soils are a more complex mixtures of minerals, water, air and organic matter, such that soils are mixtures of more than one category of particle size. Problematic soils include those soils that are intrinsically complex as well as those soils that are subject to environmental processes post deposition that create problems. The former include tropical, glacial, collapsible, expansive, fissured, organic, soluble and anthropogenic soils. The latter include the processes of weathering, seepage, contamination and temperature. Problematic soils can be grouped into four categories according to the dominant effect that creates a problem: formational effects, climatic effects, behavioural effects and environmental effects.

Examples of correlations developed for pressuremeter test results for specific soil types, including complex soils include over-consolidated clay (Huang, 1995; Kavur et al., 2019; Horvath-Kalman, 2015); weak rocks (Huang et al., 1999; Isik et al., 2008); Ballina Clay (Kelly at al., 2017); cement paste backfill (Le Roux et al., 2002); crystalline rocks (Kuvik et al., 2019); Lisbon Miocene clays (Laranjo et al., 2015); glacial till (Larsson, 2001; Cao et al., 2015); Bangkok soft marine clay (Likitlersuang et al., 2013); Dublin Boulder Clay (Long and Menkiti, 2007); Troll Clay (Lunne et al., 2006); residual soil (Silva et al., 2016; Mayne and Brown, 2003; Orhan et al., 2006; Pietrangeli et al., 2013; Wang and Borden, 1996; Zhang et al., 2019); Calvert Formation, Richmond, Virginia (Martin et al., 1995); municipal landfill (Matasovic et al., 2010; Chatzigogis et al., 2006); Thanet Sands (Menkiti et al., 2015); decomposed granite (Ng and Wang, 2001; Ng et al., 2000; Schnaid et al., 2000; Fonseca, 2003); normally consolidated clay (Powell, 2001); semi-cohesive to cohesive soils, hard soils and soft rocks, Athens (Ritsos et al., 2005); fluvial deposits, Venice (Simonini et al., 2007); varved clays, Poland (Zawrzykraj, 2017); post-glacial silt, Norway (Blaker et al., 2019); marine clay, Singapore (Bo et al., 2012; Chu et al., 2002); Connecticut Valley Varved Clay, Massachusetts (DeGroot and Luteneger, 2005); volcanic soils (Fernandez-Baniela et al., 2021); and volcanic residual soils (Fernadez-Baniela et al., 2021).

Prust et al. (2005) reported on an extensive investigation as part of the construction of the Mass Rapid Transit system, Bangkok, which included self-boring pressuremeter tests. The soil profile consists of soft clay overlying stiff clay and then dense sand.

Figure 7.63 is a comparison of an empirical prediction of K_o (Mayne and Kulhawy, 1982; Shohet, 1995; and Ladd et al., 1977) compared with that derived from SBP tests using a variety of methods (lift off, Marsland and Randolph, 1977; Hawkins et al., 1990; and curve fitting).

Figure 7.64 shows a comparison of undrained shear strength derived from triaxial, vane and pressuremeter tests; and, in the case of the pressuremeter tests, the strength was determined from the loading curve (Gibson and Anderson, 1961 and Palmer, 1972) and the unloading curve (Jefferies, 1988). The figures show that the values of strength and earth pressure at rest are dependent on the test method and the method of analysis. The results of the pressuremeter tests were used to predict the deformation of

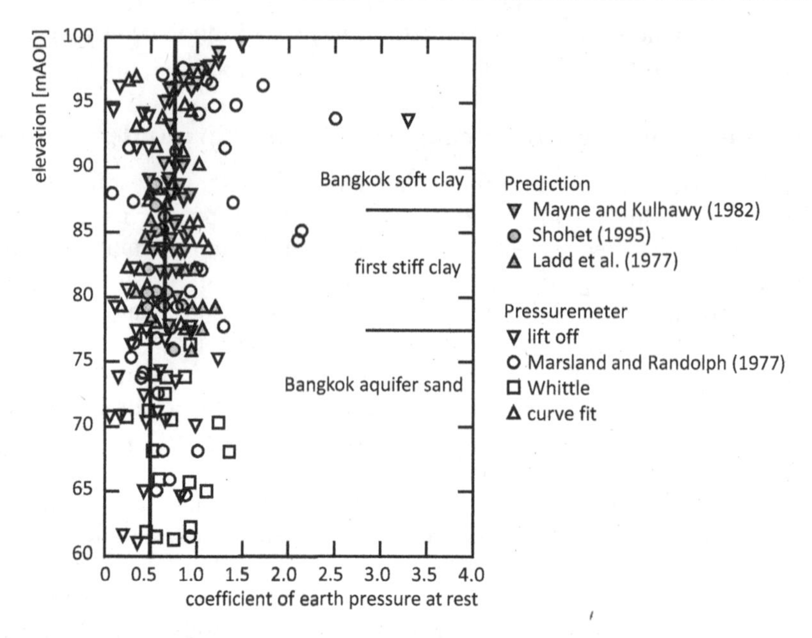

Figure 7.63 A comparison between empirical predictions of K_o with those derived from self-boring pressuremeter tests.

Source: After Prust et al., 2005.

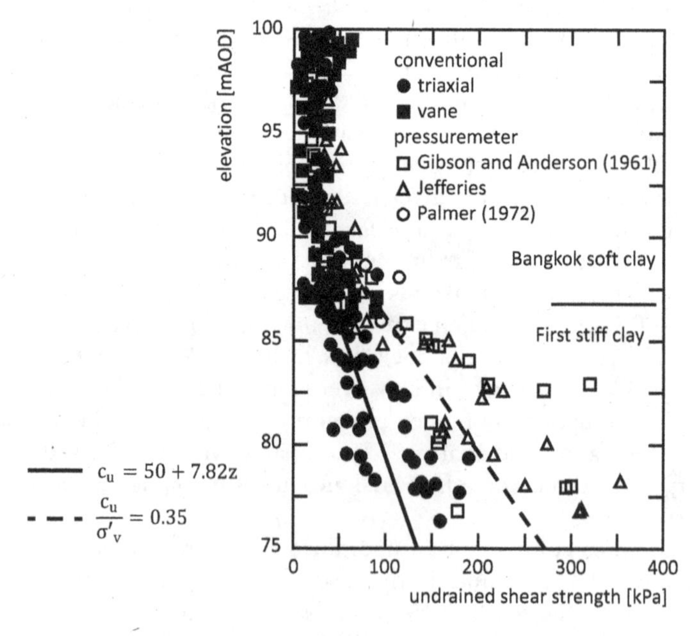

Figure 7.64 A comparison of undrained shear strength measured with vane and triaxial tests and those determined from loading and unloading curve of self boring pressuremeter tests.

Source: After Prust et al., 2005.

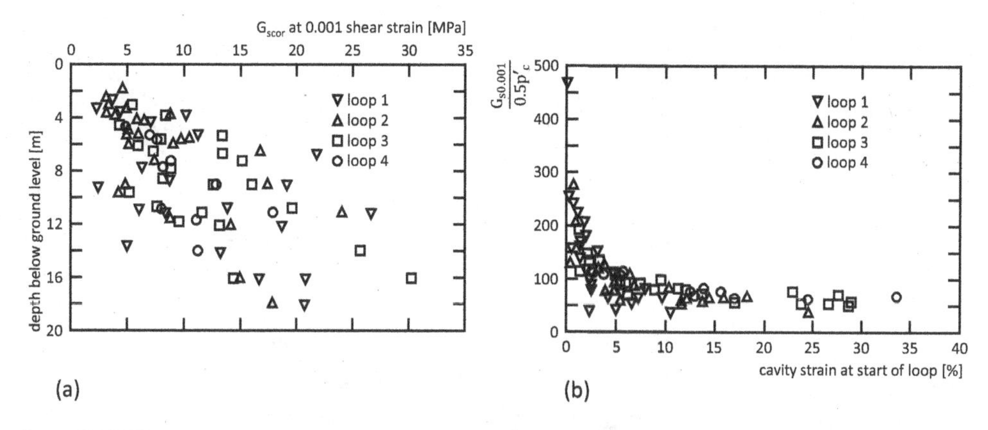

Figure 7.65 The shear modulus at 0.001 shear strain for a number of MASWs showing (a) the variation with depth and (b) the degradation curve for shear modulus.

Source: After Dixon et al., 2006.

a retaining wall supporting 25 m of stiff clay, soft clay and fill. The wall deformation measurements were similar to those predicted.

Dixon et al. (2006) provided an example of pressuremeter tests in municipal solid waste (MSW) as part of the design of the landfill. They were able to determine the variation in shear moduli with the strain from unload/reload cycles. They found that it was necessary to undertake tests up to 50% cavity strain to determine the strength of the waste: 1.5 m long, 96 mm diameter test pockets were created in the MSW, using a barrel auger. The HPD, a radial displacement prebored pressuremeter, was pushed into the test pocket. Tests were assumed to be drained with up to five unload/reload cycles to determine the shear modulus degradation curve using the power law developed by Bolton and Whittle (1999). They used a video camera to inspect the test pockets to show that the diameter was relatively uniform. Figure 7.65 shows the relationship between secant shear modulus calculated for a reference strain and depth and the degradation curve for shear modulus.

By studying the method of installation and using unload/reload cycles from a large displacement pressuremeter, Dixon et al. (2006) were able to produce MASW properties to give some confidence in the design of the barriers. Further, they showed that the properties of the waste change with time, though this has to be moderated because of the change in composition over the years.

Galera et al. (2007) undertook an extensive study of the relationships between the RMR (rock mass rating) and deformation modulus derived from pressuremeter and dilatometer tests. Tests were carried out over a ten-year period on igneous, metamorphic, and sedimentary and carbonate sedimentary rocks yielding 702 data seta to produce Figure 7.66.

They suggested that the deformation modulus, E_m, is

$$E_m = E_i e^{\left(\frac{RMR-100}{36}\right)} \tag{7.90}$$

where E_i is the deformation modulus of intact rock.

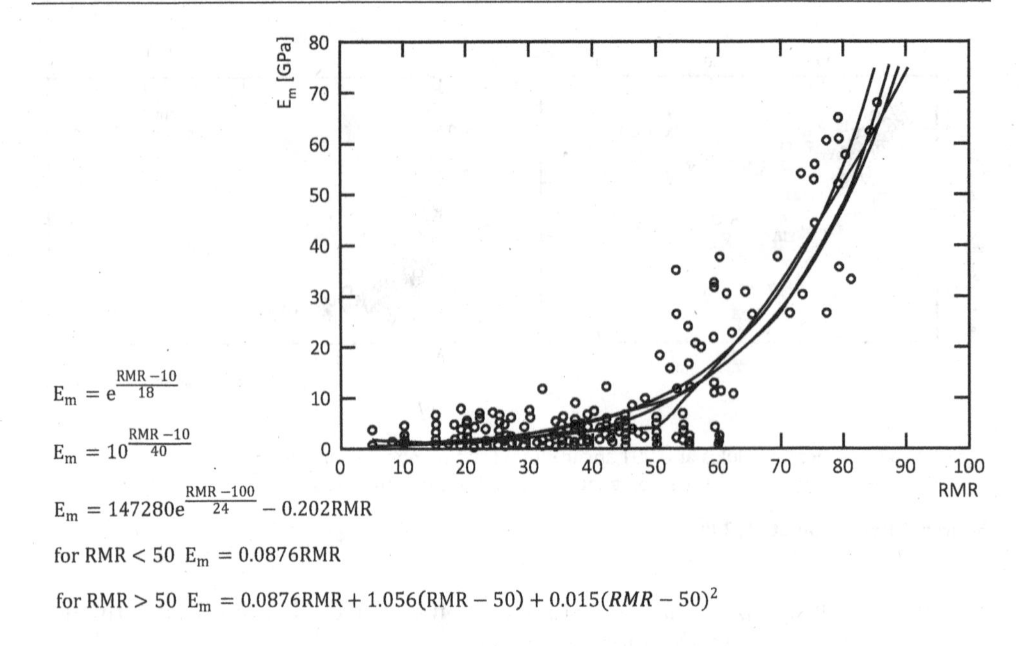

Figure 7.66 Correlations between rock mass rating (RMR) and the shear modulus (E_m) derived from prebored pressuremeters tests.

Source: After Galera et al., 2007.

It is estimated that 23% of the earth's land surface is in a permafrost zone. Springman et al. (2012) suggest that drilling techniques thermally disturb the surrounding frozen ground. It can take months for equilibrium conditions to be re-established. The thermal disturbance decreases with increasing distance from the borehole wall. Therefore, at the time of a test, the temperature in the frozen ground varies, which means that the ground response may not be completely representative of in situ conditions. However, the thermal disturbances will be similar throughout the depth of a borehole, thus it is possible to gain an insight into the relative indication of the response of the frozen ground. Ladanyi and Johnston (1973) suggested that the pressuremeter could be used to assess the in situ creep properties of permafrost, which are needed to assess the stiffness and strength properties. Arenson et al. (2003) describe multistage creep pressuremeter tests using an HPD, which had been carried out in an Alpine rock glacier. They applied the model proposed by Ladanyi and Johnston (1973) to predict the pressuremeter test curve and concluded that pressuremeter tests were feasible taking into account the thermal and installation disturbance. They also observed that if the cavity pressure was less than 1.5MPa, the frozen soil shows an elastic response. Yu et al. (2002) and Zhang et al. (2012) found that the limit pressure and pressuremeter modulus were a function of the water content or temperature in the short term. Zhang et al. (2012) used the hyperbolic model proposed by Ladanyi (1972) to confirm the findings of Arenson et al. (2003).

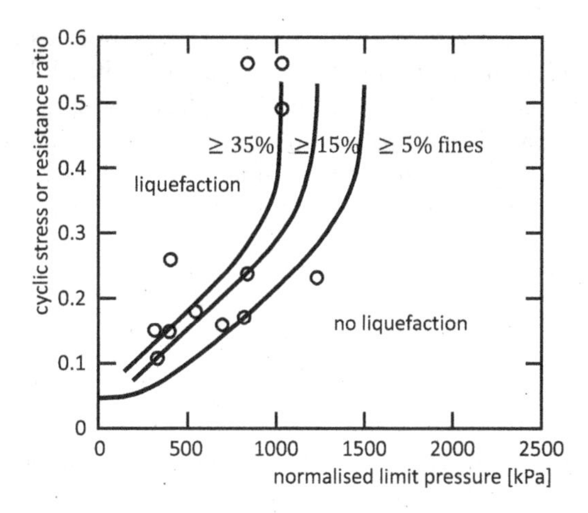

Figure 7.67 The use of the pressuremeter limit pressure to determine the liquefaction potential of soils.

Source: After Briaud, 2013.

Briaud (2013) suggested that the limit pressure from a prebored pressuremeter test could be used to assess the liquefaction potential (Figure 7.67), depending upon the fines content.

Kayabasi and Gokceoglu (2018) suggested that the cyclic resistance ratio, $CRR_{7.5}$, for magnitude 7.5 earthquakes is

$$CRR_{7.5} = \frac{1}{34 - p_{lc}} + \frac{0.2p_{lc}}{135} + \frac{50}{(2p_{lc} + 45)^2} - \frac{1}{200} \tag{7.91}$$

where p_{lc} is the corrected net limit pressure:

$$p_{lc} = p_l^* \left(\frac{p_a}{\sigma'_{vo}} \right)^{0.5} \tag{7.92}$$

This equation was developed from the relationship proposed by Youd et al. (2001), based on SPT results.

Glacial tills are complex soils because the modes of deposition results in a spatially variable soil, which does not conform to classic theories of soil mechanics based on gravitationally compacted soils (Clarke, 2017). Further, it is often difficult to create pressuremeter test pockets in glacial tills that contain significant quantities of coarse and very coarse grained particles.

Powell and Uglow (1985) compared the results of the three types of pressuremeter tests in glacial till to observe that the limit pressures were similar and the shear strength was sensitive to the installation disturbance. The difference in moduli were a function of the installation disturbance and the pressure at which they were measured and the strain over which they were measured. Measurements of horizontal stress were difficult because of the stone content.

Balachandran et al. (2015) undertook a statistical exercise as part of geotechnical investigation for the Eglinton Crosstown Light Rail Transit (LRT) Project in Toronto. The glacial till deposits in Toronto include low plasticity cohesive glacial tills (silty clay to clayey silt glacial till) and cohesionless glacial tills (sandy silt to silty sand glacial till). The data were processed by removing all those where the SPT met refusal ($N_{60} = 100$) and any outliers. The relationship for clast dominated till was

$$E_{pm} = 6.37 N_{60}^{0.59} \tag{7.93}$$

and for matrix dominated tills

$$E_{pm} = 1.15 N_{60} + 7.38 \tag{7.94}$$

However, the quality of the data was poor because of the difficulty in creating test pockets in glacial till and the influence coarse-grained particles had upon the SPT results. Balachandran et al. (2015) extended this study to show that the undrained shear strength of matrix dominated tills is

$$c_u = 0.39 p_{lm}^{0.78} \tag{7.95}$$

7.6 SUMMARY

In order to predict the response of the ground subject to change, it is necessary to develop a ground model which takes into account the topographical, geological, hydrogeological and geotechnical properties. This is interpreted from desk studies, non-intrusive and intrusive investigation, field and laboratory tests and application of theoretical, empirical and numerical methods, which is underwritten by expert opinion. Ground is complex, spatially variable and a potential hazard. Therefore to reduce risk attributed to the ground, it is essential that a realistic ground model is created. Investment in investigation is limited and methods of investigation are diverse, producing method dependent results. For that reason, it is prudent to apply a number of techniques to develop the ground model to give greater confidence in the model and the results it produces when designing a ground-related structure.

A pressuremeter can be used in any type of ground, which means it is a very versatile tool. The results of pressuremeter tests are used to produce design properties directly, to produce intrinsic properties by applying constitutive models to interpret the data, to validate more complex constitutive models used in numerical methods and as a means of control in ground improvement schemes.

The Ménard method is a theoretically based design method modified following observations of the performance of full-scale structures. The pressuremeter, installation, test procedure and interpretation must conform to a standard specification if the method is to be used. Bearing capacity and settlements of shallow and deep foundations can be predicted with a degree of confidence, provided that tests are carried out in ground conditions similar to those used in the development of the method. The increasing international use of this method implies that the database upon which the design rules are based will be extended.

The direct method of design can also make use of PBP, SBP and FDP test results though there is less experience than that for MPM design procedure. The pressuremeter installation and test procedure closely resemble the installation and loading of horizontally loaded piles. Design rules have been developed for piles using different pressuremeter types. Further developments of this method are likely to take place either by correlating parameters with the pressuremeter modulus, E_m, and modified limit pressure, p_{lm}, and using the Ménard method or by developing further design rules.

All pressuremeter tests can be interpreted to give properties of the ground, though the parameters obtained will depend on the type of probe and on the test procedure. Results of the tests have been compared with results of other tests to demonstrate that they are the same or different, though there is no reason for this, since the loading path in a pressuremeter test is different from those in other tests. However, it is possible to relate the results from different tests by taking into account the boundary conditions and test procedure. Thus, results of pressuremeter tests can be used to create site-specific empirical correlations, validate constitutive models, and produce profiles of mechanical properties to be used in design. Ideally, the pressuremeter should be used in conjunction with other tests to produce a comprehensive ground model.

Generally, it is accepted that the total horizontal stress determined from a SBP test will be the best estimate of any test. The pressuremeter was developed to measure stiffness profiles. Average and incremental stiffnesses obtained from unload/reload cycles agree very favourably with observations of the performance of full-scale structures, confirming that the pressuremeter is an ideal test by which to obtain deformation parameters. These parameters have been used to predict displacements, and the predictions agree with observations, suggesting that the use of pressuremeter results, in the indirect method of design, is valid.

Comparisons between undrained shear strength from pressuremeter tests and other tests are not conclusive, but it is likely that the average strength, a post-peak strength, of soft clay will be greater than that from a laboratory test, and that from stiff clay will be equal. Limited information on the angle of friction from SBP tests confirms that the SBP can be used to give realistic values at the in situ density.

A form of pressuremeter can be used in all ground conditions, and the parameters obtained – either directly, theoretically or empirically – represent realistic estimates of ground properties which give reasonable predictions of foundation behaviour. The selection and specification of the pressuremeter is described in chapter 8.

REFERENCES

Abchir, Z., Burlon, S., Frank, R., Habert, J. and Legrand, S. (2016) T–z curves for piles from pressuremeter test results. *Géotechnique*, *66*(2), pp. 137–148.

Anderson, J-B, Townsend, F.C., Horta, E. & Sandoval, J. (2005) Effects of modified probes and methods on PENCEL pressuremeter tests, Symposium International ISP5/PRESSIO 2005, Gambin, Magnan & Mestat, eds., pp. 21–30.

Angelim, R.R., Cunha, R.P. and Sales, M.M. (2016) Determining the elastic deformation modulus from a compacted earth embankment via laboratory and Ménard pressuremeter tests. *Soils and Rocks*, *39*(3), pp. 285–300.

Akkaya, İ., Özvan, A. and Özvan, E.E. (2019) A new empirical correlation between pressuremeter modules (EM) and shear wave velocity (Vs) for clay soils. *Journal of Applied Geophysics*, *171*, p. 103865.

Amar, S., Baguelin, F. and Canepa, Y. (1984) Etude expérimentale du comportement des fondations superficielles, *Annales de l'ITBTP*, *Sols Soils* (427), pp. 82–108.

Angelim, R.R., Cunha, R.P. and Sales, M.M. (2016) Determining the elastic deformation modulus from a compacted earth embankment via laboratory and Ménard pressuremeter tests. *Soils and Rocks*, *39*(3), pp. 285–300.

Arenson, L.U., Hawkins, P.G. and Springman, S.M. (2003) Pressuremeter tests within an active rock glacier in the Swiss Alps. In *Eighth International Conference on Permafrost. AA Balkema,* Zurich (pp. 33–38).

Arulrajah, A., Bo, M.W., Piratheepan, J. and Disfani, M.M. (2011) In situ testing of soft soil at a case study site with the self-boring pressuremeter. *Geotechnical Testing Journal*, *34*(4), p. 1.

Baguelin, F. (1982) Rules of foundation design using self boring pressuremeter test results, *Proc. Int. Symp. Pressuremeter Marine Appl.,* Paris, pp. 347–360.

Baguelin, F., Jézéquel, J.F. and Shields, D.H. (1978) *The Pressuremeter and Foundation Engineering,* Trans. Tech. Publication.

Balachandran, K., Liu, J., Cao, L. and Peaker, S. (2015) Statistical correlations between pressuremeter modulus and SPT N-value for glacial tills. In *Proc., 68th Canadian Geotechnical Conf. and 7th Canadian Permafrost Conf.* Richmond, BC, Canada: Canadian Geotechnical Society.

Balachandran, K., Liu, J., Cao, L. and Peaker, S. (2015) Statistical correlations between pressuremeter modulus and SPT N-value for glacial tills. In *Proc., 68th Canadian Geotechnical Conf. and 7th Canadian Permafrost Conf.* Richmond, BC, Canada: Canadian Geotechnical Society.

Barksdale, R.D., Ferry, C.T. and Lawrence, J.D. (1986) Residual soil settlement from pressuremeter moduli, *Proc. In situ'86: Use of In Situ Tests in Geot. Engng,* Blacksburg VA, pp. 447–461.

Battaglio, M., Ghionna, V., Jamiolkowski, M. and Lancellotta, R. (1981) Interpretation of selfboring pressuremeter tests in clays, *Proc. 10th Int. Conf. SMFE,* Stockholm, 2, pp. 433–438.

Baud, J.P., Gambin, M.P. and Schlosser, F. (2012 September) Stress-strain hyperbolic curves with Ménard PMTs. In *4th Int. Conf. on Geotechnical and Geophysical Site Characterization (ISC'4), P. de Galinhas,* Brazil (pp. 18–21).

Bellotti, R., Ghionna, V., Jamiolkowski, M., Robertson, P.K. and Peterson, R.W. (1989) Interpretation of moduli from self-boring pressuremeter tests in sand, *Geotechnique, 39*(2), 269–292.

Benoit, J. and Lutenegger, A.J. (1993) Determining lateral stress in soft clays, *Predictive Soil Mechanics, Proc. Wroth Memorial Symp.*, Oxford, pp. 56–74.

Bergado, D.T., Daria, P.M., Sampaco, C.L., Alfaro and Marolo, C. (1991) Prediction of embankment settlements by in situ tests, *Geotech. Test. J., ASTM, 14*(4), 425–439.

Blaker, Ø., Carroll, R., Paniagua Lopez, A.P., DeGroot, D.J. and L Heureux, J.S. (2019) Halden research site: Geotechnical characterization of a post glacial silt.

Bo, M.W., Chu, J. and Choa, V. (2005) The Changi east reclamation project in Singapore. In *Elsevier Geo-Engineering Book Series* (Vol. 3, pp. 247–276). Elsevier.

Bo, M.W., Chang, M.F., Arulrajah, A. and Choa, V. (2012) Ground investigations for Changi East reclamation projects. *Geotechnical and Geological Engineering*, *30*(1), pp. 45–62. Bolton and Whittle, 1999.

Bolton, M.D. and Whittle, R.W. (1999) A non-linear elastic/perfectly plastic analysis for plane strain undrained expansion tests. *Géotechnique*, *49*(1), pp. 133–141.

Boumedi, J.Y., Baud, J.P. and Radiguet, B. (2009) LNG tanks at Damietta on drilled shafts designed and tested using Ménard PMT. In *Contemporary Topics in In Situ Testing, Analysis, and Reliability of Foundations* (pp. 103–110).

Bourne-Webb, P.J., Amatya, B. and Soga, K. (2013) A framework for understanding energy pile behaviour. *Proceedings of the Institution of Civil Engineers-Geotechnical Engineering*, *166*(2), pp. 170–177.

Bozbey, I. and Togrol, E. (2010) Correlation of standard penetration test and pressuremeter data: a case study from Istanbul, Turkey. *Bulletin of engineering geology and the environment*, *69*(4), pp. 505–515.

Briaud, J.-L. (1986a) Pressuremeter and deep foundation design, *Proc. 2nd Int Symp. Pressuremeter Marine Appl., Texam,* United States, ASTM STP 950, pp. 376–405.

Briaud, J.-L. (1986b) Pressuremeter and foundation design, *Proc. In situ '86: Use of In Situ Tests in Geot. Engng,* Blacksburg VA, pp. 74–115.

Briaud, J.-L. (1992) *The Pressuremeter,* Balkeema, Rotterdam.

Briaud, J.L. (2013) September. Ménard Lecture: The pressuremeter Test: Expanding its use. In *Proceedings of the 18th International Conference on Soil Mechanics and Geotechnical Engineering,* Paris (pp. 107–126).

Briaud, J.L. and Gibbens, R. (1999) Behavior of five large spread footings in sand. *Journal of Geotechnical and Geoenvironmental Engineering, 125*(9), pp. 787–796.

Briaud, J.-L., Smith, T.D. and Meyer, B. (1983) Using the pressuremeter curve to design laterally loaded piles, *Proc. Offshore Technology Conf,* Houston, Vol. 1, pp. 495–502.

Briaud, J.-L., Tand, K.E. and Funegard, E.G. (1986) Pressuremeter and shallow foundations on stiff clay, *Transportation Research Record*, No. 1105.

Bruzzi, D., Ghionna, V., Jamiolkowski, M., Lancellotta, R. and Manfredini, G. (1986) Self-boring pressuremeter in Po River sand in the pressuremeter and its marine applications, *Proc. 2nd Int. Symp. Pressuremeter Marine Appl., Texam,* United States, ASTM STP 950, pp. 57–74.

Burland, J., 1973. Shaft friction of piles in clay--a simple fundamental approach. Publication of: Ground Engineering/UK/, *6*(3), 30–42.

Burland, J.B. (1990) Thirtieth Rankine Lecture: On the compressibility and shear strength of natural clays, *Geotechnique, 40*(3), pp. 327–378.

Burlon, S., Frank, R., Baguelin, F., Habert, J. and Legrand, S. (2014) Model factor for the bearing capacity of piles from pressuremeter test results–Eurocode 7 approach. *Géotechniqué*, *64*(7), pp. 513–525.

Bustamante, M. and Doix, B. (1985) Une méthode pour le calcul des tirants et des micropieux injectés, *Bull, de Liaison de LCPC,* Paris, *140*, 75–92.

Bustamante, M. and Gianeselli, L. (1981) Observed and predicted bearing capacity of isolated piles under axial loads – pressuremeter method, *Rev. Fr. Geotech.,* No. 16, 17–33.

Bustamante, M., Gambin, M. and Gianeselli, L. (2009) Pile design at failure using the Ménard pressuremeter: an update. In *Contemporary Topics in In Situ Testing, Analysis, and Reliability of Foundations* (pp. 127–134).

Byrne, P.M. Salgado, F.M. and Howie, J.A. (1990) Relationship between the unload shear modulus from pressuremeter tests and the maximum shear moduli for sand, *Proc. 3rd Int. Symp. Pressuremeters,* Oxford, pp. 231–242.

Campanella, R.G., Robertson, P.K., Davies, M.P. and Sy, A. (1989) Use of in situ tests in pile design, *Proc. 12th Int Conf. SMFE,* Rio de Janeiro , Vol. 1, pp. 199–203.

Cao, L.F., Peaker, S. and Ahmad, S. (2015) Engineering characteristic of glacial tills in GTA. In *68th Canadian Geotechnical Conference,* Québec.

Cardoso Bernardes H, Martines Sales M, Rodrigues Machado R, José da Cruz Junior A, Pinto da Cunha R, Resende Angelim R, Félix Rodríguez Rebolledo J. (2021) Coupling hardening soil model and Ménard pressuremeter tests to predict pile behavior. *European Journal of Environmental and Civil Engineering.* Feb 5:1–20.

Carter, J.P., Booker, J.R. and Yeung, S.K. (1986) Cavity expansion in cohesive frictional soils, *Geotechnique, 36*(3), 349–358.

Cassan, M. (1968) Les essais in situ en mécanique des sols, *Construction* (10), 337–347.

Cassan, M. (1969a) Les essais in situ en mécanique des sols, *Construction* (5), 178–187.

Cassan, M. (1969b) Les essais in situ en mécanique des sols, *Construction* (7–8), 244–256.

Chandler, R.J. (1969) Effect of weathering on the shear strength properties of Keuper Marl, Geotechnique, *19*(3), 321–334.

Chang, Y.L. (1937) Discussion on lateral pile loading tests, Feggin, L.B., *Trans. ASCE, 102*, 272.

Chatzigogos, N., Christaras, V., Makedon, T. and Tsotsos, S. (2006) Implementation of a ground investigation strategy on urban fills. In *Proceedings of the 10th Congress of the International Association for Engineering Geology and the Environment (IAEG2006)* (No. IKEECONF-2010-020). Aristotle University of Thessaloniki.

Cheshomi, A. and Ghodrati, M. (2015) Estimating Menard pressuremeter modulus and limit pressure from SPT in silty sand and silty clay soils. A case study in Mashhad, Iran. *Geomechanics and Geoengineering, 10*(3), pp. 194–202.

Cheshomi, A. and Khalili, A. (2021) Comparison between pressuremeter modulus (EPMT) and shear wave velocity (Vs) in silty clay soil. *Journal of Applied Geophysics, 192*, p. 104399.

Christoulas, S. and Frank, R. (1991) Deformation parameters for pile settlement, *Proc. 10th Eur. Conf. Soil Mech. and Found. Engng*, Florence, Italy, pp. 373–376.

Chu, J., Bo, M.W., Chang, M.F. and Choa, V. (2002) Consolidation and permeability properties of Singapore marine clay. *Journal of Geotechnical and Geoenvironmental Engineering, 128*(9), pp. 724–732.

Clarke, B.G. (1993) The interpretation of pressuremeter tests to produce design parameters, *Predictive Soil Mechanics, Proc. Wroth Memorial Symp.*, Oxford, pp. 75–88.

Clarke, B.G. and Allan, P.G. (1990) Self-boring pressuremeter tests from a gallery at 220 m below ground, *Proc. 3rd Int. Symp. Pressuremeters,* Oxford, pp. 73–84.

Clarke, B.G. and Wroth, C.P. (1984) Analysis of Dunton Green retaining wall based on results of pressuremeter tests, *Geotechnique, 34*(4), 549–561.

Clarke, B.G., Carter, J.P. and Wroth, C.P. (1979) In-situ determination of the consolidation characteristics of saturated clays, *Proc. 7th Eur. Conf. SMFE*, Brighton, Vol. 2, pp. 207–211.

Clarke, B.G., Newman, R. and Allan, P. (1990) Experience with a new high pressure self-boring pressuremeter in weak rock, *Ground Engng, 22*(5), 36–39; (6), 45–51.

Clarke, B.G. (1990) Consolidation characteristics of clays from self-boring pressuremeter tests, *Proc. 24th Ann. Conf. of the Engng Group of the Geological Soc.: Field Testing in Engineering Geology*, Sunderland, pp. 19–35.

Clarke, B.G. (1993) The interpretation of pressuremeter tests to produce design parameters, *Predictive Soil Mechanics, Proc. Wroth Memorial Symp.*, Oxford, pp. 75–88.

Clarke, B.G. (1994) Peak and post rupture strengths from pressuremeter tests, *Proc. 13th Int. Conf. SMFE*, Delhi, India, Vol. 1, pp. 125–128.

Clarke, B.G. (2017) *Engineering of glacial deposits*. CRC Press.

Clough, G.W. and Denby, G.M. (1980) Self-boring pressuremeter study of San Francisco Bay mud, *J. Geotech. Engng Div., ASCE, 106*(NGT1), 45–63.

Davidson, R.R. and Bodine, D.G. (1986) Analysis and verification of Louisiana pile foundation design based on pressuremeter results, *Proc. 2nd Int. Symp. Pressuremeter Marine Appl., Texam*, United States, ASTM STP950, pp. 423–439.

DeGroot, D.J. and Lutenegger, A.J. (2005) Characterisation by sampling and in situ testing – Connecticut Valley varved clay. *Studia Geotechnica et Mechanica, 27*, 107–120.

Dixon, N., Whittle, R.W., Jones, D.R.V. and Ng'ambi, S. (2006) Pressuremeter tests in municipal solid waste: Measurement of shear stiffness. *Geotechnique, 56*(3), pp. 211–222.

Doherty, J.P., Gourvenec, S. and Gaone, F.M. (2018) Insights from a shallow foundation load-settlement prediction exercise. *Computers and Geotechnics, 93*, pp. 269–279.

Eden, W.J. and Law, K.T. (1980) Comparison of undrained shear strength results obtained by different test methods in soft clays, *Can. Geotech. J., 17*(3), 369–381.

ENV 1997-1. (2004) Eurocode 7: Geotechnical design – part 1: General rules. Brussels: European Committee for Standardisation (CEN).

Eslami, H., Rosin-Paumier, S., Abdallah, A. and Masrouri, F. (2017) Pressuremeter test parameters of a compacted illitic soil under thermal cycling. *Acta Geotechnica*, *12*(5), pp. 1105–1118.

Fahey, M. and Carter, J.P. (1993) A finite element study of the pressuremeter test in sand using a nonlinear elastic plastic model. *Canadian Geotechnical Journal*, *30*(2), pp. 348–362.

Fawaz, A., Boulon, M. and Flavigny, E. (2002) Parameters deduced from the pressuremeter test. *Canadian Geotechnical Journal*, *39*(6), pp. 1333–1340.

Fawaz, A., Hagechehade, F. and Farah, E. (2014) A study of the pressuremeter modulus and its comparison to the elastic modulus of soil. *Study of Civil Engineering and Architecture (SCEA)*, *3*, pp. 7–15.

Fernández-Baniela, F., Arias, D. and Rubio-Ordóñez, Á. (2021) Geotechnical settings of volcanic residual soils and derived engineering problems in El Hierro Island (Spain). *Arabian Journal of Geosciences*, *14*(1), pp. 1–10.

da Fonseca, A.V. (2003) Characterising and deriving engineering properties of a saprolitic soil from granite, in Porto. *Characterisation and Engineering Properties of Natural Soils*. Tan et al. (eds.), Swets & Zeitlinger, Lisse.

Fontaine, E., Cunha, R. P., & Carvalho, D. (2005). A simplified analytical manner to obtain soil parameters from Ménard pressuremeter tests on unsaturated soils. In *M. P. Gambin, J. P. Magnan, and P. Mestat* (Eds.), *50 Years of Pressuremeters International Symposium - ISP5* (pp. 289–295). Presses des Ponts et Chaussees.

Frank, R.A. and Zhao, S.R. (1982) Estimation par les paramètres pressiométriques de l'enfoncement sous charge axiale de pieux forés dans les sols fins, *Bull, de Liaison de LCPC*, Paris, No. 119, 17–24.

Frank, R., Kalteziotis, N., Bustamante, M., Christoulas, S. and Zervogiannis, H. (1991) Evaluation of performance of two piles using pressuremeter method, *J. Geotech. Engng*, *117*(5), 695–713.

Galera, J.M., Álvarez, M. and Bieniawski, Z.T. (2007) Evaluation of the deformation modulus of rock masses using RMR: comparison with dilatometer tests. In *Workshop: Underground Works under Special Conditions*. ISP5-PRESSIO 2005 International Symposium.

Gambin, M. (1963) Calcul du tassement d'une fondation profonde en fonction des résultats pressiométriques, *Sols Soils*, *2*(7), 11–23.

Gambin, M. (1969) Calculation of laterally loaded piles, *Offshore Special Equipment*, No. 2, Paris.

Gambin, M., Flavigny, E., and Boulon, M. (1996) Le module pressiométrique. Historique et modélisation. *In XIe colloque Franco-Polonais de Mécanique des sols et des roches appliquées*. E. Dembicki and W. Cichy (Eds.). University of Gdansk, Poland, 12 September, pp. 53–60.

Gibson, R.E. and Anderson, W.F. (1961) In situ measurements of soil properties with the pressuremeter, *Civ. Engng Public Wks. Rev.*, *56*, 615–618.

Gaone, F.M., Doherty, J.P. and Gourvenec, S. (2019) An optimization strategy for evaluating modified Cam clay parameters using self-boring pressuremeter test data. *Canadian Geotechnical Journal*, *56*(11), pp. 1668–1679.

Goh, K.H., Jeyatharan, K. and Wen, D. (2012) Understanding the stiffness of soils in Singapore from pressuremeter testing. *Geotechnical Engineering Journal of the SEAGS & AGSSEA*, *43*(4), pp. 21–29.

Graham, J.P. and Jefferies, M.G. (1986) Some examples of in situ lateral stress determination in hydraulic fills using the self-boring pressuremeter, *39th Can. Geotechnical Conf. In Situ Testing and Field Behaviour*, Ottawa, pp. 191–200.

Haberfield, C.M. and Johnston, I.W. (1993) Factors influencing the interpretation of pressuremeters tests in soft rocks, *Proc. Conf. Geotechnical Engng Hard Soils-Soft Rocks*, Athens, Vol. 1, pp. 525–531.

Hamidi, B., Nikraz, H., Yee, K., Varaksin, S. and Wong, L.T. (2010) Ground improvement in deep waters using dynamic replacement. In *The Twentieth International Offshore and Polar Engineering Conference*. OnePetro.

Hamidi, B., Varaksin, S. and Nikraz, H. (2011) Predicting Ménard modulus using dynamic compaction induced subsidence. In *Proceedings of the International Conference on Advances in Geotechnical Engineering* (pp. 221–226). Australian Geomechanics Society.

Hamidi, B., Nikraz, H. and Varaksin, S. (2012) The application of dynamic compaction on Marjan Island. In *Proceedings of the 11th Australia-New Zealand Conference on Geomechanics (ANZ 2012)* (pp. 1202–1207). Australian Geomechanical Society and New Zealand Geotechnical Society.

Hawkins, P.G., Mair, R.J., Mathieson, W.G. and Muir Wood, D. (1990) Pressuremeter measurement of total horizontal stress in stiff clay, *Proc. 3rd Int. Symp. Pressuremeters,* Oxford, pp. 321–330.

Hight, D.W., Pickles, A.R., De Moor, E.K., Higgins, K.G., Jardine, R.J., Potts, D.M. and Nyirenda, Z.M. (1993) Predicted and measured tunnel distortions associated with construction of Waterloo International Terminal, *Predictive Soil Mechanics, Proc. Wroth Memorial Symp.,* Oxford, pp. 216–234.

Hobbs, N. B., Dixon, J-C. (1969) "In situ testing for Bridge Foundation in the Devonian Marl". In: Proceedings of the Conference on in situ investigations in soils and rocks. British Geotechnical Society, London, May 13–15, pp 31–38.

Hoopes, O. and Hughes, J. (2014) In situ lateral stress measurement in glaciolacustrine Seattle clay using the pressuremeter. *Journal of Geotechnical and Geoenvironmental Engineering, 140*(5), p. 04013054.

Horvath-Kalman, E. (2015) In-situ Measurements in Overconsolidated Clay. *YBL Journal of Built Environment, 3*(1–2), pp. 68–76.

Houlsby, G.T. and Nutt, N.R.F. (1992) Development of the cone pressuremeter, *Predictive Soil Mechanics, Proc. Wroth Memorial Symp.,* Oxford, pp. 254–271.

Houlsby, G.T. and Withers, N.J. (1988) Analysis of the cone pressuremeter test in clay, *Geotechnique, 38*(4), 575–587.

Houlsby, G.T. and Wroth, C.P. (1989) Influence of soil stiffness and lateral stress on the results of in-situ soil tests, *Proc. 12th Int. Conf. SMFE, Rio de Janeiro,* Vol. 1, pp. 227–232.

Huang, A.B., Pan, I.W., Liao, J.J., Wang, C.H. and Hsieh, S.Y. (1999) Pressuremeter tests in poorly cemented weak rocks. In *Vail Rocks 1999, The 37th US Symposium on Rock Mechanics (USRMS)*. OnePetro.

Huang, A.B. (1995) In situ testing in overconsolidated clays. *Transportation Research Record, 1479*, p. 35.

Hughes, J.M.O. and Ervin, M.C. (1980) Development of a high pressure pressuremeter for determining the engineering properties of soft to medium strength rocks, *Proc. 3rd Aus.-NZ Conf. Geomechanics,* Wellington, Vol. 1, pp. 243–247.

Hughes, J.M.O., Wroth, C.P. and Windle, D. (1977) Pressuremeter tests in sands, *Geotechnique, 27*(4), 455–477.

Hughes, J.M.O., Goldsmith, P.R. and Fendali, H.D.W. (1979) *Predicted and measured behavior of laterally loaded piles for the Westgate Freeway Bridge,* Victoria Geomechanics Society, Australia.

Isik, N.S., Ulusay, R. and Doyuran, V. (2008) Deformation modulus of heavily jointed–sheared and blocky greywackes by pressuremeter tests: Numerical, experimental and empirical assessments. *Engineering Geology, 101*(3–4), pp. 269–282.

Jain, S. and Nanda, A. (2009) Constitutive modeling of San Francisco Bay mud. *International Journal of Geotechnical Engineering, 3*(4), pp. 527–533.

Jamiolkowski, M., Ladd, C.C., Germaine, J.T. and Lancellotta, R. (1985) New developments in field and laboratory testing of soils, *Proc. 11th Int. Conf.* SMFE, San Francisco, Vol. 1, pp. 57–154.

Jamiolkowski, M., Ricceri, G. and Simonini, P. (2009) Safeguarding Venice from high tides: site characterization and geotechnical problems. In *Proceedings of the 17th International Conference on Soil Mechanics and Geotechnical Engineering (Vols. 1, 2, 3 and 4)* (pp. 3209–3227). IOS Press.

Jardine, R.J. (1991) Discussion on "Strain dependent moduli and pressuremeter tests," *Geotechnique, 41*(4), 621–626.

Jardine, R.J. (1992) Non-linear stiffness parameters from undrained pressuremeter tests, *Can. Geotech. J., 29*, 436–447.

Jefferies, M.G. (1988) Determination of horizontal geostatic stress in clay with self-bored pressuremeter, *Can. Geotech. J., 25*(3), 559–573.

Jefferies, M.G., Crooks, J.H.A., Becker, D.E. and Hill, P.R. (1987) Independence of geostatic stress from overconsolidation in some Beaufort Sea clays, *Can. Geotech. J., 24*(3), 342–356.

Jefferson, I., Rogers, C., Evstatiev, D. and Karastanev, D. (2005). Treatment of metastable loess soils: Lessons from Eastern Europe. In *Elsevier Geo-engineering Book Series* Vol. 3 (pp. 723–762). Elsevier.

Jewell, R.J. and Fahey, M. (1984) Measuring properties of rock with a high pressure pressuremeter, *Proc. 4th Aus.-NZ Conf. Geomechanics,* Perth, Vol. 2, pp. 535–539.

Jézéquel, J.F. and Le Méhauté, A. (1982) Cyclic tests with the self boring pressuremeter, *Proc. Int. Symp. Pressuremeter and Its Marine Appl.,* Paris, pp. 209–222.

Kalteziotis, N.A., Tsiambao, G., Sabatakakis, N. and Zervogiannis, H. (1990) Prediction of soil dynamic parameters from pressuremeter and other in situ tests, *Proc. 3rd Int. Symp. Pressuremeters,* Oxford, pp. 391–400.

Kassou, F., Benbouziyane, J., Ghafiri, A. and Sabihi, A. (2017) Settlements and consolidation rates under embankments in a soft soil with vertical drains. *International Journal of Engineering*, *30*(7), pp. 972–980.

Kavur, B., Dodigovic, F., Jug, J. and Strelec, S. (2019) The Interpretation of CPTu, PMT, SPT and Cross-Hole Tests in Stiff Clay. In *IOP Conference Series: Earth and Environmental Science*, Vol. 221, No. 1 (p. 012009). IOP Publishing.

Kayabasi, A., 2012. Prediction of pressuremeter modulus and limit pressure of clayey soils by simple and non-linear multiple regression techniques: a case study from Mersin, Turkey. *Environmental Earth Sciences*, *66*(8), pp. 2171–2183.

Kayabasi, A. and Gokceoglu, C. (2018) Liquefaction potential assessment of a region using different techniques (Tepebasi, Eskişehir, Turkey). *Engineering Geology*, *246*, pp. 139–161.

Kelly, R.B., Pineda, J.A., Bates, L., Suwal, L.P. and Fitzallen, A. (2017) Site characterisation for the Ballina field testing facility. *Géotechnique*, *67*(4), pp. 279–300.

Kirstein, J.F., Ahner, C., Uhlemann, S., Uhlich, P. and Röder, K. (2013) Ground improvement methods for the construction of the federal road B 176 on a new elevated dump in the brown coal region of MIBRAG. *Proc 18th Int. Conf. on Soil Mechs. And Found. Engng*, Paris.

Konrad, J.M., Bozozuk, M. and Law, K.T. (1985) Study of in-situ test methods in deltaic silt, *Proc. 11th Int. Conf. SMFE,* San Francisco, Vol. 2, pp. 879–886.

Kruizinga, J. (1989) Bearing capacity of a test pile compared with predictions from pressuremeter rules, *Proc. 12th Int. Conf. SMFE, Rio de Janeiro*, Vol. 2, pp. 1155–1158.

Kuvik, M., Kopecký, M. and Frankovská, J. (2019, February). Deformation modulus determination from pressuremeter and dilatometer tests for crystalline rock. In *IOP Conference Series: Materials Science and Engineering* (Vol. 471, No. 4, p. 042010). IOP Publishing.

Lacasse, S., Jamiolkowski, M., Lancellotta, R. and Lunne, T. (1981) In-situ characteristics of two Norwegian clays, *Proc. 10th Int. Conf. SMFE,* Stockholm, Vol. 2, pp. 507–511.

Ladanyi, B. (1972) In-situ determination of undrained stress-strain behaviour of sensitive clays with the pressuremeter, *Can. Geotech. J.*, *9*(3), 313–319.

Ladanyi, B. and Johnston, G.H. (1973) Evaluation of in-situ creep properties of frozen soils with pressuremeter, *Proc. 2nd Int. Conf. Permafrost North American Contributions, Yakutsk,* pp. 310–318.

Ladd, C.C., Foot, R., Ishihara, K., Schlosser, F. and Poulos, H.G. (1977) Stress-deformation and strength characteristics: state-of-the-art report, *Proc. 9th Int. Conf. SMFE,* Tokyo, Vol. 2, pp. 421–454.

Lade, P. V. & Lee, K. L. (1976). Engineering properties of soils, pp. 145. Report UCLA-ENG-7652. University of California at Los Angeles.

Laranjo, M.L., Carvalho, J., Fernandes, M.M. and Fonseca, A.V. (2015) Very small strain stiffness of Lisbon Miocene clayey formation from in situ tests.

Larsson, R. (2001) Investigations and load tests in clay till. Results from a series of investigations and load tests in the test field at Tornhill outside Lund in southern Sweden. Swedish Geotechnical Institute.

Lashkaripour, G.R. and Ajalloeian, R. (2003) Determination of silica sand stiffness. *Engineering Geology*, *68*(3–4), pp. 225–236.

LCPC-SETRA (1985) Règles de justification des fondations sur pieux à partir des résultats des essais pressiométriques, *Ministre de l'Urbanisme du Logement et des Transports,* Paris.

Lee, K.M. and Rowe, R.K. (1989) Deformation caused by surface loading and tunnelling: the role of elastic anisotropy, Geotechnique, *39*(1), 125–140.

Lehane, B.M. (2003) Vertically loaded shallow foundation on soft clayey silt. *Proceedings of the Institution of Civil Engineers-Geotechnical Engineering*, *156*(1), pp. 17–26.

Lehane, Barry M. (2008) Relationships between axial capacity and CPT q_c for bored piles in sand. In *Deep Foundations on Bored and Auger Piles-BAP V*, pp. 73–86. CRC Press.

Leischner, W. (1966) Die bautechnische Baugrundbeurteilung mittels horizontaler Belastungsversuche im Bohrloch nach dem Koglerverfahren, *Der Bauingenier,* 12.

Leroueil, S., Tavenas, F., Brucy, F., La Rochelle, P. and Roy, M. (1979) Behaviour of destructed clays, *J. Geotech. Engng Div.,* ASCE, *105*(GT6), 759–778.

Liang, Y., Cao, L. and Liu, J. (2015) Statistical Correlations between SPT N-Values and Soil Parameters. *Department of Civil Engineering, Ryerson University.*

Likitlersuang, S., Teachavorasinskun, S., Surarak, C., Oh, E. and Balasubramaniam, A. (2013) Small strain stiffness and stiffness degradation curve of Bangkok Clays. *Soils and Foundations*, *53*(4), pp. 498–509.

Long, M. and Menkiti, C.O. (2007) Geotechnical properties of Dublin boulder clay. *Géotechnique*, *57*(7), pp. 595–611.

Lunne, T., Long, M. and Uzielli, M. (2006) November. Characterisation and engineering properties of Troll Clay. In *Proceedings of the Second International Workshop on Characterisation and Engineering Properties of Natural Soils,* Singapore (Vol. 3, pp. 1939–1972).

Manassero, M. (1989) Stress-strain relationships from drained self-boring pressuremeter tests in sands, *Geotechnique*, *39*(2), 293–307.

Marchai, J. (1971) Calcul du tassement des pieux à partir des méthodes pressiométriques, *Bull, de Liaison de LCPC,* Paris (52), 22–25.

Marsland, A. and Randolph, M.F. (1977) Comparisons of the results from pressuremeter tests and large in situ plate tests in London Clay. *Geotechnique*, *27*(2), pp. 217–243.

Martin, R.E., Drahos, E.G. and Pappas, J.L. (1995) Characterization of preconsolidated soils in Richmond, Virginia. *Transportation Research Record* (1479).

Matasovic, N., El-Sherbiny, R. and Kavazanjian, Jr, E. (2010) In-situ measurements of MSW properties. *Geotechnical Characterization, Field Measurement, and Laboratory Testing of Municipal Solid Waste*, pp. 153–194.

Mayne, P.W. (1995) Profiling yield stresses in clays by in situ tests. *Transportation Research Record* (1479).

Mayne, P.W. and Kulhawy, F.H. (1982) K_o-OCR relationships in soil, *J. Geotech. Engng Div.*, ASCE, *108*(GT6), 851–872.

Mayne, P.W. and Mitchell, J.K. (1988) Profiling of overconsolidation ratio in clays by field vane. *Canadian Geotechnical Journal*, *25*(1), pp. 150–157.

Mayne, P.W. and Brown, D.A. (2003) Site characterization of Piedmont residuum of North America. *Characterization and Engineering Properties of Natural Soils*, 2, pp. 1323–1339.

McCabe, B.A. and Phillips, D.T. (2008) Design lessons from full-scale foundation load tests. In *Proc. 2nd Int. Conf. Geotechnical and Geophysical Site Characterization–ISC'3,* Taiwan (pp. 615–620).

Ménard, L. (1963) Calcul de la force portante des fondations sur la base des résultats des essais pressiométriques, *Sols – Soils, 5*(5), 9–24.

Ménard, L. and Rousseau, J. (1962) L'évaluation des tassements-tendances nouvelles, *Sols Soils,* No. 1, 13–28.

Ménard, L., Bourdon, G. and Gambin, M. (1975) Méthode générale de calcul d'un rideau ou d'un pieu sollicité horizontalement en fonction des résultats pressiométriques, *Sols Soils,* No. 22/23, 16–29.

Menkiti, C.O., Davis, J.A., Semertzidou, K., Abbireddy, C.O.R., Hight, D.W., Williams, J.D. and Black, M. (2015) The geology and geotechnical properties of the Thanet Sand Formation – an update from the Crossrail Project.

Meyerhof, G.G. (1981). The bearing capacity of rigid piles and pile groups under inclined loads in clay. *Canadian Geotechnical Journal, 18*(2), pp. 297–300.

Meyerhof, G.G., Mathur, S.K. and Valsangkar, A.J. (1981). The bearing capacity of rigid piles and pile groups under inclined loads in layered sand. *Canadian Geotechnical Journal, 18*(4), pp. 514–519.

Meyerhof, G.G. and Sastry, V.V.R.N. (1985) Bearing capacity of rigid piles under eccentric and inclined loads, *Can. Geotech. J.*, *22*, 267–276.

Meyerhof, G.G. and Sastry, V.V.R.N. (1987) Full-displacement pressuremeter method for rigid piles under lateral loads and moments, *Can. Geotech. J.*, *24*, 471–478.

Ministry of Housing and Urban-Rural Development of China (MOHURD) (2001) *Code for Investigation of Geotechnical Engineering, Ministry of Housing and Urban-Rural Development of China,* Beijing, GB 50021-2001.

Moon, J.S., Jung, H.S., Lee, S. and Kang, S.T. (2019) Ground improvement using dynamic compaction in sabkha deposit. *Applied Sciences*, *9*(12), p. 2506.

Muir-Wood, D. (1990) Strain dependent moduli and pressuremeter tests, *Geotechnique, 40*(26), 509–512.

Narimani, S., Chakeri, H. and Davarpanah, S.M. (2018) Simple and non-linear regression techniques used in sandy-clayey soils to predict the pressuremeter modulus and limit pressure: A case study of Tabriz subway. *Periodica Polytechnica Civil Engineering, 62*(3), pp. 825–839.

Ng, C.W.W. and Wang, Y. (2001) Field and laboratory measurements of small strain stiffness of decomposed granites. *Soils and Foundations*, *41*(3), pp. 57–71.

Ng, C.W., Pun, W.K. and Pang, R.P. (2000) Small strain stiffness of natural granitic saprolite in Hong Kong. *Journal of Geotechnical and Geoenvironmental Engineering*, *126*(9), pp. 819–833.

O'Brien, A.S. and Newman, R.L. (1990) Self-boring pressuremeter testing in London Clay, *Proc. 24th Ann. Conf. Engng Group of the Geological Soc.: Field Testing in Engineering Geology*, Sunderland, pp. 39–54.

Ohya, S., Imai, T., Matsubara, M. (1982) "Relationship between N value by SPT and LLT pressuremeter results." In: Proceeding 2nd Europe Symposium on Penetration Testing, vol 1, Amsterdam, The Netherlands, 24–27 May, pp. 125–130.

Orhan, M., Işık, N.S., Topal, T.A.M.E.R. and Özer, M.U.S.T.A.F.A. (2006) Effect of weathering on the geomechanical properties of andesite, Ankara–Turkey. *Environmental Geology*, *50*(1), pp. 85–100.

Özvan, A., Akkaya, İ. and Tapan, M. (2018) An approach for determining the relationship between the parameters of pressuremeter and SPT in different consistency clays in Eastern Turkey. *Bulletin of Engineering Geology and the Environment*, *77*(3), pp. 1145–1154.

Palmer, A.C. (1972) Undrained plane-strain expansion of a cylindrical cavity in clay: a simple interpretation of the pressuremeter test, *Geotechnique*, *22*(3), 451–457.

Pietrangeli, G., Cacciarini, A. and Calabrese, M. (2013) Geomechanical characterization of residual soils for foundation design of a bituminous faced rockfill dam.

Poulos, H.G. and Davis, E.H. (1980) *Elastic Solutions for Soil and Rock Mechanics*, Wiley, New York.

Powell, J.J.M. (1990) A comparison of four different pressuremeters and their methods of interpretation in a stiff heavily overconsolidated clay, *Proc. 3rd Int. Symp. Pressuremeters*, Oxford, pp. 287–298.

Powell, J.J. (2001) In situ testing and its value in characterising the UK National soft clay testbed site, Bothkennar. In *Proceedings International Conference on In Stu Measurement of Soil Properties and Case Histories, In Situ*. Bali, Indonesia.

Powell, J.J.M. and Uglow, I.M. (1985) A comparison of Ménard self-boring pressuremeter and push-in pressuremeter tests in a stiff clay till, *Proc. Conf. Advances in Underwater Technology and Offshore Engineering*, London, No. 3, pp. 201–217.

Powell, J., Lunne, T. and Frank, R. (2001) Semi empirical design procedures for axial pile capacity in clays. In *15th International Conference on Soils Mechanics and Geotechnical Engineering*, Vol. 2 (pp. 991–994).

Prévost, J.-H. (1979) Undrained shear tests on clays, *J. Geotech. Engng Div.*, ASCE, *105*(NGT1), 49–64.

Prust, R.E., Davies, J. and Hu, S. (2005) Pressuremeter investigation for mass rapid transit in Bangkok, Thailand. *Transportation Research Record*, *1928*(1), pp. 206–217.

Rashed, A., Bazaz, J.B. and Alavi, A.H. (2012) Nonlinear modeling of soil deformation modulus through LGP-based interpretation of pressuremeter test results. *Engineering Applications of Artificial Intelligence*, *25*(7), pp. 1437–1449.

Ritsos, A., Migiros, G, Kollios, A and Kolovaris, E. (2005) Evaluation of pressuremeter tests performed within the formations of the Athens Basin. *M. P. Gambin, J. P. Magnan, and P. Mestat (Eds.), 50 years of pressuremeters international symposium – ISP5* (pp. 289–295). Presses des Ponts et Chaussees.

Robertson, P.K. and Ferreira, R.S. (1993) Seismic and pressuremeter testing to determine soil modulus, *Predictive Soil Mechanics, Proc. Wroth Memorial Symp.*, Oxford, pp. 434–448.

Robertson, P.K. and Hughes, J.M.O. (1986) Determination of properties of sand from self-boring pressuremeter tests, *Proc. 2nd Int. Symp. Pressuremeter Marine Appl., Texam*, United States. ASTM STP 950, pp. 283–302.

Robertson, P.K., Hughes, J.M.O., Campanella, R.G. and Sy, A. (1984) Design of laterally-loaded displacement piles using a driven pressuremeter, *Laterally Loaded Deep Foundations*, ASTM STP 835, pp. 229–238.

Robertson, P.K., Hughes, J.M.O., Campanella, R.G., Brown, P., and McKeown, S. (1986) Design of laterally loaded piles using the pressuremeter, *Proc. 2nd Int. Symp. Pressuremeter Marine Appl., Texam,* United States, ASTM STP 950, pp. 443–457.

Rouainia, M., Elia, G., Panayides, S. and Scott, P. (2017) Non-linear finite element prediction of the performance of a deep excavation in Boston Blue Clay. *Journal of Geotechnical and Geoenvironmental Engineering, 143*(5), p. 13.

Rouainia, M. and Muir Wood, D. (2000). A kinematic hardening constitutive model for natural clays with loss of structure. *Géotechnique, 50*(2), pp. 153–164.

Roy, D., Hughes, J.M.O. and Campanella, R.G. (1999). Reliability of self-boring pressuremeter in sand. *Canadian Geotechnical Journal, 36*(1), pp. 102–110.

le Roux, K.A., Bawden, W.F. and Grabinsky, M.W.F. (2002) Comparison of the material properties of in situ and laboratory prepared cemented paste backfill. In *Annual Conference, BC,* Canada (pp. 201–209).

Roy, D. and Campanella, R.G. (1997) Interpretation of pressuremeter unloading tests in sands. *Géotechnique, 47*(5), pp. 1069–1071.

Schnaid, F. (1990) "A Study of the Cone-pressuremeter Test in Sand." PhD Thesis, University of Oxford.

Schnaid, F., Ortigao, J.A., Mántaras, F.M., Cunha, R.P. and MacGregor, I. (2000) Analysis of self-boring pressuremeter (SBPM) and Marchetti dilatometer (DMT) tests in granite saprolites. *Canadian Geotechnical Journal, 37*(4), 796–810.

Seed, H.B., Wong, R.T., Idriss, I.M. and Tokimatsu, K. (1986) Moduli and damping factors for dynamic analyses of cohesionless soils, *J. Geotech. Engng Div.,* ASCE, Vol. 112, No. 11, pp. 1016–1032.

Sellgren, E. (1981) Pressuremeter and pile foundations, *Proc. 10th Int. Conf.* SMFE, Stockholm, Vol. 2, pp. 843–846.

Shan, S., Pei, X. and Zhan, W. (2021) Estimating deformation modulus and bearing capacity of deep soils from dynamic penetration test. *Advances in Civil Engineering, 2021.*

Shields, D.H. and Bauer, G.E. (1975) Determination of the modulus of deformation of sensitive clay using laboratory and in situ tests, *Proc. ASCE Spec. Conf In Situ Measurement of Soil Properties,* Raleigh, Vol. 1, pp. 395–421.

Shohet, D.C. (1995) Prediction of in situ horizontal stresses in clay soils from the measurement of undrained shear strength, plasticity index and vertical effective stress. *Proceedings of the Institution of Civil Engineers-Geotechnical Engineering, 113*(4), pp. 206–214.

Silva, T.Q., Cândido, E.S., Marques, E.A.G. and Minette, E. (2016) Determination of Em from Ménard pressuremeter tests for gneiss residual soils. *Young, 4*(13.62), pp. 6–11.

Simonini, P., Ricceri, G. and Cola, S. (2007). Geotechnical characterization and properties of Venice lagoon heterogeneous silts. *Characterisation and engineering properties of natural soils, 4*, pp. 2289–2327.

Springman, S.M., Arenson, L.U., Yamamoto, Y., Maurer, H., Kos, A., Buchli, T. and Derungs, G. (2012) Multidisciplinary investigations on three rock glaciers in the Swiss Alps: Legacies and future perspectives. *Geografiska Annaler: Series A, Physical Geography, 94*(2), pp. 215–243.

St John, H.D. (1975) Field and theoretical studies of the behaviour of ground around deep excavations in London Clay. PhD thesis, University of Cambridge.

St John, H.D., Potts, D.M., Jardine, R.J. and Higgins, K.G. (1993) Prediction and performance of ground response due to construction of a deep basement at 60 Victoria Embankment, *Predictive Soil Mechanics, Proc. Wroth Memorial Symp.,* Oxford, pp. 581–608.

Suyama, K., Imai, T. and Ohya, S. (1982) Development of LLT pressuremeter and its application in prediction of pile behavior under horizontal load, *Proc. Int. Symp. Pressuremeter and its Marine Appl.,* Paris, pp. 61–76.

Suyama, K., Ohya, S., Imai, T., Matsubara, M. and Nakayama, E. (1983) Ground behaviour during pressuremeter testing, *Proc. Int. Symp. Soil and Rock Investigations by In-situ Testing,* Paris, Vol. 2, pp. 397–402.

Symons, I.F. and Carder, D.R. (1992) The behaviour in service of a propped retaining wall embedded in stiff clay, *Proc. 10th Eur. Conf. Soil Mech. and Found. Engng,* Florence, Italy, pp. 761–766.

Tarawneh, B., Sbitnev, A. and Hakam, Y. (2018) Estimation of pressuremeter modulus and limit pressure from cone penetration test for desert sands. *Construction and Building Materials, 169*, pp. 299–305.

Terzaghi, K. and Peck, R.B. (1948) *Soil Mechanics in Engineering Practice,* Wiley, New York.

Thompson, R.P. and Leach, B.A. (1991) Settlement prediction and measured performance of Heysham II Power Station, *Proc. 10th Eur. Conf. Soil Mech. and Found. Engng,* Florence, Italy, pp. 609–614.

Tomlinson, M.J. (1957) The adhesion of piles driven in clay soils, *Proc. 4th Int. Conf SMFE,* Vol. 2, pp. 66–71.

Varaksin, S., Hamidi, B. and D'Hiver, E. (2005) Pressuremeter techniques to determine self bearing level and surface strain for granular fills after dynamic compaction. In *International Symposium 50 Years of Pressuremeters* (ISP5-Pressio 2005).

Varaksin, S. and Hamidi, B. (2013) Pressuremeter for design and acceptance of challenging ground improvement works. In *Proceedings of the 18th International Conference on Soil Mechanics and Geotechnical Engineering*, Paris (pp. 1283–1286).

Varaksin, S. and Hamidi, B. (2015) Analysis of soil-structure interaction by Ménard pressuremeter tests and ground improvement case histories. In *Soil-Structure Interaction, Underground Structures and Retaining Walls: Proceedings of the ISSMGE Technical Committee 207 International Conference on Geotechnical Engineering,* Vol. 4 (p. 220). IOS Press.

Viggiani G., Atkinson J.H. (1995) Stiffness of fine-grained soil at very small strains. *Géotechnique.* 45(2): 249–265.

Vijayvergiya, V.N. and Focht, J.A. (1972) A new way to predict the capacity of piles in clay, *Proc. 4th Ann. Offshore Technology Conf,* Houston, Vol. 2, pp. 865–874.

Wang, C.E. and Borden, R.H. (1996) Deformation characteristics of Piedmont residual soils. *Journal of Geotechnical Engineering, 122*(10), pp. 822–830.

Waschkowski, E. (1976) "Comparaisons entre les resultats des essais pressiometriques et le SPT". Rapport de Recherche du Laboratoire Regional des Ponts Et Chaussées de Blois, F.A.E.R.1.05.23.5, Juin. Not Published.

Wen, D. (2005) Use of jet grouting in deep excavations. In *Elsevier Geo-Engineering Book Series* (Vol. 3, pp. 357–370). Elsevier.

Whittle, R.W. and Liu, L. (2013) A method for describing the stress and strain dependency of stiffness in sand, In *Proceedings of the 18th International Conference on Soil Mechanics and Geotechnical Engineering*, Paris (pp. 2–6).

Whittle, R., Palix, E. and Donaghy, D. (2017) The influence of insertion process on determining the stiffness characteristics of chalk, using pre-bored, self-bored and pushed pressuremeters. In *Offshore Site Investigation Geotechnics 8th International Conference Proceeding*, Vol. 308, No. 315 (pp. 308–315). Society for Underwater Technology.

Wilson, W. and Corke, D.J. (1990) A comparison of modulus values of sandstone derived from high pressure dilatometer, plate loading, geophysical and laboratory testing, *Proc. 3rd Int. Symp. Pressuremeters,* Oxford, pp. 351–360.

Windle, D. and Wroth, C.P. (1977) Use of self-boring pressuremeter to determine the undrained properties of clays, *Ground Engng, 10*(6), 37–46.

Wroth, C.P. (1982) British experience with the self-boring pressuremeter, *Proc. Int. Symp. Pressuremeter and its Marine Appl.,* Paris, pp. 143–164.

Yagiz, S., Akyol, E. and Sen, G. (2008) Relationship between the standard penetration test and the pressuremeter test on sandy silty clays: a case study from Denizli. *Bulletin of engineering geology and the environment*, *67*(3), pp. 405–410.

Youd, T.L. and Idriss, I.M. (2001) Liquefaction resistance of soils: Summary report from the 1996 NCEER and 1998 NCEER/NSF workshops on evaluation of liquefaction resistance of soils. *Journal of Geotechnical and Geoenvironmental Engineering*, *127*(4), pp. 297–313.

Yu, H.S. (1994) State parameter from self-boring pressuremeter tests in sand. *Journal of Geotechnical Engineering*, *120*(12), pp. 2118–2135.

Yu, H.S. (1996) Interpretation of pressuremeter unloading tests in sands. *Geotechnique*, *46*(1), pp. 17–31.

Yu, H.S. (2000) *Cavity expansion methods in geomechanics*. Springer Science & Business Media.

Yu, W., Lai, Y., Zhu, Y., Li, H., Zhang, J., Zhang, X. and Zhang, S. (2002) In situ determination of mechanical properties of frozen soils with the pressuremeter. *Cold Regions Science and Technology*, *34*(3), pp. 179–189.

Zaki, M.F.M., Ismail, M.A.M. and Govindasamy, D. (2020) Correlation between SPT and PMT for sandy silt: A Case study from Kuala Lumpur, Malaysia. *Arabian Journal for Science and Engineering*, 45(10), 8281–8302.

Zawrzykraj, P.Z. (2017) Deformability parameters of varved clays from the Iłów (Central Poland) area based on the selected field tests. *Studia Geotechnica et Mechanica*, *39*(1), pp. 89–100.

Zhang, H., Zhang, J., Su, K. and Liu, S. (2012) In-situ pressuremeter test in warm and ice-rich permafrost. *Cold Regions Science and Technology*, 83, pp. 115–121.

Zhang, W.G., Zhang, R.H., Han, L. and Goh, A.T.C. (2019) Engineering properties of the Bukit Timah Granitic residual soil in Singapore. *Underground Space*, 4(2), pp. 98–108.

Chapter 8

Choosing and specifying a pressuremeter

8.1 INTRODUCTION

The preceding chapters have shown that there are different forms of pressuremeter such that a form of pressuremeter can be found for any ground condition to obtain foundation design parameters or ground properties either directly, theoretically, numerically or empirically. The choice of pressuremeter will depend on availability, ground conditions and parameters required. Whichever pressuremeter is chosen, the data will only be as good as the quality with which the site work and interpretation are carried out. Thus, it is important to specify the correct equipment, site operations, and acceptance criteria so that the operator is able to produce the results the engineer requires. It is important that the operators demonstrate that they have sufficient experience to undertake the work to the satisfaction of the engineer. The selection and specification of the pressuremeter are discussed here, and guidelines to a specification are given in the appendix. Table 8.1 is a summary of the advantages and disadvantages of pressuremeters in a range of ground conditions.

Pressuremeter testing has evolved from the prebored pressuremeter developed by Ménard in 1957, used to produce design parameters directly to the more sophisticated, instrumented devices, which aim to resemble most closely the ideal pressuremeter test, which can then be analysed as an infinitely long, expanding cavity to produce ground properties directly. As pressuremeters have developed, so have methods of design. In 1957, design methods were semi-empirical methods validated against observations of ground response. This approach is still valid, but now it is feasible to model ground response more accurately using numerical methods with increasingly sophisticated constitutive models. The parameters that define these models are derived from a number of sources, including the results of pressuremeter tests. However, given the boundary conditions imposed by a pressuremeter test, it is also feasible to use results of pressuremeter tests to validate numerical methods. This versatility means that pressuremeters are increasingly being used in ground investigation, which is facilitated by the development of international standards.

8.2 CURRENT STATE OF PRESSUREMETER TESTING

There are five (Figure 8.1) main reasons for using pressuremeters: to measure ground properties; to produce design parameters; to produce parameters for constitutive models; to validate numerical methods; and for research. Ground properties are

DOI: 10.1201/9781003028925-8

Table 8.1 Advantages and disadvantages of pressuremeters in a range of ground conditions

Ground Type	*Type of Pressuremeter*	*Advantages*	*Disadvantages*
Strong rock	Ménard	• Stable test pocket • Sample of rock to be tested	• Limited to stiffness
	Prebored	• Stable test pocket • Sample of rock to be tested	• Limited to stiffness
	Self-bored		• Not possible
	Full-displacement		• Not possible
Weak rock	Ménard	• Stable pocket with drilling mud • Use of split casing • Sample of rock to be tested	• Limited to stiffness unless very weak when strength may be determined • Potential collapse of borehole unless supported with casing or drilling mud • Disturbance of surrounding ground
	Prebored	• Stable pocket with drilling mud • Use of split casing • Sample of rock to be tested	• Limited to stiffness unless very weak when strength may be determined • Potential collapse of borehole unless supported with casing or drilling mud • Disturbance of surrounding ground
	Self-bored	• Pocket stabilised by probe • In very weak rocks horizontal stress, stiffness and strength determined	• No sample • Potential collapse of borehole unless supported with casing or drilling mud
	Full-displacement		• Not possible
Very coarse-grained soils	Ménard	• Driven into place with split casing	• Ground is significantly disturbed during installation • Limited to stiffness unless in a loose state • Potential collapse of borehole unless supported with casing or drilling mud • No sample • Disturbance of surrounding ground
	Prebored	• Driven into place with split casing	• Ground is significantly disturbed during installation • Limited to stiffness unless in a loose state • Potential collapse of borehole unless supported with casing or drilling mud • No sample • Disturbance of surrounding ground
	Self-bored		• Not possible
	Full-displacement		• Not possible

Table 8.1 Cont.

Ground Type	*Type of Pressuremeter*	*Advantages*	*Disadvantages*
Coarse-grained soils	Ménard	• Sample of soil to be tested • Design properties pressuremeter modulus and limit pressure	• Potential collapse of test pocket unless supported with split casing or drilling mud • Potential collapse of borehole unless supported with casing or drilling mud • Disturbance of surrounding ground
	Prebored	• Sample of soil to be tested • Assessment of stiffness • Empirical assessment of horizontal stress and strength	• Potential collapse of test pocket unless supported with split casing or drilling mud • Potential collapse of borehole unless supported with casing or drilling mud • Disturbance of surrounding ground
	Self-bored	• Minimal disturbance of surrounding ground • Test pocket supported by probe • Assessment of horizontal stress, stiffness and strength	• Potential collapse of borehole unless supported with casing or drilling mud • No sample of soil to be tested
	Full-displacement	• Test pocket supported by probe • Assessment of stiffness • Empirical assessment of horizontal stress and strength	• Potential collapse of borehole unless supported with casing or drilling mud • No sample of soil to be tested • Disturbance of surrounding ground
Fine-grained soils	Ménard	• Sample of soil to be tested • Design properties pressuremeter modulus and limit pressure	• Disturbance of surrounding ground
	Prebored	• Sample of soil to be tested • Assessment of stiffness • Empirical assessment of horizontal stress and strength	• Disturbance of surrounding ground
	Self-bored	• Minimal disturbance of surrounding ground • Test pocket supported by probe • Assessment of horizontal stress, stiffness and strength	• No sample of soil to be tested
	Full-displacement	• Test pocket supported by probe • Assessment of stiffness • Empirical assessment of horizontal stress and strength	• No sample of soil to be tested • Disturbance of surrounding ground

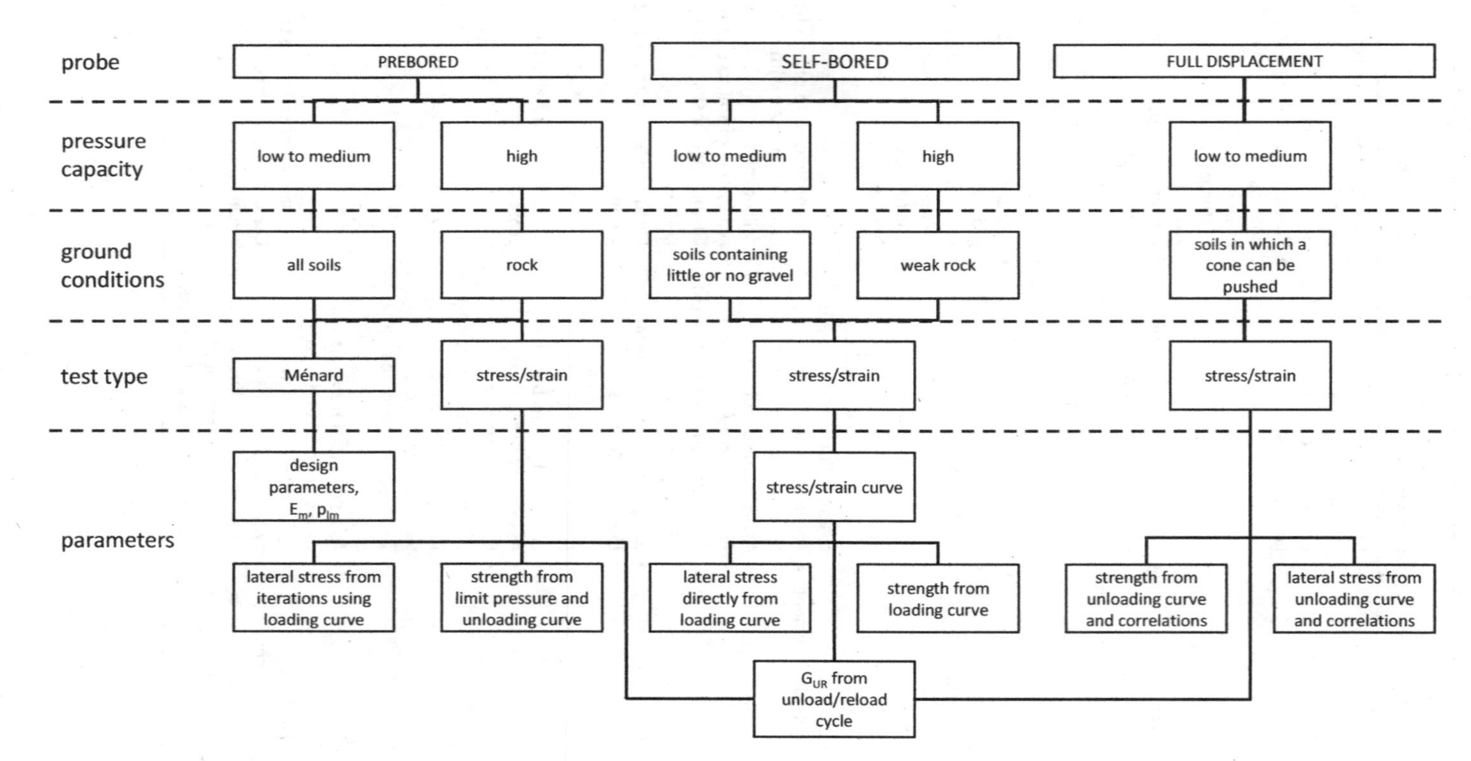

Figure 8.1 The use of pressuremeters in different ground conditions to obtain ground properties.

measured either for use in a design method or to demonstrate changes in the ground properties following some modification such as compaction or to allow comparisons to be made with other test results to justify methods of interpretation.

There are three groups of pressuremeters in commercial practice. Prebored pressuremeters include volume-displacement Ménard type tricell probes (Figure 2.5) and radial-displacement type monocell probes (e.g., Figures 2.9, 2.10, 2.11, 2.17, 2.24). These probes are used in all ground conditions, though the sensitivity and robustness of the probes are different for different ground conditions. Rotary rigs are used to advance the borehole, create the pocket and install the probe into the pocket. Pockets are created either by sampling, coring or drilling and, in all cases, a mud or polymer flush is recommended (Tables 3.3, 3.4 and 3.7).

Self-boring probes include the volume-displacement type French probes (Figure 2.15) and the radial-displacement Cambridge type probes (Figures 2.17). They are drilled into place either from the base of a borehole created using conventional drilling and sampling techniques or directly from the surface in soils containing little or no gravel and, in some cases, weak rock. They can be operated with purpose-built rotary rigs or from conventional rotary rigs, which can also advance the borehole. It is now more common to use a conventional rig since this rig is used to clear obstructions, case the borehole in unstable ground and advance the borehole between test positions when the distance exceeds the minimum required (Figure 3.10).

Full-displacement probes such as cone pressuremeters (Figure 2.24) are pushed into the ground, either from the base of a borehole or from the surface. If the pressuremeter is attached to a cone, which is used to measure the cone resistance, then the probe should be pushed into the ground at 2cm/s. This probe can only be used in soils into which it is possible to push a cone.

Tests are either stress- or strain-controlled loading and unloading sequences, which can include unload/reload cycles and, in some cases, constant stress or strain before unloading. The Ménard test (Figures 3.26 and 3.27) is a form of stress-controlled loading test carried out with a volume-displacement type PBP that has a displacement capacity of at least 50%. The parameters E_m and p_{lm} (section 6.3) obtained from this test curve are used in design formulae developed from observations of full-scale structures. It is usual with radial-displacement type probes to use strain-controlled tests in soils (Figure 3.28) and stress-controlled tests in rocks. Stress-controlled tests are the most common method of testing when using volume-displacement type pressuremeters.

PBP tests, including Ménard tests, can be analysed to give an initial stiffness (section 6.5.1). This is not a property of undisturbed ground but is a function of the disturbance created during installation. It is possible, if there is sufficient strain, to interpret the loading curve to give a value of post-peak strength. This can be found by using a semi-empirical method based on limit pressure (Table 6.10) or by selecting the reference datum (Figure 6.7) to give a reasonable assessment of the in situ horizontal stress and then applying a theory of cavity expansion.

An SBP test is the only type of pressuremeter test that gives a complete stress-strain curve from which it is possible to evaluate the initial stiffness of the ground (section 6.5.1), the peak strength (Figure 6.33 and sections 6.6 and 6.7) and the post-peak strength (Figure 6.34). The horizontal stress is found directly from the pressure at which the membrane lifts off from the body of the probe (section 6.4.1).

The installation of an FDP is a form of cavity expansion such that, theoretically, the pressure acting on the membrane is equal to the limit pressure. In fact, the actual pressure is less than that because of unloading that occurs as the soil particles move past the shoulder of the cone. It is normal to include a friction reducer behind the cone, and this allows further unloading to take place. The post-peak undrained shear strength of clays can be estimated from the final unloading curve (section 4.5.2, 4.5.3 and 4.6.3). The limit pressure is given directly, and this is used, together with the cone resistance, to estimate the angle of friction of sands and the in situ horizontal stress (section 6.7).

An unload/reload cycle can be carried out in any test and is usual in all except the Ménard test. The cycle can be carried out at any stage during a SBP test, and this cycle can be interpreted to give an average stiffness (section 6.5.2), the variation of average stiffness with strain (section 6.5.3) and the variation of elemental stiffness with strain. The stiffness obtained represents the stiffness of ground unaffected by installation. Normalising the stiffness with respect to stress level permits comparisons to be made between tests at different depths in the same deposits and cycles at different strain levels in the same test. The cycle in a PBP test must be carried out at as large a strain as possible to ensure that ground unaffected by installation is tested. The cycle can be interpreted to give the same data as that from the SBP test but it may be more difficult to compare the results since the in situ stress can only be estimated. The cycle in an FDP test should be carried out once the limit pressure is reestablished. The cycle can be interpreted in the same manner as the SBP test and, provided an estimate can be made of the in situ stress, it may be possible to compare stiffnesses.

Simple rheological models are used to interpret tests, since the improvements in the results obtained by using more sophisticated methods are not justified given the natural variability of the ground and the uncertainty of the condition of the ground at the start of testing (Wroth, 1984). Improvements in installation techniques are reducing this uncertainty, but there is still the natural variability. Horizontal stress can only be obtained directly from good-quality SBP tests (section 7.4.1). It is possible to use iterative and semi-empirical techniques to estimate horizontal stress from PBP and FDP tests (section 7.4.1). The pocket diameter at the horizontal stress is the reference datum used to interpret the loading curve to obtain strength. Generally, strength can only be found directly from the loading portion of good-quality SBP tests. Other tests, including PBP, FDP and poor-quality SBP tests, are interpreted either semi-empirically using correlations or by analysing the unloading curve (sections 7.4.3 and 7.4.4).

Predictions of foundation performance using ground properties determined from pressuremeter tests are generally good. There is extensive experience of the use of the Ménard method to predict capacity and movement of shallow foundations (section 7.2.1), axially loaded piles (section 7.2.2) and horizontally loaded piles (section 7.2.3). The FDP test has been used to predict, with limited success, the displacement of laterally loaded piles (section 7.3). The Ménard test has also been used to predict the capacity of ground anchors (section 7.2.4) and to design road pavements. Data taken from SBP tests have been correlated with E_m and p_{lm} (section 7.2.6) so that the direct method of design developed for Ménard tests can be used with data from SBP tests but, generally, the SBP test is used to produce ground properties.

Current research is devoted to improving the analysis of the pressuremeter test to gain an insight into the factors that influence the test, including the effects of installation, probe geometry, measuring systems, drainage, anisotropy and soil structure. Databases of foundation behaviour and comparisons with predictions from pressuremeter test results are continually being updated, leading to further refinements of the design formulae and confidence in the use of pressuremeter test results in design. There are international standards for all types of pressuremeter (e.g., BS 5930:2015+A1:2020; NF P 94-110 2000; GOST 20276–12:2012; ASTM D4719–20:2020; BS EN ISO 22476-4:2021; BS EN ISO 22476-5; BS EN ISO 22476-6; and BS EN ISO 22476-8:2018). Semi-empirical correlations are being improved as further laboratory tests are completed on different types of soil. Instrumentation, test control and automatic data processing continue to be developed.

8.3 CHOOSING A PRESSUREMETER

The choice of pressuremeter depends on the ground conditions and end product required, that is, design parameters or ground properties. There is a limitation in that not all pressuremeters are available throughout the world, though it is possible to mobilise the devices to a site possibly at some expense.

The decisions that have to be taken when selecting a pressuremeter are as follows:

1. What pressure capacity is required for the ground to be tested?
2. Are the tests to be interpreted to give design parameters or ground properties?
3. If ground properties are required, which ones?
4. Does the operator have the necessary experience and equipment?
5. Are tests to be carried out in test pockets below a borehole?

Figure 8.2 refers to the selection process.

Only high-pressure PBPs can be used in moderately strong to strong rock. The weak rock SBP and high-pressure PBP can be used in weak rock. All types of pressuremeter can be used in soil, though if the soil contains too much gravel then only the PBP can be used.

If a design method based on the pressuremeter curve rather than interpreted ground properties is to be used, then the displacement capacity of the probe is fixed (Figure 8.2b). The Ménard method is based on volume-displacement type tricell probes that can expand by up to 50%. It may be possible to use radial-displacement type monocell probes, but the effects of the non-cylindrical expansion due to the finite length of the membrane means that the modified limit pressure measured may be in error. The effect on the initial stiffness is minimal. Further work is required to justify the use of these probes in the Ménard method. Correlations between the SBP test curve and the PBP curve have been proposed so that the SBP test can be analysed to give E_m and p_{lm} provided that the expansion of the probe is at least 20%. There is limited experience of this approach using SBP data so it should be treated with caution. Methods of predicting the performance of horizontally loaded piles from a FDP test curve have been developed.

If a complete stress-strain curve is required then it is necessary to carry out an SBP test. PBP tests can be interpreted and corrected to produce a stress-strain curve but the

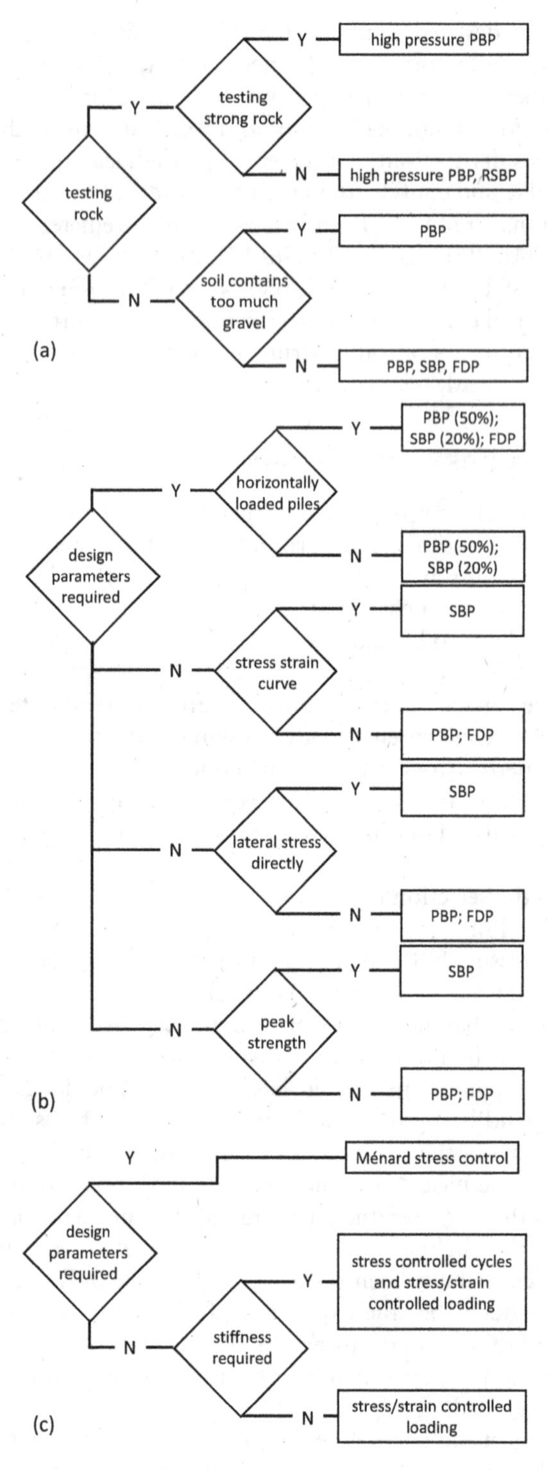

Figure 8.2 Choosing a pressuremeter: (a) the pressure capacity and type; (b) the displacement capacity and type; and (c) the test method.

installation effects are such that any interpretation has to be treated with caution. It is not possible to obtain a stress-strain curve from an FDP test because of the significant disturbance during installation. The disturbance created during the installation of a PBP and a FDP are such that the horizontal stress can only be assessed by iterative procedures based on an assumed soil model and semi-empirical correlations. The horizontal stress can be determined directly from good-quality SBP tests.

All pressuremeter tests can be analysed to give stiffness and strength, but a test should be designed for the parameters required (Figure 8.2c). E_m and p_{lm} are obtained from a stress-controlled test in which the expansion is at least 50%. Stiffness is best obtained from stress-controlled cycles, ideally after the membrane has been held at a constant stress or strain before unloading. If a cycle is included in a test then the cycling process will affect the derived strength unless the cycle is carried out at small strains, which is only possible with good-quality SBP tests. Stiffness is likely to be more important than strength since foundation performance is more important than failure. However, there are instances when strength will be required and, in these cases, it is recommended that no cycles are carried out or the cycles are carried out at a large strain once post-peak conditions have been established. Stress- and strain-controlled tests will give different results, but all cycles are likely to be stress-controlled and therefore the choice of stress- or strain-control is governed by the chosen probe and its control system, and it will only affect strength.

8.4 A TYPICAL SPECIFICATION

A specification should be written with the end product in mind, which is either going to be ground properties or design parameters. As in all quality ground investigations there is a need for the engineer to liaise closely with the contractor, keeping them informed of the purpose of the investigation and changes to that purpose, and to monitor and assess data when they become available. Table 8.2 summarises the key points of a specification and acceptance criteria required.

There are a number of national and international standards covering pressuremeter testing (e.g. BS 5930:2015+A1:2020; NF P 94-110 2000; GOST 20276–12:2012; ASTM D4719–20:2020; BS EN ISO 22476-4:2021; BS EN ISO 22476-5; BS EN ISO 22476-6; and BS EN ISO 22476-8:2018). It is recommended that the relevant standard is incorporated in the contract documents.

Different pressuremeters and different site operations will give different results. Specifying a pressuremeter by name can be too specific, leading to increased costs. It is better to specify the type of pressuremeter required and its pressure and displacement capacity and ensure that the operator has sufficient experience. Most probes have a length/diameter ratio in excess of 5. The length of the expanding section does affect the results, but this can be taken into account in the interpretation. Calibration procedures must be clearly defined, since results obtained depend on those calibrations. The number and type of calibrations will depend on the probe, the ground conditions and the accuracy of the results required. For instance, it is unnecessary to specify membrane compression calibration for tests in soft ground or for volume-displacement type probes. It is recommended that operators maintain instrument registers to demonstrate that the transducer calibrations remain the same, within reason, over a period of time.

Table 8.2 Guidelines for preparing a specification for pressuremeter testing

Group	*Detail*	*General*	*Acceptance criteria*
Probe	Dimensions	1. Create cylindrical expanding cavity	1. L/D > 6 2. Cylindrical 3. Straight to within 1 mm/m
Probe	Pressures	1. Sensitive enough to measure pressure required to reach maximum displacement or maximum capacity of probe	1. 20MN/m^2 for rock 2. 4 MN/m^2 for soil
Probes	Types and displacement	1. Prebored	1. Tricell or monocell 2. 50% capacity for Ménard test 3. 25% capacity for all other tests
		2. Self-bored	1. Minimum disturbance drilling systems integral with expanding section 2. 10% capacity for testing weak rock 3. 15% capacity for testing soil
		3. Pushed in	1. Cone integral with expanding section 2. 25% capacity
Operator		1. Experience	1. Three months' experience under supervision of technician with two years' experience
Transducers			1. Resolution 0.05% of full working range
Calibrations	Register	Register	1. Instrument register to show calibrations
Calibrations	Procedure	Pressure	1. 10 to 20 increments to max capacity 2. Increments maintained until displacement constant 3. Readings at 10 s
		Displacement	1. 10 increments up to maximum displacement 2. Reading at each increment
		System compression	1. Probe, testing equipment and connecting tubes 2. 20 to 40 increments up to maximum capacity 3. Increments maintained until displacement constant 4. Readings at 10 s
		Membrane stiffness	1. Inflate membrane in air 2. New membranes inflated at least 3 times 3. Stress-controlled tests to define curve
		Membrane compression	1. Increase pressure to maximum capacity 2. Reduce to 10% of maximum 3. Carry out pressure calibration
Calibrations	Frequency	1. Transducers 2. Membrane stiffness 3. Membrane and system compression	1. Beginning and end of project and after any major repair 1. Beginning and end of project and when membrane changed 1. Specify

Table 8.2 Cont.

Group	*Detail*	*General*	*Acceptance criteria*
Installation		1. Prebored	1. Minimum disturbance to pocket walls 2. Specify sampling 3. Specify limits to mud 4. 1 <[pocket/probe diameter] <1.1 5. Pocket length = 4 × length of expanding section 6. Probe in centre of pocket 7. Install probe within 60 min of completing pocket
		2. Self-bored	1. Minimum disturbance drilling beyond base of hole or from surface 2. Drill twice distance between centre of expanding section and base of probe
		3. Pushed in	1. 2 cm/s
Test depths		1. Depths 2. Minimum spacing	
Testing	Time between installation and testing	1. Prebored	1. Start testing as soon as probe in place
		2. Self-bored	1. 30 min wait in cohesive soils, no wait in granular soils
		3. Pushed in	1. Start testing as soon as probe in place
Testing	Termination		1. Maximum pressure capacity 2. Maximum displacement 3. Possible damage to equipment
Testing	Procedure	1. Time intervals to record data	1. Between 10 and 30 s
		1. Strain-controlled test (PBP)	1. Increment of 5% of maximum capacity every 1 min
		1. Strain-controlled test (SBP)	1. Stress control at first 2. 1%/min
		1. Strain-controlled test (PIP)	1. Stress control at first 2. 5%/min
		1. Ménard	1. 7 to 15 loading increments 2. First increment <0.2 MN/m^2 3. Increment held for 1 min
Testing	Additions	1. Unload/reload cycles	1. Specify number 2. Specify approximate strain level
		1. Time before unloading	1. Specify time or minimum strain rate
		1. Dissipation tests	1. Specify approximate strain level

Source: Based on BS 5930:2015+A1:2020; NF P 94-110: 2000; GOST 20276–12:2012; ASTM D4719–20:2020; BS EN ISO 22476-4:2021; BS EN ISO 22476-5: 2012; BS EN ISO 22476-6: 2018; and BS EN ISO 22476-8: 2018.

If they do, it is only necessary to prove the transducer calibrations conform to the instrument register. Therefore, unless there are changes to the probe, it is only necessary to calibrate the transducers at the beginning and end of a project. Calibrations for transducers are generally linear and therefore the acceptance criteria should be clearly defined in terms of the coefficient of determination and repeatability. Calibrations for membrane stiffness and compression vary with the membrane since the material properties change with age and use. Calibrations are carried out every time a membrane is changed and at frequent intervals during testing.

Installation procedures are critical since they affect the surrounding ground. A prebored pressuremeter should be installed within a pocket as soon as possible after it is drilled. The pocket must be drilled with minimum disturbance to the surrounding ground. The drilling requirements will be partly a function of the ground type and depth and therefore will be the responsibility of the drilling contractor. However, it must be very clearly stated that the pocket should be drilled in such a manner as to reduce the disturbance to the ground – that is, the drill type, speed of advance, mud type and mud pressure must all be adjusted accordingly. The length of pocket will depend on the length of the expanding section and the zone of disturbance due to drilling the borehole. An allowance has to be made for debris collecting in the base of the pocket – either debris from drilling or debris from the sides of the pocket being scraped as the probe is lowered to the test position. There is a limit to the diameter of the pocket, since it must not be smaller than the probe to avoid pushing the probe into the cavity and, hence, displacing the soil, and not too large to ensure that the ground can be pressurised sufficiently to obtain the parameters required. Once a pocket is complete the probe should be lowered to the test position as quickly as possible to prevent further deterioration of the pocket walls and reduce the possibility of collapse.

It is usual to specify that a pocket should be drilled with minimum disturbance to the surrounding ground, to specify the diameter and length of the pocket and the time between completing the pocket and starting the test. It is often difficult to establish that the criteria have been met until a test is started. The measured displacement will give an indication of the pocket diameter. The test curve will give an indication of the zone of disturbance and the test depth, in relation to the pocket depth, will give an indication of the amount of debris. A test in a pocket that fails to meet the criteria need not necessarily be abandoned. Any data from the ground is useful. In some circumstances it may be very difficult to meet the criteria and in those cases the engineer should discuss with the operator any possible changes that could be made.

The quality of the results from SBP tests is dependent on the quality of the drilling, which is the responsibility of the operator and drilling contractor. The speed of drilling, the position of the cutter, the mud pressure and the return flow all affect the degree of disturbance. The quality of drilling can only be assessed once a test has started. If the results of a test suggest that the drilling process has affected the surrounding ground, then the engineer must discuss with the operator changes that could be made to reduce the disturbance.

The full-displacement FDP, or cone pressuremeter, should be installed at the same speed every time in order to produce the same amount of disturbance in the same soils. The cone data are also used in the interpretation of tests therefore, the cone should be pushed in at 2 cm/s, the same speed as the static cone.

Testing should start immediately a PBP or FDP probe is in place. This reduces further deterioration of the pocket walls when using a PBP; it prevents further changes to the soil due to consolidation and relaxation when using a FDP. The SBP is used to obtain ground properties directly, since it most closely resembles the ideal test. Drilling the SBP can give rise to changes in pore pressure in clays, therefore it is usual to specify a minimum period between drilling and testing (typically half an hour).

Alternatively, the test could be started once the excess pore pressures have dissipated (if pore pressure transducers are fitted), though this can lead to excessive and uneconomic periods between drilling and testing. SBP tests in granular soils can be started once the probe is in place.

The depths of tests are usually specified. The minimum distance between tests depends on the probe but, typically, for most probes it is 1 metre. The spacing between test locations will vary according to the elevation of the test relative to the elevation of the proposed structure. The spacing between tests will increase with depth for foundations at the surface. SBP tests are often used to obtain representative parameters of a stratum, and in that case tests should be required at 1-metre intervals to ensure that a full profile of results is obtained. FDP and PBP tests used to obtain representative parameters will be carried out at the same frequency, but if they are to be used for design parameters the spacing will vary with distance from the formation level of the proposed foundation.

Derived ground properties and design parameters are affected by the type and speed of test. There are two main test procedures: the stress-controlled test (including the Ménard test) and the strain-controlled test. In addition to the loading sequence, it may be necessary to specify additions or modifications to the test procedure, such as unload/reload cycles, time before unloading and changes in stress or strain rates. Data can be recorded manually or automatically, though there is a tendency towards automatic recording. The time at which the data should be scanned and the format of the data should be specified. The latter is especially important if the engineer is to undertake further processing at a later stage.

A stress-controlled test is usually specified in terms of the number of increments, the time to maintain those increments and the minimum and maximum size of increments. A strain-controlled test can either be specified in terms of the cavity or volumetric strain rate or, in the case of volume-displacement type probes, in terms of the increment of volume and the time for maintaining that increment.

It is usual to specify an unload/reload cycle in all tests exept for the Ménard test. The strain level at which the unloading should start and the stress amplitude of the cycle should be indicated but not specified. The reason for this is that it is often difficult to start unloading at a specific point, especially if the membrane is not expanding uniformly. The stress amplitude is indicated so that the cycle is confined to the elastic range, though at the time of the test, the amplitude, which is related to strength, is unknown. However, the strain level and stress amplitude are not critical, since theoretical corrections can be applied to adjust the modulus to take into account the variations between tests – that is, a secant modulus can be determined over a specified stress or strain range.

In strain-controlled tests it is becoming common to specify a period before unloading to allow consolidation and relaxation to occur. If this is not done, the unloading curve will include a component of expansion due to consolidation and creep which, in the extreme case, can lead to a negative stiffness. If this is specified then the loading test curve can no longer be used to derive strength. If it is not specified then the reloading portion of the curve will be used to determine the stiffness of the ground. This period is unnecessary in stress-controlled tests, since the strain rate at the end of each increment is likely to be small.

Data reduction can be carried out once a test is complete. Transducer outputs are converted to stress and strain. Corrections are applied for membrane stiffness and compression, system compliance and membrane thinning. The stress-strain curve should be submitted to the engineer within 24 hours of completing the test so that decisions can be taken on the installation and testing procedure. Further analysis may be required but this can be carried out at any time.

The interpretation of the data is often the responsibility of the operator, but the engineer may want to undertake his own interpretation at a later stage. Thus, the data should be stored in some easily read format.

Interpretation includes theoretical and semi-empirical methods and empirical correlations. A Ménard test is interpreted to give E_m and p_{lm}. A good-quality SBP test with an unload/reload cycle can be interpreted directly to give horizontal stress, shear modulus and strength. A good-quality PBP test with an unload/reload cycle can be interpreted directly to give shear modulus and semi-empirically to give horizontal stress and strength. A good-quality FDP test with an unload/reload cycle can be interpreted theoretically to give shear modulus and, empirically, horizontal stress and strength. Further analyses allow the test curves to be interpreted to give more refined ground properties, but this is usually the responsibility of the engineer.

8.5 COSTS

An example Bill of Quantities is given in the appendix, showing the items that should be considered when specifying pressuremeter tests. The equipment has to be delivered to site and, since it is often separate from the drilling rig, it should be itemised separately. When on site, the equipment has to be moved between boreholes and, while this may not be itemised separately, it is a separate activity which ensures that the pressuremeter equipment and rig are treated separately, allowing more flexibility on site.

The installation of all pressuremeters involves some activity with depth. The pocket for a PBP is created by the drilling rig. The drilling items for the rig include creating the pocket, advancing the borehole and overdrilling the pocket. Overdrilling is the same item as drilling the borehole, therefore the full depth of borehole and the extra depth to create the pockets should be paid for. Lowering and raising the drill rods is within the rate for drilling. The SBP drills its own pocket but, if it is drilled from the base of a borehole and that borehole is advanced between test positions, then overdrilling the pocket should be allowed for in the drilling item. The FDP is pushed in but, as with the SBP, when it is used in a borehole the total depth of the borehole should be paid for.

A pressuremeter has to be calibrated before and after use when there are any changes to the equipment and if specified. There should be separate items for each of the calibrations. There should also be an item for the standard test which includes any specified relaxation time between creating the pocket and starting the test. If a test lasts more than the time specified – for example, when carrying out dissipation tests – then an hourly rate applies. Additional unload/reload cycles are an extra item. The test rate should include the standard minimum interpretation to produce the parameters required. Additional processing to produce further information is an extra item. The further activity of raising and lowering the probe is usually within the rate of testing.

8.6 FUTURE DEVELOPMENTS

There are a number of international standards for creating boreholes, installing a pressuremeter, undertaking tests and interpreting the results (BS 5930:2015+A1:2020; NF P 94-110 2000; GOST 20276–12:2012; ASTM D4719–20:2020; BS EN ISO 22476-4:2021; BS EN ISO 22476-5; BS EN ISO 22476-6; and BS EN ISO 22476-8:2018). Specifications range from those that specify every aspect of pressuremeter testing, including the installation and interpretation, to those that are written as guidelines allowing a degree of flexibility. Therefore, to realise the benefits of pressuremeter testing, it is important that the engineer is familiar with all aspects of pressuremeter testing and the use of results in design, and that the contractor has the necessary experience to conduct the testing in accordance with the principles and specification for pressuremeter testing.

Burlon and Reiffsteck (2015) argued for a re-evaluation of the use of pressuremeters in ground investigation to improve the accuracy of displacement calculations for geotechnical structures and assessment of the bearing capacities of shallow and deep foundations with the development of numerical methods. This includes the following:

- The development of testing devices and test procedures to increase reliability and performance of the test;
- Development of constitutive models that incorporate data from pressuremeter tests to complement advances made in the field of numerical modeling of geotechnical structures;
- Creation of a database for calibration and validation of calculation methods using results obtained from pressuremeter tests;
- Development of guidelines for the correct use of pressuremeters and the application of their results in design.

Most forms of pressuremeter evolved from research at universities and research institutions to such an extent that they are commercially viable. They have become more robust and reliable with use. The principles that define a pressuremeter probe are now fixed, but manufacturers will continue to improve the equipment within the scope of those principles.

Failures occur because of burst membranes and electrical problems. Improved membranes and better instrumentation are continually being developed, making the pressuremeter even more reliable.

The results of a pressuremeter test depend to some extent on the system used to install the probe. The aim is to produce a consistent method, which minimises the disturbance to the ground. Developments in sampling techniques and laboratory testing are leading to a better understanding of the behaviour of the ground during sampling and drilling. Hence, it is possible to control the amount of disturbance and take any disturbance into account when analysing a test. Prebored pressuremeters are installed in test pockets of a specific size; self-bored pressuremeters create their own minimally disturbed test pockets; and full-displacement pressuremeters displace the soil uniformly prior to testing.

Pressuremeter tests are automatically logged allowing rapid interpretation and data storage for further analysis at a later stage. It is likely that the information stored will be standardised following the development of data-management systems which will allow easy transfer of data. Computers are used to control tests allowing standard procedures to be followed exactly, removing the dependency on the operator. It also allows test procedures to be adjusted – for example, to allow additional unload/reload cycles or pore-pressure dissipation.

The increasing use of pressuremeters and the publication of comparisons of predictions based on pressuremeter test data and observations have led to developments of design methods for the all types of pressuremeter; and the database upon which the Ménard results are based has been extended.

The analysis of pressuremeter tests to take into account disturbance and improvements in the soil model has led to increasing confidence in the results to the extent that pressuremeter tests are used to validate more complex constitutive models, However, for routine geotechnical design, simple interpretations of standard tests are adequate.

The importance of stiffness measurements has increased with improvements in analytical techniques to predict foundation performance. The most consistent parameters obtained from a pressuremeter test are the shear moduli and this, together with the speed with which they are obtained, makes the pressuremeter test ideal for foundation design.

For the same results, the cost of a pressuremeter test and all associated activities is less than the cost of other tests and their associated activities. Thus, as confidence develops in the use of the pressuremeters and as, increasingly, results are required more quickly it is likely that pressuremeter testing will become even more common.

REFERENCES

ASTM D4719–20 (2020) Standard Test Methods for Prebored Pressuremeter Testing in Soils, ASTM International.

BS 5930:2015+A1:2020, *Code of practice for ground investigations*, British Standards Institution.

BS EN ISO 22475-1:2021, *Geotechnical investigation and testing – Sampling methods and groundwater measurements – Part 1: Technical principles for the sampling of soil, rock and groundwater*, British Standards Institution.

BS EN ISO 22476-4:2021, *Geotechnical investigation and testing. Field testing – Prebored pressuremeter test by Ménard procedure*, British Standards Institution.

BS EN ISO 22476-5:2012, *Geotechnical investigation and testing. Field testing - Flexible dilatometer test*, British Standards Institution.

BS EN ISO 22476-6:2018, *Geotechnical investigation and testing. Field testing – Self-boring pressuremeter test*, British Standards Institution.

BS EN ISO 22476-8:2018, *Geotechnical investigation and testing. Field testing – Full displacement pressuremeter test*, British Standards Institution.

Burlon, S. and Reiffsteck, P., 2015, ARSCOP: a French national project to continue with the development of the pressuremeter. In *ISP7-Symposium International pour le 60ème anniversaire du pressiomètre* (p. 6).

GOST 20276–12: 2012 updated version of GOST 20276–85 (translated by Foque, J.B. and Sousa Coutinho, G.F.) *Soils Methods for Determining Deformation Characteristics*, Interstate Council for Standardization, Metrology and Certification, Russia.

NF P 94-110 2000, 2000, *Sols: Reconnaissance et essais – Essai pressiométrique Ménard*, AFNOR

Wroth, C.P., 1984, The interpretation of in situ soil tests. *Geotechnique*, *34*(4), pp. 449–489.

Appendix

Specifications and quantities

A.1 INTRODUCTION

The following is based on the Ménard recommendations for carrying out stress-controlled prebored pressuremeter tests to obtain the pressuremeter modulus E_m and the modified limit pressure p_{lm}; the ASTM D4719-20 standard test method for stress- and strain-controlled prebored pressuremeter tests to obtain the pressuremeter modulus E_m and modified limit pressure p_{lm}; and the standards BS EN ISO 22476-4:2021; BS EN ISO 22476-5:2012; BS EN ISO 22476-6:2018; and BS EN ISO 22476-8:2018. This is a set of guidelines that could be used to produce a specification. The engineer is required to specify certain requirements and acceptance criteria. The figures in bold and square brackets refer to sections in the main text that give further details.

A.2 GENERAL

A.2.1 The equipment

The pressuremeter shall be a cylindrical device designed to apply uniform pressure to the walls of the pocket by means of a flexible membrane. The length/diameter ratio of the expanding section shall be greater than 5. The vertical axis of the pressuremeter shall be straight to within 1 mm per metre over the length of the major diameter.

The probe shall either be a probe that is lowered into a prebored pocket (prebored pressuremeter) or a probe that is bored into the ground (self-bored pressuremeter) or a probe that is pushed into the ground (full displacement pressuremeter). The particular probe to be used shall comply with the specification and be robust enough to be used in the ground conditions. The working range of the pressure transducer or gauge shall comply with the specification or be equal to the working capacity of the probe. Whichever probe is used it must be properly calibrated. The engineer must agree that the probe to be used is correct and will produce the results required.

A.2.2 The operator

The operator of the equipment shall have at least three months' experience of the pressuremeter under the full-time supervision of a technician with a minimum of two years' experience in the use of the pressuremeters specified. The operator shall be responsible to a supervising engineer who has a minimum of five years' experience in geotechnical engineering and is fully conversant with the operation of pressuremeters and the interpretation of the data obtained.

A.3 PROBES

A.3.1 General

A.3.1.1 Prebored pressuremeters [2.3]

A prebored pressuremeter shall be lowered in to a test pocket created specifically for the pressuremeter test. This probe can either be a tricell or monocell.

A.3.1.2 Self-bored pressuremeters [2.4]

A self-bored pressuremeter shall be drilled into the ground using an integral cutting head at its lower end such that the probe replaces the material it removes. The drilling techniques and flushing medium shall be consistent with achieving minimum disturbance to the ground surrounding the probe such that a direct assessment of the *in-situ* horizontal stress can be made.

A.3.1.3 Full displacement pressuremeters [2.5]

A full displacement pressuremeter shall be pushed into the ground using an integral cone at its lower end. A friction reducer, if required, shall be placed between the cone and the pressuremeter section.

A.3.2 Volume-displacement type probes

Volume changes shall be measured either at the surface or in the probe. The volume-measuring unit shall be capable of measuring changes in volume equal to 50 or 25% of the initial volume of the expanding section depending on the test method specified (50% for the Ménard method). The total pressure shall be measured either at the surface or in the probe, and the gauge or transducer shall have a working range greater than or equal to the maximum capacity of the probe.

The resolution of the measuring devices shall be within 0.05% of their full working range. The data-acquisition system used with the probe shall be capable of recording the measured outputs to that resolution.

A.3.2.1 Low- to medium-pressure pressuremeters (prebored, self-bored and full displacement) (for soils) [2.3.1, 2.3.2, 2.4.1, 2.4.2, 2.5.1]

This pressuremeter shall have a maximum pressure capacity of 4 MN/m^2.

A.3.2.2 High-pressure prebored pressuremeters (for rocks) [2.3.1, 2.3.3, 2.4.3]

This pressuremeter shall have a maximum pressure capacity of 20 MN/m^2.

A.3.3 Radial displacement type probes

Displacement shall be measured directly at the midpoint of the membrane and at least three or more equidistant points around the circumference by displacement transducers mounted within the probe. The displacement measuring system shall be capable of measuring an increase in the diameter of the probe as specified below or 50% if the Ménard method is required. The total pressure shall be measured by a transducer within the probe and shall have a working range greater than or equal to the maximum capacity of the probe.

The resolution of any transducer shall be to within 0.05% of its full working range. The data-acquisition system used with the probe shall be capable of recording the transducer outputs to that resolution.

A.3.3.1 Low- to medium-pressure prebored pressuremeters (for soils) [2.3.1, 2.3.2, 2.4.1, 2.4.2, 2.5.1]

This pressuremeter shall have a maximum pressure capacity of 4 MN/m^2 and a displacement-measuring system capable of measuring an increase in the diameter of at least 25% of the original diameter of the probe.

A.3.3.2 High-pressure prebored pressuremeters (for rocks) [2.3.1, 2.3.3]

This pressuremeter shall have a maximum pressure capacity of 20 MN/m^2 and a displacement-measuring system capable of measuring an increase in the diameter of at least 25% of the original diameter of the probe.

A.3.3.3 Soft ground self-bored pressuremeter [2.4.1, 2.4.2]

This pressuremeter shall have a maximum pressure capacity of 4 MN/m^2 and a displacement-measuring system capable of measuring an increase in diameter of at least 15% of the original diameter of the probe. Two transducers, if required, shall be mounted diametrically opposite on the central section of the membrane such that pore pressures can be determined by those transducers.

A.3.3.4 Weak rock self-bored pressuremeter [2.4.3]

This pressuremeter shall have a maximum pressure capacity of 20 MN/m^2 and a displacement-measuring system capable of measuring an increase in the diameter of at least 10% of the original diameter of the probe.

A.3.3.5 Full displacement pressuremeters [2.5.1]

This pressuremeter shall have a maximum pressure capacity of 4 MN/m^2 and a displacement-measuring system capable of measuring an increase in the diameter of at least 25% of the original diameter of the probe.

A.4 CALIBRATIONS

A statement from the manufacturer shall be given, if required, showing that the probe specified meets the requirements of the specification for straightness and sensitivity.

Calibrations for each probe shall be kept in a register for that probe and be available for inspection. The mean and standard deviation for the correlations of each transducer shall be given. Calibrations shall be carried out in shaded, clean, dry conditions. The record of each calibration shall include the ambient temperature at which the calibration was carried out and a description of the location where it was carried out.

The output from transducers shall be read to within 0.05% of the output over the full working range. The data for each transducer shall be tabulated and presented as a graph of measured displacement or pressure against corresponding output. Regression analysis shall be used to determine the correlation between output and displacement or pressure and the coefficient shall be expressed to at least three significant figures. The coefficient of determination for each calibration shall be greater than 0.95. The correlation coefficients for each calibration shall be within 5% of the mean correlation given in the register. Calibrations which do not meet these criteria will not be acceptable without the approval of the engineer. All calibrations shall be traceable to National Standards.

A.4.1 Types

The procedures described below for calibrating the probe shall be carried out:

Volume displacement type probes

(1) pressure transducers (if fitted)
(2) displacement transducers (if fitted)
(3) system compression
(4) membrane stiffness

Radial displacement type probes

(1) pressure transducers
(2) displacement transducers
(3) membrane stiffness
(4) membrane compression

A.4.2 *Total-pressure transducer* (1) [3.3.4]

The probe shall be placed in a cylinder strong enough to restrain safely the expansion at full working pressure. The pressure in the probe shall be continuously increased in equal increments, such that there are 10 to 20 increments up to the total working capacity of the installed pressure transducer. The pressure in the probe shall be reduced by the same number of increments used for pressurisation. The pressure at the end of each increment or decrement shall be held constant until the readings from the transducer remain substantially constant, readings being recorded at 10 s intervals throughout. The calibration of a pressure transducer shall be the change in output from the transducer against the pressure applied to the transducer measured by a calibrated Bourdon gauge. The maximum capacity of the Bourdon gauge shall be between 100 and 120% of the maximum capacity of the transducer and shall read to within 0.25% of the maximum capacity of the gauge.

A.4.3 *Displacement transducer* (2) [3.3.3]

The calibration of a displacement transducer shall be the change in output from the transducer against the movement of the transducer measured by a micrometer. The micrometer shall read to 0.01 m. The displacement shall be continuously increased up to the maximum working range of the transducer and then reduced to zero at increments of displacement of no greater than 10% of the maximum displacement. Each transducer shall be calibrated independently.

A.4.4 *Pore-pressure transducer (if fitted)* (1a) [3.3.5]

The procedure and equipment for calibrating the pore-pressure transducers shall be the same as that for the total-pressure transducer. If the pore pressure is measured relative to the pressure within the probe then the transducer shall be calibrated by internally pressurising the probe within a cylinder strong enough to restrain safely the expansion at full working pressure. If the pore pressure is measured relative to atmospheric pressure then the transducer shall be calibrated by externally pressurising the probe within a cylinder strong enough to restrain safely the applied pressure.

A.4.5 *System compression* (3) [3.3.9]

The probe shall be placed in a cylinder strong enough to restrain safely the expansion at full working pressure. The readout and control units, and connecting hoses shall be the same as those to be used in the testing programme. The pressure in, or the volume of, the probe shall be continuously increased in equal increments, such that there are 20–40 increments up to the total working capacity of the probe. The pressure in, or volume of, the probe shall be reduced by the same number of increments used for pressurisation. The pressure or volume at the end of each increment or decrement shall be held constant until the readings of volume or pressure remain substantially constant, readings being recorded at 10 s intervals throughout. The readings at the end of each increment or decrement shall be plotted to give a pressure versus volume graph, which is the system compression calibration.

A.4.6 *Membrane stiffness* (4) [3.3.6]

The probe shall be clamped in a vertical position. A new membrane shall be inflated up to the maximum volume or movement of the displacement transducers and deflated to the deflated probe diameter at least three times before calibration.

The membrane shall be inflated in free air by applying pressure in increments until the expansion commences. There shall be sufficient increments of pressure to accurately define the initial portion of the expansion. Upon expansion a cavity strain rate or pressure rate shall be selected to give sufficient data to accurately define the expansion curve. The expansion shall continue until either the maximum movement of one of the displacement transducers or the maximum volume change is achieved. The membrane shall then be deflated by reducing the diameter of the membrane in the same decrements as the increments used for inflation. The output from all the measuring systems shall be recorded at 10 s intervals.

The data for each transducer shall be converted to engineering units by applying the coefficients determined in type 1 and 2 calibrations and presented as a graph of measured total pressure against corresponding cavity strain or displacement for each displacement transducer or for volume displacement type probes, as a graph of measured total pressure against corresponding volumetric strain or change. The membrane stiffness calibration coefficients shall be expressed as the pressure required to lift the membrane away from the body of the probe and the pressure required to reach the maximum volume or movement of the displacement transducers. The coefficients shall be expressed to at least two significant figures.

A.4.7 *Membrane compression* (5) [3.3.8]

The probe shall be placed in a cylinder of known elastic properties strong enough to restrain safely the expansion at full working pressure. A close- fitting cylinder with end restraints shall be used for self-bored pressuremeters. A cylinder with a ratio of internal diameter of the cylinder to deflated probe diameter of no greater than 1.1 shall be used with prebored pressuremeters.

The probe shall be pressurised up to its working pressure capacity. The pressure in the probe shall be reduced to 10% of the working capacity of the probe and then increased up to the working capacity in equal increments. The maximum increment or decrement shall be no greater than 5% of the working capacity. The pressure in the probe shall then be reduced to zero. The pressure at the end of each increment or decrement shall be held constant until the readings from the transducer remain sensibly constant, readings being recorded at 10 s intervals throughout.

The data for each transducer shall be converted to engineering units by applying the coefficients determined in type 1, 2 and 4 calibrations. The data shall be corrected for the expansion of the cylinder and presented as a graph of total pressure corrected for membrane stiffness against average cavity strain. The membrane compression calibration coefficient shall be expressed as the average movement of the displacement transducers with pressure to at least two significant figures.

A.4.8 *Pressure loss* (6) [3.3.10]

This applies to volume-displacement systems. The probe is placed next to and, ideally, level with the control unit. The membrane shall be inflated in free air by applying pressure in increments until the expansion commences. There shall be sufficient increments of pressure to accurately define the initial portion of the expansion. In the case of Ménard pressuremeters, the stress increment shall be 10kPa. Upon expansion a cavity strain rate or pressure rate shall be selected to give sufficient data to accurately define the expansion curve. The expansion shall continue until either the maximum movement of one of the displacement transducers or the maximum volume change is achieved. In the case of Ménard pressuremeters, the maximum volume change shall be 700 cm^3. The membrane shall then be deflated by reducing the diameter of the membrane in the same decrements as the increments used for inflation. The output from all the measuring systems shall be recorded at 10 s intervals.

A.4.9 *Volume loss* (7) [3.3.11]

This applies to volume-displacement systems. The probe is placed next to and, ideally, level with the control unit. The probe shall be placed in a cylinder of known elastic properties strong enough to restrain safely the expansion at full working pressure. A close-fitting cylinder with end restraints shall be used for self-bored pressuremeters. A cylinder with a ratio of internal diameter of the cylinder to deflated probe diameter of no greater than 1.1 shall be used with prebored pressuremeters.

The probe shall be pressurised up to its working pressure capacity. The pressure in the probe shall be reduced to 10% of the working capacity of the probe and then increased up to the working capacity in equal increments. The maximum increment or decrement shall be no greater than 5% of the working capacity. The pressure in the probe shall then be reduced to zero. The pressure at the end of each increment or decrement shall be held constant until the readings from the transducer remain sensibly constant, readings being recorded at 10 s intervals throughout.

A.4.10 *Frequency* [3.3.13]

The transducers shall be calibrated prior to the commencement and following the completion of the testing on site. A copy of the calibrations shall be presented to the engineer. Membrane stiffness shall be carried out prior to the commencement of testing on site and after the completion of every borehole. A new membrane shall be calibrated before use and after it is first removed from the borehole. Type 1 and 2 calibrations shall not be required on change of a membrane unless there are other changes to the instrument, including the probe, control/readout unit and connecting hoses.

The displacement and pressure transducers shall be recalibrated on site following any repair of the transducers readout/control unit and connecting hoses. The engineer shall be able to inspect the probe calibration register or shall be provided with copies as specified. The frequency of membrane compression calibrations to be carried out shall be specified.

A.5 INSTALLATION

The probe must be installed in such a manner that the disturbance to the surrounding ground is kept to a minimum for the type of probe. The initial volume of a deflated volume displacement type probe shall be determined at the surface before it is lowered into the borehole. The output of pore-pressure transducers shall be determined at the surface before the probe is lowered into the borehole.

A.5.1 *Prebored pressuremeters* [3.2.3]

The hole shall be advanced to a test position using boring and drilling equipment and techniques and, where necessary, a flushing medium. Particular boring, drilling and sampling requirements shall be specified. Provision shall be made during construction of the borehole for supporting the sides of the borehole by temporary casing, in which case it shall not extend beyond the base of the borehole.

Drilling to form a test pocket below the base of a hole for the insertion of a prebored pressuremeter shall be carried out using equipment, techniques and a flushing medium consistent with achieving minimum pocket disturbance. Any particular test pocket sampling requirements shall be specified.

The test pocket diameter shall not be less than the deflated probe diameter and the ratio of the test pocket diameter to the deflated probe diameter shall be no greater than 1.1. Test pockets with a ratio of pocket diameter to deflated probe diameter greater than 1.1 will not normally be acceptable unless the engineer is satisfied that a ratio less than 1.1 is impractical under the prevailing ground conditions. The length of the test pocket shall not be less than four times the length of the expanding section. In cases where the ratio of the test pocket diameter to the deflated probe diameter exceeds 1.1, the contractor shall discuss with the engineer any changes to the equipment, techniques and flushing medium necessary to achieve an acceptable pocket diameter.

If the orientation of the probe is to be recorded prior to commencement of insertion into the borehole and before it is removed therefrom, it shall be specified. The probe shall be inserted in the test pocket within 60 min of completion of construction of test pockets to depths of 30 m. For pockets in excess of 30 m deep, additional time shall be allowed following discussions with the contractor. The probe shall be at least a distance below the top and above the base of the test pocket equal to half the length of the expanding section.

A.5.2 *Self-bored pressuremeter* [3.2.4]

The self-bored pressuremeter shall be drilled continuously from the ground surface or from the base of a predrilled hole. It shall be drilled beyond a previous test position or beyond the base of a predrilled hole such that the ground to be tested has not already been tested or disturbed when drilling the borehole. The minimum distance shall be equal to twice the distance between the centre of the expanding section and the base of the probe. If a length greater than that distance is required then the length to be self-drilled shall be specified. If the engineer is satisfied that it is impractical under the prevailing ground conditions to drill a length greater than 80% of the minimum distance of a length specified, an expansion test may be carried out.

Drilling or boring to advance the borehole to a test location shall be carried out using equipment, techniques and a flushing medium selected to ensure minimum disturbance to the ground. Provision shall be made during construction of the borehole for supporting the sides of the borehole by temporary casing, in which case it shall not extend beyond the base of the borehole.

During self-drilling of the pressuremeter the setting, type and rotational speed of the cutter, type and characteristics of drilling fluid, the drilling fluid pressure, rate of advance and ram pressure shall all be adjusted to ensure minimum disturbance without causing undue risk of damage to the equipment.

If the orientation of the probe is to be recorded prior to commencement of self-drilling and before it is removed from the drilled test pocket, it shall be specified.

The frequency of recording the output from the pore-pressure transducers during self-drilling and relaxation, if required, shall be specified.

In free-draining soils, an expansion test shall be commenced as soon as practically possible following the completion of self-drilling. In all remaining soils and rocks there shall be a minimum of 30 min and a maximum of 1 h between completion of self-drilling and the commencement of the expansion test.

A.5.3 *Full displacement pressuremeters* [3.2.5]

The probe shall be pushed continuously into the soil at 2 cm/s either from the base of a borehole or from the surface. Drilling or boring to advance the borehole to a test location shall be carried out using equipment, techniques and a flushing medium selected to ensure minimum disturbance to the ground. Provision shall be made during construction of the borehole for supporting the sides of the borehole by temporary casing, in which case it shall not extend beyond the base of the borehole.

If the orientation of the probe is to be recorded prior to commencement of pushing and before it is removed from the drilled test pocket, it shall be specified.

An expansion test shall be commenced as soon as practically possible following the completion of pushing.

A.6 TESTING PROCEDURE

The pressuremeter shall be pressurised until either the maximum pressure capacity is reached, or any one of the displacement transducers has reached its full working range or the volume capacity is reached or there is undue risk of damage to the equipment or to a specified displacement which is less than or equal to the maximum capacity of the probe.

The pressuremeter shall be unloaded, if required, both during the unload/ reload cycle and the final unloading at the same rates as the loading stage.

Prior to unloading, there shall be, if required, a period during which the pressure is held constant. This period shall be specified.

The reduction in stress during any unload/reload cycle shall be limited to ensure that it remains within the elastic range of the ground under test. The strain at the start of any additional unload/reload cycles shall be specified.

The output from the transducers shall be recorded at a minimum frequency of 10 s intervals throughout the test if automatic recording is used or 30 s if manual readings are taken.

The duration of the test shall be no longer than 90 min.

A.6.1 *Strain-controlled tests* [3.5.4]

A.6.1.1 Prebored probes

Strain-controlled tests shall be carried out at a constant rate of volumetric or cavity strain increase. The increment of volume or strain shall be 5% of the maximum capacity of the probe and each increment should be maintained for 1 min. An unload/

reload cycle, if required, shall be carried out when either the pocket has increased in diameter by between 1 and 3% or when the pressure has reached 10 MN/m^2, whichever comes first.

A.6.1.2 Self-bored probes

Strain-controlled tests shall be carried out at a constant rate of stress increase during the early stage of the test until expansion commences and then at a constant rate of strain of 1%/min, with one unload/reload cycle, if required, included in the loading sequence between 1 and 3% total cavity strain. There shall be a sufficient number of readings taken to define accurately the pressure at which expansion starts.

A.6.1.3 Full-displacement probes

Strain-controlled tests shall be carried out at a constant rate of stress increase during the early stage of the test until expansion commences and then at a constant rate of strain of 5%/min, with one unload/reload cycle, if required, included in the loading sequence once constant pressure conditions are approximately achieved.

A.6.2 Stress-controlled tests [3.5.3]

Stress-controlled tests shall be carried out at a constant rate of pressure increase, with one unload/reload cycle, if required, included in the loading sequence when either the cavity has increased in diameter by between 1 and 3%, or when the pressure has reached 10 MN/m^2, whichever occurs first. The increments shall be no greater than 5% of the maximum capacity of the probe and shall be adjusted to ensure that there are at least 15 increases in pressure throughout the loading stage (excluding any unload/reload cycles). The size of the pressure increment shall be adjusted during the early stage of the test to ensure that a sufficient number of readings are taken to define accurately when the membrane comes in contact with the side of the pocket, and shall be no greater than 0.2MN/m^2.

Each pressure increment shall be held constant for 1 min.

A.6.3 Stress-controlled tests: Ménard method [3.5.2]

Stress-controlled tests shall be carried out at a constant rate of pressure increase. The increments shall be no greater than 5% of the maximum capacity of the probe and shall be adjusted to ensure that there are between 7 and 15 increases in pressure throughout the loading stage. There shall be no unload/reload cycles. The size of the pressure increment shall be adjusted during the early stage of the test to ensure that a sufficient number of readings are taken to define accurately when the membrane comes in contact with the side of the pocket, and shall be no greater than 0.2MN/m^2.

Each pressure increment shall be held constant for 1 min.

A.7 ON-SITE DATA PROCESSING

A.7.1 *Volume displacement type probes* [3.7.2]

The data for each transducer, when fitted, shall be converted to engineering units by applying the coefficients determined in calibration types 1 and 2; the applied pressure shall be corrected for membrane stiffness (type 4), system compliance (type 3), pressure loss (type 6) and volume loss (type 7).

A.7.2 *Radial displacement type probes* [3.73, 3.7.4, 3.75]

The data for each transducer shall be converted to engineering units by applying the coefficients determined in calibration types 1 and 2; the applied pressure shall be corrected for membrane stiffness (type 4) and, if necessary, the calculated cavity strains shall be corrected for membrane compression (type 5) and thinning (change in thickness of the membrane due to expansion).

A.8 INTERPRETATION

The stiffness from the unload/reload cycle and, if required, the initial stiffness shall be determined using the analyses developed by Windle and Wroth (1977a) [4.4]. If the stiffness degradation curve is required, the analyses described in [4.8, 6.5.3] shall be used.

The strength from self-bored pressuremeter tests in clays shall be determined using the analyses developed by Windle and Wroth (1977a) [4.5.2] and in sands using Hughes et al. (1977) [4.6.2].

The strength from prebored tests in clays can be determined from the modified limit pressure using the relationships in Table 6.10.

The strength from full displacement pressuremeter tests in clays shall be determined from the unloading curve using the analysis developed by Houlsby and Withers (1988) [4.8.2].

The strength from full displacement pressuremeter tests in sands shall be determined from the empirical correlations developed by Houlsby and Nutt (1993) [6.7].

The horizontal stress (self-bored pressuremeter tests only) shall be determined using the lift-off procedure proposed by Wroth (1982) and Hawkins et al. (1990) [6.4.2].

The pressuremeter modulus and modified limit pressure shall be determined using the method proposed by Ménard [6.3].

Any additional data processing and analyses required, including reporting, shall be specified.

A.9 INFORMATION TO BE SUBMITTED

All information shall be presented in a digital format with graphs presented at a suitable scale.

A.9.1 Prior to commencing work on site

- Full details of the pressuremeter and testing equipment.
- Details of the proposed drilling equipment and flushing medium to be used for drilling boreholes.
- Description of the methods of carrying out all the drilling operations for forming a test pocket.
- Details of the experience of the proposed operator and supervising engineer.
- Typical test data sheets and forms for presenting final results.

A.9.2 Preliminary results for each test

The following information shall be submitted for each test within one working day of the completion of that test:

- Contract name and number, borehole number, depth of the top and bottom of the test pocket, depth of the displacement measurement axes.
- Names of drilling and testing personnel.
- Details of equipment used.
- Details of boring, drilling (and self-drilling if applicable), including date and time of start and finish of all drilling, description and estimate of any drilling fluid returns and depth and size of casing used, if any.
- The output from the transducers recorded prior to and during installation and on removal from the borehole for any prebored pressuremeter or pressuremeter fitted with pore-pressure transducers.
- Tabulated output of the transducers during a test, if required, time of start and finish of test and rates of stress and/or strain in a digital format.
- The calibration coefficients used to convert the test data to engineering units.
- Tabulated calibrated test data in digital format.
- A plot of applied pressure against the volumetric or average cavity strain expressed as a percentage shall be submitted, and the additional plots given in Table A.1 shall also be submitted, if required.
- A preliminary assessment of the parameters as outlined in the previous section.

A.9.3 Report data processing and analysis

The calibrations shall be reviewed by the supervising engineer. The data for each transducer shall be recalibrated, if required, by applying all the appropriate calibration coefficients determined following this review.

The supervising engineer shall reassess the calibrated data to give the parameters outlined in section A.8. These, together with the corrected plots, detailed in section A.9.2, comprise the definitive interpretation and presentation of the data for formal reporting purposes.

A.9.4 Information to be submitted in the report

The information submitted in the report shall be as specified.

Table A.1 Typical plots from pressuremeter tests

Probe	*Ground type*	*Abscissa*	*Ordinate*
Radial displacement type			
Self-bored, full displacement	All	Cavity strain for each arm	Applied pressure
Prebored	All	Cavity strain for each pair of arms	Applied pressure
Self-bored	All	Initial cavity strain for each arm	Applied pressure
All	All	Cavity strain for unload/reload cycle for each arm	Applied pressure
All	Clay	Logarithm of cavity strain for each arm	Applied pressure
All	Sands	Natural logarithm of current cavity strain for each arm	Natural logarithm of effective applied pressure
Volume displacement type			
Prebored	All	Volume change	Applied pressure
Prebored	All	Rate of change of volume	Applied pressure

A.10 BILL OF QUANTITIES

Item	*Item description*	*Unit*
	Prebored pressuremeter	item
1	Establishment on site of the drilling rig and crew	
2	Establishment on site of prebored pressuremeter equipment and operator	item
3	Moving of the drilling rig to the site of each exploratory borehole	no.
4	Moving of prebored pressuremeter equipment to the site of each exploratory borehole	no.
5	Open hole drilling/boring to advance the borehole to a test location	
5.1	Between existing ground level and not exceeding 10 m depth	m
5.2	Between 10 m and not exceeding 20 m depth	m
5.3	Between 20 m and not exceeding 30 m depth	m
6	Rotary core drilling to advance the borehole to a test location	
6.1	Between existing ground level and not exceeding 10 m depth	m
6.2	Between 10 m and not exceeding 20 m depth	m
6.3	Between 20 m and not exceeding 30 m depth	m
7	Manufacturer's statement for straightness and sensitivity	item
8	Calibration of prebored pressuremeter prior to establishment on site and following the completion of testing on site	item
9	Carry out additional calibrations as instructed by the engineer	
9.1	Displacement transducers	no.
9.2	Total-pressure transducer	no.
9.3	Membrane stiffness	no.
9.4	System compression or membrane compression	no.
9.5	Pressure loss	no.
9.6	Volume loss	no.

Item	*Item description*	*Unit*
10	Provide copies of the probe calibration register	item
11	Drilling as specified to form a test pocket	
11.1	Between existing ground level and not exceeding 10 m depth	m
11.2	Between 10 m and not exceeding 20 m depth	m
11.3	Between 20 m and not exceeding 30 m depth	m
12	Provide orientation data for pressuremeter tests	no.
13	Expansion test of duration not exceeding 1.5 h	
13.1	Between existing ground level and not exceeding 10 m depth	no.
13.2	Between 10 m and not exceeding 20 m depth	no.
13.3	Between 20 m and not exceeding 30 m depth	no.
14	Extra over items 13 for expansion tests in excess of 1.5 h	H
15	Extra over items 13 for carrying out additional unload/reload cycles as specified	per cycle
16	Standing time of prebored pressuremeter equipment	h
17	Backfilling to prebored pressuremeter borehole	
17.1	Between existing ground level and not exceeding 10 m depth	m
17.2	Between 10 m and not exceeding 20 m depth	m
17.3	Between 20 m and not exceeding 30 m depth	m
18	Additional data processing, analysis and reporting as specified	per test
19	Supply?? copies of the report	item
	Self-boring pressuremeter	
1	Establishment on site of the drilling rig and crew	item
2	Establishment on site of self-bored pressuremeter equipment and operator	item
3	Moving of the drilling rig to the site of each exploratory borehole	no.
4	Moving of self-bored pressuremeter equipment to the site of each exploratory borehole	
5	Drilling or bored to advance the borehole to a test location	
5.1	Between existing ground level and not exceeding 10 m depth	M
5.2	Between 10 m and not exceeding 20 m depth	M
5.3	Between 20 m and not exceeding 30 m depth	M
6	Manufacturer's statements for straightness and sensitivity	Item
7	Calibration of self-bored pressuremeter prior to the establishment on site and item following the completion of testing on site	Item
8	Carry out additional calibrations as instructed by the engineer	
8.1	Displacement transducers	no.
8.2	Pore-pressure transducers	no.
8.3	Total-pressure transducer	no.
8.4	Membrane stiffness	no.
8.5	Membrane compression	no.
9	Provide copies of the probe calibration register as specified	Item
10	Self-drilling to form test pocket	
10.1	Between existing ground level and not exceeding 10 m depth	M
10.2	Between 10 m and not exceeding 20 m depth	M

Item	*Item description*	*Unit*
10.3	Between 20 m and not exceeding 30 m depth	M
11	Provide orientation data for pressuremeter tests	no.
11.1	Between existing ground level and not exceeding 10 m depth	no.
11.2	Between 10 m and not exceeding 20 m depth	no.
11.3	Between 20 m and not exceeding 30 m depth	no.
12	Provide pore-pressure cell data during installation	
12.1	Between existing ground level and not exceeding 10 m depth	no.
12.2	Between 10 m and not exceeding 20 m depth	no.
12.3	Between 20 m and not exceeding 30 m depth	no.
13	Expansion test of duration not exceeding 1.5 h including 30min standing	
13.1	Between existing ground level and not exceeding 10 m depth	no.
13.2	Between 10 m and not exceeding 20 m depth	no.
13.3	Between 20 m and not exceeding 30 m depth	no.
14	Extra over item 13 for expansion tests in excess of 1.5 h	H
15	Extra over item 13 for carrying out additional unload/reload cycles as specified	per cycle
16	Standing time of self-bored pressuremeter equipment	H
17	Backfilling self-bored pressuremeter borehole	
17.1	Between existing ground level and not exceeding 10 m depth	M
17.2	Between 10 m and not exceeding 20 m depth	M
17.3	Between 20 m and not exceeding 30 m depth	M
18	Additional data processing, analysis and reporting as specified	per test
19	Supply?? copies of the report	Item
	Full displacement pressuremeter	
1	Establishment on site of the drilling rig and crew	Item
2	Establishment on site of the cone truck and crew	Item
3	Establishment on site of full displacement pressuremeter equipment and operator	Item
4	Moving of the drilling rig to the site of each exploratory borehole	no.
5	Moving of the cone truck the site of each exploratory borehole	no.
6	Moving of full displacement pressuremeter equipment to the site of each exploratory borehole	no.
7	Drilling or bored to advance the borehole to a test location	
7.1	Between existing ground level and not exceeding 10 m depth	M
7.2	Between 10 m and not exceeding 20 m depth	M
7.3	Between 20 m and not exceeding 30 m depth	M
8	Manufacturer's statements for straightness and sensitivity	Item
9	Calibration of full displacement pressuremeter prior to the establishment on site and item following the completion of testing on site	
10	Carry out additional calibrations as instructed by the engineer	
10.1	Displacement transducers	no.
10.2	Pore-pressure transducers	no.
10.3	Total-pressure transducer	no.

Item	*Item description*	*Unit*
10.4	Membrane stiffness	no.
11	Provide copies of the probe calibration register as specified	Item
12	Pushing the cone to form test pocket	
12.1	Between existing ground level and not exceeding 10 m depth	M
12.2	Between 10 m and not exceeding 20 m depth	M
12.3	Between 20 m and not exceeding 30 m depth	M
13	Provide orientation data for pressuremeter tests	no.
13.1	Between existing ground level and not exceeding 10 m depth	no.
13.2	Between 10 m and not exceeding 20 m depth	no.
13.3	Between 20 m and not exceeding 30 m depth	no.
14	Expansion test of duration not exceeding 1 h	
14.1	Between existing ground level and not exceeding 10 m depth	no.
14.2	Between 10 m and not exceeding 20 m depth	no.
14.3	Between 20 m and not exceeding 30 m depth	no.
15	Extra over item 12 for expansion tests in excess of 1.5 h	H
16	Extra over item 14 for carrying out additional unload/reload cycles as specified	per cycle
17	Standing time of full displacement pressuremeter equipment	H
18	Backfilling full displacement pressuremeter borehole	
18.1	Between existing ground level and not exceeding 10 m depth	M
18.2	Between 10 m and not exceeding 20 m depth	M
18.3	Between 20 m and not exceeding 30 m depth	M
19	Additional data processing, analysis and reporting as specified	per test
20	Analysis of cone data	per test
21	Supply?? copies of the report	Item

Index

Note: Page numbers in **bold** refer to tables and those in *italic* refer to figures.